THE ROUTLEDGE HANDBOOK OF DISASTER RISK REDUCTION INCLUDING CLIMATE CHANGE ADAPTATION

The Routledge Handbook of Disaster Risk Reduction Including Climate Change Adaptation aims to provide an overview and critique of the current state of knowledge, policy, and practice, encouraging engagement, and reflection on bringing the two sectors together. This long-awaited and welcomed volume makes a compelling case that a common research agenda and a series of practical policies and policy recommendations can and should be put in place.

Over 40 contributions explore DRR including CCA in five parts. The first part presents and interrogates much of the typical vocabulary seen in DRR including CCA, not only pointing out the useful and not-so-useful dimensions, but also providing alternatives and positive examples. The second part explains how to move forward creating and supporting positive cross-overs and connections, while the third one explores some aspects of multi-dimensional approaches to knowing and understanding. The fourth part argues for a balanced approach to governance, taking both governmental and non-governmental governance, as well as different scales of governance, into consideration. The final part of the *Handbook* emphasises DRR including CCA as an investment, rather than a cost, and connects its further implementation with livelihoods of people around the world.

This handbook highlights the connections amongst the processes of dealing with disasters and dealing with climate change. It demonstrates how little climate change brings which is new and emphasises the strengths of placing climate change within wider contexts in order to draw on all our strengths while overcoming limitations with specialities. It will prove to be a valuable guide for graduate and advanced undergraduate students, academics, policy makers, and practitioners with an interest in disaster risk reduction and climate change.

Ilan Kelman is a Reader in Risk, Resilience, and Global Health at University College London, England and a researcher at the University of Agder, Kristiansand, Norway. His overall research interest is linking disasters and health. More details from: http://www.ilankelman.org and Twitter @IlanKelman

Jessica Mercer is a consultant with Secure Futures (www.secure-futures.net) focused on risk reduction for communities worldwide. Previously, she has worked with academia, United Nations agencies, government and non-governmental organisations in the areas of DRR including CCA for over 10 years.

JC Gaillard is Associate Professor at the University of Auckland, New Zealand. His work focuses on developing participatory tools for engaging minority groups in disaster risk reduction with an emphasis on ethnic and gender minorities, prisoners and homeless people. More details from: https://jcgaillard.wordpress.com

THE ROUTLEDGE HANDBOOK OF DISASTER RISK REDUCTION INCLUDING CLIMATE CHANGE ADAPTATION

Edited by Ilan Kelman,
Jessica Mercer, and JC Gaillard

Routledge
Taylor & Francis Group

LONDON AND NEW YORK

First published 2017
by Routledge
2 Park Square, Milton Park, Abingdon, Oxon OX14 4RN

and by Routledge
52 Vanderbilt Avenue, New York, NY 10017

First issued in paperback 2020

Routledge is an imprint of the Taylor & Francis Group, an informa business

British Library Cataloguing-in-Publication Data
A catalogue record for this book is available from the British Library

Library of Congress Cataloging-in-Publication Data
Names: Kelman, Ilan, editor. | Mercer, Jessica, editor. | Gaillard, J. C., editor.
Title: The Routledge handbook of disaster risk reduction including climate
change adaptation / edited by Ilan Kelman, Jessica Mercer and JC Gaillard.
Other titles: Handbook of disaster risk reduction including climate change adaptation
Description: Abingdon, Oxon ; New York, NY : Routledge, 2017. |
Includes bibliographical references and index.
Identifiers: LCCN 2016048450 | ISBN 9781138924567 (hardback : alk. paper) |
ISBN 9781315684260 (ebook)
Subjects: LCSH: Natural disasters—Risk assessment—Handbooks, manuals, etc. |
Climatic changes—Risk assessment | Hazard mitigation—Handbooks, manuals, etc. |
Emergency management—Handbooks, manuals, etc.
Classification: LCC GB5014 .R679 2017 | DDC 363.34/6—dc23
LC record available at https://lccn.loc.gov/2016048450

ISBN 13: 978-0-367-58128-2 (pbk)
ISBN 13: 978-1-138-92456-7 (hbk)

Typeset in Bembo
by Apex CoVantage, LLC

CONTENTS

Contents

FIGURES

TABLES

BOXES

CONTRIBUTORS

Bayes Ahmed is affiliated as a Lecturer at the Department of Disaster Science and Management, University of Dhaka, Bangladesh. He has a background in disaster risk reduction, urban planning, and geoinformatics. His PhD research at University College London is focused on developing a framework in transforming indigenous cultural knowledge to address community vulnerability to landslides in the Chittagong Hill Districts of Bangladesh. Recently, he worked with the earthquake vulnerable refugee-people in Ecuador, and with the cyclone affected coastal communities in Bangladesh.

Amina Aitsi-Selmi is a public health physician and consultant in International Public Health based at Public Health England. Her work focuses on evidence translation at international level to advocate for health outside the health sector. She has a special interest in health equity and emerging non-communicable disease risks in countries undergoing economic transition, having completed a PhD in 2013 with Professor Sir Michael Marmot at UCL in 2013 on the epidemiology of the social determinants of obesity in low- and middle-income countries.

Bob Alexander is an independent researcher, trainer, and consultant working in issues related to participatory integrated disaster, climate change, and development vulnerability reduction and resilience strengthening for over twenty years. His foci include CBDRR/CBA; livelihood/nutrition/BSS security; rural socio-economic vulnerability identification/analysis/assessment/adaptation; effective risk communication; and factors affecting choices of displaced persons.

David Alexander is Professor of Risk and Disaster Reduction at the Institute for Risk and Disaster Reduction at University College London, UK. He is the author of several books, including Natural Disasters, Confronting Catastrophe, Principles of Emergency Planning and Management, and How to Write an Emergency Plan. He founded and edits the International Journal of Disaster Risk Reduction and authored a paper on 'The Geography of Italian Pasta'.

Camillo Boano is an architect and urbanist. He is Senior Lecturer at The Bartlett Development Planning Unit, UCL, where he directs the MSc in Building and Urban Design in Development. He is also co-director of the UCL Urban Laboratory. Camillo has over twenty years of experience in research, DRR and development work in South America, the Middle East, Eastern Europe and

South East Asia. He researches the encounters between critical theory and radical philosophy with urban and architectural design processes.

Lee Bosher is a Senior Lecturer in Disaster Risk Reduction at Loughborough University, UK. He has a background in disaster risk management and his research and teaching embraces disaster risk reduction and the multi-disciplinary integration of proactive hazard mitigation strategies into the decision-making processes of key stakeholders, involved with the planning, design, construction and operation of the built environment. Lee's books include *Hazards and the Built Environment* (2008) and *Disaster Risk Reduction for the Built Environment: An introduction* (2017).

David Bourguignon holds a PhD in Geography (University of Montpellier 3) and specialises in evaluating the cost of insured damage caused by disasters, especially floods. Since 2010, he has worked as a research analyst for the 'Mission Risques Naturels', which is an association created by the French insurance trade associations and dedicated to the prevention of disasters involving environmental hazards in France.

Sarah Bradshaw is a Professor of Gender and Sustainable Development (@DrSarahBradshaw) at Middlesex University, London. She has worked in the general field of gender and development for twenty years and combines her academic work with practitioner activities, working with a number of NGOs in Nicaragua and with the UN Sustainable Development Solutions Network. Living through Hurricane Mitch led to an interest in gender post-disasters, and her book *Gender, Development and Disasters* was published in 2013 by Edward Elgar.

Christophe Buffet is a researcher and consultant in climate change adaptation and disaster risk reduction. Under the direction of Amy Dahan at the Centre Alexandre Koyré (EHESS/CNRS) in Paris, his PhD thesis dealt with the construction, framing, and actors of adaptation from the Conference of Parties meetings to vulnerable populations in Bangladesh. He studies adaptation and official development assistance, the interface between climate sciences and policies, and community-based adaptation.

Tom R. Burns (BS in Physics and PhD in Sociology, both at Stanford; Professor Emeritus, University of Uppsala, Sweden; CIES, Lisbon University Institute) has published more than twenty books and numerous articles in the areas of governance and democracy, the sociology and politics of environment and technology, studies of markets, administration and management. Recent publications include *Sustainable Development: Agents, Systems and the Environment* (2015); *The Sustainability Revolution: A Societal Paradigm Shift* (2012); and with Peter M. Hall and others *The Meta-Power Paradigm* (2012).

Jake Rom D. Cadag is a Senior Lecturer at the Department of Geography of the University of the Philippines Diliman. His professional specialties include disaster risk reduction and management and development studies. He is particularly skilled in community work and conduct of participatory methods and tools involving different actors.

John Campbell is an Associate Professor of Geography at the University of Waikato, New Zealand. He has a PhD from the University of Hawaii and has worked on issues of environment and development, and in particular, the human dimensions of disasters and climate change in Pacific island countries for the past forty years.

Jean Connolly Carmalt is an Assistant Professor at John Jay College of Criminal Justice (CUNY) and the Earth and Environmental Sciences Program at the CUNY Graduate Center. She holds a PhD in Geography (University of Washington) and a J.D. (Cornell University). Dr Carmalt focuses on public international law and society, with a particular interest in the right to health, UN human rights processes, and environmental disasters.

Ksenia Chmutina is a lecturer in sustainable and resilient urbanism at Loughborough University, UK. Her main research interest is in synergies of resilience and sustainability in the built environment, including holistic approach to enhancing resilience to natural and human-induced threats, and a better understanding of the systemic implications of sustainability and resilience under the pressures of urbanisation and climate change.

Ellie Cosgrave is a Research Associate on the EPSRC-funded Liveable Cities Programme, a cross disciplinary research programme that aims to develop realistic and radical approaches to engineering that support the development of low carbon, resource secure cities that maximise citizen wellbeing. Her work explores transformative approaches to policy making and governance of cities.

Rachel Cowell is a mobile product manager specialising in the use of ICT for development. In early 2015 she launched mobile agriculture service iShamba in Kenya, which empowers farmers with market prices, weather information and bespoke farming tips for their farm. She has worked with both UNEP and UNDP to explore the use of mobile technologies in early warning systems and ways to increase private sector engagement in disaster response.

Kate Crowley is a Hazard and Risk Researcher at the National Institute of Water and Atmospheric Research (NIWA) in Wellington, New Zealand. She is part of the RiskScape team developing and enabling the application of a natural hazards impact and loss modelling tool. Previously, Dr Crowley worked as the Disaster Risk Reduction Adviser for an international development and humanitarian agency called CAFOD, based in London, United Kingdom. In addition to RiskScape her applied research includes examining social vulnerability, culture and disasters, risk science communication and knowledge exchange.

Lydia Cumiskey is a PhD researcher at the Flood Hazard Research Centre, Middlesex University London for the System-Risk ETN project. She is a guest researcher at Deltares, The Netherlands where she worked for four years on disaster risk management research and consultancy projects. She coordinates the Disaster Risk Reduction (DRR) team at the Water Youth Network (WYN) and acts as the Regional Focal Point for Europe within UN Major Group for Children and Youth, DRR working group.

Soledad Natalia M. Dalisay is a faculty member of the Department of Anthropology, University of the Philippines in Diliman, the Philippines. She is also currently the Chair of the Department. Her research interests include the Anthropology of Disaster, Gender and Sexuality, Cultural Ecology of Health, and Nutritional Anthropology.

Rajarshi DasGupta is a Post-Doctoral Research Fellow in the Graduate School of Agriculture and Life Sciences, The University of Tokyo. He received his PhD from Kyoto University and has previously worked with governmental and non-governmental organisations in the field of environment and disaster management. Current research interests are ecosystem approaches for DRR, resilience planning and utilising ecosystem services. He has authored a number of articles and contributed to edited volumes on DRR, CCA, sustainable development, and natural resource conservation.

Annika Dean holds a Bachelor of Development Studies with first class honours from the University of Newcastle, Australia and a PhD from the Climate Change Research Centre at the University of New South Wales. Her PhD research explored the effects of climate finance on social vulnerability, risk and capacity to adapt to climate change in the Republic of Kiribati. She is currently lecturing in climate change and energy policy at UNSW.

Loreine B. dela Cruz has been the Executive Director of the Center for Disaster Preparedness Foundation since July 2012. She is a development specialist with more than 30 years of professional experience in various fields of development work. Her specialisation is in gender-inclusive community-based disaster risk reduction and management; peace and human rights; and psychosocial development work. Gender perspectives are integrated into her methodology and work methods. She has a Masters in Development Management.

Zenaida Delica-Willison is currently the Chairperson of the Board of Trustees of the Center for Disaster Preparedness Foundation. She has over forty years experience in development and disaster risk reduction in the Philippines and in other Asian countries. She has held senior posts in NGOs and advisory posts in the UNDP Office for South–South Cooperation, leading teams and programmes, developing strategies, and undertaking training, research, and consultancy. She developed the concept of and implemented the South–South Community Based Development Academy in South and Southeast Asia. She holds a Masters of Public Health and Masters of Development Practice.

Deborah P. Dixon is a Professor of Geography at the University of Glasgow, Scotland, UK. Her research draws out the interstices between geopolitics, geoaesthetics, and feminist theory, with a substantive focus on the complex, wicked problems and possibilities associated with the Anthropocene.

Nathalie Doswald is an independent biodiversity and climate change consultant in Geneva, Switzerland. She previously worked as a Programme Officer at UNEP-WCMC where she became an expert on ecosystem-based adaptation. Her PhD looked at the impacts of climate change on the distribution and migration of migrant birds. She holds an MRes Distinction in Ecology and Environmental Management from University of York, UK, and a BSc First class in Environmental Sciences from Stirling University, UK.

Melanie Duncan is a hazard and risk researcher within the volcanology team at the British Geological Survey (BGS) – a Natural Environment Research Council Institute – in their Edinburgh office, United Kingdom. Her research interest lies in the multi-hazard characteristics of volcanic environments, particularly cascading and interacting hazards. Melanie is a member of the Global Volcano Model network and is currently working with other members to deliver its core goal of reducing risk from volcanic eruptions. Before joining BGS, she was a disaster risk reduction consultant for the United Nations and World Bank.

Marisol Estrella is currently Programme Coordinator for Disaster Risk Reduction in the Post-Conflict and Disaster Management Branch of the United Nations Environment Programme (UNEP), based in Geneva, Switzerland. Ms Estrella has previous field experience in community-based disaster risk management in Southeast Asia. She holds an MA Distinction in Environment,

Development and Policy from University of Sussex, UK, and BA First Class Honours in Anthropology and Environmental Studies from the University of McGill, Canada.

Carina Fearnley is an interdisciplinary researcher and Lecturer in Science and Technology Studies at the Department of Science and Technology Studies at University College London, UK. Her research draws on concepts of scientific uncertainty, risk, and complexity to be reframed and communicated within the context of disaster risk reduction.

Nava Fedaeff is a Climate Scientist at the National Institute of Water and Atmospheric Research (NIWA) in Auckland, New Zealand. She is a regular contributor to several climate products including National Climate Summaries, the Seasonal Climate Outlook and the Island Climate Update. Nava has also carried out consultancy work relating to climate drivers for the Bay of Plenty and Hawke's Bay Regional Councils and is a key member of the Forecasting Services team responsible for a range of NiwaWeather products.

Heather Fehr is a long-time member of the Red Cross Red Crescent Movement and is currently the Disaster Risk Reduction Advisor at the British Red Cross working with National Societies globally to prepare for and respond to disasters. She holds a Master's degree in Urban Planning specialising in Disaster Risk Management from the University of British Columbia School of Community and Regional Planning.

Elizabeth Ferris currently serves as research professor at Georgetown University's Institute for the Study of International Migration and non-resident senior fellow at the Brookings Institution. She previously served as co-director of the Brookings-LSE Project on Internal Displacement, as programme director at the World Council of Churches in Geneva, Switzerland and as a professor at several US universities.

JC Gaillard is Associate Professor at the University of Auckland and a former member of the faculty of the University of the Philippines Diliman. His present work focuses on developing participatory tools for disaster risk reduction and in fostering the participation of minority groups in disaster-related activities with an emphasis on ethnic minorities, gender minorities, prisoners and homeless people. JC also collaborates in participatory mapping and trainings with NGOs, local governments and community-based organisations. More details from: https://jcgaillard.wordpress.com

Virginia García-Acosta's research has focused on food history and on disaster and risk from an historical-anthropological perspective. Her interests are around earthquakes and disasters as processes associated to hydrometeorological hazards throughout Mexican and Latin American history. She is currently preparing a catalogue on hurricanes which have occurred in Mexico during the last five centuries, similar to those published before by her on earthquakes and agricultural disasters, with particular emphasis on preventive and adaptive strategies.

Vincent T. Gawronski is Professor of Political Science and Program Director for Latin American Studies at Birmingham-Southern College. His research and teaching interests include the politics of disaster and how disasters, violence, crime, and migration interact. With co-authors, recent publications include 'From Disaster Event to Political Crisis: A "5C+A" Framework for Analysis' in *International Studies Perspectives* (2010); 'Communities of Practice and Hazard Reduction' (2012); and 'Disasters as Crisis Triggers for Critical Junctures? The 1976 Guatemala Case' in *Latin American Politics and Societies* (2013).

Terry Gibson was Operations Director at the Global Network of Civil Society Organisations for Disaster Reduction (GNDR) from 2008–2016, leading on the design and development of the 'Views from the Frontline' and 'Frontline' action research programmes. He has completed doctoral studies in learning and communications within international networks – taking GNDR as the case study – and researches and publishes on these themes, and on the role and development of transnational networks.

Christopher Gomez is Associate Professor at Kobe University in the Faculty of Maritime Sciences. He specialises in volcanoes at sea, metrology and disaster risks. He works mostly on the island nations of the Ring of Fire, where multi-cultural experience and being polyglots are central to shaping meaningful and culturally relevant research.

Mary Joy K. Gonzales is an Environmental Planner and a DRR Training and Advocacy Officer for Tearfund Philippines. She is a graduate student at the School of Urban and Regional Planning of the University of the Philippines Diliman, and is currently doing her thesis on small islands resilience. She specialises in building the capacities of local government units in DRR and climate change adaptation planning.

Lorenzo Guadagno works with the International Organization for Migration, managing its capacity building programme on 'Reducing the vulnerability of migrants in emergencies'. He has worked on DRR, human mobility and ecosystem management with several international and non-governmental organisations. He holds a PhD in sociology from the University of Sannio, Benevento, Italy.

Annelies Heijmans is currently lecturer-researcher at Van Hall Larenstein University of Applied Sciences, Velp, and coordinates the Major Disaster Risk Management programme. She further designs and implements 'Living Labs', a form of interactive research with humanitarian aid agencies, governments, disaster-affected populations, and the private sector in the field of DRR, CCA, and conflict transformation. She has more than twenty years of working experience in the field of DRR in disaster- and conflict-affected areas, promoting the view of strengthening local capacities in dealing with risk. Her email is annelies.heijmans@hvhl.nl

Moa M. Herrgård is a law student at Stockholm University and medical student at the Karolinska Institute in Sweden. She is a former intern at the World Health Organization. Moa coordinates the Disaster Risk Reduction working group within the UN Major Group for Children and Youth.

Mischa Hill is an Emergency Management Advisor at the Wellington Region Emergency Management Office in New Zealand. Mischa works as part of the Community Resilience Team, engaging with communities to build their capacity to be able to respond to and recover from a disaster. Her interests lie in pre-disaster recovery planning, social wellbeing post-disaster, and working with Marae/iwi.

Lisa Hiwasaki is an environmental anthropologist with extensive experience leading interdisciplinary research and development projects on environmental conservation, water resources management, and disaster risk reduction, including climate change adaptation. She is passionate about promoting sustainable development using participatory approaches, and increasing the resilience of marginalised populations, in particular indigenous peoples and ethnic minorities. Lisa has been living and working in Southeast Asia for the past six years, and is currently based in Vietnam with the World Agroforestry Centre (ICRAF).

Rohit Jigyasu is a conservation architect and risk management professional from India, currently working as UNESCO Chair-holder professor at the Institute for Disaster Mitigation of Urban Cultural Heritage at Ritsumeikan University, Kyoto, Japan and Senior Advisor at the Indian Institute for Human Settlements (IIHS). He has been the President of ICOMOS International Scientific Committee on Risk Preparedness (ICORP) since 2010. Rohit has been working with various organisations for consultancy, research and training on Disaster Risk Management with special focus on cultural heritage.

Lindsey Jones is a Research Fellow at the Overseas Development Institute. His research focuses primarily on issues related to adaptation to climate change, disaster risk reduction and the uptake of climate information in decision making. Lindsey has previously worked in a number of countries across Asia and Africa for the CGIAR Climate Change Agriculture and Food Security programme (CCAFS), World Food Programme (WFP) and United Nations Development Programme (UNDP).

Ilan Kelman http://www.ilankelman.org and Twitter @IlanKelman is a Reader in Risk, Resilience and Global Health at University College London, England and a researcher at the University of Agder, Kristiansand, Norway. His overall research interest is linking disasters and health, including the integration of climate change into disaster research and health research. That covers three main areas: (i) disaster diplomacy and health diplomacy, http://www.disasterdiplomacy.org; (ii) island sustainability involving safe and healthy communities in isolated locations, http://www.islandvulnerability.org; and (iii) risk education for health and disasters, http://www.riskred.org

Andrea Lampis is associate professor at and, at present, also the director of the Department of Sociology of the Universidad Nacional de Colombia, Bogotá. His research line, Global Risks and Local Vulnerabilities, combines the study of urban vulnerability and poverty with environmental risk and the socio-institutional implications of climate change adaptation. He has recently co-edited *Untamed Urbanisms* (2016) and edited *Cambio Ambiental Global, Estado y Valor Público: La Cuestión Socio-Ecológica en América Latina entre Justicia Ambiental y Legítima Depredación*.

Carl Lavery is a Professor of Theatre and Performance at the University of Glasgow, Scotland. His research interests are in the fields of ecology and environment, contemporary French theatre and performance and performance writing. Much of this work is informed by a crossdisciplinary interest in site, politics, and aesthetics.

Judy Lawrence is Research Fellow at the Climate Change Research Institute, Victoria University of Wellington, New Zealand. Judy is Co-Chair of the New Zealand Government's Climate Change Adaptation Technical Working Group and previously was Convenor of the National Science Strategy Committee on Climate Change and Director of the New Zealand Climate Change Office. Judy's research interests include governance and institutional design for decision making at local and national scales for addressing uncertainty and changing climate, including application of dynamic adaptive pathways planning. http://www.victoria.ac.nz/sgees/research-centres/ccri/resources

Loïc Le Dé is a lecturer in Emergency and Disaster Management from the School of Public Health and Psychosocial Studies at Auckland University of Technology (AUT) in New Zealand.

His research focuses on disaster risk reduction and management, migration and remittances, livelihoods assessment and strengthening disaster risk reduction, and participatory tools for disaster risk reduction. https://www.aut.ac.nz/profiles/health-sciences/lecturers/loic-le-de

Virginie Le Masson is a Research Fellow working for the Social Development and Risk and Resilience programmes at the Overseas Development Institute in London. With a PhD in human geography, her research focuses on the gender dimension of disaster risk reduction and climate change mitigation and adaptation. Before joining ODI, Virginie worked with local Non-Governmental Organisations in the Philippines and with the French Red Cross's disaster risk management programme in Reunion Island, Indian Ocean.

James Lewis, Datum International, http://www.datum-international.eu, author of 'Development in Disaster-prone Places: Studies of Vulnerability', 1999 (IT Publications [Practical Action], London) is an independent researcher and writer on causes of vulnerability to natural hazards, corruption and poverty, island vulnerability, climate change, resilience and capacity, and interconnections with socio-economic development; inspired by field assignments of former international consultancy in North Africa, South Asia and island states of the South Pacific, Caribbean, and Indian Oceans.

Brian Linneker is an independent scholar and senior research fellow in economic geography at London Universities (@DrBrianLinneker). He has worked for over twenty-five years in academia, government and NGOs on poverty, vulnerability and social exclusion, both in developed and developing world contexts.

Emma Lovell is a researcher in the Risk and Resilience Programme at ODI. She has an MA in Disasters, Adaptation and Development from King's College London, and her areas of work include disaster risk reduction (DRR), climate change, resilience, and gender and social inclusion in DRR. Before ODI, Emma was based in Bangkok, Thailand, where she worked for the United Nations Economic and Social Commission for Asia and the Pacific (UNESCAP) and the Asian Disaster Preparedness Center (ADPC).

Darren Lumbroso is a chartered civil engineer and technical director at the independent research organisation HR Wallingford in the UK. He has over twenty years experience carrying out studies on risk assessments, early warning systems and disaster risk reduction for weather-related hazards. He has worked in some forty countries worldwide including ones in sub-Saharan Africa, Asia, North America, the Pacific and the Caribbean.

Emmanuel M. Luna is a Professor of Community Development at the College of Social Work and Community Development, University of the Philippines. He has been in the field of Community-Based Disaster Management and Disaster Risk Reduction studies and practice for twenty-five years. He is a Co-Editor of the journal *Disaster Prevention and Management*.

Gregor Macara has been employed as a Climate Scientist at the National Institute of Water and Atmospheric Research (NIWA) in Wellington, New Zealand since 2013. Prior to this, he completed his Master of Science degree (Geography) at the University of Otago, with his thesis exploring the spatial and temporal variability of snowpack stability. His areas of interest include climate and weather extremes, climate change and New Zealand's cryosphere.

Nora Machado des Johansson's (CIES, Lisbon University Institute; CLEPUL, University of Lisbon; and Sociology at the University of Gothenburg) main areas of research and teaching are ethics, cultural sociology, social psychology and institutional and governance analysis. Among her publications are articles in *Public Administration, Social Science and Medicine, Death Studies* and *Human Systems Management* plus a book *Using the Bodies of the Dead: Ethical and Organizational Dimensions of Organ Transplantation*. She is currently doing research on the sociology of sacrality and religious movements.

Stavros Mavrogenis is a researcher at the European Centre for Environmental Research and Training, Panteion University of Athens, Greece. His main field of interest is climate change adaptation law and theory with focus on the case of Small Island Developing States, disaster risk reduction in the Global South, and disaster diplomacy.

Jessica Mercer is currently an independent consultant with Secure Futures (www.secure-futures.net) where she works predominantly in disaster risk reduction including climate change adaptation. Prior to establishing 'Secure Futures' for the last ten years Jessica has worked with NGOs, academia, UN agencies and other international organisations across Asia, Africa and the South Pacific in disaster risk reduction including climate change adaptation.

Daria Mokhnacheva works in the Migration, Environment and Climate Change Division at the International Organization for Migration and focuses on the impacts of disasters and environmental change on human mobility. Before joining IOM, she worked with UNDP in Russia, and with IDDRI in Paris. She has co-authored *The Atlas of Environmental Migration* and several papers on migration, environment, and climate change. She holds an undergraduate degree from the University of Cambridge and a Masters at Sciences Po Paris and Columbia University.

Fatima Gay J. Molina serves as Head of the Advocacy, Partnership and Networking Program of the Center for Disaster Preparedness Foundation. She has implemented projects on advocacy, research, and training for disaster risk reduction (DRR) and climate change adaptation (CCA) with UN agencies, international NGOs, and humanitarian groups since 2008. She specialises in disaster and environmental anthropology, child-centred DRR and CCA, and policy, governance, and local and indigenous knowledge studies. She has a Postgraduate Diploma in Children, Youth and Development. She is currently completing her Masters in Anthropology.

Virginia Murray was appointed as Consultant in Global Disaster Risk Reduction for Public Health England in April 2014. This appointment is to take forward her work as vice-chair of the UNISDR Scientific and Technical Advisory Group Qualified in medicine. She has over forty years' experience in disaster reduction and was a coordinating Lead Author for chapter 9 Case Studies for the IPCC Special Report on Managing the Risks of Extreme Events and Disasters to Advance Climate Change Adaptation, 2012.

Livhuwani David Nemakonde worked for the National Government of South Africa in the past fifteen years before joining the African Centre for Disaster Studies in February 2015. He holds a Master's degree in Public Administration (University of Pretoria) and a Master's degree in Development Studies (University of Free State). He submitted his PhD thesis in May 2016 with the title: Integrating parallel structures for disaster risk reduction and climate change adaptation in the Southern African Development Community. Livhuwani's research interests are disaster risk governance, Eco-disaster risk reduction, climate change adaptation, climate smart agriculture, and natural resource management.

Calum T.M. Nicholson (BA Cantab; MPhil Oxon) holds a PhD in Human Geography and is the author of the forthcoming *Of People: A Discourse about the Conflict Between Narratives and Relationships*. Philosophically, his work considers the relationship between politics, knowledge, and education. Thematically, his work is illustrated through case studies from the applied social sciences, specifically the debate on climate-induced migration.

Dewald van Niekerk is a Professor in Disaster Risk Reduction and the founder and Director of the African Centre for Disaster Studies at North-West University, Potchefstroom Campus. Dewald is a C2 NRF graded researcher. His interests include community based disaster risk management, disaster risk assessment, disaster risk governance, building institutional capacities for disaster risk reduction, and transdisciplinary disaster risk reduction.

Ian O'Donnell leads on research and knowledge management at the Red Cross Red Crescent Global Disaster Preparedness Center. He has developed key components for the One Billion Coalition for Resilience, a multi-partner initiative to engage people and communities around the world in resilience action. Previously, he managed programs in urban resilience, risk financing, and preparedness planning with the Asian Development Bank, ProVention Consortium, and American Red Cross. He worked early in his career at NeXT Software and was a Peace Corp volunteer in Mali.

Richard S. Olson, PhD is Professor of Politics and International Relations and Director of the Extreme Events Institute and the International Hurricane Research Center at Florida International University. With co-authors, recent publications include 'From Disaster Event to Political Crisis: A "5C+A" Framework for Analysis', *International Studies Perspectives* (2010); 'Establishing Public Accountability, Speaking Truth to Power, and Inducing Political Will for Disaster Risk Reduction: "Ocho Rios+25"', *Environmental Hazards* (2011); and 'Disasters as Crisis Triggers for Critical Junctures? The 1976 Guatemala Case', *Latin American Politics and Societies* (2013).

Sarah Opitz-Stapleton works at the intersection of water resources, climate services and social vulnerability and risk analysis and solutions for climate adaptation and disaster risk reduction programs in Asia (China, India, Nepal, Thailand and Vietnam) and Latin America (Colombia, Ecuador and El Salvador). She is a consultant with Plan8 Risk Consulting and Senior Affiliate Scientist with ISET-International.

Mark Pendleton is a cultural historian and Lecturer in Japanese Studies in the School of East Asian Studies at the University of Sheffield, UK. His research interests include twentieth-century Japanese history and the politics of memory in East Asia.

Sandrine Revet is a social anthropologist and Research Fellow at the Sciences Po's Centre for International Studies (CERI) in Paris. Her work focuses on the anthropology of disasters in Venezuela and Latin America (*Anthropologie d'une catastrophe*, 2007) and on the international governance of 'natural' risks and disasters (*Disasterland*, forthcoming). She is a co-founder of the Anthropological Association for Research on Disasters and Risks (ARCRA) and of the Disaster and Crisis Anthropological Network (DICAN).

Manik Saha is working as the Senior Programme Manager of the British Red Cross in the Bangladesh Country Office. He has a background in Project Management specialising in Food Security and Livelihoods and DRR. He has expertise in planning, designing, and implementing, with monitoring and evaluation, projects. Recently, he worked as the focal point of integrated Vulnerability-to-Resilience projects covering both rural and urban communities in Bangladesh. He holds a Master's degree from the University of Dhaka, Bangladesh in Disaster Management.

Juan Pablo Sarmiento, MD, MPH, is a Research Professor at the Florida International University's Extreme Events Institute. He is a Medical Doctor with post-graduate studies in Disaster Management, and Public Administration; he has a specialisation degree in Medical Education, and a residency in Nutrition. He holds Master's degrees in Project Management, and Public Health, Specialty Promotion and Community Development. He has worked for the Pan American Health Organization (PAHO/WHO), and the United States Agency for International Development's Office of Foreign Disaster Assistance (USAID/OFDA).

Wendy Saunders is the Senior Natural Hazards Planner at GNS Science in Lower Hutt, New Zealand. Wendy specialises in assisting local, regional and central government to improve their land use planning policies and plans for risk reduction. In recent years Wendy has developed a risk-based approach to manage the consequences of natural hazards for land use planning, which has been implemented at regional and district levels (http://gns.cri.nz/Home/RBP/Risk-based-planning/A-toolbox).

Erica Seville is co-Leader of Resilient Organisations and Director of Resilient Organisations Ltd. Resilient Organisations is a public-good research programme working collaboratively to improve the resilience of organisations so they can both survive adversity and thrive in a world of uncertainty. Erica has authored over 100 research articles and is a regular international speaker on resilience. She is a member of the leadership team, and leads the Pathways to Resilience Flagship within QuakeCoRE, the New Zealand Center of Research Excellence dedicated to improving earthquake resilience. She is also a member of the Resilience Expert Advisory Group (REAG), providing advice and support to the Australian Federal Government on organisational resilience issues. Erica is an Adjunct Senior Fellow with the Department of Civil and Natural Resources Engineering at the University of Canterbury and has a PhD in risk management.

Rajib Shaw is the Executive Director of the Integrated Research on Disaster Risk (IRDR) programme. He is also the Senior Fellow of the Institute of Global Environmental Strategies (IGES) Japan, and the Chairperson of SEEDS Asia, a Japanese NGO. Previously, he was a Professor in the Graduate School of Global Environmental Studies of Kyoto University, Japan. His expertise includes community-based disaster risk reduction, climate change adaptation, urban risk management, and disaster and environmental education.

Asha Sitati is a postgraduate student at University College London (UCL). Prior to joining UCL, she worked as an Associate Programme Officer on an Emerging Issues Project at the United Nations Environment. She also worked for two years on a CLIM-WARN project at the UN Environment researching ways to design efficient, integrated and actionable early warning systems for climate-related hazards. She has a BSc in Environmental Science with Information Technology and an MA in Environmental Planning and Management.

Joanne R. Stevenson is a Senior Analyst with Resilient Organisations Ltd and an Adjunct Senior Fellow with the Department of Management, Marketing, and Entrepreneurship at the University of Canterbury. As a social systems researcher, her work explores vulnerability, recovery, and resilience of social-ecological and socio-economic systems; organisation and social network theory; and the role of place and space in organisational decision-making and outcomes. Joanne has her PhD in Geography and Commerce from the University of Canterbury.

Karen Sudmeier-Rieux is a senior researcher at the University of Lausanne, Institute for Earth Science, where she manages a research project on landslide risk, bio-engineering, migration, resilience and vulnerability of mountain communities in Western Nepal. She is also an education and training consultant with the United Nations Environment Programme, with whom she conducts training workshops on ecosystem-based disaster risk reduction.

Petros Theodorou is an independent researcher currently working with Sustainable Futures, a London-based advisory firm on climate change adaptation, risk reduction, and resilience. He has an academic background in economics, sustainable development, and environmental governance. Petros has working and research experience on the climate change adaptation-migration nexus in Southeast Asia and Europe.

Freddy Vinet is a professor of geography at the University of Montpellier (France) with twenty years' experience in teaching and leading research on the assessment of damage due to disasters. He participated in numerous post-disaster field studies; e.g., after the 2004 tsunami in Indonesia and after the storm surge 'Xynthia' in western France in 2010. He leads a Master's degree course on disaster prevention at the University Paul Valery in Montpellier (France), teaching students on methods and tools (especially GIS) to reduce the consequences of disasters.

Elisabeth Vogel is a doctoral student at the University of Melbourne and the Potsdam Institute for Climate Impact Research. Her research interests are centred on the impacts of climate extreme events, with a particular focus on the influence of weather extremes on global food production. As part of her research, Elisabeth worked with the CLIM-WARN project of the United Nations Environment Programme (UNEP) in Nairobi, Kenya, researching ways and methods to design effective early warning systems for climate-related hazards.

Elizabeth Wagemann is an architect currently conducting doctoral research at the University of Cambridge on post-disaster accommodation in Latin America. She graduated from the Catholic University of Chile with a Bachelors in Architecture (2001) and a Masters in Architecture (2005), and received an MPhil in Architecture (2012) from the University of Cambridge. She has experience in designing low-cost and post-disaster housing in Chile, Peru, Ecuador, Brazil, and the Philippines, and has been a lecturer at the Catholic University of Chile.

Rory Walshe is a PhD Student in the Geography department of King's College London. He has a special interest in Small Island Developing States, climate change adaptation, disaster risk reduction, and traditional knowledge, particularly for indigenous communities. His PhD is examining the assumptions surrounding climate change impacts on small islands.

Chadia Wannous is a Senior Advisor at the UN Office for Disaster Risk Reduction (UNISDR), coordinating implementation of the health components of the Sendai Framework for Disaster Risk Reduction and supporting the Science and Technology Secretariat. Prior to this she was

Senior Policy Advisor to the UN Secretary General Special Envoy on Ebola and the UN System Influenza Coordination. Chadia is a public health professional with a PhD in International Health and Development from Tulane University with twenty years of work experience.

Gideon Wentink has worked at the African Centre for Disaster Studies since 2010 as a researcher. He has been involved in various research projects, national and international, where he contributed as a researcher. In 2013 he obtained his Master's degree in Development and Management. He is also a lecturer for the Postgraduate diploma in Public and Non-Profit Management (Disaster risk studies). His research interests are in Communication, Disaster Risk Reduction, legislation, and governance.

Gustavo Wilches-Chaux devoted his life to environmental education and public information, participatory risk management and climate adaptation, having in his hands post-disaster reconstruction processes after the earthquakes that destroyed his city, Popayán, in 1983, and the region of Tierradentro in 1994. He is based in Colombia and works as an independent consultant, professor in Universidad Externado de Colombia, and writer, seeking to include communities at risk with their knowledge and their local tools.

Emily Wilkinson is a Research Fellow at the Overseas Development Institute in London, UK. She has fifteen years' experience working on collective responses to environmental hazards. Emily is an internationally renowned expert in disaster and climate risk governance and has published extensively on these topics.

Ben Wisner has worked on people-focused, rights-driven development in Africa, Latin American, the Caribbean, and Asia since 1966. He is a pioneer using participatory action research for community based disaster reduction, leading these efforts for the ProVention Consortium (http://www.proventionconsortium.net) and is advisor to the GNDR (http://www.globalnetwork-dr.org) and Earthquakes without Frontiers (http://ewf.nerc.ac.uk).

Zehra Zaidi is a sustainable development expert, specialising in disaster risk reduction, and is currently Marie Curie Research Fellow at Fondazione Eni Enrico Mattei (FEEM). Her work focuses on institutional risk governance, organisational capacity assessment, and the development of risk management strategies and adaptive capacity to improve resilience at both the institutional and community levels. Areas of expertise include national disaster risk management strategies and institutional capacity building, including the role of government and social institutions in influencing adaptive behaviour and decision making.

Zinta Zommers is a DRR, climate change and resilience technical specialist with the Office of Food for Peace, USAID. She has also worked with the UN Secretary-General's Climate Change Support Team and as a Programme Officer with the UN Environment Programme. Zinta led UNEP's Climate Change Early Warning Project, working directly with communities in Kenya, Ghana and Burkina Faso. She co-authored the book, *Preventing Disaster: Early Warning Systems for Climate Change*. Prior to joining the UN, Zinta was a Junior Research Fellow at the University of Oxford.

FOREWORD

Dancing with Donors, Dicing with Death

Ben Wisner

When I was asked to write this foreword and contemplated the rich material in this book and thought about what is at stake, my mind turned to folk dancing. There are such clear and apparent synergies to be gained by embedding CCA within DRR or at least coordinating efforts, it is puzzling why donors, government institutions, non-profits and researchers continue to pursue DRR and CCA separately. Equally obvious is the urgency of evidence-based action in the face of increasing disaster losses. Deeply entrenched interests and competition among donors, government departments, NGOs and, alas, career-aware researchers have impeded the integration of CCA and DRR. This is the dance. Most of the dancers hear the same music. The *desastrólogos* (disasterologists) are aware that roughly 80 per cent of disaster deaths and 70 per cent of economic losses occur in climate-related events. Climate scientists and policy advocates are perhaps less aware of, but many appreciate, the partial success made by diverse DRR actors in mainstreaming disaster awareness into everyday governance, especially at the local scale – the county and municipality. Yet the steps in this complicated dance separate partners, swing them around and introduce them to yet new rhythms as the 'calls for research proposals' echo around the dance floor.

This long-awaited and welcomed book makes a compelling case that a common research agenda and a series of practical policy and policy recommendations should and *can* be put in place. If they are not, the death toll, especially among people at the margin of society, and the destruction of their livelihoods will accelerate. This is what I mean by 'dicing with death'. Human existence should not depend on the roll of dice or the accident of birth. A farm family in Tanzania, Niger or Malawi should enjoy social protection and access to resources, infrastructure and technical knowledge as the equally drought-exposed farm family in Australia (Burton *et al.* 1993; Smucker and Wisner 2008). A roll of the dice and accident of birth should not determine whether a school child has lessons in a structure susceptible to destruction by flood, landslide, fire, high wind or earthquake (Petal *et al.* 2015), or whether while studying or at play, the school child is vulnerable to electrocution by lightning.

A report by the UN published in 2015 looked back twenty years at *The Human Cost of Weather Related Disasters* (CRED 2015). It claimed that 90 per cent of major disaster losses are associated with 6,457 floods, storms, heatwaves, droughts and other weather-related events during that period. More than 600,000 people lost their lives in these events that also accounted for 71 per cent of economic loss from all natural hazard-triggered disasters. During that same period there were approximately 130,000 deaths in disasters triggered by earthquakes and volcanoes;

thus climate-related events account for about 78 per cent of all natural hazard-related disasters (EM-DAT 2016; Livescience 2005).

The one most useful lesson that DRR has to offer is the fruit of more than a hundred years of field research: disasters are not natural. Whenever and wherever people suffer and assets and livelihoods are lost in a hazard event, one has to look at the vulnerability of these people and at underlying risk factors or root causes. Applied to CCA, this means that 'adaptation' with a big 'A' by the state and corporations should not submerge and distort or block adaptation with a little 'a' by ordinary farmers, fishers, herders, foresters, traders and craft workers. In Tanzania, government has used the rhetorical call of 'the climate imperative' to justify huge land grabs so that foreign agribusiness companies can 'feed the country', displacing small farmers and pastoralists who, studies have shown, are quite adept at adapting (with a small 'a') spontaneously to climate change (Smucker *et al.* 2015). The root causes or underlying risk factors in such a case are not addressed by conventional CCA programmes such as REDD+, however well meaning the intentions of the Norwegians and other donors. A long series of critical studies of food security that date at least from Meillassoux's classic work with a committee on the Sahel famine (Comité d'Information Sahel 1974) have exposed such root causes. Reading (or, perhaps, re-reading) this literature would help to refocus CCA and root it more deeply in the daily reality of resource-dependent primary producers (Wisner 1988).

The one most useful lesson that CCA has got to offer is a generous approach to time and the notion that there is no going back, that change is a constant and the most prudent approach is transformation. This is an almost Buddhist conception of existence as impermanence. Applied to DRR, it means that the conventional notion of a 'disaster management cycle' has to be abandoned. A return to the status quo ante in terms of rights, power, access to resources, land use and mobility – and alas, it is the status quo that usually results despite however much rhetoric there is about 'building back better' – can only result in another disaster (Boano 2014; Susman *et al.* 1983).

But what of the geological hazards, the earthquakes for instance, that capture headlines and send humanitarian agencies scrambling? In fact, even here fruitful collaboration between DRR and CCA is possible and desirable. Nepal is a case in point. Recovery from the April 2015 earthquake has barely begun as I write in August 2016. Survivors have weathered two monsoon seasons and a harsh winter. Now another winter approaches and most of the 700,000 homes damaged or destroyed are yet to be rebuilt (Shelter Cluster 2015). In Nepal both monsoon rains and earthquakes may trigger landslides. Earthquake related landslides block rivers and cause flooding, as do the monsoon rains. One would think that repair and replacement of roads, bridges, location of health centres, rural schools and homes would benefit from the longer-term perspective that climate adaptation brings to planning. Yet in Nepal donor-funded DRR and CCA projects run along in parallel with little or no connection.

A final reason why this book is so timely and welcome is this. Ordinary people experience, perceive, respond and adapt to what goes on around them holistically. This is well known by those researchers and practitioners who work with civil society and local governments in community based DRR (CBDRR). Recent research has highlighted the significance of everyday threats in the social psychology of risk perception. Crime, domestic violence, polluted water, power outages and other social, economic and environmental risks came up in more than 6,000 open-ended conversations with people in fifteen low and medium income countries (Gibson and Wisner 2016). The poorest, most marginalised rural people and their urban cousins who inhabit the world's burgeoning megacities live in a condition of informality, complexity, insecurity and fragility (GNDR 2015, p. 5). In order to interest such people in CCA, it may be necessary to begin with helping them deal with a perceived priority that has apparently nothing to do with risk – climate-related or not. Equally, a study of more than eighty applications of CBDRR revealed that those that made the step from outputs (a hazard map, a plan, etc.) to outcomes (action) often had a link with a non-disaster risk related problem perceived as a priority in the community (Wisner *et al.* 2008).

All concerned with DRR and with CCA need to work together – researchers, policy makers, advocates, administrators and officials. Time is too short and the stakes in terms of human life and wellbeing are too high for continuing competition over terminology, buzzwords, grant money and academic prestige. I finish with two equally salient situations – one squarely to the so-called 'climate' side of things, one very clearly belonging to the domain of 'non-climate' hazards. Ten per cent of humanity lives in the two per cent of the earth on coasts less than ten metres above sea level. Sixty per cent of them are urban (IIED n.d., p. 3), and at least half of these are slum dwellers. Thus hundreds of millions of people are potentially exposed to the hazards of sea level rise with few resources to allow them to adapt. At the same time, hundreds of millions of children go to school in seismically active areas, and relatively few schools worldwide have been built to reduce earthquake risk (Petal *et al.* 2015; Plan International 2016; Wisner 2006). The methods and lessons learned by DRR and by CCA can help deal with all risk, and that cooperation should begin *now*!

Ben Wisner
University College London, London, UK and
Oberlin College, Oberlin, Ohio, USA

References

Boano, C. (2014) 'Post-disaster recovery planning: Introductory notes to its challenges and potentials', in A. Lopez-Carresi, M. Fordham, B. Wisner, I. Kelman and JC Gaillard (eds), *Disaster Management: International Lessons in Risk Reduction, Response and Recovery*, London: Earthscan, pp. 191–210.

Burton, I., Kates, R. and White, G. (1993) *The Environment as Hazard*, 2nd edition, New York: Guilford.

Comité d'Information Sahel, C. (1974) *Qui se nourrit de la famine en Afrique?*, Paris: Maspero.

CRED (Center for Research on Epidemiology of Disasters). (2015) *The Human Costs of Weather Related Disasters*, Brussels and Geneva: CRED and United Nations International Strategy for Disaster Reduction (UNISDR). Online http://cred.be/HCWRD (accessed 20 September 2016).

EM-DAT. (2016) *Cred Crunch 42*, Brussels: Center for Research on the Epidemiology of Disasters. Online https://twitter.com/creducl/status/720934305745235968 (accessed 20 September 2016).

Gibson, T. and Wisner, B. (2016) 'Lets talk about you . . . Opening space for local experience, action and learning in disaster risk reduction', *Disaster Prevention and Management: An International Journal*, 25, 5: 664–684.

GNDR (Global Network of Civil Society Organisations for Disaster Reduction). (2015) *We Need a Reality Check*, London: GNDR. Online http://www.gndr.org/learning/resources/gndr-publications/item/1461-we-need-a-reality-check.html (accessed 20 September 2016).

IIED (International Institute for Environment and Development). (nd) *Climate Change and the Urban Poor: Risk and Resilience in 15 of the World's Most Vulnerable Cities*. London: IIED. Online http://pubs.iied.org/G02597/ (accessed 20 September 2016).

Livescience. (2005) *Global Natural Disaster Deaths 1990–2005*. Online http://www.livescience.com/416-global-natural-disaster-deaths-1990-2004.html (accessed 20 September 2016).

Petal, M., Wisner, B., Kelman, I., Alexander, D., Cardona, O., Benouar, D., Bhatia, S., Bothara, S., Dixit, A., Green, R., Kandel, R., Monk, T., Pandey, B., Rodgers, J., Sanduvaç, Z. and Shaw, R. (2015) 'School seismic safety and risk mitigation', in M. Beer, I. Kougioumtzoglou, E. Patelli and I. Au (eds), *Encyclopedia of Earthquake Engineering*, Berlin: Springer, pp. 2450–2460.

Plan International. (2016) *Safe Schools Global Programme*. Online https://plan-international.org/publications/safe-schools-programme (accessed 20 September 2016).

Shelter Cluster. (2015) *Needs Analysis: Nepal Earthquake 2015*, Global Shelter Cluster: Coordinating humanitarian shelter. Online https://www.sheltercluster.org/response/nepal-earthquake-2015 (accessed 20 September 2016).

Smucker, T. and Wisner, B. (2008) 'Changing household responses to drought in Tharaka, Kenya: Persistence, change, and challenge', *Disasters* 32, 2: 190–215.

Smucker, T., Wisner, B., Mascarenhas, A., Munishi, P., Wangui, E., Sinha, G., Weiner, W., Bwenge, C. and Lovell, E. (2015) 'Differentiated livelihoods, local institutions and the adaptation imperative: Assessing climate change adaptation policy in Tanzania', *Geoforum* 59: 39–50.

Susman, P., O'Keefe, P. and Wisner, B. (1983) 'Global Disasters: A radical interpretation', in K. Hewitt (ed.), *Interpretations of Calamity from the Viewpoint of Human Ecology*, Boston: Allen & Unwin, pp. 263–283.

Wisner, B. (1988) *Power and Need in Africa*, London: Earthscan.

Wisner, B. (2006) *Let our Children Teach Us! A Review of Education and Knowledge in Disaster Risk Reduction*, London and Geneva: ActionAid and United Nations International Strategy for Disaster Reduction (UNISDR). Online https://www.unisdr.org/we/inform/publications/609 (accessed 20 September 2016).

Wisner, B., Haghebaert, B., Schaerer, M. and Arnold, M. (2008) *Community Risk Assessment Toolkit*, Geneva: ProVention Consortium. Online http://www.proventionconsortium.net/?pageid=39 (accessed 20 September 2016).

PART I

Vocabularies and Interpretations

1

EDITORIAL INTRODUCTION TO THIS HANDBOOK

Why Act on Disaster Risk Reduction Including Climate Change Adaptation?

Ilan Kelman, Jessica Mercer, and JC Gaillard

Why a handbook on disaster risk reduction (DRR) including climate change adaptation (CCA)? Part of the answer lies in the definitions and application of these terms.

DRR Includes CCA

According to the 2014 glossary of the Intergovernmental Panel on Climate Change (IPCC), CCA is 'The process of adjustment to actual or expected climate and its effects. In human systems, adaptation seeks to moderate harm or exploit beneficial opportunities. In natural systems, human intervention may facilitate adjustment to expected climate and its effects' (http://ipcc-wg2.gov/AR5/images/uploads/WGIIAR5-Glossary_FGD.pdf). In simple terms, it is about dealing with climate.

The key word is 'adjustment', noting that 'adaptation' is, in effect, defined as 'adjusting'. In fact, 'adjustment' is the term that the disasters field used for several decades prior to climate change being accepted as a global issue. Not all languages differentiate between 'adapt' and 'adjust', making it difficult to explain the difference, especially for policies and on-the-ground action. The forms of action that CCA suggests are sensible: reducing harm from any potential hazards which manifest as the climate changes while taking advantage of emergent opportunities.

Examples are dealing with floods, storms, landslides, and droughts to avoid disasters occurring; having reliable water and food supplies; ensuring healthy ecosystems; and developing overall livelihood strategies that are suitable under a wide range of climatic conditions. These activities are basic development and sustainability processes. No activities new or exclusive to climate change are enacted, instead focusing on needed development approaches. CCA, as a standalone approach to dealing with climate processes, therefore reasserts well-known and long-accepted development priorities.

DRR is defined by the UN Office for Disaster Risk Reduction (UNISDR, formerly the secretariat of the United Nations International Strategy for Disaster Reduction) as 'The concept and practice of reducing disaster risks through systematic efforts to analyse and manage the causal factors of disasters' (http://www.unisdr.org/we/inform/terminology). This definition is a bit circular, in that 'disaster risk reduction' is defined as 'reducing disaster risk', plus many terms such as 'risk', 'disaster', and 'disaster risk' can be and have been deconstructed and critiqued. On the positive side, this definition incorporates theory and practice, links to understanding and acting, and highlights the importance of root causes to avoid problems in the first place.

The definitions of DRR and CCA make it evident that they have numerous similarities. Despite much debate and separation, there should never be any construction of DRR and CCA as being opposites, as opposing one another, or as precluding each other. Instead, to a large degree, DRR's definition embraces the elements of CCA's definition, generalising it beyond climate and making it more straightforward. DRR's inclusion of CCA can be examined further to highlight differences between them (Table 1.1).

For factor 1 in Table 1.1 (hazards), CCA, by definition, deals with only climate and hence with only climate-related environmental hazards. Examples are too much or too little precipitation, fog, storms, and wind. Some environmental hazards such as wildfires and landslides can be climate-driven, climate-influenced, or with limited connection to climate. Some biological and geological environmental hazards – such as epidemics, earthquakes, volcanic eruptions, and tsunamis – have links to climate, and on occasion influence or are influenced by climate, but are generally considered to be comparatively separate from climate change. One notable exception is climate change's influence on many disease vectors, such as by expanding their geographic ranges, speeding up their life cycles, or increasing precipitation intensity which washes away vector eggs and larvae. Climate influencers other than climate change, which can also substantially influence environmental hazards, include the El Niño-Southern Oscillation cycle, the North Atlantic Oscillation, and the Indian Ocean Dipole.

While CCA is about changes to the climate, DRR by definition deals with all environmental hazards and environmental hazard influencers, including climate and its trends, changes, variabilities, and cycles. All the environmental hazards mentioned in the previous paragraph – and all the environmental hazards and hazard influencers connected with climate change – are considered by DRR. Consequently, in terms of hazards considered, factor 1 in Table 1.1 (hazards) shows that DRR deals with the hazards relevant to CCA and many more.

Consequently, factor 2 in Table 1.1 (timescale) emerges. Because DRR deals with all hazards, by definition, DRR must deal with all timescales, by definition, from sudden-onset hazards such as earthquakes through to decades-long trends in climate. DRR actions also encompass all timescales, from learning the basics of first aid which can be achieved in a few hours to formulating, promulgating, and enforcing design, building, and planning codes which could require decades. Whether short-term or long-term, DRR covers environmental hazards, hazard influencers, and actions to deal with those hazards across all timescales.

Conversely, CCA by definition tackles climate, which means long-term interests. Climate, by the IPCC's definition (http://ipcc-wg2.gov/AR5/images/uploads/WGIIAR5-Glossary_FGD.pdf), is the long-term average of weather parameters (e.g., air temperature, wind speed and direction, and precipitation volume and rate) usually over a period of thirty years. Therefore, CCA considers dealing with long-term influences, usually seeking long-term actions. Short-term phenomena and responses are not typically the first priority of CCA.

Table 1.1 Comparing DRR and CCA

Factor	DRR	CCA
1. Hazards	All hazards	Only climate-related hazards
2. Timescale	Long and short term	Long term
3. Society–environment interactions	Society ←→ environment	Environment → society

Since CCA covers the long-term while DRR covers all timescales, factor 2 in Table 1.1 (time-scale) demonstrates that DRR deals with the timescales relevant to CCA and much more.

Factor 3 in Table 1.1 (society–environment interactions) illustrates the final potential divergence between CCA and DRR: society–environment interactions. DRR's definition explicitly states that its purpose is to address 'the causal factors of disasters'. Implicit is that, irrespective of the origin of the root causes, they will be addressed, whether from society, from the environment, or from their interaction. The key is to make decisions that avoid disasters manifesting in the first place or that reduce disaster impacts.

In contrast, CCA aims for society to react 'to actual or expected climate and its effects' (see the IPCC definition above). That is, await climatic stimuli or changes, or analyses about climatic stimuli or changes, and respond to this phenomenon, actual or expected. CCA is waiting for the environment to change, or to have expectations of the environment changing, and then human systems respond.

The approaches of both CCA and DRR are needed. We do not and cannot know exactly what the environment will do, so it is sometimes appropriate to wait, to run models, to project outcomes, and to estimate consequences in order to react to that information, as per CCA. DRR also involves these elements, which is often integrated into disaster preparedness and emergency planning, as well as being foundational for many aspects of damage mitigation and disaster prevention. DRR, however, does more than respond to actual or expected environmental stimuli, deliberately setting out to reduce or eliminate risks from environmental hazards and hazard influencers based on societal needs, not just on what the environment does or is expected to do. As such, factor 3 (society–environment interactions) illustrates that DRR encompasses CCA's activities and covers much more than CCA.

Climate change mitigation, to some extent, seeks to reduce or eliminate risks due to climate change by stopping climate change from happening – or, at least, by reducing the ensuing changes to the climate and environment. Climate change mitigation has been deliberately separated from CCA to a large degree by, more or less, setting up climate change mitigation and CCA as different fields with different institutions, different vocabularies, and different strategies. The origins of this separation are obscure, especially given that many were advocating for keeping them joined as the IPCC was being founded.

Three key factors of the DRR and CCA definitions are examined in Table 1.1, demonstrating that, in each case, DRR does what CCA does – and much more. Consequently, DRR encompasses CCA meaning that CCA sits as a subset within DRR.

Not accepting that DRR includes CCA could lead to problems, such as successfully adapting to climate change while increasing disaster risk. Consider a school that is built in a location outside the expected floodplain under climate change, demonstrating CCA (as well as DRR for floods). The school might also be built with its own renewable energy systems for local electricity generation alongside natural ventilation reducing the need for electricity; that is, implementing climate change mitigation. If the school is built without fire-resistance measures, and could be in a seismic zone without earthquake-resistance measures, then CCA has been achieved without DRR.

Conversely, if a school is meant to be disaster-resistant, then all hazards will be addressed. It will have fire-resistance measures, earthquake-resistance measures, and will be built outside the expected floodplain under climate change. DRR is achieved and, by consequence, CCA is achieved. DRR also means maintaining essential services after an environmental hazard appears. Reduced electricity demand and local electricity generation can assist in maintaining electricity needs by not relying on external sources. Implementing DRR for a school could also mean having its own renewable energy systems for local electricity generation alongside natural ventilation design.

In summary, pursuing CCA or climate change mitigation does not inevitably yield full DRR. Pursuing DRR, by definition, means implementing all CCA measures that are needed and often many climate change mitigation measures. Based on basic definitions from IPCC and UNISDR, DRR includes CCA.

CCA's Separation

Despite the importance of working together and joining forces to achieve a safer world, the definitions of DRR and CCA are often ignored in order to seek separation of climate change from wider development contexts. The view of climate change as being a separate issue is epitomised by the separate UN processes for it. IPCC seeks government member consensus on the synthesis and assessment of the state of climate change science. The United Nations Framework Convention on Climate Change (UNFCCC) seeks international regimes to address climate change, originally focused on climate change mitigation but now increasingly addressing CCA. Neither institution is well connected to DRR.

Attempts at linking climate change to wider processes such as DRR are made. IPCC commissions special reports on targeted topics. In 2012, a Special Report was published entitled 'Managing the Risks of Extreme Events and Disasters to Advance Climate Change Adaptation (SREX)'. The idea was to bring together DRR and CCA concepts to seek a melding of the ideas. The result was an expostulation of CCA's views of DRR without providing equal balance to DRR's views of CCA. Vocabulary, citations, and approaches from CCA dominated, with the report's ethos etching in the DRR–CCA separation by claiming some differences, rather than using basic definitions and all literature to explain how CCA sits within DRR.

UNFCCC accords, most notably the widely covered Paris Agreement from December 2015, often acknowledge the existence of DRR and its international processes, but rarely engage directly with them. Part of the reason is the legalities. UN processes for DRR and CCA are entirely divorced, with the CCA regime aiming for legally binding international treaties whereas DRR seeks voluntary agreements. Merging them would encroach on so much established territory and would threaten so many vested interests within the UN system and national governments, that from a practical perspective, it would be unlikely to succeed even if the suggestion were discussed at the highest levels. The DRR agreements place climate change increasingly high on the DRR agenda, emphasising the jurisdiction of UNFCCC over climate change related processes.

The slow takeover by climate change of DRR and other development concerns while enshrining the separation has three main consequences. First, CCA can be a distraction from root causes and wider concerns. The earlier example of the school illustrates. If climate change dominates, then new schools might be built for only CCA rather than for wider DRR and development. It makes much more sense to consider development including DRR so that CCA is automatically included, rather than being distracted by CCA so that other needed efforts fall by the wayside.

Additionally, focusing on only long-term priorities often distracts from short-term initiatives that ultimately underpin sustainable CCA. For example, strengthening the livelihoods of those vulnerable to environmental hazards, including changes in climate patterns, to enable them to meet their short-term needs will ultimately reinforce their ability to adapt over the long-term. Everyday needs tend to dominate people's priorities in all societies, unlike long-term climate change. There might be limited point in anticipating the relocation of those living on atolls threatened by sea-level rise and other changes to the ocean if the islanders do not have enough food to put on their plates or enough cash income to pay for their bills and their children's school tuition fees. In fact, focusing on long-term climate change often leads to eroding traditional response mechanisms to cope with environmental hazards, including changing climate patterns, as observed amongst many Pacific societies. Instead, all timescales ought to be considered simultaneously to tackle all the challenges, as DRR does, but not CCA.

Part of the distraction is the second consequence that climate change is proving to be a scapegoat for many other DRR and development challenges. After the 2004 and 2011 tsunamis, several scientific and non-governmental organisation (NGO) commentators suggested those events as being examples of the disasters expected under climate change, despite limited (not no) connection between climate change impacts and tsunami impacts. Hurricane Katrina was blamed on climate change, yet it was on the borderline between categories three and four at landfall, rather than being as extreme as a hurricane could be. Blaming climate change for Hurricane Katrina's devastating impacts further distracts from the disaster's long-term root causes. Climate change did not build New Orleans below sea level without adequate mechanisms to deal with flooding. Climate change did not elect George W. Bush as President of the USA or Ray Nagin as mayor of New Orleans. Climate change did not appoint Michael Brown as the head of the Federal Emergency Management Agency. Climate change is used as a scapegoat for pre-existing vulnerabilities that caused the Hurricane Katrina disaster.

Meanwhile, in the Maldives, the government is suggesting that people from outer islands must move closer to the capital before climate change inundates their settlements and destroys their livelihoods. Yet for several decades, the Maldivian government – first under a totalitarian dictatorship and then under elected presidents, although with disputes about the level of democracy in the country – has been suggesting that people from outer islands must move closer to the capital in order to make it easier (and presumably cheaper) for the government to provide basic services. Climate change is used as a scapegoat for chronic development challenges.

Part of the combination of climate change as a scapegoat and distraction is the third consequence of identifying climate change as a diffuse, top-down problem that technology can solve. Climate change is causing problems for people in Malawi, Bangladesh, and Florida who are not used to the climate regime into which we are moving and who lack the resources and options to deal with the challenges themselves without experiencing adverse consequences. Yet, all three jurisdictions have also suffered from internal problems including corruption, poor governance, mismanagement of development processes, and power imbalances, which keep many people in those locations poor and marginalised. Instead of fully admitting and comprehensively tackling those root causes, we now hear the need for rapid technological development to support Malawi, Bangladesh, and Florida in dealing with climate change.

Technological innovation and application will, nonetheless, contribute to assisting the Malawians, Bangladeshis, and Floridians who are vulnerable to climate change. It is also true that full CCA for those locations will have limited impact on people's lives and livelihoods because internal problems remain including corruption, poor governance, mismanagement of development processes, and power imbalances. Separating CCA from these wider contexts can be detrimental to implementing suitable development measures.

None of this discussion denies climate change impacts or obviates the need to tackle them. Climate change is already being highly damaging to many people and is a major, serious impediment to development – as are many other topics. Inequity, volcanic eruptions, poor governance, water resources, wealth imbalances, large dams, and pollution, amongst many other concerns, are being highly damaging to many people and are major, serious impediments to development. The relative weighting is contextual. Some people in Alaska and Papua New Guinea are experiencing severe livelihood problems and are moving almost exclusively due to climate change. Some people in Antigua, Russia, and Saudi Arabia are barely being influenced by climate change, instead experiencing severe livelihood problems due to the host of other factors mentioned including corruption and discrimination.

Climate change should be neither ignored nor permitted to dominate. It should complement development endeavours, not distract from or be a scapegoat for them. CCA is needed, although it is one of many needed development processes, so it should not be permitted to create or exacerbate problems. Despite the continual separation of climate change from many other contexts, it should always be viewed as sitting within, not separate from, wider contexts.

Beyond DRR Including CCA

There is no doubt that climate change has grabbed the political and common consciousness around the world. This energy and attention should be harnessed in order to bring the wider contexts into play and to indicate the importance of placing CCA within DRR so that we collaborate rather than compete. We can learn from and teach each other by melding our work, rather than spending energy constructing and maintaining fences.

DRR does not and cannot cover all topics, since DRR itself is a subset of wider development and sustainability processes. A school perfect with respect to DRR including CCA might not help in achieving development if only boys are permitted to attend or if only those who can afford to pay fees are permitted to attend. Wider concepts including justice and equity need to be considered to ensure that development problems are not created or exacerbated.

When we define DRR and CCA, and when we seek to apply them, these broader concerns emerge leading to broader approaches, much more encompassing than the single topic of climate change. This handbook puts together a balance of breadth and depth alongside a balance of theory, policy, and practice to describe and analyse CCA without losing sight of factors beyond, most notably DRR but also development and sustainability. The definitions give the direction for application. We are told directly what needs to be done, which this handbook details along with advice on how to do it.

Revisiting the question opening this Editorial Introduction, why this handbook? Because we need it to ensure that DRR including CCA succeeds.

2

EDITORIAL INTRODUCTION TO PART I: VOCABULARIES AND INTERPRETATIONS

Say What We Mean, Say What We Do

Ilan Kelman, Jessica Mercer, and JC Gaillard

Words, languages, vocabularies, meanings, and interpretations change. Terminology that was commonplace a century ago is racist and discriminatory today. In the meantime, new words and phrases emerge, becoming commonplace. These shifts happen across languages and cultures.

The same holds true for science, policy, and practice. Even the key phrases in this handbook's title, 'disaster risk reduction' (DRR) and 'climate change adaptation' (CCA) are jargon that evolved from previous vocabulary and which are not always easy to translate into different disciplines, different professions, different languages, or different cultures.

The field of DRR including CCA has also spawned its own buzzwords that are frequently challenging to explain, translate, interpret, understand, and apply. Even over the last decade, phrases have waxed and waned, notably social-ecological systems, resilience, human security, integral theory, transformation, and quantum social theory. Often, they purport to be new and innovative, but in reading history, it is fascinating to learn how much contemporary work is re-packaged, re-named, and re-labelled from what has long been published and, frequently, applied.

The words 'panarchy' and 'consilience', for instance, have birthed conferences, books, and what is purported to be new paradigms. Both words have been used in English for over 150 years, yet they are still not common vocabulary in English or other languages. They require extensive hand-waving and complex diagrams to convey their core points and meanings. Yet, bizarrely, their fundaments are as old as humanity: many cultures have long accepted the meanings of these concepts engrained within their cultures.

Why not learn from other cultures and use their vocabularies to express and apply these ideas? We could bring long-standing practice into academia rather than fabricating academic fields to foist them onto policy makers and practitioners.

The same applies to 'social-ecological systems' (SES) and its numerous variants including 'socio-ecological systems', 'socio-environmental systems', 'social-environmental systems', and 'coupled human and natural systems' (CHANS). Why SES and CHANS evolved separately is an interesting exploration in itself, but let's explore the phrases in more detail.

'Social' and 'human' in this context broadly refer to society. 'Ecological', 'environmental', and 'natural' in this context broadly refer to biotic and abiotic elements of the natural environment. 'Systems' basically means a connected collection or combination. SES and CHANS are

9

consequently connected collections or combinations of society and the natural environment. In English, we tend to use the term 'reality'. Especially in DRR including CCA, what would not involve different aspects of people, nature, and their interactions?

Within reality, we are often told that we should seek 'adaptive capacity' and 'adaptive management'. In effect, they mean: Can we change for the better? Surely it is much more engaging and much more useful to seek change for the better rather than complicated jargon?

Which brings us to 'transformation'. As with much other jargon, 'transformation' has multiple definitions, some of which are almost mutually exclusive! Vagueness manifests with the birth of phrases such as 'Transformational Change' and 'Transformative Change'. How could those be differentiated, especially when they do not translate readily into many other languages and cultures?

In English, 'transformation' and 'change' are almost synonyms, although 'transformation' is frequently seen as a subset of 'change'. What about 'changeable transformation' or 'changeative transformation' instead? Where do 'transition', 'transmutation', and 'transmogrification' sit, especially in languages that do not have separate words for them or for 'transformation'? What will come next once 'resilience' (or is it 'resiliency'?) and 'transformation' (or even 'transfiguration') lose their lustre?

Connotations of vocabulary need to be considered as well. 'Vulnerability' is frequently viewed as being negative, often frightening, and disempowering. Call people 'vulnerable' and they might believe it, leading to an assumption that no further option exists except to be vulnerable – and thus external people and organisations need funds to play their heroic roles in saving the vulnerable. Conversely, some disaster-affected people sometimes purposefully label themselves as 'vulnerable' to garner attention and to gain external resources. There can be conflicting views on how to use 'vulnerability' between 'insiders' and 'outsiders' as well as within 'insiders' and 'outsiders'.

Yet, labelling people as 'resilient' might also mean that they believe it. By virtue of being resilient, we need not worry about changing anything. Even more pernicious is the power-brokers labelling others as 'resilient' meaning that no resources need to be provided for DRR including CCA. We are so resilient that we should take care of ourselves.

Consequently, it becomes clear that not all resilience, transformation, or change is good, useful, or needed. Even with laws and cultural norms that have transformed some societies against discrimination, remarkable resilience is seen in racism, sexism, homophobia, and how people with disabilities are treated. Transformation could mean a major shift of resources away from DRR because people are resilient, so why help them further? Vulnerabilities could be swiftly created and re-established, enacting transformation. How do we separate the destructive from the constructive in order to focus on the latter?

The wise words of Allan Lavell should be considered, coming from a brief for a handbook chapter that he proposed but then was unable to pursue (quoted with permission):

> Concepts are the building blocks for understanding, relating to and guiding practical or on-the-ground actions. Whether known or not, explicit or not, concepts always guide our actions. Theory and practice are based on these. The best concepts are those that in simple terms capture essential aspects of a problem and causality, projecting the needed direction of interventions.

We might all be aiming for a similar notion of Sustainable Holistic Integrated Transformation, but does it really help to call it as such? If we need a detailed essay to explain a concept, then we should question the utility of that concept.

DRR including CCA is about action. By saying what we mean, we can inspire and galvanise to appropriate action. By saying what we do, we can spend much more time acting and much less time debating the subtle intricacies of complex formulations.

Vocabularies and interpretations contribute to shaping policy and practice, positively or detrimentally. The chapters in this part of the *Handbook* present and interrogate much of the typical vocabulary seen in DRR including CCA, pointing out the useful and not-so-useful dimensions. It is not just critique, but also providing alternatives, positive examples, and ways forward.

It is not just that words, languages, and vocabularies change. It is also about us choosing to transform them so that the phrases serve us rather than us serving them.

3

DISASTER RISK REDUCTION: A CRITICAL APPROACH

Rajarshi DasGupta and Rajib Shaw

Introduction

From ancient civilizations to modern 'technological-driven' societies, human history has shaped and been constantly reshaped by natural hazards. One of the several hypotheses behind the collapse of the great Indus Valley Civilization during 1800–1700 bc, is of a series of droughts followed by an eastward shift of the monsoon. Consequently, historians and palaeo-climatologists argued that the vibrant ancient cities of Mohenjo-Daro and Harappa came to a complete halt due to extreme paucity of rain that virtually forced its dwellers to abandon these ancient cities. Likewise, a series of natural hazards – e.g., earthquakes and drought – between 1225 bc and 1177 bc led to the downfall of ancient societies, including the great Egyptian civilizations, heralding the beginning of the 'dark age' (Cline 2014). As people and communities continue to deal with a plethora of natural hazards, they also learn from, and adapt to, these experiences. This collective experience largely indicates, that, at least to some extent, the adverse consequences of natural hazards can be avoided through careful and timely planning; and that the solution lies in either containing the forces of nature, which in itself leads to problems and often transference of the hazard elsewhere, or by altering our own behaviour and addressing vulnerability. This has led to significant societal transformation over the years and as a result, the concept of 'Disaster Risk Reduction' (DRR) emerged and received recognition.

The Evolution of 'Disaster Risk Reduction' and Links with Climate Change Adaptation

According to the English dictionary, the word 'disaster' can be traced to its origin from a sixteenth-century Italian term 'disastro', which generally refers to an 'ill-starred event' [dis- (expressing negation) + astro 'star' (from Latin *astrum*)]. The word has been retained in the English vocabulary ever since and has been indiscriminately used to denote a wide sense of purpose ranging from the occurrences of large natural hazards (e.g., earthquakes, typhoons, etc.), undesirable physical events (e.g., power cuts, disease outbreaks, etc.) to social and political blunders. Nevertheless, in a traditional sense, the word 'disaster' is closely connected with the word 'accident', since any event that led to loss of life and/or property was either referred to as a 'disaster' or an 'accident'. Since the English dictionary definitions of these two words are quite close, it was necessary to set up the criteria by which we may distinguish a physical or human caused incident as 'disaster'.

Over the last three decades, there have been several attempts to draw a distinctive definition of 'disaster' and to set up the criteria by which an incident can be termed as a 'disaster'. For example, a definition of disaster, posed by the International Worker's Party and cited in Rutherford and Boer (1983, p. 10), described disasters as a 'destructive event, which, relative to the resources available, causes many casualties, usually occurring within a short period of time'. This definition implied three boundary conditions to separate disasters from accidents: (a) 'resource availability', (b) 'many casualties' and (c) 'short period of time'. De Boer (1990) later justified that if a sudden shock could be managed by utilizing own resources and minimizing casualties, this could be referred to as an 'accident', rather than a 'disaster'. For instance, they argued that 'even a serious explosion or fire need not be a disaster in the presence of adequate facilities for rescue and treatment' (De Boer 1990, p. 592).

Nevertheless, the general assumption even during the early sixties was that natural hazards were more or less an 'Act of God', and that they could potentially trigger an extreme event that would require humanitarian assistance to recover. During the late sixties, this assumption changed and DRR strategies such as sea walls, dams etc., were used to 'tame the force of nature'. This changed again in the late seventies and early eighties when the 'vulnerability' approach to disasters gained momentum by rejecting the assumption that disasters are an outcome of a natural hazard(s) alone and outlining links with the surrounding social environment (see Box 3.1). The realization of the fact that disasters result from complex interactions between human and natural systems also led to the hypothesis, that even with a high probability of natural hazards occurring, not all communities remain equally prone to damage. Consequently, it was identified, that disasters are functions of 'Hazard' and 'Vulnerability' (Disaster Risk (R) = Hazard (H) × Vulnerability (V)), and in particular, since 'Hazards' are generally considered as sudden shock events (although also slow-onset, e.g., droughts), it is rather imperative to manage the 'vulnerability' component, which is inclusive of several complex and interlinked social, human, economic, environmental and physical variables (Blaikie *et al.* 1994). For example, amongst other things, 'vulnerability' may include poor design and construction of buildings, social isolation, economic incapacity, inadequate protection of assets, lack of public information and awareness, limited official recognition of risks and preparedness measures, and disregard for wise environmental management (UNISDR 2009). For instance, taking into consideration solely the earthquake of the Great East Japan Earthquake and Tsunami that occurred in 2011, then this cannot be referred to as a 'disaster' in the true sense of the term due to Japan's excellence in earthquake engineering (Kelman and Glantz 2015). However, sadly the tsunami generated by the earthquake did cause a disaster, demonstrating failure to reduce vulnerability and prevent the tsunami from becoming a disaster (Kelman and Glantz 2015).

Box 3.1   Disaster Risk Reduction – A Brief History

Rajarshi DasGupta[1] and Rajib Shaw[2]

[1] Graduate School of Agriculture and Life Sciences, The University of Tokyo, Japan
[2] Integrated Research on Disaster Risk, Beijing, China

DRR alongside other related terms including emergency assistance, disaster response and relief, humanitarian assistance, emergency management, disaster mitigation and prevention, and disaster risk management were originally incorporated under the umbrella term 'Disaster Management' (UNISDR 2004). The past forty years have seen DRR disassociate itself from 'disaster management', moving away from the technocratic paradigm that viewed natural hazards as something to contain and control through engineering structures and sophisticated monitoring technology. The technocratic paradigm primarily focused upon the hazard and single, event-based scenarios under command-and-control mechanisms with established hierarchical relationships. However, despite use of such technology the impact of natural hazards upon populations worldwide has only continued to increase. That is due to the over-focus upon the 'naturalness' of disaster events when it is the interaction between a hazard(s) and humans that could potentially result in a disaster. This has denied the wider historical and social dimensions of the hazard (Bankoff 2001). That contributed to the development of the alternative paradigm that viewed disasters as products of vulnerability; i.e., socio-economic and political in origin rather than natural (Hewitt 1983; Torry 1978, 1979; Blaikie *et al.* 1994). The alternative paradigm focuses largely on vulnerability and risk issues, and responds to dynamic and multi-risk scenarios utilising multiple actors and adapting to differing situations. Hence, DRR in its present day form is now seen as 'the systematic development and application of policies, strategies and practices to minimise vulnerabilities, hazards and the unfolding of disaster impacts throughout a society, in the broad context of sustainable development' (UNISDR 2004, p. 3).

In recent disaster literature, the above relationship has been modified further with the incorporation of two terms, 'exposure' and 'coping capacity'. Within hazard research, 'exposure' is largely defined by the entities exposed and prone to hazard impact(s). In particular, 'exposure' is defined mostly in temporal and spatial terms. For example, according to the World Risk Report (2011), exposure is related to the potential average number of people who are exposed to earthquakes, storms, droughts and floods, etc. On the contrary, societal dynamics may change the vulnerability of particular groups or individuals over time (Shaw *et al.* 2010). For instance, experiential learning from mega disasters such as the Great East Japan Earthquake and Tsunami in 2011 essentially indicates that a high awareness of communities and effective, efficient resource management also leads to substantial reduction of disaster risks and effective post-disaster humanitarian response. For the same reason, the consequences of a powerful earthquake in Japan and Nepal are distinctly different. These factors, collectively known as 'coping capacity', are the intrinsic properties of a community that can substantially reduce disaster impacts. According to these conceptual amendments, 'Disaster Risk' is now represented as:

$$\text{Disaster Risk (R)} = \frac{\text{Hazard (H)} \times \text{Exposure (E)} \times \text{Vulnerability (V)}}{\text{Coping Capacity (C)}}$$

Many researchers, predominantly over the last decade, have attempted to clarify this equation. In particular, it has become essential to look further into the variables of Vulnerability (V), Exposure (E) and Coping Capacity (C), in order to implement strategies to reduce vulnerability and exposure, and increase coping capacity with respect to particular individuals, communities and regions. A host of theoretical research has been conducted to identify the key factors that affect these three components. For example, Blaikie *et al.* (1994) categorized the component vulnerability into root causes (such as limited access to power and resources, poverty, health, education or lack of human development) and dynamic pressures (such as lack of local institutions, population growth, etc.). The root causes outlined by Blaikie *et al.* (1994) are linked to a lack of human development that potentially leads to high vulnerability levels. For instance, poverty, in all its forms and means, is the single largest denominator of high vulnerability to natural hazards irrespective of geographic locations. In addition, given anticipated changes in the climate, potentially more and more people will live under new or escalated risk of natural hazards and extreme weather events. As in the case of natural hazards, natural hazard drivers including climate change will hit the poor hardest. However, observed and predicted consequences of climate change are highly uncertain and are essentially not linear. For instance, changes in rainfall may lead to flood in one area and drought in another, and may be of benefit to some whilst of negative consequence to others.

Since the publication of the third Intergovernmental Panel on Climate Change (IPCC) assessment report, there has been strong consensus among the international community that climate change will bring certain adverse consequences despite our best possible mitigation measures. In response to that, governments need to prepare systematic plans to adapt to these changes. At a global policy level the domain of Climate Change Adaptation (CCA) received recognition during the Conference of the Parties (COP) 11 in 2000 for the United Nations Framework Convention for Climate Change (UNFCCC) in Nairobi, emerging alongside the earlier advocated Green House Gas (GHG) mitigation strategies. In particular, the lack of international commitment observed in the Kyoto protocol led less wealthy countries to shift more towards adaptation rather than mitigation. Hence, significant attention was given to CCA issues in successive COPs, particularly in the 'Bali Road Map (2007)' as well as during COP 15 in Copenhagen (2009).

CCA, therefore, has become a developmental priority over the last decade or so. However, the main question is whether CCA is a standalone policy entity or whether it should be embedded within the scope of DRR, given the many overlaps that exist both in policy and in practice. The IPCC Special Report on Managing the Risks of Extreme Events and Disasters to Advance Climate Change Adaptation (SREX) (IPCC 2012) recommends addressing both the issues in coherence rather than in isolation. However, critiques of this report outline the clear tension within it between authors who place CCA within wider contexts and those that prefer to view climate change in isolation as an aside from other disaster and vulnerability concerns (Kelman *et al.* 2015). The emphasis of the report as per the mandate of the IPCC is still to advance CCA (IPCC 2012) without due consideration to wider disaster and vulnerability concerns.

Others argue that the domain of DRR essentially relies on experiential learning, while CCA largely revolves around the meticulous identification of possible future scenarios (Shaw *et al.* 2010). However, DRR by definition and by nature must look to the future as well as the past to determine the most cost effective DRR strategies. It therefore remains imperative to identify the overlap areas between these two domains, particularly from a practitioner perspective. The following section attempts to narrate the commonalities between CCA and DRR and argues how best the ideas of CCA can be embedded within the scope of DRR.

Commonalities and Coherences between Disaster Risk Reduction and Climate Change Adaptation

It is imperative to understand that both CCA and DRR, alongside the Sustainable Development Goals (SDGs), are essentially developmental issues with many of the targets highly overlapping. For example, the eradication of poverty, which is one of the SDGs, will essentially contribute to achieving the target of reducing disaster including climate risk as well. Moreover, in reality, it is extremely difficult to segregate the issues of climate change as a hazard driver and hazards, since communities do not feel the impact of natural hazards and climate change separately. Hence, practitioners and policy planners need to consider both these issues coherently. On the contrary, CCA itself is often considered as a subset of DRR from an academic perspective, with DRR being a subset of development. As has been mentioned, historically human civilizations have adapted to large-scale natural hazards and gained experiential learning. This, in turn, has helped them to prepare for the next big event – e.g., the Moken indigenous people of Thailand and Jarawa tribes of India's Andaman and Nicobar islands managed to anticipate the Indian Ocean Tsunami during 2004. The powerful ability to interpret the signs and symbols of nature is a great example of the exceptional ability of these indigenous people, who, over thousands of years, have learnt to live by the sea and adapt to change experienced.

CCA has the potential to substantially reduce many of the adverse impacts of climate change, thereby reducing future 'vulnerability'. Similarly to DRR, CCA also requires substantial access to information, mobilization, active management of resources and knowledge sharing. Hence, principal CCA activities are identified as similar to DRR approaches such as vulnerability reduction, strengthening coping capacity and reducing direct exposure through long-term, self-sustaining community based or ecosystem based measures (see below). Additionally, implementation of CCA and DRR approaches starts with a similar process through building a common understanding, undertaking a structured review of potential strategies and cost–benefit analysis (Shaw *et al.* 2010). Further, it is often argued that for both CCA and DRR, active community involvement remains imperative for successful implementation of DRR including CCA policies and action plans.

Institutionalization of Disaster Risk Reduction

Although the components of disaster risk (hazard, vulnerability, exposure and coping capacity) have been identified since the early seventies, in many cases national governments have continued to focus on relief centric approaches. That is, given the difficulties in tackling the many facets of vulnerability. The thematic identity of DRR, however, received its desired attention when the United Nations declared the 1990s as the 'International Decade for Natural Disaster Reduction' (IDNDR). Consequently, the 'Yokohama Strategy and Plan for Action' were adopted at the first United Nations World Conference on Disaster Risk Reduction (WCDR) in 1994 as the first international policy guidelines for DRR. On the termination of IDNDR, the United Nations General Assembly established the secretariat of the United Nations International Strategy for Disaster Risk Reduction (UNISDR) to facilitate implementation of the International Strategy for Disaster Reduction (ISDR), a successor mechanism of the IDNDR. The second WCDR held in Kobe, Japan, in 2005, was attended by 168 member countries and resulted in the 'Hyogo Framework for Action (HFA), 2005–2015: Building the Resilience of Nations and Communities to Disasters' (UNISDR 2005). This was the first milestone of the institutionalization of DRR and considered as a monumental shift. Through five sets of priorities, the concepts of DRR, as put forth in the HFA, reflected a stronger focus upon risk preparedness and prevention, as compared with response and recovery.

The HFA resulted in some policy alteration at national government level, with many countries coming up with Disaster Management Acts, policies and long-term risk reduction plans. Yet, despite a significant mobilization of resources and development of new capacities for DRR at the national government level, contemporary progress reviews (e.g., Global Assessment Report on Disaster Risk Reduction (UNISDR 2015) and the Global Network of Civil Society Organisations for Disaster Reduction (GNDR) Frontline Report (2013)), largely indicated the exclusion of local governments and communities in local level DRR policy framing and implementation. In addition, little has been done to reduce underlying risks, such as environmental degradation, contributing to vulnerability.

The HFA framework was replaced by the Sendai Framework for Disaster Risk Reduction (SFDRR), which was put into place following the third WCDR held in Sendai City, Japan, in 2015. The SFDRR has an operational period of 15 years (2015–2030) and outlines its goals in four priorities: 1) Understanding disaster risk; 2) Strengthening Disaster Risk Governance; 3) Investing in Disaster Risk Reduction for Resilience; and 4) Enhancing disaster preparedness for effective response, and to build back better in recovery, rehabilitation and reconstruction. Although it is too early to measure the utility and effectiveness of the SFDRR, many argue that the absence of baselines and lack of distinctive responsibilities and quantitative targets are the major drawbacks of the SFDRR (Chatterjee *et al.* 2015).

The year 2015 was equally important from the climate change perspective, particularly due to the adoption of the Paris Agreement of UNFCCC, which sets forth new targets for greenhouse gas emissions mitigation, adaptation and finance starting in the year 2020. The agreement is now open for signature and as of 2 August 2016, there are 179 signatories. Similar to the SFDRR, the Paris Agreement has also been criticized due to a lack of binding targets. However, beside the traditional GHG mitigation, one of the thrust areas of the Paris agreement is enforcing CCA and adaptation financing for resilient societies. It is, however, imperative to mention that within the scope of CCA (as formulated in Paris agreement) and DRR (as formulated in SFDRR), a whole lot of similarities and broad overlap exist between the targets. In particular, both these documents revolve around enhancing community resilience against short-, medium- and long-term hazards by incorporating effective adaptations through proper institutionalization and global partnership.

Contemporary Approaches for Disaster Risk Reduction and its Synergies in Climate Change Adaptation

The key issues for both DRR and CCA are to identify the approaches that can translate the theoretical notion and policies into appropriate risk reduction measures. In this section, we discuss the existing approaches of DRR that may also have simultaneous applications in CCA. Over the last two decades, especially since the adoption of HFA, several theoretical developments of specific DRR approaches have been formulated against a variety of physical and economic backgrounds. What is interesting to note, is that, compared with the traditional hard engineering based risk reduction approaches, which, at a point in time, were considered the only way to reduce the exposure of communities to specific hazards, several alternative approaches evolved to meet the objectives of controlling exposure, reducing vulnerability and enhancing coping capacities. These approaches remain equally applicable for a climate resilient society and form the very basis of long-term CCA planning.

Some of the well-researched approaches that have been applied include Community-based Disaster Risk Reduction (CBDRR), Ecosystem based Disaster Risk Reduction (Eco-DRR) and Restrictive Planning. Collectively known as 'Soft-approaches or No-Regrets approaches for Disaster Risk Reduction', these specifically aim to capitalize on existing human and natural resources for proactive risk reduction from hazards including hazard drivers such as climate change. These approaches, in many cases, are complementary to traditional hard engineering based DRR approaches, and often, based on the hazard profile and the community capacity, can be used in tandem. It is, however, imperative to understand the potential scope and opportunities of these specific approaches. Table 3.1 provides a summary of the contemporary approaches of DRR alongside their specific contribution to reducing vulnerability and exposure, and increasing coping capacities. The following section describes in short the specific applicability of these different approaches.

Table 3.1 Approaches for DRR

DRR Approaches	Reducing Exposure	Reducing Vulnerability	Enhancing Coping Capacity	Applicability and Utility
Engineering based disaster risk reduction	(Major Role) For example, sea dikes, earthquake resilient buildings, dams and reservoirs	(Semi-Major Role) Mainly reducing physical vulnerability	(Major Role) Advanced early warning, scientific modelling of risks	• Adequate financial capacity of the national and local government • In-depth understanding of hazards and uncertainty • Reliability
Community-based disaster risk reduction (CBDRR)	(Minor Role) Not directly related, however, better community understanding of risks leads to reduction of exposure; such as not settling by the sea etc.	(Major Role) Creating local assets and mutual understanding, enhances social capital	(Major Role) High disaster awareness, efficient evacuation, culture of preparedness	• Existence of weak local governments • Applicable in local risk management • Net social benefits with long-term sustainability
Ecosystem based disaster risk reduction (Eco-DRR)	(Major Role) For example, storm surge attenuation, soil accumulation, erosion control etc.	(Major Role) Asset creation in terms of livelihood and physical resources	(Minor Role) Creating environmental awareness for sustainable ecosystem management	• Low cost adaptive approach. • High utility in CCA • Generates net environmental benefits
Restrictive Planning	(Major Role) For example, planned retreat, coastal regulation zones	(Semi-Major Role) Mainly reducing physical vulnerability, however, may increase social vulnerability	(No Role)	• High relative cost • Requires proper legislation and policy reforms • May lack social acceptability

Engineering Based Disaster Risk Reduction Including Climate Change Adaptation

Having emerged during the 1960s, the traditional approaches for DRR have been, by and large, hard engineering approaches, which, to a significant extent, have been successful in reducing risk from natural hazards including climate-related hazards. For example, the extensive embankment network of the Netherlands serves as the most prominent illustration of this. This system of robust dikes with extensive mechanization has been tremendously effective in mitigating storms and surges. The system is constantly upgraded to meet changing exposures and anticipated impacts of climate change. However, as outlined in the introduction the implementation of engineering based approaches for DRR including CCA also comes with a negative side; e.g., encouraging people to build in flood plains or in areas that may possibly be inundated in the future or transference of the hazard elsewhere. For example, nearly 9,600 km of Japan's 35,000 km coastline are protected by sea dikes that are designed to protect the adjacent human habitation from the Level-1 Tsunami [according to Kaigan hou (Japanese Sea-Coastal Law of 1953, amended 1999), Level-1 Tsunami are the events that may occur once in a hundred years]. These hard engineering risk reduction measures demand major capital investments and recurring maintenance costs. Yet, these hard and strong coastal defence mechanisms have not always been as productive as planned, since the sea dikes only provided a false sense of security during the East Japan Earthquake and Tsunami in 2011. In addition, sea dikes are also considered to have negative environmental impacts, such as disruption of natural shoreline processes and destruction of shoreline habitats such as wetlands and intertidal beaches. Nevertheless, despite certain ambiguities, engineering solutions for DRR remain a trusted approach in many wealthier countries.

Community-based Disaster Risk Reduction (CBDRR)

CBDRR is defined as a process in which affected communities are at the centre of any risk reduction strategy. This is often referred to as a participatory and bottom-up process that is initiated, led and/or managed by the community itself (often with the assistance of outsiders such as NGOs) and is not a request/order from higher authorities. The principal arguments of this approach lie in the understanding of the fact that communities know their situation best and as such are best placed to initiate a process of identifying strategies to address their risk. In particular, CBDRR attempts to reduce the risk of disasters within a community, by focusing on the root causes of risks, addressing it through the most appropriate knowledge, whether this be local, internal knowledge or external knowledge and/or assistance – or, usually, a combination (see Box 3.2). This approach has been adopted in many countries within the last decade and has been one of the principal developmental doctrines since the adoption of the HFA. Although it still lacks legal recognition in many countries, international agencies, including UNISDR, strongly encourage active participation of communities in managing local disaster risks.

Box 3.2 Community-based Disaster Risk Reduction and External Vulnerabilities in Coastal Bangladesh

Bayes Ahmed[1], Ilan Kelman[1,2], Heather K, Fehr[3], and Manik Saha[4]

[1] University College London, London, UK
[2] University of Agder, Kristiansand, Norway
[3] British Red Cross, London, UK
[4] British Red Cross, Dhaka, Bangladesh
Based on Ahmed *et al.* (2016).

From 2013 to 2016, two Bangladeshi coastal communities in Patuakhali district, Nowapara and Pashurbunia, participated in the Vulnerability to Resilience (V2R) programme implemented by the Bangladesh Red Crescent through funding and technical support made available by the British Red Cross and Swedish Red Cross.

With special attention given to the poorest households, activities included cash-for-work for constructing and improving community access roads; cash for new livelihoods and livelihood diversification; building latrines and tube-wells for freshwater while providing hygiene advice; and providing safety equipment for fishers. Training was provided for business development, market access,

---

community mapping, first aid, search and rescue, and early warning. Community members across socio-economic classes were engaged in CBDRR activities, whereas little occurred before, particularly through using sustainable livelihoods and livelihood diversification to implement DRR.

Just after V2R ended on 30 April 2016, Cyclone Ruanu made landfall on Bangladesh's south coast on 21 May 2016. The people in the V2R communities received warnings and evacuated, so no casualties were reported. They returned afterwards to find that many of the new initiatives had withstood the storm and subsequent flooding from a broken embankment. Their new freshwater supply and latrines continued to function. Livelihoods received little interruption, apart from agricultural fields being salinated by the storm surge, so many people switched to fishing while rehabilitating the land to continue cultivation. CBDRR succeeded by focusing on root causes of vulnerability through shoring up sustainable livelihoods and basic services.

A new seaport might be constructed nearby, meaning that the communities might experience forced eviction and major landscape changes which would significantly disrupt their livelihood patterns and undermine V2R's work. Although the people now have more skills and resources to address this challenge, CBDRR can be undermined by external creators of vulnerability and disaster risk.

Ecosystem-based Disaster Risk Reduction (Eco-DRR)

One of the much referred to international policy documents of the last decade was the Millennium Ecosystem Assessment (2005), which reemphasized the need for harvesting the unbounded relations between humans and nature, especially by promoting the concept of 'Ecosystem Services'. With poor environmental practices, many of the world's ecosystems remain critically degraded. This, in turn, increases risk, not only by increasing exposure to natural hazards, but also by enhancing the potential vulnerability, especially in terms of livelihoods and access to resources. For example, many communities in less wealthy countries are directly dependent on ecosystem services for their livelihoods, and, therefore, remain particularly vulnerable to changes in environmental conditions (Renaud *et al.* 2013). Efficient management of ecosystems, that aims to revitalize or enhance ecosystem services, is an essential approach for DRR including CCA. The HFA also recognized environmental degradation as a major contributing factor for disaster risk (Doswald and Estrella 2015).

The synergies between sound ecosystem management and DRR including CCA have received wide recognition since the Indian Ocean Tsunami, as various reports and case studies of the potential wave attenuation capabilities of mangroves were referred to after the catastrophic tsunami (e.g., EJF 2006; Kathiresan and Rajendran 2005). This, in particular, renewed interest in ecosystems and their services in DRR, giving rise to a new concept of 'Ecosystem-based Disaster Risk Reduction' or Eco-DRR. The 'Eco-DRR' concept has received wide recognition since the United Nations Environment Programme (UNEP) adopted this concept as their doctrine for ecosystem conservation. In general, these approaches are hypothesized as low-cost, futuristic risk reduction approaches that are aimed at generating net social and ecological benefits.

For example, Hiraishi and Harada (2003), based on a theoretical study, suggested that a coastal forest of 30 trees per 100 m^2 in a 100-m wide belt may reduce the maximum tsunami flow pressure by more than 90 per cent. However, clearly this is not always the case and is dependent on many factors, not least the extent of the hazard(s) concerned. In monetary terms, Eco-DRR approaches are typically considered as highly cost-effective. For example, Gilman *et al.* (2008) outlined that the replacement cost of existing mangroves with rock walls in Malaysia has been estimated at USD 300,000 per km. Further, the cost of current mangrove restoration in Thailand is estimated as USD 946 per hectare, while the cost for protecting existing mangroves is capped to only USD 189 per hectare (Gilman *et al.* 2008). In comparison to engineered seawalls, this cost is negligible.

Restrictive Planning

Another approach for DRR including CCA largely revolves around the principles of restrictive planning. Evolved mostly in the early 1990s, restrictive planning largely attempts to reduce disaster risks by decreasing direct physical exposures, such as by demarking specific zones in coastal areas and planned retreats of human settlements from highly vulnerable and disaster prone areas. Collectively, known as 'soft' approaches for DRR including CCA, restrictive planning such as planned retreat / managed realignment of coastal zones under threat of rising sea levels provides a potential approach where building hard engineering structures is beyond the capacity of the governments (Abel *et al.* 2011). In addition, it also provides the space for ecological succession that is imperative to reduce the exposure further. In this process derelict lands are often left to natural regeneration of vegetation, such as mangroves in coastal areas. This then forms the basis of long-term risk reduction including the risks of hazard drivers such as climate change.

Conclusion and Way Forward

All the above-discussed DRR approaches, individually and collectively, remain imperative from the perspective of minimizing the risks of natural hazards that may intensify or diminish with climate change. Thus, they are imperative for CCA approaches also. As has been mentioned, governments and policy planners can choose one or a combination of approaches based on their understanding of risks as well as the social, economic and technical capacity of the concerned communities. This understanding of risk needs to be inclusive as well as futuristic to accommodate the probable uncertainties of climate change.

Nevertheless, it is imperative to recognize the underlying causes of disaster risks and a thematic assessment of the 'scale' and 'context' for policy planners needs to consider a specific DRR including CCA approach or a suitable combination. 'Scale' – i.e., the extent of coverage area – is among the basic requirements for adopting a specific DRR including CCA approach. For example, engineering based DRR including CCA may be applicable on a small scale or for a certain region, while restrictive planning such as planned retreat from susceptible coastlines can be advocated for at a national level. Similarly, the 'context', especially high or low technical and economic capacity of the government, also remains a major determinant for choosing a specific DRR including CCA approach. CBDRR has become overwhelmingly popular in less wealthy countries due to the lesser requirement for capital investments. Additionally, policy researchers have profoundly recommended this approach as being ameliorative for fostering planned and spontaneous adaptation to the adverse impacts of climate change. Similarly, Eco-DRR is considered as a potential alternative to engineered DRR in the backdrop of limited technical and economic capacity. These approaches not only have long-term sustainability, but also are often self-sustaining and cost effective for reducing disaster including climate change risk.

Decision-making for the appropriate combination of risk reduction approaches is not easy and must remain site specific. Local knowledge is an essential component that needs to be harnessed and integrated for drafting effective DRR including CCA strategies. A multitude of other factors such as economic and technical capacity, thematic understanding of risks, social acceptance (e.g., it is probably impossible to prohibit a fishing community from living near the coast), community participation and perception, and long-term environmental and social impacts are some of the key parameters that lead to effective decision-making in this regard. In particular, community-based and ecosystem-based approaches have high social return that will benefit even in the absence of major hazards. One of the substantial challenges for appropriate decision-making in combining DRR including CCA approaches is the lack of previous experience and shortage of case studies, which, largely, restricts evidence-based choices. These gaps will minimize over time as global communities, including governments and academia, put more emphasis on local and regional DRR including CCA through innovative, action-led, proactive risk reduction measures.

Another substantial problem in decision-making is that, despite the many similarities, the domains of CCA and DRR have not been integrated at international and national levels due to political and governance reasons. At the international level, the parallel existence of CCA and DRR forums and platforms working to achieve similar targets is, at times, highly confusing for practitioners. While these platforms are working separately on many similar issues, it remains imperative to merge and accomplish the common targets in order to simplify the complexities. Segregating CCA measures from DRR strategies is illogical and deceptive in many cases. A community-led sectoral approach, for example, improving livelihood resilience utilizing DRR including CCA strategies is perhaps a more cognitive risk reduction approach than segregated action planning for DRR or CCA.

References

Abel, N., Gorddard, R., Harman, B., Leitch, A., Langridge, J., Ryan, A. and Heyenga, S. (2011) 'Sea level rise, coastal development and planned retreat: Analytical framework, governance principles and an Australian case study', *Environmental Science and Policy* 14, 3: 279–288.

Ahmed, B., Kelman, I., Fehr, H.K. and Saha, M. (2016) 'Community resilience to cyclone disasters in coastal Bangladesh', *Sustainability* 8: 805–834.

Bankoff, G. (2001) 'Rendering the world unsafe: "Vulnerability" as Western discourse', *Disasters* 25, 1: 19–35.

Blaikie, P., Cannon, T., Davis, I. and Wisner, B. (1994) *At Risk,* London: Routledge.

Chatterjee, R., Shiwaku, K., Das Gupta, R., Nakano, G. and Shaw, R. (2015) 'Bangkok to Sendai and beyond: Implications for disaster risk reduction in Asia', *International Journal of Disaster Risk Science* 6, 2: 177–188.

Cline, E.H. (2014) *1177 BC: The Year Civilization Collapsed,* USA: Princeton University Press.

De Boer, J. (1990) 'Definition and classification of disasters: Introduction of a disaster severity scale', *The Journal of Emergency Medicine* 8, 5: 591–595.

Doswald, N. and Estrella, M. (2015) *Promoting Ecosystems for Disaster Risk Reduction and Climate Change Adaptation: Opportunities for Integration,* Geneva: United Nations Environment Programme.

EJF (Environmental Justice Foundation). (2006) *Nature's Defence against Tsunamis: A Report on the Impact of Mangrove Loss and Shrimp Farm Development on Coastal Defences,* London: Environmental Justice Foundation.

Gilman, E.L., Ellison, J., Duke, N.C. and Field, C. (2008) 'Threats to mangroves from climate change and adaptation options: a review', *Aquatic Botany* 89, 2: 237–250.

GNDR (Global Network of Civil Society Organisations for Disaster Reduction). (2013) *Views from the Frontline (VFL) 2013*. Online http://www.gndr.org/programmes/views-from-the-frontline.html (accessed 9 February 2016).

Hewitt, K. (ed.) (1983) *Interpretations of Calamity from the Viewpoint of Human Ecology,* Boston: Allen & Unwin.

Hiraishi, T. and Harada, K. (2003) 'Greenbelt tsunami prevention in South-Pacific region', *Report of the Port and Airport Research Institute* 42, 2: 1–23.

IPCC (Intergovernmental Panel on Climate Change) (2012) *Special report on managing the risks of extreme events and disasters to advance climate change adaptation (SREX),* Cambridge: Cambridge University Press.

Kathiresan, K. and Rajendran, N. (2005) 'Coastal mangrove forests mitigated tsunami', *Estuarine Coast Shelf Sciences* 65, 3: 601–606.

Kelman, I., Gaillard, J.C. Mercer, J., Lewis, J. and Carrigan, A. (2015) 'Island vulnerability and resilience: Combining knowledges for disaster risk reduction including climate change adaptation', in E. DeLoughrey, J. Didur and A. Carrigan (eds.), *Global Ecologies and the Environmental Humanities: Postcolonial Approaches,* London: Routledge, pp. 162–185.

Kelman, I. and Glantz, M.H. (2015) 'Analyzing the Sendai framework for disaster risk reduction', *International Journal of Disaster Risk Science* 6, 2: 105–106.

Millennium Ecosystem Assessment (2005). *Ecosystems and Human Wellbeing: Synthesis*, Washington, DC: Island Press.

Renaud, F.G., Sudmeier-Rieux, K. and Estrella, M. (2013) *The Role of Ecosystems in Disaster Risk Reduction*, Tokyo: United Nations University Press.

Rutherford, W.H. and De Boer, J. (1983) 'The definition and classification of disasters', *Injury* 15, 1: 10–12.

Shaw, R., Pulhin, J.M. and Pereira, J.J. (2010) 'Climate change adaptation and disaster risk reduction: Overview of issues and challenges', in R. Shaw, J.M. Pulhin and J.J. Pereira (eds.), *Climate Change Adaptation and Disaster Risk Reduction: Issues and Challenges*, Bingley: Emerald, pp. 1–22.

Torry, W.I. (1978) 'Bureaucracy, community and natural disasters', *Human Organization* 37, 3: 302–307.

Torry, W.I. (1979) 'Anthropological studies in hazardous environments: Past trends and new horizons', *Current Anthropology* 20, 3: 517–540.

UNISDR (United Nations International Strategy for Disaster Reduction) (2004) *Living with Risk: A Global Review of Disaster Reduction Initiatives*, Geneva: UNISDR.

UNISDR (2005) *Hyogo Framework for Action 2005–2015: Building the Resilience of Nations and Communities to Disasters*, Geneva: UNISDR.

UNISDR (2009) *UNISDR Terminology on Disaster Risk Reduction*, Geneva: UNISDR.

UNISDR (2015) *Global Assessment Report (GAR) on Disaster Risk Reduction: Loss Data and Extensive Risk Analysis*, Geneva: UNISDR. Online http://www.preventionweb.net/english/hyogo/gar/2015/en/gar-pdf/Annex2-Loss_Data_and_Extensive_Risk_Analysis.pdf (accessed 9 February 2016).

World Risk Report (2011) *World Risk Report*, Tokyo: Institute of Environment and Human Security, United Nations University.

4

CLIMATE CHANGE ADAPTATION
A Critical Approach

Stavros Mavrogenis, Petros Theodorou, and Rory Walshe

Introduction

This chapter critically reviews the theory, policy and practice of climate change adaptation (CCA). 2015 was an important year for sustainable development, as three world conferences have taken place aiming to compose and reach long-term agreements for sustainability. The first was the third United Nations (UN) World Conference on Disaster Risk Reduction (third WCDRR), held from 14 to 18 March 2015 in Sendai, Japan, wherein the participants consulted on and formed the successor to the Hyogo Framework for Action (HFA) which is called the Sendai Framework for Disaster Risk Reduction (SFDRR). Following this, the UN ratified the Sustainable Development Goals (SDGs) in September in New York, which are replacing the Millennium Development Goals established in 2000 (UN 2000). The United Nations Framework Convention on Climate Change (UNFCCC) Conference of the Parties (COP) rounded off 2015 in December, in Paris. The purpose of the Paris COP was to produce a binding agreement for the reduction of greenhouse gas emissions to address climate change.

These policy processes straddle the concerns of CCA, disaster risk reduction (DRR) and sustainable development. As a discipline DRR is older than CCA, but both were preceded by activities grouped thematically as 'poverty reduction', now known broadly as 'sustainable development'. The aims of DRR and CCA are linked since for both, poverty reduction and sustainable development are essential components of vulnerability reduction. Indeed, CCA and DRR are integral to several of the recently ratified SDGs, particularly goals 13 (climate action) and 11 (sustainable cities and communities). Kelman *et al.* (2015) use the Paris COP to highlight the interconnectedness among these three processes as well as the inter-dependency of the outcome of each process on the others. Despite these similarities and overlaps the three processes are separate and independent, which as Mercer *et al.* (2012) note, is limiting and illogical given that they aim to address and examine similar processes regarding vulnerability and resilience, and to improve methods to anticipate, resist, cope with and recover from hazard impact.

As well as their aims, DRR and CCA also share 'mutual benefits', in that DRR already attempts to reduce climate-related disaster risk and, increasingly, to offset long-term impacts of climate change (Kelman *et al.* 2015). Therefore, in order for CCA to be effective, it must build on and be embedded within DRR, and should not be undertaken in isolation from this wider agenda (Sperling and Szekely 2005). Nevertheless, including all three issues, namely DRR, CCA and sustainable development, in a unique, common agenda is a difficult task due to academic and institutional entrenchment (Kelman *et al.* 2015).

The SDGs may not be achieved if disaster risk is ignored. In an interconnected world, the losses incurred from disasters affect individuals, critical public infrastructure and ecosystems as well as the broader economic and increasingly complex environment in which the public and private sectors operate. Consequently, a priority for policy could also be to incorporate both DRR and CCA within core strategy and programme objectives in a more holistic approach for achieving sustainability. Hewitt (1983) and Lewis (1999), among others, laid the foundation for justifying this approach when evidencing that all disasters, climate related or otherwise, hinder development. Critically, they assert that it is the root causes contributing to these disasters that need to be adequately addressed.

Adaptation Theory

The concept of a 'disaster' is often associated with the insufficient capacity, or disinterest, of existing national and local systems to reduce a population's vulnerability (Glantz 1976; Hewitt 1983; Lewis 1999; Wisner *et al.* 2004). Too many disasters are rooted in corruption, the misuse of available resources, the looting of natural and economic resources to benefit the most powerful and a lack of enforcement or disobedience of laws such as building codes.

The Intergovernmental Panel on Climate Change (IPCC 2014, pp. 119–120) defines climate as 'the average weather, or more rigorously, as the statistical description in terms of the mean and variability of relevant quantities over a period of time ranging from months to thousands or millions of years' and climate change as an alteration in 'the state of the climate that can be identified (e.g., by using statistical tests) by changes in the mean and/or the variability of its properties and that persists for an extended period, typically decades or longer'. It goes on to specify that climate change 'may be due to natural internal processes or external forcings such as modulations of the solar cycles, volcanic eruptions and persistent anthropogenic changes in the composition of the atmosphere or in land use'. It is important to note that this differs from the UNFCCC (1992) definition, which specifically attributes climate change to direct or indirect human activity and distinguishes this from climate variability attributable to natural causes.

Climate change is increasingly utilised as a 'catch all' phrase assigning blame for many disasters (Kelman *et al.* 2015; Mercer 2010). Yet, climate change is just one factor amongst many that has the capacity to influence hazard parameters. Furthermore, it is relatively easy to accept climate change as a problem, and to blame climate change for problems experienced, without accepting responsibility for one's own actions at an individual, community, institutional, or governmental level. The existing debate supports the assertion that a vast wealth of knowledge in relation to 'adapting to change' and 'dealing with disaster' is not currently utilised to maximum effect (Mercer 2010). Therefore, encompassing successful CCA within the wider contexts of DRR, development and sustainability remains a challenge (Gaillard 2013; Kelman *et al.* 2015; Mercer 2010).

It is also acknowledged by Gaillard (2013) that a focus on climate change has been used by governments of less wealthy countries as a scapegoat for the root causes of vulnerability to disasters. Gaillard (2013, p. 222) explains that marginalisation (geographic, social, economic, and political) is instead the crucial element of vulnerability.

CCA in the IPCC's Fourth Assessment Report was defined as (2007, p. 869):

- Adaptation – Adjustment in natural or human systems in response to actual or expected climatic stimuli or their effects, which moderates harm or exploits beneficial opportunities. Various types of adaptation can be distinguished, including anticipatory, autonomous and planned adaptation.

- Anticipatory adaptation – Adaptation that takes place before impacts of climate change are observed. Also referred to as proactive adaptation.
- Autonomous adaptation – Adaptation that does not constitute a conscious response to climatic stimuli but is triggered by ecological changes in natural systems and by market or welfare changes in human systems. Also referred to as spontaneous adaptation.
- Planned adaptation – Adaptation that is the result of a deliberate policy decision, based on an awareness that conditions have changed or are about to change and that action is required to return to, maintain, or achieve a desired state.

More recently, the IPCC (2014, p. 118) provided a complicated set of vocabulary regarding adaptation to climate change. It defines adaptation as

> [t]he process of adjustment to actual or expected climate and its effects. In human systems, adaptation seeks to moderate or avoid harm or exploit beneficial opportunities. In some natural systems, human intervention may facilitate adjustment to expected climate and its effects.

This represents a departure from the 2007 IPCC definition in terms of the breadth and focus. Namely, the 2014 definition differentiates between adaptation for human and natural systems; however, the separation of natural and human systems is misleading because they are already connected. Nature has been 'socialised' and there is no nature without human presence (for the concept of socio-nature see Castree and Braun 2001).

The definitions for anticipatory and planned adaptation that were present in the 2007 report were removed from the 2014 report. However, the definition for autonomous adaptation remains:

> Autonomous adaptation – Adaptation in response to experienced climate and its effects, without planning explicitly or consciously focused on addressing climate change. Also referred to as spontaneous adaptation.
>
> *(IPCC 2014, p. 1759)*

This is a significant change in the definition of autonomous adaptation from 2007, which moves beyond only considering the 'ecological changes in natural systems and by market or welfare changes in human systems' (IPCC 2007, p. 869). This narrow definition excluded a variety of reactions related to human conservation and other societal changes.

Another issue that shapes CCA is its relationship with mitigation. Article 3.3 of the UNFCCC (1992) states that '[t]he Parties should take precautionary measures to anticipate, prevent or minimize the causes of climate change and mitigate its adverse effects'. Adaptation is not defined in a clear way but is implied by the term mitigation. Eventually, these two terms were separated in the UNFCCC process and came to have different meanings. This was a development that harmed the main goal of the climate change agenda since most CCA projects, such as reforestation, installing photovoltaic panels, and education can be both adaptation and mitigation actions by climate change definitions. On the other hand, hard engineering adaptation measures such as coastal walls and breakwaters require a lot of energy in order to be constructed and maintained, so CCA can interfere with climate change mitigation.

The top-down approach of CCA policies is one of the main reasons for these kinds of malpractices. Local communities are excluded from the decision-making process and their voice is not heard regarding the expression of their needs and their knowledge on CCA. Irrespective of various terms, overlaps and confusion it is clear that CCA is needed to deal with human induced climate change (Mavrogenis and Kelman 2013).

International Governance of Adaptation: The UNFCCC Process

CCA emerged in the UNFCCC, which was signed in 1992. Articles 2 and 3 of the UNFCCC (1992) clearly state that the parties to the Convention are obliged to stabilise greenhouse gas emissions in order to 'prevent dangerous anthropogenic interference with the climate system' (Article 2). Thus, the UNFCCC forms the pillars of international climate change governance, focusing on mitigation but nonetheless referring to adaptation four times throughout the text (UNFCCC 1992; see also Bodansky 1993). In fact, Article 4 places specific obligations on the countries labelled as 'developed' to share the cost of adaptation measures in the countries labelled as 'developing'. Article 3.1 states that developed countries and emerging economies should undertake all the necessary measures to combat the adverse impacts of climate change, while Article 3.2 describes how these measures should take the vulnerabilities of developing countries under consideration.

Article 4.1 is especially important for adaptation governance. It requires that member states have to plan, implement, publish, and amend national plans for combating climate change, including adaptation measures. Article 4.3 of the UNFCCC (1992) refers to the financing of adaptation measures in which developed countries must provide 'new and additional financial resources to meet the agreed full costs incurred by developing country Parties'. The onus is especially on developed countries to act given 'the developing country Parties that are particularly vulnerable to the adverse effects of climate change' according to Article 4.4.

Despite identifying the more affluent countries as the cause of human–induced climate change, the language of the UNFCCC lacks clarity and specificity. In this context, the COPs have resulted in a series of agreements (Table 4.1). To some extent, the lynchpin of the UNFCCC was the Kyoto Protocol (UNFCCC 1998), an internationally binding agreement that committed wealthier countries 'to reducing their overall emissions of such gases by at least 5 per cent below 1990 levels in the commitment period 2008 to 2012'. The Kyoto Protocol entered into force in February 2005, yet many of the countries that have ratified it have still not met the targets to which they agreed.

Table 4.1 Summary of Milestones from UNFCCC COPs and the EU

Milestone	Year
UNFCCC's article 4 makes direct reference to adaptation	1992
COP3 resulted in the Kyoto Protocol	1997
COP7 the Marrakesh Accords established the National Adaptation Programmes of Action (NAPAs), the Least Developed Country (LDC) Expert Group, the LDCs Fund, and the Adaptation Fund	2001
COP12 established the Nairobi Work Programme in Impacts, Vulnerability and Adaptation to climate change (NWP)	2006
COP13 established the Bali Action Plan (BAP) and officially launched the Adaptation Fund	2007
The European Commission launched the Green Paper on Climate Change Adaptation	2007
The European Commission published the White Paper 'Adapting to climate change: Towards a European framework for action'	2009
COP15 tried to secure unsuccessfully a successor to the Kyoto Protocol	2009
COP16 established the Cancun Adaptation Framework and the Green Climate Fund	2010
The European Commission adopted the 'EU Strategy on adaptation to climate change'	2013
COP21 Paris Agreement	2015

CCA started emerging as a key concern within the UNFCCC during the previous decade (2000–2010). COP7 in Marrakesh delivered the opportunity for LDCs to develop National Adaptation Programmes of Action (NAPAs). This was built upon by COP12 in Nairobi, with the introduction of the 5-year (2005–2010) Nairobi Work Programme (NWP) on Impacts, Vulnerability and Adaptation to Climate Change for the LDCs and the UN-designated Small Island Developing States (SIDS). Therefore, both the Marrakesh and Nairobi COPs were significant first steps that have tried to ensure the scientific basis of adaptation actions and to catalyse action on adaptation.

In December 2007, COP13 in Bali resulted in a few notable developments regarding adaptation. The Adaptation Fund was identified as a particularly important financial mechanism for CCA, as it gave the recipient countries more confidence in its operations, reducing bureaucracy and facilitating bottom-up CCA (Kelman and Mavrogenis 2014). Additionally in Bali, in an attempt to coordinate CCA with DRR approaches, a number of adaptation plans were implemented both in Africa and the SIDS. This effort essentially acknowledges DRR and risk management as major elements of CCA. Yet, it was done without fully recognising what DRR could offer CCA, and how CCA sitting within DRR would bring together the two concerns more coherently and effectively.

Following the highly anticipated yet ultimately ineffectual COP15 in Copenhagen, COP16 in Cancun in 2010 resulted in the establishment of the Cancun Adaptation Framework. Cancun also delivered the Green Climate Fund, a multilateral financing entity with the mandate to serve the UNFCCC in delivering equal amounts of funding to less wealthy countries for both adaptation and mitigation. However, the fund has hitherto received little support and developed states seem reluctant to transfer their money to the fund.

Additionally, the agreement from COP16 introduced the possibility for compensation in case of losses attributed to climate change, continuing the debate about whether or not compensation for impacts could count as adaptation. Finally, for the first time in international adaptation governance, an Adaptation Committee (AC) was founded in order to coordinate the implementation of international adaptation. In October 2014, at its 6th meeting, the AC agreed to include several recommendations in its report to the COP20 in Lima – in particular, to 'encourage the Adaptation Fund, the Global Environment Facility and the Green Climate Fund to consider and integrate local, indigenous and traditional knowledge and practices into adaptation planning and practices, as well as procedures for monitoring, evaluation and reporting' (Adaptation Committee 2014, p. 2).

In COP21 the international community adopted the Paris Agreement, a legally binding text with the main goal of keeping the increase in global average temperature (above pre-industrial levels) to well below 2 °C and at the same time attempting to limit the increase to 1.5 °C (UNFCCC 2015). Parties presented their 'intended nationally determined contributions' – INDCs which are effectively voluntary targets – covering 5- or 10-year periods starting in 2020. Starting from 2023, governments will come together every five years in a 'global stocktake', based on the latest science and implementation progress to date. The stocktake will set the context for the raising of ambition by all Parties by looking at what has been collectively achieved and what more needs to be done to achieve the below 2 °C objective. There is no punitive mechanism in the Paris Agreement but a set of procedures such as technical expert reviews, a multilateral peer review process, and a standing committee on implementation and compliance will evaluate the progress of the implementation. In conclusion, the Paris Agreement sets ambitious goals but it is up to states to fulfil their commitments and pledges.

National and Multinational Governance Regimes for Adaptation

In 2001 at COP7 in Marrakesh, the NAPAs (specifically for LDCs) entered into force in the context of the 'Marrakesh Accords'. The basic feature of NAPAs is that they provide the opportunity for LDCs to develop their own CCA priorities and to propose solutions based on their own needs and capacities. In this sense, they are designed to summarise and build on exist-ing strategies and knowledge, identifying adaptation actions that are based in communities. In practice however, many NAPAs were produced by external consultants with varying degrees of local, on-the-ground input (Kelman and Mavrogenis 2014). Post-Marrakesh COPs continued to support NAPAs, mainly by mainstreaming NAPAs into developing planning and revising NAPAs.

As of August 2013, the UNFCCC Secretariat had received NAPAs from 49 out of the 50 LDCs that received funding for preparing the document. Despite the fact that the Least Devel-oped Countries Fund (LDCF) is currently the primary source that LDCs can tap into for NAPAs, and although 97 NAPA-related projects had been approved for funding as of April 2013, there remains a lack of clarity regarding by whom and how they will be implemented (Huq and Khan 2006). In fact, the progress report on the LDCF in 2013 pointed out that the supply of resources for adaptation continued to fall far short of the current and projected demand, and adaptation finance remained unpredictable. This consequently provides vulnerable countries with limited opportunities and incentives to invest in longer-term capacity building, institutional frameworks, planning and investments. Furthermore, as Kalame *et al.* (2011) note, few NAPAs have been implemented, with existing successes due to effective inter-ministerial mainstreaming of adap-tation into development planning. Lastly, institutional barriers as well as the exclusion of mar-ginalised and more vulnerable groups from participatory processes have been highlighted as key constraints in the NAPA processes (Huq and Khan 2006).

In most of the LDC countries that have submitted NAPAs, the UNFCCC has been influential in pushing the adaptation agenda forward (Juhola 2010). In Europe, the EU members have been developing their own CCA strategies. In some countries, such as the UK and France, extreme weather events have influenced the policy agenda, underscoring the vulnerabilities of societies to current weather. In contrast, more long-term concerns, such as sea-level rise, have been a driving force in the Netherlands (Peltonen *et al.* 2010). Nevertheless, the EU is shifting towards a more coherent regional approach, including connections with other common policies, such as the Common Agricultural Policy (CAP) and the Common Fisheries Policy (CFP), in order to support and promote multinational governance of adaptation.

Community-based Adaptation and Ecosystem-based Adaptation in SIDS: Putting the Communities in the Driving Seat

SIDS are said to be one of the most vulnerable groups in the world to the adverse impacts of climate change (IPCC 2014). IPCC (2014) suggests that SIDS will experience significant sea level rise over the next 100 years and some portions of land could be inundated. Other major impacts of climate change might occur such as changing weather, saltwater intrusion into groundwater that leads to land erosion, and freshwater shortages. SIDS societies will face impacts of climate change and if migration scenarios play out, which is not certain, then potential loss of languages, identities, and cultures could result.

Adapting to climate change without integrating traditional and local knowledge might lead to failure and malpractices (Mercer *et al.* 2012). The most recent literature on DRR and CCA converges to this point, with DRR continuing to integrate bottom up, 'grassroots' strategies with appropriate top-down approaches, as it has long advocated. While CCA emerged from top-down policy and was initially largely disconnected from communities (Mercer 2010), CCA is also increasingly appreciating the value of community-based and bottom-up approaches and therefore could learn valuable lessons from DRR. Consequently, a more holistic approach should be adopted in terms of enhancing the synergies between DRR and CCA, preferably through placing CCA as a subset of DRR, and also combining scientific and local knowledge. Community-based Adaptation (CbA) and Ecosystem-based Adaptation (EbA) might address this point.

Reid *et al.* (2009, p.13) define CbA as 'a community-led process, based on communities' priorities, needs, knowledge and capacities which should empower people to plan for and cope with the impacts of climate change'. Therefore, CbA focuses largely on supporting people to help themselves towards CCA. EbA is an approach that helps people adapt to the adverse impacts of climate change by using biodiversity and ecosystem services to their advantage (Marshall *et al.* 2010; Pérez *et al.* 2010). EbA promotes sustainable management and conservation and restoration of ecosystems, taking into account anticipated climate change impacts, to increase the resilience of ecosystems and people (Renaud *et al.* 2013; Mercer *et al.* 2012).

CbA tools for climate change only recently entered the academic discourse and focus largely on empowerment within communities. Kelman *et al.* (2009, p. 52) adopt the 'guided discovery' framework as a four-step process that leads to establishing long-term cooperative partnerships between communities and collaborators outside the community at national, regional and international levels. Its main strength is that it recognises scientific and local knowledge as resources for successful strategies for vulnerability reduction. Thus, the main issue that CbA methodologies should address is the possibility of integrating bottom-up and top-down activities. A major gap in CCA (e.g., Mercer 2010 amongst others) is that development practitioners can reject or be unaware of scientific knowledge, while scientists that consider themselves experts often do not engage with practitioners.

Many NGOs are using CbA and EbA to try to fill that gap by conducting research on and implementation of CbA and EbA tools (see Marshall *et al.* 2010; Pérez *et al.* 2010). Marshall *et al.* (2010, pp. 28–29) provide a synthesis of CbA toolkits. In their gap analysis of EbA in the Caribbean, Mercer *et al.* (2012, p. 1924) point out that 'EbA activities are often not differentiated from non-EbA activities, instead recognising adaptation as happening or being needed, with some aspects involving or related to ecosystems and other aspects not'. Boxes 4.1, 4.2, and 4.3 summarise the applicability of CCA measures in SIDS and non-SIDS countries.

Box 4.1   Ecosystem-based Adaptation in Tonga

Stavros Mavrogenis[1], Petros Theodorou[2], and Rory Walshe[3]

[1] European Center for Environmental Research and Training, Panteion University of Athens, Greece
[2] Independent researcher, Utrecht, the Netherlands
[3] King's College London, London, UK
Based on Mavrogenis and Kelman (2013).

For Tonga, successful community-based EbA means linking ecosystems and local livelihood benefits. A key message is that showing the community the benefits from EbA creates local buy-in leading to behavioural change and sustainability. When the benefits derived by the community from mangroves were shown (e.g., reduced storm damage and livelihoods support), people dumped less waste into the mangroves. Locals saw that their livelihoods would gain from ecosystem restoration, so they did the work themselves. The youth were motivated because the revenues from EbA helped them to improve their own livelihoods. Meanwhile, women were supported through an empowerment project encouraging women to lead, which then motivated the women to continue the EbA activities.

Box 4.2   Community-based and Ecosystem-based Adaptation in Seychelles

Stavros Mavrogenis[1], Petros Theodorou[2], and Rory Walshe[3]

[1] European Center for Environmental Research and Training, Panteion University of Athens, Greece
[2] Independent researcher, Utrecht, the Netherlands
[3] King's College London, London, UK

The government of the Seychelles has demonstrated a strong commitment to fighting climate change, and in 1992 was the second country in the world to sign the UNFCCC. Two months later, the country established a national commission for coordinating, developing, and implementing a national plan on climate change, for acting as an intermediary between the national plan and the government, and for preparing national communications to the UNFCCC. The country's national strategy for climate change has the main goal of minimising climate change impacts through coordinated and preventative action at all levels of society – deliberately connecting the local, national, and international. The Seychelles' national adaptation strategy has already achieved institutional governance and community engagement through a series of open public consultations. Integrating top–down and bottom-up approaches has ensured progress on CCA despite the problems of funding, slow exchange of knowledge and technology, and the continued marginalisation of SIDS.

Box 4.3   Climate Change Adaptation in the EU

Stavros Mavrogenis[1], Petros Theodorou[2], and Rory Walshe[3]

[1] European Center for Environmental Research and Training, Panteion University of Athens, Greece
[2] Independent researcher, Utrecht, the Netherlands
[3] King's College London, London, UK

In 2007, the European Commission launched a consultation by publishing a Green Paper on CCA in Europe (European Commission 2007). Two years later, building on this consultation, it published a transitional policy paper – the White Paper 'Adapting to climate change: Towards a European framework for action' (European Commission 2009). In April 2013, the Commission adopted the 'EU strategy on adaptation to climate change' (European Commission 2013) aiming at the provision of specific steps for implementing EU adaptation policies. This strategy aims to provide more specific steps for implementing EU adaptation policies. Whereas the EU members have been developing their own national CCA strategies, the EU is now moving towards a more coherent regional approach including connections with other common policies such as the Common Agricultural Policy (CAP), the Cohesion Policy, the Common Fisheries Policy (CFP) and the Blue Growth Policy to better support multinational governance of adaptation. See also Table 4.1.

A good example of integrating CbA and EbA into the national CCA agenda is what the Seychelles accomplished in recent years (Box 4.2).

EbA is a major opportunity for adaptation at the community level – especially for the world's most vulnerable communities. It addresses many of the existing concerns and priorities of these communities. Consequently, CbA may be a useful vehicle for implementing EbA at the community level.

Local communities will have to adapt to climate change due to the slow pace of the international climate negotiations, as they have always had to adapt to changes in the environment, irrespective of international environmental negotiations and treaties. Adapting to a changing climate requires that communities be put in the driving seat and lead the CCA actions. CbA and EbA are useful tools in implementing this approach, although these actions should not be narrowly applied at the local level, but should also attempt to cross over into national and regional levels as well.

In non-SIDS countries CCA strategies are often national policies that follow the top-down or managerial approach. The EU countries, for example, develop their own adaptation strategies that promote mainstreaming of adaptation into other national policies, from the national level to local level. However, EbA is not considered as a separate adaptation policy but usually part of biodiversity conservation strategies.

Conclusions

This chapter has reviewed theory, policy, and practice for CCA, demonstrating how it sits well within DRR. In theory, this process can be heavily top-down. First, developing the theoretical framework, then mainstreaming it into international and national decision-making processes, and finally practising various models on the ground. Yet, people and communities have been adapting for centuries in climates that have always had variability, trends, and changes. It is true that today's climate change is a result of human intervention but this does not change the fact that environmental change is an ongoing process, which has forced humankind to cope with it throughout history.

The international community of scientists, decision-makers and practitioners over the last few years has increasingly rejected the assumption that implementing adaptation is either a top-down or a bottom-up approach, but instead is a process that should combine both as well as internal/external knowledge and local/external support. Community-based and ecosystem-based adaptation overlap and target the same goals: full and fair access to resources in order to secure livelihoods for local communities to thrive in and live with their environment. Communities are the most genuinely viable element of sustainable development. Therefore, communities should be at the centre of CCA policies, both external and on the ground. This means that in order to achieve the ideal of 'putting communities in the driving seat' the communities themselves should have a say in defining and applying CCA according to their needs.

Global environmental change affects everyone's lives and therefore CCA actions can and should start now. The urgency of mainstreaming and implementing CCA might bridge the gap between theory and practice and in this manner may help to overcome local and external elites who use climate change as a scapegoat in order to perpetuate their power and malpractices that eventually result in environmental degradation. Climate change is part of the wider environmental change processes that already provoke numerous challenges and opportunities at global and local scales, including with respect to food, energy, water, and natural hazards. Therefore, CCA is part of the development process that addresses the aforementioned challenges and is not a separate domain.

Furthermore, in view of the similarities to exploit and the differences to account for between CCA and DRR, there have been increased calls to integrate CCA into DRR (e.g., Kelman *et al.* 2015). Kelman *et al.* (2015) further suggest that while CCA should be subsumed within DRR, DRR should be integrated within sustainable development. Both DRR and CCA prioritise their particular framing of issues, and neglect those outside of it, yet the largest development challenges are often the result of neither disasters nor climate change and are instead a complex combination of development and poverty issues on many scales across space and time.

The IPCC and the UNFCCC processes are the pillars of CCA that represent a science and policy driven approach of dividing climate change from the development discourse into something separate. The recent milestones of the SDGs, the Paris Agreement, and the SFDRR all point in the same direction. Integration of the three realms based on the people's needs will assist in achieving sustainable livelihoods on the people's own terms without causing problems for others.

References

Adaptation Committee (AC) (2014) *Sixth Meeting of the AC Bonn, Germany 29 September to 1 October 2014,* AC: Germany.

Bodansky, D. (1993) 'The United Nations Framework Convention on Climate Change: a commentary', *Yale Journal of International Law* 18, 2: 451–558.

Castree, N. and Braun, B. (eds.) (2001) *Social Nature: Theory, Practice, and Politics,* Oxford, Malden, MA: Blackwell Publishers.

European Commission (2007) *Green Paper – From the Commission to the Council, the European Parliament, the European Economic and Social Committee and the Committee of the Regions. Adapting to Climate Change in Europe – Options for EU Action (No. COM(2007) 354 final {SEC(2007) 849}),* Brussels: Commission of the European Communities.

European Commission (2009) *White Paper – Adapting to Climate Change: Towards a European Framework for Action (COM(2009) 147 final),* Brussels: Commission of the European Communities.

European Commission (2013) *An EU Strategy on Adaptation to Climate Change (No. COM/2013/0216 final),* Brussels: European Commission.

Gaillard, JC (2013) 'Missing the target: Western disaster management policies versus root causes of calamities in the developing world', in A. Sarkar and S.R. Sensarma (eds.), *Disaster Risk Management: Conflict and Cooperation,* New Delhi: Concept Publishing, pp. 1–18.

Glantz, M.H. (ed.) (1976). *The Politics of Natural Disaster: The Case of the Sahel Drought,* New York: Praeger Publishers.

Hewitt, K. (1983). 'The idea of calamity in a technocratic age', in K. Hewitt (ed.), *Interpretations of Calamity: From the Viewpoint of Human Ecology,* Boston: Allen & Unwin, pp. 3–32.

Huq, S. and Khan, M.R. (2006) *Equity in national adaptation programs of action (NAPAs): the case of Bangladesh,* Cambridge, Massachusetts: MIT Press, pp. 131–153.

IPCC (Intergovernmental Panel on Climate Change) (2007) *Climate Change 2007: Impacts, Adaptation and Vulnerability. Contribution of Working Group II to the Fourth Assessment Report of the Intergovernmental Panel on Climate Change,* Cambridge: Cambridge University Press.

IPCC (2014) *Climate Change 2014: Impacts, Adaptation, and Vulnerability. Contribution of Working Group II to the Fifth Assessment Report of the Intergovernmental Panel on Climate Change,* Cambridge: Cambridge University Press.

Juhola, S. (2010) 'Mainstreaming climate change adaptation: The case of multilevel governance in Finland', in E.C.H. Keskitalo (ed.), *Development of Adaptation Policy and Practice in Europe: Multi-level Governance of Climate Change,* Berlin: Verlag Springer, pp. 1–37.

Kalame, F.B., Kudejira, D. and Nkem, J. (2011) 'Assessing the process and options for implementing National Adaptation Programmes of Action (NAPA): A case study from Burkina Faso', *Mitigation and Adaptation Strategies for Global Change* 16, 5: 535–553.

Kelman, I. and Mavrogenis, S. (2014) 'Theory, policy and practice for climate change adaptation', in *Environmental Change: Adaptation Challenges,* Czech Republic: Global Change Research Centre, the Academy of Sciences of the Czech Republic, pp. 12–20.

Kelman, I., Mercer, J. and West, J.J. (2009) 'Combining different knowledges: Community-based climate change', in H. Reid, M. Alam, R. Berger, T. Cannon, S. Huq and A. Milligan (eds.), *Community-based Adaptation to Climate Change 60,* London: International Institute for Environment and Development, pp. 41–53.

Kelman, I., Gaillard, J.C. and Mercer, J. (2015) 'Climate change's role in disaster risk reduction's future: Beyond vulnerability and resilience', *International Journal of Disaster Risk Science,* 6, 1: 21–27.

Lewis, J. (1999) *Development in disaster-prone places: Studies of vulnerability*, London: Intermediate Technology Publications.

Marshall, N.A., Marshall, P.A., Tamelander, J., Obura, D., Malleret-King, D. and Cinner, J.E. (2010) *A Framework for Social Adaptation to Climate Change: Sustaining Tropical Coastal Communities and Industries*, Switzerland: International Union for Conservation of Nature.

Mavrogenis, S. and Kelman, I. (2013) 'Lessons from local initiatives on ecosystem-based climate change work in Tonga', in F. Renaud, M. Estella and K. Sudmeier (eds.), *The Role of Ecosystems in Disaster Risk Reduction: From Science to Practice*, Tokyo: United Nations University Press, pp. 191–218.

Mercer, J. (2010) 'Disaster risk reduction or climate change adaptation: Are we reinventing the wheel?' *Journal of International Development*, 22, 2: 247–264.

Mercer, J., Kelman, I., Alfthan, B. and Kurvits, T. (2012). 'Ecosystem-based adaptation to climate change in Caribbean small island developing states: Integrating local and external knowledge', *Sustainability* 4, 8: 1908–1932.

Peltonen, L., Juhola, S. and Schuster, P. (2010) *Governance of climate change adaptation: policy review*. Online http://www.baltcica.org/documents/baltcica_policyreview_090310_SJ.pdf (accessed 31 August 2016).

Pérez, A.A., Fernández, B.H. and Gatti, R.C. (2010) *Building Resilience to Climate Change: Ecosystem-based Adaptation and Lessons from the Field*, Switzerland: IUCN.

Reid, H., Alam, M., Berger, R., Cannon, T., Huq, S. and Milligan, A. (2009) 'Community-based adaptation to climate change: an overview', *Participatory Learning and Action*, 60, 1: 11–33.

Renaud, F., Sudmeier-Rieux, K. and Estrella, M. (2013) 'The relevance of ecosystems for disaster-risk reduction', in F. Renaud, M. Estrella and K. Sudmeier (eds.), *The Role of Ecosystems in Disaster Risk Reduction: From Science to Practice*, Tokyo: United Nations University Press, pp. 3–25.

Sperling, F. and Szekely, F. (2005) *Disaster Risk Management in a Changing Climate – Discussion Paper Prepared for the World Conference on Disaster Reduction on behalf of the Vulnerability and Adaptation Resource Group (VARG)*. Washington, DC: World Bank.

UNFCCC (United Nations Framework Convention on Climate Change) (1992) *United Nations Framework Convention on Climate Change*, New York: United Nations.

UNFCCC (United Nations Framework Convention on Climate Change) (1998) *The Kyoto Protocol to the United Nations Framework Convention on Climate Change*, New York: United Nations.

UNFCCC (United Nations Framework Convention on Climate Change) (2015) *Paris Agreement*, New York: United Nations.

United Nations (2000) *United Nations Millennium Declaration*, New York: United Nations.

Wisner, B., Blaikie, P., Cannon, T. and Davis, I. (2004) *At Risk: Natural Hazards, People's Vulnerability and Disasters*, London: Routledge.

5

CLIMATE AND WEATHER HAZARDS AND HAZARD DRIVERS

Kate Crowley, Nava Fedaeff, Gregor Macara, and Melanie Duncan

Introduction

Understanding natural hazards is, alongside understanding our capacities and vulnerabilities, a critical foundation for Disaster Risk Reduction (DRR) and Climate Change Adaptation (CCA). There can be little progress in reducing risk and adapting to hazards without fully understanding where, when and how a hazard may impact an area. Our understanding of these hazards may originate from experience, knowledge as well as scientific research. However, climate and weather hazards are dynamic owing to natural climate variability and, more recently, anthropogenic climate change. Our current understanding of these hazards combined with our changing vulnerabilities may no longer provide us with the knowledge we need to adapt, survive and thrive.

Shifting hazard-scapes continue to change hazard frequency, magnitude, and other parameters, with impacts from these hazards dictated by vulnerabilities and capacities. Communities around the world are experiencing some weather hazards for the first time in living and cultural memory, whilst other regions are expecting to see continuing shifts in the magnitudes, frequencies, and impacts of extreme weather events and regular weather patterns. Such experiences are not unique to weather and climate: there are a number of documented cases of geophysical hazards occurring in areas where there was no prior local knowledge or experience of these events (e.g., the 1963 volcanic eruption of Mt Agung in Indonesia). Additionally, climate and weather can influence the triggering and impact of a number of geophysical hazards (e.g., rainfall triggered mass movements, volcanic ash dispersion by winds, etc.). Therefore, identifying and improving our understanding of both slow and fast onset, frequent and infrequent, as well as single and multi-hazard events – alongside their impacts – requires an understanding of the processes that lead to the occurrence of a hazard.

The complexity of the climate system combined with relatively recent coordinated data-gathering at local, regional and international scales means that our scientific understanding of the climate (long-term patterns of atmospheric behaviour) and weather (short-term patterns of atmospheric behaviour) is still relatively new and evolving, although land use practices (such as farming) demonstrate people's long-standing understanding of their local climate. Understanding the present and future climate, and forecasting the weather, remains heavily reliant on studying the past. But this understanding is hampered by the varying quality, format, breadth, consistency and availability of relevant scientifically collected data, and the capability to analyse and interpret data (Bronnimann *et al.* 2008). Despite these challenges, significant efforts have been made to collect and analyse data.

This chapter explores the natural processes that 'drive' climate and weather hazards: the long-term behaviour of the earth's atmosphere and the short-term atmospheric variations respectively. We introduce large-scale processes currently triggering and fuelling the hazards we face. We then explore the complexities of these hazards and highlight the challenges for DRR including CCA.

Invisible Giants: The Patterns and Behaviours of Large-scale Climate Oscillations

Natural global warming and cooling trends occur at different temporal scales. Trends can be seasonal, annual, decadal, centurial or longer, but they all tend to be intricately linked.

Solar radiation is the main source of heat for the planet, with the average amount of energy received from the sun (i.e., the solar irradiance) varying at different locations on Earth. Low-latitude locations receive higher solar irradiance than high-latitude locations, and the resulting heat imbalance is a key driver of the atmosphere's general circulation. Specifically, atmospheric circulation shifts the relatively high energy received near the equator poleward. This is achieved by a variety of mechanisms.

One such circulation mechanism is the Hadley Cell. The Hadley Cell characterises the meso-scale atmospheric circulation pattern that occurs between approximately 30°N and 30°S latitudes. Here, warm air rises near the equator. This rising of air in the tropics forms areas of relatively low pressure at the Earth's surface and enhances convection, resulting in abundant rainfall in these parts. As the air rises, it flows towards higher latitudes, cools, and subsides back towards the Earth's surface. This subsidence is characterised by a belt of subtropical anticyclones (high pressure). These anticyclones suppress convection, such that many of the world's deserts are located along these latitudes. Once the air descends over the sub-tropics, it flows back towards the equator, as air will preferentially flow from areas of high pressure to areas of low pressure. This equator-ward flow of air completes the Hadley Cell, and is characterised by the north-easterly trade winds in the Northern Hemisphere, and south-easterly trade winds in the Southern Hemisphere. In addition, the belt of subtropical anticyclones is a key driver of atmospheric circulation in the mid-latitudes, where westerly winds predominate.

As described above, meridional (north–south) imbalances of energy (and air pressure) are a key component of atmospheric circulation. In addition, there are zonal (west–east) atmospheric circulation features that also play an important role in the weather and climate observed on Earth.

One example is the Walker Circulation. This conceptual model describes the typical atmospheric circulation over the tropical Pacific Ocean, specifically, the predominance of easterly trade winds in the tropics. These trade winds 'push' warmed ocean surface waters from the tropical eastern Pacific to the tropical western Pacific, which enables an upwelling of relatively cool deep waters in the tropical eastern Pacific Ocean. The Walker Circulation periodically strengthens, or weakens/reverses, and these periodic changes are associated with sea surface temperature (SST) variations in the tropical eastern Pacific Ocean.

Such SST variations characterise the El Niño Southern Oscillation (ENSO) phenomenon. ENSO has three phases: neutral, warm (El Niño) and cold (La Niña). During El Niño, the Walker Circulation weakens or reverses, resulting in anomalously high SSTs in the tropical eastern Pacific Ocean. Conversely, during La Niña the Walker Circulation strengthens, resulting in anomalously low SSTs in the tropical eastern Pacific Ocean. Both the El Niño and La Niña phases of ENSO contribute to seasonal variability of processes such as temperature, rainfall, pressure, winds, and sea level in many regions of the world, which in turn may result in alterations to numerous climate and weather hazards such as flooding and drought.

Many zonal and meridional atmospheric circulation patterns are now well understood and a selection of the most influential and common are provided in Table 5.1. The implications of

Table 5.1 Summary of Natural Climate Fluctuations (Adapted from Goosse (2015), International Research Institute for Climate and Society (2016), and World Meteorological Organisation (2015))

Natural Climate Fluctuations	Explanation	Impact
El Niño Southern Oscillation (ENSO)	• The Southern Oscillation is a fluctuation of atmospheric and oceanic variations over the tropical Indo-Pacific region. • The ENSO swings between warm (El Niño) and cold (La Niña) conditions.	• Changes the likelihood of particular climate patterns around the globe. • The outcomes of each event are never exactly the same. • El Niño can lead to increased rainfall along the western coast of the Americas, and drought conditions around southeastern Asia, India, southeastern Africa, Amazonia and northeast Brazil with fewer tropical cyclones around Australia and in the North Atlantic and Australia. • La Niña conditions are more or less the opposite of El Niño.
Madden-Julian Oscillations (MJO) or Intra-Seasonal Oscillation	• Over the Indian and Pacific Oceans tropical convection periodic oscillation (30–60 days). • Forecasting its behaviour is difficult. • Strongest from September to May.	• Associated with rainfall and formation of tropical cyclones. • Can stimulate the rapid development of El Niño.
South Pacific Convergence Zone (SPCZ)	• Warm sea surface temperatures produce lower surface air pressures resulting in atmospheric instability and uplift resulting in convection. • Runs from Solomon Islands to and beyond Samoa. • Position is linked to ENSO. During El Niño it shifts east and north. During La Niña it shifts west.	• Rising air produces clouds and rainfall. • Position of SPCZ can influence tropical cyclone genesis, locations and tracks.
Intertropical Convergence Zone (ITCZ)	• Belt of intense convective activity and rising air. • Shifts with seasonal latitude of maximum solar heating.	• During the northern hemisphere summer, the ITCZ leads into a sustained expansion of the monsoon over Southern Asia. • Activity around the ITCZ can lead to cyclone development. • Southern hemisphere summer impacts are similar to the northern hemisphere but are less pronounced. • In the tropics ITCZ produces short-lived convective clouds and intense rainfall.

(Continued)

Table 5.1 Continued

Natural Climate Fluctuations	Explanation	Impact
North Atlantic Oscillation (NAO)	• Measure of sea surface westerly winds across the Atlantic.	• The positive phase is associated with cold winters over the northwest Atlantic and mild winters over Europe, as well as wet conditions from Iceland to Scandinavia and drier winter conditions over southern Europe. • A negative phase indicates weaker westerlies, a more meandering circulation pattern, often with blocking anticyclones occurring over Iceland or Scandinavia, and colder winters over northern Europe.
Interdecadal Pacific Oscillation (IPO)	• Interdecadal (20–30 years) fluctuation in atmospheric pressure.	• When the IPO is negative, sea surface temperatures are cooler than average in the central North Pacific. • When the IPO is positive, the opposite occurs.
Antarctic Oscillation (AAO) or Southern Annular Mode (SAM) Arctic Oscillation or Northern Annular Mode	• Oscillation of atmospheric pressure in the far southern or northern latitudes. • Alternates between phases approximately every month. • Positive phases have been tentatively linked to ozone depletion in the southern hemisphere.	• A high SAM index is associated with anomalously dry conditions over southern South America, New Zealand and Tasmania and wet conditions over much of Australia and South Africa.
Indian Ocean Dipole (IOD)	• Characterised by cooling of the sea surface in the southeast equatorial Indian Ocean and warming of the sea surface in the western equatorial Indian Ocean.	• Brings heavy rainfall over east Africa and severe droughts and wildfires over the Indonesian region.

climate change on these circulation patterns have not yet been fully resolved, and as such are areas of ongoing research (IPCC 2013). Similarly, considerable resources are focused on seasonal–scale (3–6 months) climate forecasting (linked to variability of atmospheric circulation patterns) in order to improve its efficacy. This stems from recognition that seasonal forecasting is an important tool that can enable communities to anticipate potential climate and weather hazards. Climate change influenced by human actions adds another layer, superimposed on top of, and being influenced by, weather and shorter scale natural climatic variations. Weather hazards have a short timescale such as storms, whilst climate hazards have a longer time frame and can develop over months to years such as drought. Consequently, all temporal scales need to be considered when assessing both climate and weather hazards.

Climate fluctuations can also be driven by processes originating from the sub-surface of the earth. Volcanic eruptions can eject significant volumes of ash, gas and aerosols, which have been demonstrated to affect global temperatures during eruptions with mass flow rates sufficient to eject significant volumes into the stratosphere. Volcanic aerosols, in particular, block solar radiation causing global cooling that can last for several years, for instance the years following the eruptions of Tambora (1815), Krakatoa (1883), and Pinatubo (1991). Further back in geological time, large-scale volcanism caused global warming owing to the release of substantial concentrations of greenhouse gases into the atmosphere.

Climate and Weather

Our knowledge of the global circulation helps to explain the seasonal changes we observe and the more extreme weather we experience. Local to regional extreme weather includes drought, surface and river flooding, cyclones, wind and snow storms. Each of these environmental phenomena has a distinctive set of natural physical characteristics that when exposed to a vulnerable system can have negative consequences and so is often termed a hazard. Whether a hazard is extreme depends on the context, with 'extreme' being defined as 'The occurrence of a value of a weather or climate variable above (or below) a threshold value near the upper (or lower) ends of the range of observed values of the variable' (IPCC 2013, p. 3).

Beyond the individual characteristics, observations of the climate and weather indicate that many, but not all hazards are influenced by large-scale processes. These processes, such as ENSO, can lead to weather extremes impacting several different regions simultaneously.

But, examining just the physical triggers and statistics can be misleading. For example, according to the International Disaster Database (Guha-Sapir *et al.* 2015), the Great Chinese Drought of 1958–1962 is the most lethal drought disaster in recorded history, killing between 30 and 45 million people (Dikötter 2010). However, the disaster was not fuelled by natural processes alone. The political system in China at the time was systematically starving its own people into submission. An estimated 8 per cent of those who died were tortured to death, whilst unknown numbers died due to regulated access to food and enforced hard labour (Dikötter 2010). It is possible that this disaster could have been avoided through appropriate management of resources. The second and third most lethal droughts on record, in Bangladesh (1942) and India (1943), triggered famine and resulted in 1.9 million and 1.5 million deaths respectively; these were also influenced by political decisions as much as natural processes.

Between 1972–73 and 1982–83, clusters of droughts and floods were observed across the globe. Researchers believe that these clustered events were linked to changes in sea surface temperatures in the central and eastern Pacific Ocean (El Niño) and changes in atmospheric sea surface pressure across the Pacific basin (Southern Oscillation). Despite uncertainties in the data, it seems that ENSO provides a climatic setting that is relatively predictable with drought conditions statistically more likely in certain regions during an El Niño (Goddard and Dilley 2005) and the reverse during La Niña.

Knowing that these larger-scale processes may result in the conditions that trigger certain hazards over a number of months provides an opportunity for disaster managers to plan ahead and prioritise mitigation and preparedness strategies. In essence, understanding these processes may provide a way of 'forecasting', or at least of being aware of a heightening frequency of, certain hazard events.

Although there are many errors in global disaster data, droughts, tropical cyclones, and floods are often ranked the most deadly and costly disasters (e.g., Guha-Sapir *et al.* 2015) – at least, so far.

Drought is a complex, slow-onset and long-term hazard that can lead to significant local to global scale impacts. Meteorological drought occurs when there are long periods of below-normal rainfall. There are a range of other definitions for drought reflecting different interests, sectors and causes (Smucker 2012). The potential for large geographic coverage and endurance of drought accentuates socioeconomic conditions that can lead to widespread lack of water and poor agricultural productivity.

Identifying the early warning signs of variations in climatic and oceanic patterns paves the way for early action. Adapting to hazards influenced by large-scale climate variations, such as ENSO-related drought, is crucial not only for individually impacted communities, countries and regions but also for our global society. Chapter 43 indicates how hazard drivers can have severe global repercussions, but also why it is important not to attribute political outcomes directly to only weather or climate.

The impacts of the 2010–2012 La Niña were felt by several countries and territories in the Pacific. Most notably, a state of emergency was declared in Tuvalu on 28 September 2011. Prolonged dry weather due to La Niña led to severe water shortages in the capital atoll, Funafuti, as well as the outer islands of Nukulaelae and Nanumaga. Parts of Tuvalu had less than a week's worth of water supply before international aid arrived.

Officially declaring drought conditions has social, economic and political repercussions and motivations (e.g., Bastos and Miller 2013; Keller 1992; Smucker 2012), as does taking action to reduce drought impacts. One trigger of drought hazards could be long-term decreases in precipitation or could be due to changes in water use and management while precipitation remains stable (or even increases). The driver of drought disasters is vulnerability, in other words a lack of DRR including CCA measures available to communities combined with a lack of action at all scales. The political responsibility does not lie with one governance regime but requires global or regional collaboration. Box 5.1 illustrates the importance of collaboration and partnership for monitoring and forecasting seasonal climate.

Box 5.1   Climate Early Warning Systems in the Pacific

Kate Crowley[1], Nava Fedaeff[2], Gregor Macara[1], and Melanie Duncan[3]

[1] National Institute of Water and Atmospheric Research (NIWA), Wellington, New Zealand
[2] National Institute of Water and Atmospheric Research (NIWA), Auckland, New Zealand
[3] British Geological Survey, Edinburgh, UK

Climate monitoring and the communication of warnings are important components for DRR including CCA. Since 2008, the Samoa Meteorology Division (SMD) within the Samoan Ministry of Natural Resources and Environment, in collaboration with New Zealand's National Institute for Water and Atmospheric Research (NIWA), and the Australian Bureau of Meteorology, have been installing and implementing a climate data service known as a Climate Early Warning System (CLEWS). Funded by the Global Environmental Facility (GEF), CLEWS provides near real-time climate services crucial for risk reduction and adaptation decision making (Porteous *et al.* 2013).

The CLEWS (Figure 5.1) is a combined framework to provide data analysis and support for weather and climate forecasts, which can be integrated with local knowledge to provide tailored early warning

Figure 5.1   Climate Early Warning System Components

(By Authors)

information for specific economic and social sectors (e.g., agriculture). It is leveraged on the deployment of an open source relational database for storing and managing climate data called Climate Data for the Environment, Services Client (CliDE) software (Martin *et al.* 2015). The system comprises five main processes: (1) data capture and logging at automatic or manual climate stations; (2) data transfer by electronic or manual processes; (3) automatic or manual ingest into CliDE; (4) data storage, curation and delivery in a systematic database system; and (5) the interactive CliDEsc product generator and user services.

Using the CLiDE software as the long-term storage for the automated weather observations and linking with NIWA's FloSys, software that enables 'on demand' collection of data via radio or telephone networks (Porteous *et al.* 2013), Samoa now has a near real-time database of scientific meteorological observations (Martin *et al.* 2015).

SMD is applying CLEWS within a wider programme called Integrating Climate Change Risks in the Agriculture and Health Sectors, and Forestry Sector in Samoa (ICCRAHS and ICCRFS), which includes community engagement, capacity building and strong partnerships with other core government departments, such as the National Disaster Management Office and those working in the health sectors. In 2014, for example, the Fire Warning System (FWSYS) was launched, which provides real-time online fire risk alerts. The successful use of this system stems from the demand for such a tool through policy.

But where some regions experience drought, other areas are more likely to suffer from flood during ENSO. In order to reduce flood risk, events must be considered in context, particularly in terms of vulnerabilities. In other words, it is crucial to fully understand what causes and accentuates flood damage at a local level in order to mitigate against it and reduce vulnerability over the long term. So, while floods and droughts will continue to occur, they do not necessarily lead to disasters.

Flooding is a generic term that can be broken down according to the trigger mechanism, for example: storm surge, tsunami, dam burst, debris flow, riverine flooding, surface water flooding, and storm water flooding. Many of these are interlinked and it is not uncommon to experience more than one type of flood during a single event. Heavy precipitation may accompany a storm system that creates a storm surge leading to coastal flooding combined with riverine flooding and urban surface water flooding. This leads to a complexity for DRR including CCA given that each flood type may require a distinctive adaptation or mitigation measure.

During the 2010–2012 La Niña, Australia experienced its biggest flood events in the country's recorded history. The floods resulted in Australia's fifth largest insured loss (Insurance Council of Australia 2016). The increased rainfall followed a significant period of drought. Despite expectations that La Niña would increase the number of tropical cyclones in the Australian region, this did not occur in 2010–2012. But despite the numbers being close to average, the severity of the cyclones was above average – which is in line with projections under climate change (Knutson *et al.* 2010). For example, out of the three tropical cyclones that crossed Queensland during the 2010–2011 season, Tropical Cyclone Yasi was the most powerful to make landfall in Queensland, Australia, since 1918 (Australian Government Bureau of Meteorology 2012).

The impact of climate change is altering the environmental conditions in areas where tropical cyclones form. Although anthropogenic climate change appears to be reducing the frequency of tropical cyclones, they are tending towards a pattern of increased intensity (Knutson *et al.* 2010; Van Aalst 2006). It is currently difficult to distinguish whether ENSO is driving this 'fuelling' of cyclones or climate change – or more likely a combination of both.

Small Island Developing States (SIDS) are experiencing the brunt of climate oscillations and climate change (Connell 2013), owing principally to wider political and social processes that create and maintain these locations' vulnerability (Lewis 1999; Box 5.2).

Box 5.2 Vanuatu and Cyclone Pam in 2015

Kate Crowley[1], Nava Fedaeff[2], Gregor Macara[1], and Melanie Duncan[3]

[1] National Institute of Water and Atmospheric Research (NIWA), Wellington, New Zealand
[2] National Institute of Water and Atmospheric Research (NIWA), Auckland, New Zealand
[3] British Geological Survey, Edinburgh, UK

SIDS are located at the interface of multiple hazards and are in most cases highly vulnerable (Méheux *et al.* 2007; Pelling and Uitto 2001). In March 2015, Vanuatu in the Pacific was struck by Tropical Cyclone Pam. Pam was a category 5 cyclone with estimated wind speeds of 250km/h when it travelled over the islands. The official post-disaster needs assessment (Esler 2015) estimated that 60,000 people were displaced from their homes and 17,000 buildings were damaged or destroyed. The cyclone seriously harmed the livelihoods of over 40,000 households with low-income households and subsistence farmers suffering the most. Total recovery and reconstruction needs were estimated at USD 316 million with USD 95 million required over the following 1–4 years (Esler 2015). However, before the humanitarian effort had ceased, warning signs of drought were emerging with below normal rainfall across Vanuatu identified from June 2015 (Vanuatu Meteorological Services 2016). By December 2015, it was clear that a strong El Niño was likely to lead to a poor wet season for the islands and led to drought lasting until early to mid-2016. Could this be the beginning of Vanuatu and its neighbouring states being locked into a cycle of multi-hazards and potential disasters because vulnerabilities continue to be augmented?

Significant research has been conducted on the negative impacts of ENSO; however, it can also have a positive impact. Munich Re reported that in 2015 there was, in fact, a reduction of insured losses due to reduced hurricane activity (Munich Re 2016). This corresponds with previous El Niño phases where below-average hurricane numbers were recorded along the Atlantic and Gulf coasts of the USA (Glantz 2001), although these statistics are biased towards developed country losses.

Like tropical cyclones/hurricanes, the majority of climate and weather hazards involve multiple hazards. One hazard may trigger another causing a cascading effect, or hazards may coincide due to the triggering forces at play. For example, during a tropical cyclone, strong winds combine with heavy rainfall leading to subsequent flooding.

Many locations are exposed to multiple hazards. It is therefore necessary to account for the risk related to the full range of hazards that might affect people's lives, livelihoods and assets. There have, however, been a number of instances where the entirety of hazards has not been considered; for instance, following the devastating 2004 Indian Ocean tsunami, some housing in Aceh, Indonesia, was reconstructed in highly flood-prone areas (Benson 2009).

Hazards can also interact, resulting in the overall impact being greater than if these hazards had occurred at separate times, which has major implications for risk assessment and vulnerability reduction. In spite of the growing recognition of their importance, there lacks an agreed-upon terminology of interrelated hazards (Kappes *et al.* 2012). They tend to be categorised by the process (e.g., one hazard triggering another), actual examples of interaction (e.g., an earthquake triggering landslide), and/or their influence (e.g., positive or negative impact on the subsequent hazard). The coincidental occurrence of hazards and the triggering or cascade of hazards are generally the most considered processes. Hazard interrelations can be further differentiated into (for instance) four interdependent categories (Table 5.2). Each of the interrelated hazard processes can occur during a single disaster, depending on the analytical spatial and temporal scale considered (Duncan 2014). Multi-hazard events therefore add complexity for DRR including CCA.

Table 5.2 Categories of Interrelated Hazard Processes (Duncan 2014)

Causation	Amplification or alleviation	Coincidence	Association
Hazards that generate secondary hazards, which may occur immediately or shortly after the primary hazard (including cascading hazards).	Hazards that exacerbate or reduce future hazards. (Applies to the same or different hazards).	The simultaneous occurrence of hazards in space and/or time, resulting in compounding effects or secondary hazards.	Hazards that increase or decrease the probability of a secondary event, but which are difficult to quantify.

Shifting Patterns

Although the climatic processes referred to in Table 5.1 have been identified because of their regular cyclical nature, their patterns can shift. For example, El Niño could be stimulated by solar insolation, or respond to external radiative forcing (Glantz 2001). Rapid changes in the global climate add another major driver to environmental hazards: anthropogenic climate change is a reality and is adding to natural climate change (IPCC 2013). The exact impact of climate change on hazards remains inconclusive; however 'a changing climate leads to changes in the frequency, intensity, spatial extent, duration and timing of weather and climate extremes, and can result in unprecedented extremes' (IPCC 2012, p. 115).

Climate change can influence the casual factors of hazards. Increases to the sea surface temperatures could lead to stronger and more prolonged El Niño episodes (Vecchi and Wittenberg 2010), although the IPCC report a low confidence in projected changes in ENSO variability. A shift would change the hazards we experience at all temporal scales. However, as has always been the case with weather and climate, a considerable amount of uncertainty remains in our understanding of how these changes will impact extreme events and 'everyday' hazards. Uncertainty lies both within our current understanding of climatic patterns (Burroughs 2003; Glantz 2001) and with our understanding of climate change (IPCC 2013).

The uncertainties in the climate projections relate to the natural variability of the climate; uncertainties in the climate model parameters and structure; and finally the projections of future emissions and emissions uptake (IPCC 2012). For example, in terms of tropical cyclones, the IPCC (2013, p. 1220) state 'it is likely that the global frequency of occurrence of tropical cyclones will either decrease or remain essentially unchanged, concurrent with a likely increase in both global mean tropical cyclone maximum wind speed and precipitation rates'. However, overall this IPCC report notes a low level of confidence in region-specific projections of frequency and intensity. This is particularly pertinent for low-lying islands and coastal settlements – including large cities such as New York, Miami, and Shanghai – where sea-level rise combined with more intense cyclones could result in extremely damaging multi-hazard events if steps are not taken to reduce vulnerability and to implement DRR including CCA.

It is not yet fully possible to attribute single events such as tropical cyclones (e.g., Tropical Cyclone Pam) to climate change given the poor historical record of tropical cyclones and the large number of factors leading to tropical cyclone genesis and evolution. The limited data are not sufficient to rule out other (natural) climatic influences such as the Pacific Decadal Oscillation.

In some cases, it is nonetheless possible to attribute long-term hazard trends (rather than specific events) as being influenced by anthropogenic climate change. Recent studies have been able to discern anthropogenic influences from natural climate variability (Box 5.3). It is only recently that such statements have been able to be made with conviction.

Box 5.3   Attributing Changing Hazard-scapes in New Zealand to Anthropological Climate Change

Kate Crowley[1], Nava Fedaeff[2], Gregor Macara[1], and Melanie Duncan[3]

[1] National Institute of Water and Atmospheric Research (NIWA), Wellington, New Zealand
[2] National Institute of Water and Atmospheric Research (NIWA), Auckland, New Zealand
[3] British Geological Survey, Edinburgh, UK

Attribution of extreme events to climate change is an evolving science and exceptionally challenging, but recently there has been advancement. Quantifying the impact of a drought requires an understanding of the entire system including human behaviour, such as land and water use practices, as well as the physical environment. Previously, a limited observation record of extreme events posed one of the biggest challenges for discerning an anthropogenic influence from natural climate variability. A special report by the Bulletin of the American Meteorological Society examined the role of anthropogenic climate change on several extreme events including the 2013 drought in New Zealand and heat waves in Australia (Herring *et al.* 2014).

During the summer of 2013, an 'adverse event' brought on by drought was declared for parts of the North Island of New Zealand. The New Zealand Treasury estimated that this event cost the economy at least NZD 1.5 billion. Using a suite of climate model simulations from the Coupled Model Intercomparison Project Phase 5 (CMIP5) archive that alternately included and excluded the effect of anthropogenic climate change, Harrington *et al.* (2014) found that monthly atmospheric pressure anomalies associated with the drought were higher (0.4 hPa on average) as a result of anthropogenic climate change. While it was also noted that natural variability played a key role in the severity of the drought, the results showed a significant shift in several climate indices. This suggested that drought conditions are now more likely to occur over New Zealand due to anthropogenic climate change.

Furthermore, in the 2015 edition of the same special report by the Bulletin of the American Meteorological Society (Herring *et al.* 2015), Rosier *et al.* (2015) found that the likelihood of an extreme five-day July rainfall event over Northland, New Zealand, such as was observed in early July 2014, had increased due to the anthropogenic influence on the climate.

Conclusion

The purpose of this chapter was to introduce the large-scale atmospheric drivers of climate and weather hazards in the context of DRR including CCA. These varied processes influence every aspect of our lives from the daily weather forecast to extreme weather and climate hazards. Our climate is now changing rapidly with major influences from human activities. Poor quality datasets undermine our ability to understand climatic processes and progress is needed for improved models of ocean–atmosphere–land interactions. Improved models may, in time, enable us to make sense of the interdependencies between all the components of the Earth's climate system.

Large-scale oscillations, such as ENSO, influence wide geographic and temporal scale hazards, which in turn have significant local and global impacts. Although progress has been made, how these processes are being impacted by climate change is still uncertain. But we can implement DRR including CCA in the face of such uncertainty, building on existing knowledge and experience, especially recognising that uncertainty has always existed, but this should not preclude effective action.

Lessons can be shared and action taken for improved climate monitoring and early warning programmes such as CLEWS in Samoa, but this will only work where governments have a clear understanding of the technology and science that exists and their needs. The technology and science must be placed within the context of the social conditions which create and maintain vulnerability. Changing vulnerabilities and capacities also alter the impact of hazard events, given that some weather and climate induced disasters are caused directly by human activities, such as water (mis)management and land use (mal)practices. Therefore, a balanced approach to risk assessment is vital.

Ultimately, striving to understand large-scale processes in combination with projected climate change impacts, as well as our individual through to global vulnerabilities, improves our ability to implement DRR including CCA in the short and long term.

References

Australian Government Bureau of Meteorology (2012) *Annual Climate Summary 2012*, Canberra, Australia: Australian Government Bureau of Meteorology.

Bastos, P. and Miller, S. (2013) 'Politics under the weather: Droughts, parties and electoral outcomes', IDB Working Paper, Washington, D.C.: Inter-American Development Bank.

Benson, C. (2009) *Mainstreaming Disaster Risk Reduction into Development: Challenges and Experience in the Philippines*, Geneva: ProVention Consortium Secretariat.

Bronnimann, S., Luterbacher, J., Ewen, T., Diaz, H.F., Storlarski, R.S. and Nue, U. (2008) 'A focus on climate during the past 100 years', in S. Bronnimann, J. Luterbacher, T. Ewen, H.F. Diaz, R.S. Stolarski and U. Neu (eds.), *Climate Variability and Extremes during the Past 100 Years*, Switzerland: Springer, pp. 1–25.

Burroughs, W.J. (2003) *Weather Cycles: Real or Imaginary?*, Cambridge: Cambridge University Press.

Connell, J. (2013) *Islands at Risk? Environments, Economies and Contemporary Change*, Cheltenham, UK: Edward Elgar Publishing Limited.

Dikötter, F. (2010) *Mao's great Famine: The History of China's most Devastating Catastrophe, 1958–62*, London: Bloomsbury.

Duncan, M. (2014) *Multi-hazard Assessments for Disaster Risk Reduction: Lessons from the Philippines and Applications for Non-governmental Organisations*, Engineering Doctorate, London: University College London.

Esler, S. (2015) *Vanuatu Post Disaster Needs Assessment Tropical Cyclone Pam*, Vanuatu: Government of Vanuatu.

Glantz, M.H. (2001) *Currents of Change: Impacts of El Niño and La Niña on Climate and Society*, Cambridge, UK: Cambridge University Press.

Goddard, L. and Dilley, M. (2005) 'El Niño: Catastrophe or opportunity', *Journal of Climate* 18, 5: 651–665.

Goosse, H. (2015). *Climate System Dynamics and Modelling*. New York, Cambridge University Press.

Guha-Sapir, D., Below, R. and Hoyois, P. (2015) *EM-DAT: The CRED/OFDA International Disaster Database*, Brussels, Belgium: Université Catholique de Louvain.

Harrington, L., Rosier, S., Dean, S.M. and Scahill, A. (2014) 'The role of anthropogenic climate change in the 2013 drought over North Island, New Zealand', *Bulletin of the American Meteorological Society* 95, 9: S15–S18.

Herring, S.C., Hoerling, M.P., Peterson, T.C. and Stott, P.A. (2014) 'Explaining extreme events of 2013 from a climate perspective', *Bulletin of the American Meteorological Society* 95, 9: S1–S96.

Herring, S.C., Hoerling, M.P., Kossin, J.P., Peterson, T.C. and Stott, P.A. (2015) 'Explaining extreme events of 2014 from a climate perspective', *Bulletin of the American Meteorological Society* 96, 12: S1–S172.

Insurance Council of Australia. (2016) *Catastrophe Events and the Community*. Online http://www.insurance council.com.au/issue-submissions/issues/catastrophe-events (accessed 15 March 2016).

International Research Institute for Climate and Society. (2016). 'ENSO Essentials'. Online http://iri.columbia.edu/our-expertise/climate/enso/enso-essentials/ (retrieved 5 January 2016).

IPCC (Intergovernmental Panel on Climate Change) (2012) *Special Report on Managing the Risks of Extreme Events and Disasters to Advance Climate Change Adaptation (SREX)*, Cambridge: Cambridge University Press.

IPCC (Intergovernmental Panel on Climate Change) (2013) *Climate change 2013 – The Physical Science Basis: Summary for Policy Makers*, in the Fifth Assessment Report of the Intergovernmental Panel on Climate Change, Geneva: IPCC.

Kappes, M., Keiler, M., Elverfeldt, K. and Glade, T. (2012) 'Challenges of analyzing multi-hazard risk: A review', *Natural Hazards* 64: 1925–1958.

Keller, E. J. (1992) 'Drought, war, and the politics of famine in Ethiopia and Eritrea', *The Journal of Modern African Studies* 30, 04: 609–624.

Knutson, T.R., McBride, J.L., Chan, J., Emanuel, K., Holland, G., Landsea, C., Held, I., Kossin, J.P., Srivastava, A.K. and Sugi, M. (2010) 'Tropical cyclones and climate change', *Nature Geoscience* 3, 3: 157–163.

Lewis, J. (1999) *Development in disaster-prone places: studies of vulnerability*, London, UK: ITDG Publishing.

Martin, D.J., Howard, A., Hutchinson, R., McGree, S. and Jones, D.A. (2015) 'Development and implementation of a climate data management system for western Pacific small island developing states', *Meteorological Applications* 22, 2: 273–287.

Méheux, K., Dominey-Howes, D. and Lloyd, K. (2007) 'Natural hazard impacts in small island developing states: A review of current knowledge and future research needs', *Natural Hazards* 40, 2: 429–446.

Munich Re. (2016) *El Niño Curbs Losses from Natural Catastrophes in 2015*. Online https://www.munichre.com/site/touch-naturalhazards/get/params_Len_Dattachment/1130649/2016-01-04-natcat-2015-en.pdf (accessed 6 January 2016)

Pelling, M. and Uitto, J.I. (2001) 'Small island developing states: Natural disaster vulnerability and global change', *Global Environmental Change Part B: Environmental Hazards* 3, 2: 49–62.

Porteous, A., Tait, A., Titimaea, A., Seuseu, S., Ramsay, D., Lefale, P., Wratt, D., Allen, T. and Moneo, M. (2013) *Strengthening Climate Services in Samoa*, New Zealand: National Institute of Water and Atmospheric Research (NIWA).

Rosier, S., Dean, S., Stuart, S., Carey-Smith, T., Black, M.T. and Massey, N. (2015) 'Extreme rainfall in early July 2014 in Northland, New Zealand – Was there an anthropogenic influence?', *Bulletin of the American Meteorological Society* 96, 12: S136–140.

Smucker, T. (2012) 'Drought', in B. Wisner, JC Gaillard and I. Kelman (eds.), *The Routledge Handbook of Hazards and Disaster Risk Reduction*, London, UK: Routledge, pp. 257–268.

Van Aalst, M.K. (2006) 'The impacts of climate change on the risk of natural disasters', *Disasters* 30, 1: 5–18.

Vanuatu Meteorological Services. (2016) *ENSO alert report 2015. El Nino is now well Established: Prepare for Drought*. Online http://www.meteo.gov.vu/ClimateForecastsRainfall/ENSOAlertreport/tabid/225/Default.aspx (accessed 6 January 2016).

Vecchi, G.A. and Wittenberg, A.T. (2010) 'El Niño and our future climate: Where do we stand?' *Wiley Interdisciplinary Reviews: Climate Change* 1, 2: 260–270.

World Meteorological Organisation. (2015). 'Significant Natural Climate Fluctuations'. Online https://www.wmo.int/pages/themes/climate/significant_natural_climate_fluctuations.php (retrieved 22 December 2015).

6

VULNERABILITY
AND RESILIENCE

Ilan Kelman, Jessica Mercer, and JC Gaillard

This chapter is a reprint of the following paper with the kind permission of the journal:

Kelman, I., Gaillard, JC and Mercer, J. (2015) 'Climate change's role in disaster risk reduction's future: Beyond vulnerability and resilience', *International Journal of Disaster Risk Science* 6, 1: 21–7.

It has been reformatted to suit this handbook's style and all the boxes are additions from the original paper with text adjustments in the chapter to refer to the boxes.

Common Goals and Interests: Beyond 'Normal'

Humanity has created numerous challenges for Planet Earth and, consequently, for ourselves. A seminal policy year for environment and development takes place in 2015 due to three parallel but interacting United Nations processes: (1) seeking a long-term agreement on dealing with greenhouse gases and climate change impacts; (2) aiming for the finalization and adoption of the Sustainable Development Goals; and (3) striving to develop a successor to the Hyogo Framework for Action as a global disaster risk reduction plan.

Why three separate processes? Why not join them? They all have common themes, use common approaches, and deal with common terms, including the examples of 'vulnerability' (Box 6.1) and 'resilience' (Box 6.2). In theory, there should be no need to separate them, but instead to use 2015 as an opportunity to bring them together and to learn from each other in order to improve society and build a better future. The point of these processes is to create something new, beyond the normal situation of poor development, poverty, vulnerability, and disaster.

Box 6.1 'Vulnerability': A Matter of Perception?

Annelies Heijmans[1]

[1] Van Hall Larenstein University of Applied Sciences, Velp, the Netherlands
(This text updates Heijmans 2001.)

Agencies, institutions, and organizations involved in DRR including CCA use differing definitions of 'vulnerability' to describe factors, processes, or constraints of an economic, social, physical, or geographic nature which reduce the ability of individuals and communities to prepare for and cope with disasters. Most of these groups recognize that vulnerability is a significant concern for the poor and that the most vulnerable sectors require attention. The way they explain and frame vulnerability determines the kind of DRR measures they propose.

Three different views on vulnerability and resulting strategies to address vulnerability exist, although they are not mutually exclusive:

1   Unexpected Natural Events as the Cause Of Disasters → Technological and Scientific Solutions

Vulnerability is viewed as the result of unforeseen, external natural events and risk (exposure to events measured in terms of proximity). In order to reduce vulnerability, technological solutions are designed and applied, such as early warning systems, flood control, remote sensing for drought and fire monitoring, equipment to measure seismic activity, and building code regulations focusing on 'protecting' the building or the occupants against nature.

2   Poverty as the Cause of Disasters → Economic and Financial Solutions

Technological solutions are often costly. To reduce vulnerability, social safety nets, insurance, calamity funds, financial assistance, and livelihood programmes are provided to build people's assets and abilities to better cope with hazards and disasters.

3   Powerlessness as the Cause of Disasters → Political Solutions

Socio-economic and political processes in society produce conditions that adversely affect the ability of communities or countries to respond to, cope with, or recover from the damaging effects of disasters. These conditions precede the disaster, contribute to its severity, and may continue to exist afterwards (Anderson and Woodrow 1989, 10; Blaikie *et al.* 1994, 9). Vulnerability can be reduced only when people affected by disasters politically start demanding DRR.

None of the above three views mentions explicitly how people experience and view vulnerability. In the context of recurrent disasters, people do not talk about their vulnerability or that they rely on outside aid (Heijmans 2012, 259). Although they do not necessarily use the notion of 'vulnerability' to describe their worsening situation, they feel the stress, face difficulties, talk about 'risks', and make risk-taking or risk-avoiding decisions. People are neither passive nor powerless; the options people have to deal with risks are embedded in local institutional settings. People comply with these settings, adjust them, contest norms and policies, or evade them in numerous quiet and subtle ways, which Kerkvliet (2009) refers to as the practice of 'everyday politics'.

Hence, 'vulnerability' is not a real-time condition in which people live, but rather a social construct of external groups to legitimate their aid-giving actions. However, the concept of 'vulnerability' can still be used as an analytical tool and an instrument for raising consciousness. It can unravel institutional arrangements and power relations by taking an historical perspective and by shifting away from routinely categorizing women, children, refugees, elderly, widows, indigenous peoples, and others as 'most vulnerable'. In this process, 'vulnerability' is used to make people understand their current conditions, the reasons behind those conditions, and what actions to take (Heijmans 2012, 110; Lewis 1999).

Box 6.2   A History of Resilience

Zehra Zaidi[1] and David Alexander[1]

[1] University College London, London, UK

Resilience is an ancient concept that has been subject to many different interpretations over its rich and varied history. It originated in politics and natural history and entered into the sciences in the 1600s. It found its way into mechanics in the nineteenth century and in the mid-twentieth century it was taken up by anthropology and child psychology (Alexander 2013). A social view of resilience was introduced to the disasters field by Timmerman (1981), who in effect presented the concept as the opposite of vulnerability. He analysed it in terms of both the limits to growth (respecting them would provide space for resilience to be achieved) and the systems model favoured by Holling (1973).

Resilience has gradually been accepted into both the social and physical science approaches to disaster reduction. Resilient defences against physical forces are robust and flexible. Resilient societies are endowed with resolve and adaptability. This reflects the engineer's view of resilience as an optimum combination of strength and ductility, so that a material can both resist and absorb a force applied to it (Rankine 1867).

For DRR including CCA, resilience brought the idea that resistance and adaptability could apply equally to the security field (civil defence and counter terrorism), civil protection (conventional disaster management) and other aspects of society in which people and their forms of association needed to be more robust in the face of adversity. The concept has experienced a surge in popularity and flowed from academic debates to practical applications of disaster and development agendas in the field. Efforts have been made to mainstream resilience thinking into policy and programming at all levels of disaster and development operations.

This is reflected in the preoccupation with improving resilience in the recent Sendai Framework for Disaster Risk Reduction (UNISDR 2015), as well as the Paris Agreement and Framework Convention on Climate Change (UNFCCC 2015) and the Sustainable Development Goals (UNGA 2015), with resilience directly or indirectly expressed as a desired outcome of all three frameworks. Resilience is seen as the bridging concept between DRR, CCA, humanitarian assistance, and development goals. The targets of the Sendai Framework for Disaster Risk Reduction can be seen as directly addressing the improvement of resilience at a global level, offering a practical tool for reducing vulnerability.

Regrettably, in the evolution of resilience, disillusion also rapidly set in (e.g. Zebrowski 2009; MacKinnon and Derickson 2013). The fear is that resilience does not 'deliver the goods': perhaps it fails to adequately illuminate the complex processes of vulnerability reduction and the subtle linkages between hazard and vulnerability that eventually give us disasters. Rather than outright failure, the problem with resilience seems to be that it does not offer much more insight into disasters than do any of the approaches it has replaced.

Seeking something new, rather than perpetuating the normal, seems to be at odds with many conceptualizations of 'resilience' in which the core idea is to return to 'normal' or, after a disaster, to return to the pre-disaster state. Yet returning to normal means returning to poor development, poverty, vulnerability, and disaster, not building a better future (Box 6.3). As one example from among many, with a heavy basis in ecosystem science, the Resilience Alliance (epitomized by Folke 2006) states that resilience is about 'still hav[ing] the same identity (retain[ing] the same basic structure and ways of functioning)' [editors' note: see www.resalliance.org although their definition has since been updated]. No explanation is given for why it should necessarily be an objective to retain the same identity, structure, and ways of functioning. Conversely, overcoming racial segregation and giving women equal rights is based on overturning the standard functions and controls of society; that is, permitting a disturbance to fundamentally change a system (Box 6.4).

Box 6.3   Resilience: Paralogism or Sophism?

James Lewis[1]

[1] Datum International, Marshfield, UK
Based on Lewis (2015).

Over-emphasis on resilience serves to obscure root causes of disastrous consequences it seeks to ameliorate.

Disasters expose society's vulnerabilities within inequality, poverty and ensuing social or physical inadequacies. Most people's vulnerability is not of their own making but is caused by actions and inactions of other people, those more powerfully self-interested. Vulnerability is a political issue against which resilience may be politically expedient.

Large numbers of people are caused to be at risk by processes, often corrupt and over long periods of time, working to favour the few and in opposition to the interests of the many. Examples of such processes are: discrimination, displacement, impoverishment by others' self-seeking expenditure, denial of access to resources and personal siphoning of public money meant to be spent for public good. Such behaviours prevail within and beyond most concepts of governance and operate powerfully and pervasively at domestic, group, corporate and institutional levels. How can community resilience succeed against pervasive corrupt criminality entrenched within the same communities; and why should it be expected to do so?

Resilience trades on well-meaning innate optimism, positive drive, and honourable 'common good' for the safer future of humankind. What is not expressed is that governments, societies and communities may include the negative as well as the positive, the corrupt as well as the ethical, and the powerfully active as well as the inactive, inert and the weak. Whilst initiating or permitting favourable systems and procedures, good governance has also to prevent and inhibit those other systems that, whilst advantageous to the powerful few, result in widespread deprivation for the many. Can resilience alone counter these realities?

Now a popular byword from board rooms, staterooms, ministerial departments, parliaments, and committee rooms, is 'resilience' simply a paralogism of ill-informed non-thinkers or worse, is it a conscious sophism of conspired delusion – or of intended or unintended deceit?

Box 6.4   Precondition May Govern Capacity for Resilience

James Lewis[1]

[1] Datum International, Marshfield, UK
Based on Lewis (2013).

In some contexts, neither capacity nor resources may be available without assistance and external inputs. As its definition states, resilience depends upon the *ability* of a system, community or society, which may mean it cannot be assumed; for example, in an aftermath of physical and psychological shock, the effects of which may be long term.

Resilience, individually and collectively, is a human characteristic that refers to coping and adapting in the aftermath of a stressful occurrence. How much an 'abstract judgement' can bear up to disaster aftermath depends upon physical, social and psychological vulnerabilities of the pre-existing community and how those vulnerabilities may have been exposed over time. All human characteristics require mental and physical energy to emerge and to mobilize, characteristics that may be in short supply where destruction, deaths, injuries and shock prevail. In some preceding long-term contexts, such characteristics may have been eroded over long periods of time before disaster impact. Prescribed characteristics of resilience only rarely refer to preceding contexts.

Not all disasters are sudden. Disasters of 'slow onset' usually include drought and famine but, in reality, all disasters are slow onset when evolutions of vulnerabilities are recognized, whether of natural hazards, conflicts or wars. The experience of slow-onset events is therefore prolonged.

Reality suggests that the negatives of resilience are significant. People's capacities, variously preconditioned from birth, are further affected by limited access to food, water, medical care, schools and teachers, and maybe also overall lack of accustomed infrastructure and services for transport, telephones, radios, newspapers, mail and other facilities. In much the same way, people are also affected by shock, fear, anxiety and grief and resilience may require similarly long periods of time to emerge.

Wolf (1988, p. 51) described her formative experiences under Nazi rule as living with fear, hypnotic conformity and obedience: 'A whole generation, more than one, had suffered great damage to the fundamental principles of its psychic existence on earth. That cannot be so easily repaired'. She describes the need for time 'so that one is able to live in reality, with reality, and in a way adapt to reality, that is, to act in a way that influences reality'.

Only then would resilience re-emerge.

In fact, the assumption that society has a 'normal' state could be questioned, since society always changes. Is society ever on an even, steady trajectory that could be called 'stable', 'usual', or 'normal' over the long term? The assumption that society would not wish to, or should not, change is questionable, because there are fundamental aspects of society's controls, functions, and processes that have changed in the past – such as excluding people on the basis of gender and race – and that should change in the future – such as continuing sexism and racism. How these changes may be enacted should be discussed in order to avoid accusations of external imposition for societal change and cultural imperialism, if it is desired to avoid such accusations.

Glantz and Jamieson (2000) and Tobin (1999) note that if resilience involves a return to pre-disaster conditions, then it is simply a return to the conditions, including vulnerability, which led to a disaster in the first place. Vulnerability is the chronic, 'normal' condition related to poor development and sustainability practices (Hewitt 1983; Lewis 1999; Wisner *et al.* 2004, 2012). Should a desire exist to return to that 'normal' of the vulnerability process? That would be setting up another disaster. If post earthquake Haiti rebuilt to its status prior to the 2010 disaster, then the country is deliberately constructing the conditions that killed over 200,000 people in the first place.

Furthermore, survivors carry a disaster with them for life, through emotions and reactions. An example is flood survivors feeling stressed when it rains (Tapsell *et al.* 2002). 'Recovery' can be achieved through continuing with life without letting the flood experience control all decisions – an appropriate development approach. But the experience of being flooded might never, and perhaps never should, go away and be forgotten as if it never happened. Instead, hazards can be 'normalized' through response mechanisms that are fully embedded within people's everyday life (Anderson 1968; Bankoff 2003). 'Return to normal' might never be feasible after a disaster (Fordham 1998; Hills 1998) – and it might never be desirable. Rather than 'bouncing back', resilience and sustainability could instead be demonstrated through a society that does not get 'back to normal', but instead does better, even through 'bouncing forward' (Manyena *et al.* 2011). The post-disaster development paradigm of 'Build Back Better' personifies that perspective, within the critiques of that term and process (Kennedy *et al.* 2008).

The three 2015 processes have an opportunity to embrace, promote, and make practical these notions, including using vulnerability and resilience concepts that would break out of the trajectories leading to disasters and sustainability difficulties. Applying long-term, deeper perspectives seeks a 'normal' in which hazard effects, including those from climate change, are less detrimental and more advantageous for society. Part of this strategy entails deepening our approach to vulnerability and resilience in order to step beyond standard approaches that have proven counterproductive to the common 2015 goals.

Deepening Our Approach to Vulnerability and Resilience

To facilitate improvement and integration, five points are suggested here because they embrace wider and deeper meanings that ensure a robust future for disaster risk reduction. These points are presented a priori but emerge from a long history of research, policy, and practice for which only a few example citations were given in the previous section.

(1) Vulnerability and resilience are dominated by quantitative approaches, even though they are also qualitative. Not all aspects of vulnerability and resilience can be demonstrated by calculation or quantification, even where these actions assist with some aspects. Qualitative characteristics are shown by the value of intangible items, including photographs and archaeological sites, in understanding how people and communities avoid, react to, and recover from disasters (Parker *et al.* 2007).

(2) Vulnerability and resilience are presented as being objective, when subjectivity is more realistic. For example, Russia has been saved at least three times from invading armies when the harsh winter weather, coupled with poor strategic military decisions by the invader and solid tactics from the Russians, contributed to the invaders' defeat. The invasions were by Charles XII of Sweden from 1708 to 1709, Napoleon Bonaparte at the end of 1812, and Adolf Hitler from 1941 to 1943. Similarly, the victory of the English navy in the Camperdown Campaign in 1797 has been attributed as much to weather as to military tactics (Wheeler 1991). In such

cases, one side in a military conflict saw weather damage as vulnerability, while the other saw it as resilience. The perspective depended on to whom the damage was being done. A parallel interpretation is that the environmental phenomenon itself can be a hazard or a resource/ opportunity, depending on one's perspective. If it is viewed as a hazard, then vulnerability is emphasized. Conversely, if it is viewed as a resource or opportunity, then resilience is emphasized. In this way, environmental phenomena can be intertwined with the interpretation conferred on it by society.

(3) Vulnerability and resilience are assumed to have absolute metrics, but proportional approaches are important too. Lewis (1979, 1999), amongst others, provides an alternative to the frequent focus on presenting absolute numbers to describe vulnerability and resilience. He describes why proportional impact, indicative of proportional vulnerability, provides important information for development activities. For instance, islands have small populations relative to cities. Even if 100 per cent of an island country's population is affected by a poor water supply or by a cyclone, that situation is unlikely to match the numbers of people who would be affected in a megacity with only 1 per cent of the population affected. Yet 100 per cent of a population affected can be much worse than 1 per cent of a population affected. Absolute and proportional metrics provide different characteristics of vulnerability and resilience, so both are needed.

(4) Vulnerability and resilience are assumed to be non-contextual, when contextuality or localization tends to be more realistic. Often, a method for quantifying objective vulnerability or resilience is assumed to be transferable to other contexts. That assumption might not be appropriate. Vulnerability and resilience might be predominantly Western constructs that make their understanding and application highly contextual (see also Baldacchino 2004; Bankoff 2001). In fact, some languages do not have words for 'vulnerability' or 'resilience' and the concepts can be difficult to explain within those cultural contexts.

(5) Vulnerability and resilience are often presented as being the current state, whereas examining a long-term process with a past and future is needed. Vulnerability and resilience are not only about the present state, but are also about what society has done to itself (and especially what some sectors have done to other sectors) over the long-term; why and how society has taken that set of actions in order to reach the present state; and how society might change the present state to improve in the future (see also Bankoff 2001; Garcia-Acosta 2004; Lewis 1999; Wisner *et al.* 2004, 2012).

These five points show how varying perspectives and wider contexts would contribute to fully accounting for development's long-standing contributions to vulnerability and resilience studies. Multiple theoretical and practical difficulties emerge when broader temporal and spatial perspectives and contexts are not considered, as shown by examining the suggestion of 'double exposure'.

'Double exposure' describes how vulnerability is augmented by having to deal simultaneously with problems from the impacts of global environmental change and economic globalization (Leichenko and O'Brien 2008). The history of and literature from vulnerability and resilience research and on-the-ground practice, from the 1970s to today, has highlighted 'multiple exposure' (Bankoff 2003; Cuny 1983; Glantz 1977; Hewitt 1983; Lewis 1999; Shaw *et al.* 2010a; Wisner *et al.* 2004). Climate change, globalization, poverty, earthquakes, injustice, tropical cyclones, lack of livelihood opportunities, inequity, landslides, overexploitation of natural resources, epidemics, and lack of water supply – amongst many other ongoing challenges – often converge to most affect those who have the fewest options and resources for dealing with those challenges (Box 6.5).

Box 6.5   Causes of Vulnerability, Destroyers of Resilience

James Lewis[1]

[1] Datum International, Marshfield, UK
Based on Lewis and Kelman (2012) and Lewis and Lewis (2014).

Causes of vulnerability to disasters serve to diminish capacity for resilience against their consequences.

Discrimination, displacement, impoverishment, denial of access to resources, corrupt siphoning of funds intended for public good, and denial of tax payments are globally identified as repetitive causes of inequality, poverty and ultimate vulnerability. Poverty may itself be a further cause, as well as an indicator, of low income, increased living costs, consequent debt, homelessness, poor nutrition and physical and mental ill-health.

Suffering the effects of these and other actions means lives drained of initiative and energy, the basic requirement of capacity necessary for personal and community resilience.

People's resilience is preferred by those guilty of causes of vulnerability but to 'bounce back' is invalid in the longer-term if it means a return to poverty.

Consequently, those with the fewest options and resources tend to be most vulnerable across all forms of threats, demonstrating multiple exposure to multiple threats simultaneously. To refer to 'double exposure' by assuming that only two forms of threat are especially important does not factor in the more expansive forms of the notion that have been long established in the literature. Leichenko and O'Brien (2008, p. 31) mention 'multiple stressors', but do not reference the multiple exposure approaches of prior literature and they nonetheless continue to focus on global environmental change and economic globalization as being the most important factors for their analyses. Is this simply a theoretical dispute without much practical meaning?

The practical problem arises from the fact that, in many locations, the most prominent or fundamental development challenges are neither climate change nor globalization. Decision-makers might be distracted by double exposure and forget about, or wish to ignore, the 20 per cent HIV infection rate or the upslope deforestation destroying the delta. That does not deny potential globalization inputs into these phenomena, but accepts that multi-scalar processes across time and space are influential. Elsewhere, double exposure is more sinister. Exacerbated sea flooding in certain places in Bangladesh was blamed on climate change, whereas it was actually due more to using structural sea defences (Auerbach *et al.* 2015). Villagers in Vanuatu were termed 'climate change refugees' even though the increased sea flooding was due more to tectonic subsidence than to sea-level rise (Ballu *et al.* 2011).

Research in the Maldives (Kothari 2014) shows how climate change and globalization are being used as excuses by the government to force a policy of population consolidation (resettlement) on outer islanders. Yet, the government has long been trying to resettle the outer islanders closer to the capital using other reasons, such as that it is hard to provide a scattered population with services including health, harbours, and education. Both arguments have legitimacy and can be countered, but climate change is used as an excuse to do what the government wishes to do anyway. 'Double exposure' can be used insidiously to achieve hidden agendas by obscuring the full picture of multiple exposure.

The Role of Climate Change

Yet, misuse of climate change does not obviate climate change as a significant concern that will cause major problems, not just for low-lying islands such as the Maldives and low-lying coasts such as Bangladesh, but for all of humanity (IPCC 2013–2014). Many good practice examples of resilience exist, to climate change impacts and to other hazard drivers and hazards. These practices demonstrate what can be achieved when broader concepts of vulnerability and resilience are accepted and applied (Global Network of Civil Society Organisations for Disaster Reduction 2009, 2011). An ongoing challenge is framing climate change in research, policy, and practice to try to avoid the difficulties resulting from narrow views of vulnerability and resilience or too much focus on a single phenomenon such as climate change.

It is not appropriate to disparage or to ignore climate change. Nor should a false duality be created by suggesting that the debate is climate change versus other concerns such as earthquakes, injustice, HIV/AIDS, gender equity, or water resources. Care is nonetheless needed when highlighting climate change, since addressing climate change has the potential to create or exacerbate other development concerns.

For example, large hydroelectric dams might contribute to climate change mitigation through reduced dependence on fossil fuels. Large dams might also contribute to climate change adaptation by permitting a more stable water supply, irrespective of precipitation variations. But large dams tend to increase flood risk over the long term in a process termed 'risk transference' (Etkin 1999). Structural defences including large dams stop smaller floods and permit people to live in floodplains while remaining relatively dry. As a result of this false sense of security, vulnerability to flooding increases (Fordham 1999). Most structural defences could fail at some point; often from an event that exceeds or has different characteristics from the design flood but sometimes because maintenance requirements have not been met. Then the damage incurred by the flood is much greater than it would have been without the false sense of security imposed by the structural defences. Short-term flood risk has decreased, but long-term flood risk has increased. Risk is transferred into the future and augmented, hence the term 'risk transference' (Etkin 1999). Risk can also be transferred amongst locations, sub-populations, and sectors (Graham and Weiner 1995), which makes it important to consider a multitude of challenges (similarly to 'multiple exposure') when assessing and addressing vulnerability and resilience. Other than risk transference, many other development concerns have also been identified through relying on large dams (World Commission on Dams 2000) irrespective of their potential contributions to climate change mitigation and adaptation. Rather than keeping climate change as a separate or dominating topic, the proposal from a development perspective is to enact the 'multiple exposure' perspective by viewing climate change as one challenge amongst many (Gaillard 2010; Mercer 2010). Researchers and practitioners have long published on and tried to address vulnerability and resilience to the consequences of change, positive and negative, at all time and space scales and based on many forms of change (Aysan and Davis 1992; Bankoff 2001; Etkin 1999; Glantz 1977; Hewitt 1983; Lewis 1979, 1999; Wisner *et al.* 2004, 2012). Examples are aridification and desertification, climatic changes from meteorite strikes and volcanic eruptions, local water drawdown, and availability and use of local and locally appropriate building materials. Contemporary climate change is one more to add to this well-established list – and it should be added to ensure that climate change vulnerability and resilience are addressed. Nonetheless, climate change should not be tackled at the expense of other challenges and opportunities in everyday life.

The subset within development work that is best suited for placing climate change adaptation in perspective and context is disaster risk reduction (Shaw *et al.* 2010a, 2010b). That arises due to the long history within disaster risk reduction of dealing with climate-related changes at all time and space scales and from multiple causes (Garcia-Acosta 2004; Glantz 1977; Hewitt 1983; Lewis 1999; Wisner *et al.* 2004, 2012). Therefore, research, policy, and practice should accept contemporary climate change as one challenge amongst many within disaster risk reduction.

Climate change as a subset within disaster risk reduction can be elaborated through three main points.

First, climate change is one contributor to disaster risk amongst many. Climate change should not be ignored but neither does it necessarily dominate other contributors. Those contributors include, but are not limited to, non-climate-related environmental phenomena (for example, earthquakes and volcanoes), inequities, injustices, social oppression, discrimination, poor wealth distribution, and a value system that permits exploitation of environmental resources irrespective of the long-term consequences. Climate change drives both hazards and vulnerabilities.

It drives hazards, for instance, in that a hotter atmosphere can hold more water vapour leading to increased precipitation. When and where that moisture is released can augment the intensities of floods and blizzards as they occur. Sometimes, climate change drives hazards by making the hazards less extreme such as by reducing the frequencies of Arctic storms called polar lows (Zahn and von Storch 2010), Atlantic hurricanes (Knutson *et al.* 2010), and winter floods in central Europe (Mudelsee *et al.* 2003). Climate change drives vulnerabilities by changing local environmental conditions so rapidly that local environmental knowledge cannot keep pace with and is less applicable to, for example, local food resources. Whether climate change is a more significant or a less significant contributor than other factors – such as relying on structural approaches for floods or increasing the social oppression that creates and perpetuates food-related vulnerabilities – depends on the specific context.

Second, climate change is one 'creeping environmental change' amongst many. Creeping environmental changes are incremental changes in conditions that cumulate to create a major problem, apparent or recognized only after a threshold has been crossed (Glantz 1994a, 1994b). Climate change fulfils that definition. Other creeping environmental changes not linked to contemporary anthropogenic climate change include soil erosion due to intensive farming, salinization of freshwater supplies due to excessive draw-down, and slow subsidence of land due to water or fossil fuel pumping (Glantz 1994a, 1994b; Wisner *et al.* 2012). Development work has long dealt with such topics (such as the historical descriptions provided by Crush 1995; Gaillard 2010; Glantz 1999; Mercer 2010) and climate change can readily be integrated into this set of development concerns. Third, the reality is that climate change has become politically important, within and outside of development. That should provide an opportunity, not to focus exclusively on climate change, but rather to raise the points made in this article in order to engage interest in more comprehensive development processes. Little point exists in building a new school with natural ventilation techniques that save energy and that cope with higher average temperatures, if that school will collapse in the next moderate, shallow earthquake. Similarly, if a hospital is renovated with water-resistant materials and finishes for climate change adaptation due to the projected expansion of the floodplain, but is put out of action by toxic contaminants in the floodwater, then little has been achieved. Climate change is one topic amongst many and should be dealt with in wider contexts. Since climate change drives hazards and vulnerabilities and since disaster risk reduction efforts provide more comprehensive views of vulnerability and resilience, a prudent place for climate change would be placement within disaster risk reduction.

Climate change adaptation therefore becomes one of many processes within disaster risk reduction.

Moving Beyond Climate Change, Vulnerability, and Resilience

By placing climate change within disaster risk reduction, while using the prominence of climate change to promote and achieve wider development agendas, a long-term perspective is supported in which related research better serves policy and practice – and vice versa. The long-term perspective further assists in addressing the vulnerability process and the resilience process. An historical perspective avoids being distracted by a myopic concentration on climate change, instead directing attention to root causes and the fundamentals of vulnerability and resilience as long-term processes. Research, policy, and practice would move forward in concert by accepting the widespread, long-term causes and consequences of vulnerability and resilience from multiple sources and requiring multiple, integrated interventions.

Achieving this theoretical approach in practice would set aside and move beyond vocabulary differences, instead bringing together the 2015 processes under the common banner of sustainability. Oliver-Smith (1979) referred to a 400-year earthquake in examining the 31 May 1970 earthquake and rock avalanche in Yungay, Peru, that killed thousands of the city's inhabitants. The '400 year' timeframe is not the geological return period of the seismic or avalanche event. Instead, it refers to the fact that the root causes of the vulnerability, which were exposed as a result of the event, took 400 years to build up – a long-term process. The vulnerability that caused the disaster can be traced back to the Spanish conquest of the region, in terms of demographics, settlement locations, and ways of living – exactly the aspects that the Sustainable Development Goals aim to address. No longer must ways of living and livelihoods be categorized as vulnerability and/ or resilience, but instead they are accepted as supporting multiple sustainability goals, tackling multiple exposure (and see Box 6.6).

Box 6.6   Capacities as a Standalone Concept for Disaster Risk Reduction Including Climate Change Adaptation

JC Gaillard[1]

[1] The University of Auckland, Auckland, New Zealand

Capacities, as a plural, standalone concept, was first coined by Anderson and Woodrow (1989). It explicitly emerged as a spin-off from the so-called vulnerability paradigm. It was strongly influenced by the growing and concomitant momentum gained by the idea that people, including the most marginalized, should be at the forefront of development because they are knowledgeable and resourceful. Capacities in facing hazards and disasters were then seen as a rationale for fostering people's participation in DRR. Recognizing that people have capacities therefore underpins that they should also participate in DRR. This suggests a shift in power relations to the detriment of outside institutions and organizations who were, at the time, considered the dominant, if not exclusive, stakeholder for DRR.

Despite its strong epistemological grounding, there have been very few definitions of the concept of capacities and, interestingly, most spring from publications by practitioners or international organizations rather than scholars. All definitions nonetheless converge to suggest that capacities encompass the strengths people possess to mitigate, prepare for, cope with and recover from disasters. These strengths include a wide array of cognition faculties, knowledge, skills and resources as well as the ability to claim, access and use them. In other words, capacities refers to the set of diverse knowledge, skills and resources people can claim, access and resort to in dealing with hazards and disasters.

Capacities as a 'plural' set of knowledge, skills and resources should be distinguished from the 'singular' capacity or ability to do something. Capacities as a standalone concept therefore differs from the concepts of (i) 'adaptive capacity'; i.e., the ability to adapt to environmental change, especially climate change and (ii) 'coping capacity'; i.e., the ability to survive amidst an adverse environment. In fact, capacities cover both the prospective dimension of 'adaptive capacity', or the ability to anticipate future changes in the environment, and the retrospective nature of 'coping capacity', based on the experience of past events.

Capacities apply to people, in all their diversity, who face hazards and disasters as part of their everyday life. They are both an individual and collective attribute. Everyone possesses a unique set of knowledge, skills and resources that are often shared and combined with relatives, kin, neighbours and others around them. The diversity of people's unique capacities taken all together form a collective pool of capacities tapped by both individuals and the broader 'community' that share these capacities. The use of capacities as a plural noun is important for emphasizing this diversity. Furthermore, capacities are not necessarily place-based and can involve transnational links. However, they are endogenous to the community of people who share and combine them in dealing with the same hazards and disasters.

As such, people's capacities differ from their vulnerability that is largely driven by exogenous and structural factors beyond their reach (i.e., how power and resources are shared within society). This explains why, unlike common assumptions, vulnerability and capacities are not the two ends of the same spectrum. Some highly vulnerable people may display a whole range of diverse knowledge, skills and resources while similarly vulnerable individuals may have minimal capacities. Conversely, less vulnerable individuals may have more or fewer capacities. This distinction is essential in defining priorities to be given to different initiatives on DRR including CCA at various scales.

Expanding the school and hospital examples at the end of the previous section, disaster risk reduction is not the ultimate endeavour. A school that withstands multiple hazards might not achieve development and sustainability goals if only boys are permitted to attend. A hospital built with all disaster risk reduction considerations, including with climate change adaptation factored in, but serving only the most affluent people, sets back development by expanding the rich–poor gap.

Consequently, although the role of climate change is to be positioned within disaster risk reduction, disaster risk reduction's future is to be a subset of wider development and sustainability processes. Having three separate streams for international negotiations duplicates efforts and disperses energy. But given this situation, bringing them all together would be challenging; for instance, the climate change negotiations seek a legally binding accord ratified by world parliaments while the disaster risk reduction process and the Sustainable Development Goals aim for voluntary agreements. None of the three has yet articulated a verifiable monitoring and enforcement mechanism, although that could potentially develop. With effort and will, these practical difficulties could be overcome, although territorialism and vested interests are likely to preclude such action.

The theoretical strength of climate change sitting within disaster risk reduction, which in turn sits within development and sustainability, can lead to positive policy and practice outcomes. This approach would represent a vision for disaster risk reduction's future, ending tribalism and separation in order to work together to achieve common goals. Although the prospect of this integrated approach occurring seems unlikely, not despite but because of the three 2015 processes and their long histories, the momentum of three independent but overlapping institutional paths should not stop us from doing our best to bring all areas together in order to save humanity from itself.

References

Alexander, D.E. (2013) 'Resilience and disaster risk reduction: An etymological journey', *Natural Hazards and Earth System Sciences* 13, 11: 2707–2716.

Anderson, J.W. (1968) 'Cultural adaptation to threatened disaster', *Human Organization* 27, 4: 298–307.

Anderson, M.B. and Woodrow, P. (1989) *Rising from the Ashes: Development Strategies in Times of Disasters*, Boulder: Westview Press.

Auerbach, L.W., Goodbred Jr., S.L., Mondal, D.R., Wilson, C.A., Ahmed, K.R., Roy, K., Steckler, M.S., Small, C., Gilligan, J.M. and Ackerly, B.A. (2015) 'Flood risk of natural and embanked landscapes on the Ganges–Brahmaputra tidal delta plain', *Nature Climate Change* 5, 2: 153–157.

Aysan, Y. and Davis, I. (eds.) (1992) *Disasters and the small dwelling: Perspectives for the UN IDNDR*. Oxford: James and James.

Baldacchino, G. (2004) *Moving Away from the Terms Vulnerability and Resilience in Small Islands. Wise Coastal Practices for Sustainable Human Development Forum, 11 March 2004*. Online http://www.csiwisepractices. org/?read=490 (accessed 17 January 2012).

Ballu, V., Bouin, M.N., Siméoni, P., Crawford, W.C., Calmant, S., Boré, J.M., Kanas, T. and Pelletier, B. (2011) 'Comparing the role of absolute sea-level rise and vertical tectonic motions in coastal flooding, Torres Islands (Vanuatu)', *Proceedings of the National Academy of Sciences* 108, 32: 13019–13022.

Bankoff, G. (2001) 'Rendering the world unsafe: "Vulnerability" as western discourse', *Disasters* 25, 1: 19–35.

Bankoff, G. (2003) *Cultures of Disaster; Society and Natural Hazard in the Philippines*, London: Routledge.

Blaikie, P., Cannon, T., Davis, I. and Wisner, B. (1994) *At risk: Natural hazards, people's vulnerability and disasters*, London: Routledge.

Crush, J. (ed.) (1995) *Power of Development*, London: Routledge.

Cuny, F. (1983) *Disasters and Development*, Oxford: Oxford University Press.

Etkin, D. (1999) 'Risk transference and related trends: Driving forces towards more mega-disasters', *Environmental Hazards* 1, 2: 69–75.

Folke, C. (2006) 'Resilience: The emergence of a perspective for social–ecological systems analyses', *Global Environmental Change* 16, 3: 253–267.

Fordham, M. (1998) 'Making women visible in disasters: Problematising the private domain', *Disasters* 22, 2: 126–143.

Fordham, M. (1999) 'Participatory planning for flood mitigation: Models and approaches', *The Australian Journal of Emergency Management* 13, 4: 27–34.

Gaillard, JC (2010) 'Vulnerability, capacity, and resilience: Perspectives for climate and disaster risk reduction', *Journal of International Development* 22, 2: 218–232.

Garcia-Acosta, V. (2004) 'La perspectiva histórica en la antropología del riesgo y del desastre: acercamientos metodológicos', *Relaciones* 97, 25: 125–142.

Glantz, M.H. (1977) 'Nine fallacies of natural disaster: The case of the Sahel', *Climatic Change* 1, 1: 69–84.

Glantz, M.H. (1994a) 'Creeping environmental problems', *The World & I*, June: 218–225.

Glantz, M.H. (1994b) 'Creeping environmental phenomena: Are societies equipped to deal with them?' in *Creeping Environmental Phenomena and Societal Responses to Them*, Proceedings of workshop held 7–10 February 1994 in Boulder, Colorado, ed. M.H. Glantz, 1–10. NCAR/ESIG, Boulder, Colorado.

Glantz, M.H. (ed.) (1999) *Creeping Environmental Problems and Sustainable Development in the Aral Sea Basin*, Cambridge: Cambridge University Press.

Glantz, M.H. and Jamieson, D. (2000) 'Societal response to Hurricane Mitch and intra-versus intergenerational equity issues: Whose norms should apply?', *Risk Analysis* 20, 6: 869–882.

Global Network of Civil Society Organisations for Disaster Reduction. (2009) *'Clouds but Little Rain . . .': Views from the Frontline – A Local Perspective of Progress Towards Implementation of the Hyogo Framework for Action*, Teddington: Global Network of Civil Society Organisations for Disaster Reduction.

Global Network of Civil Society Organisations for Disaster Reduction. (2011) *If we do not join hands . . . Global Network of Civil Society Organisations for Disaster Reduction Views from the Frontline: Local reports of progress on implementing the Hyogo Framework for Action, with strategic recommendations for more effective implementation*, Teddington: Global Network of Civil Society Organisations for Disaster Reduction.

Graham, J.D. and Weiner, J.B. (eds.) (1995) *Risk vs. Risk: Trade-offs in Protecting Health and the Environment*, Cambridge, MA: Harvard University Press.

Heijmans, A. (2001) *Vulnerability: A Matter of Perception, Working Paper 4*, London: Benfield Hazard Research Centre, University College of London.

Heijmans, A. (2012) *Risky Encounters: Institutions and Interventions in Response to Recurrent Disasters and Conflict*, PhD Thesis, The Netherlands: Wageningen University.

Hewitt, K. (ed.) (1983) *Interpretations of Calamity from the Viewpoint of Human Ecology*, London: Allen & Unwin.

Hills, A. (1998) 'Seduced by recovery: The consequences of misunderstanding disaster', *Journal of Contingencies and Crisis Management* 6, 3: 162–170.

Holling, C.S. (1973) 'Resilience and stability of ecological systems', *Annual Reviews of Ecological Systems* 4: 1–23.

IPCC (Intergovernmental Panel on Climate Change). *IPCC 2013–2014*, Geneva: IPCC.

Kennedy, J., Ashmore, J., Babister, E. and Kelman, I. (2008) 'The meaning of "Build Back Better": Evidence from post-tsunami Aceh and Sri Lanka', *Journal of Contingencies and Crisis Management* 16, 1: 24–36.

Kerkvliet, B.J.T. (2009) 'Everyday politics in peasant societies (and ours)', *The Journal of Peasant Studies* 36, 1: 227–243.

Knutson, T.R., McBride, J.L., Chan, J., Emanuel, K., Holland, G., Landsea, C., Held, I. and Kossin, J.P. (2010) 'Tropical cyclones and climate change', *Nature Geoscience* 3 (February): 157–163.

Kothari, U. (2014) 'Political discourses of climate change and migration: Resettlement policies in the Maldives', *The Geographical Journal* 180, 2: 130–140.

Leichenko, R., and O'Brien, K. (2008) *Environmental Change and Globalization: Double exposures.* Oxford: Oxford University Press.

Lewis, J. (1979). 'The vulnerable state: An alternative view', in L. Stephens and S.J. Green (eds.), *Disaster Assistance: Appraisal, Reform and New Approaches*, New York: New York University Press, pp. 104–129.

Lewis, J. (1999) *Development in Disaster-prone Places: Studies of Vulnerability*, London: Intermediate Technology Publications.

Lewis, J. (2013) 'Some realities of resilience: a case-study of Wittenberge', *Disaster Prevention and Management* 22, 1: 48–62.

Lewis, J. (2015) 'Cultures and contra-cultures: Social divisions and behavioural origins of vulnerabilities to disaster risk', in F. Kruger, G. Bankoff, T. Cannon, B. Orlowski and L.F. Schipper (eds.), *Cultures and Disasters: Understanding Cultural Framings in Disaster Risk Reduction*, London and New York: Routledge, pp. 109–122.

Lewis, J. and Kelman, I. (2012) 'The good, the bad and the ugly: Disaster risk creation (DRC) versus disaster risk reduction (DRR)', *PLOS Currents: Disasters.* Online http://currents.plos.org/disasters/article/the-good-the-bad-and-the-ugly-disaster-risk-reduction-drr-versus-disaster-risk-creation-drc (accessed 2 September 2016).

Lewis, J. and Lewis, S. (2014) 'Processes of vulnerability in England? Place, poverty and susceptibility', *Disaster Prevention and Management* 23, 5: 586–609.

MacKinnon, D. and Derickson, K.D. (2013) 'From resilience to resourcefulness: a critique of resilience policy and activism', *Progress in Human Geography* 37, 2: 253–270.

Manyena, S.B., O'Brien, G., O'Keefe, P. and Rose, J. (2011) 'Disaster resilience: A bounce back or bounce forward ability?', *Local Environment* 16, 5: 417–424.

Mercer, J. (2010) 'Disaster risk reduction or climate change adaptation: Are we reinventing the wheel?', *Journal of International Development* 22, 2: 247–264.

Mudelsee, M., Börngen, M., Tetzlaff, G. and Grünewald, U. (2003) 'No upward trends in the occurrence of extreme floods in central Europe', *Nature* 425, 6954: 166–169.

Oliver-Smith, T. (1979) 'Post disaster consensus and conflict in a traditional society: The 1970 avalanche of Yungay, Peru', *Mass Emergencies* 4, 1: 39–52.

Parker, D., Tapsell, S. and McCarthy, S. (2007) 'Enhancing the human benefits of flood warnings', *Natural Hazards* 43, 3: 397–414.

Rankine, W.J.M. (1867) *A Manual of Applied Mechanics*, London: Charles Griffin & Co.

Shaw, R., Pulhin, J.M. and Pereira, J.J. (2010a) *Climate Change Adaptation and Disaster Risk Reduction: Issues and Challenges*, Bingley, UK: Emerald.

Shaw, R., Pulhin, J.M. and Pereira, J.J. (2010b) *Climate Change Adaptation and Disaster Risk Reduction: An Asian perspective*, Bingley, UK: Emerald.

Tapsell, S.M., Penning-Rowsell, E.C., Tunstall, S.M. and Wilson, T.L. (2002) 'Vulnerability to flooding: Health and social dimensions', *Philosophical Transactions of the Royal Society of London* 360, 1796: 1511–1525.

Timmerman, P. (1981) *Vulnerability, Resilience and the Collapse of Society.* Environmental Monograph no. 1, Toronto: Institute for Environmental Studies, University of Toronto.

Tobin, G.A. (1999) 'Sustainability and community resilience: The holy grail of hazards planning?', *Environmental Hazards* 1, 1: 13–25.

UNFCCC (United Nations Framework Convention on Climate Change). (2015) *Paris Agreement*, New York: United Nations.

UNGA (United Nations General Assembly). (2015) *Resolution Adopted by the General Assembly on 25 September 2015, A/RES/70/1,* New York: United Nations.

UNISDR (United Nations International Strategy for Disaster Reduction). (2015) *Sendai Framework for Disaster Risk Reduction 2015–2030,* Geneva: UNISDR.

Wheeler, D. (1991) 'The influence of the weather during the Camperdown campaign of 1797', *The Mariner's Mirrors* 77, 1: 47–54.

Wisner, B., Blaikie, P., Cannon, T. and Davis, I. (2004) *At Risk: Natural Hazards, People's Vulnerability and Disasters, 2nd edn,* London: Routledge.

Wisner, B., Gaillard, JC, and Kelman, I. (eds.) (2012) *Handbook of Hazards and Disaster Risk Reduction,* Abingdon, UK: Routledge.

Wolf, C. (1988) *The Fourth Dimension: Interviews with Christa Wolf (Translation: H Pilkington),* London: Verso.

World Commission on Dams. (2000) *Dams and Development: A New Framework for Decision-making,* London: Earthscan.

Zahn, M. and von Storch, H. (2010) 'Decreased frequency of North Atlantic polar lows associated with future climate warming', *Nature* 467, 7313: 309–312.

Zebrowski, C. (2009) 'Governing the network society: A biopolitical critique of resilience', *Political Perspectives* 3, 1: 1–38.

7

A DUE DILIGENCE APPROACH TO BUZZWORDS FOR DISASTER RISK REDUCTION INCLUDING CLIMATE CHANGE ADAPTATION

Calum T.M. Nicholson

Introduction

Categories and labels are widely held to be a prerequisite for both understanding and changing the world we live in. Not surprisingly, a concern for categorisation is so deeply ingrained in this cultural and historical milieu, that it could be said to be not only its most notable characteristic, but one of its most integral values. A category that has come into particular vogue is what we term a 'buzzword'.

Many technocratic labels and 'buzzwords' presume to be apolitical: concerns with climate change, and environment; migration, and integration; social vulnerability, resilience and transformation, among many others, not to mention combinations and recombinations of such themes. Each of these umbrella themes continually sub-divides, with narrower and narrower areas of specialisation, often marked by distinct terminologies. Such buzzwords are particularly prevalent in the literature for both climate change adaptation (CCA) and disaster risk reduction (DRR), and underpin their artificial separation.

Cynically, some might conclude that the proliferation of overlapping buzzwords is, at root, an exercise in disciplinary politics and the marking of thematic territory. Alternatively, and more generously, we might presume that such buzzwords are generated in politically neutral attempts to produce tractable, certain knowledge of societal problems. However, regardless of intention, one thing is certain: the proliferation of new labels and 'buzzwords' has resulted in clutter. Everyone is an 'expert', but in the sense of knowing, to quote Butler, 'more and more about less and less' (Jeffares and Gray 1995: 121). While the explanation of in what, precisely, they are expert, is often opaque, what is certainly being lost is a clear view of the whole, into which each of these specialisations purportedly fits, and in relation to which they presumably have their relevance. As Kelman *et al.* have noted, thematic 'tribalism and separation', with regard to discussion of climate change, DRR and sustainability, is not conducive to 'work[ing] together to achieve common goals for humanity' (Kelman *et al.* 2015: 21).

With these problems in mind, this chapter provides a basic framework through which we can independently interrogate the welter of buzzwords and labels marking our present era. The provision of such a framework is important, as the alternative is to take categories at face value or on sentiment, which of course would be to uncritically defer to some norm, which in turn might serve a political agenda. As noted, categories are often held, normatively, to be prerequisites to understanding and changing the world we live in. But we need to be aware of how they can be used to obfuscate, and thus maintain a potentially pernicious status quo.

This chapter will approach 'buzzword' categories and labels from a 'bottom up' perspective of citizens confronted by them, rather than from the 'top down' perspective of political and academic actors (or those who aspire to be such, keeping in mind that the two categories are not necessarily distinct) who invest in their utility. The chapter endeavours to do our due diligence, to give us tools to sift through the language, sentiment, and rhetoric, inevitably generated as part of anyone's attempt to build a successful societal intervention. As such, the chapter is an attempt to meet a challenge set by Isabelle Stengers: 'how can we present a proposal intended not to say what is, or what ought to be, but . . . to "slow down" reasoning and create an opportunity to arouse a slightly different awareness of the problems and situations mobilizing us?' (Stengers 2005: 994).

It is surely as important to have the tools – the framework – to discern a nebulous, ill-conceived concept, as it is to know the grounds on which a strong implementable concept may be conceived. As Stewart (2011: xiii) wrote:

> the question of whether and how to intervene . . . is not a question of what we ought to do but what we can: of understanding the limits of Western institutions in the twenty-first century and of giving a credible account of the specific context of a particular intervention.

Stewart was here referring to military interventions and state-building. Yet his admonition is as applicable to interests in DRR including CCA.

In order to provide a framework for conducting due diligence, and thereby provide 'a slightly different awareness of the problems and situations mobilising us' (Stengers 2005: 994) when confronted with a buzzword, this chapter has a three-section structure. In Section I, a three-step framework is presented for how we should approach the interrogation of a buzzword. Section II then illustrates this framework by applying it to a case study of the Nansen Initiative on 'Disaster Displacement'. Section III will then present the implications of this framework for how we conduct our work, in both research and policy contexts.

Section I: Framework

When faced with a new category – or 'buzzword' – holding implicit a concern for what, societally speaking, is, or ought to be, it is easy for our critical faculties to be overwhelmed. This is particularly the case when the buzzword in question is attended by a nimbus of well-meaning sentiment, clouding our capacity for healthy dispassion. How, in such cases, are we able to go about doing due diligence, interrogating the buzzwords presented as nostrums for our impoverished understanding of the world around us?

This section provides a framework allowing us to conduct such due diligence. It is in three steps. Step A outlines how we should orient ourselves when faced with a discussion of a new buzzword; Step B reminds us of what we should prioritise when examining a new buzzword; Step C details how we should examine a new buzzword. The framework here is presented in an abstract manner (Section II will then illustrate its utility through three case studies).

Step A – Orientation

When exposed to a new buzzword, the first step is to bring order to the chaos, allowing us to orient ourselves. It is first important to realise that the significance of a buzzword is in its utility, regardless of whether that utility is explicit or implied. Language does many things, but what we are concerned with here are assertoric terms to what is, and what is said ought to be (terms such as 'environmental migrant', 'climate change' or 'social resilience'), where the intention is to inform some sort of societal intervention. In short, those terms that are used against a particular political backdrop, with all its attendant agendas.

In light of this, there are three types of concerns that might be raised in the context of a buzzword. These are to do with its subject (to what the buzzword refers), inclusive of concepts and data; purpose (why it is being used), inclusive of norms, laws and institutions; and practice (how it would be implemented), inclusive of financial and political considerations.

While any discussion of a buzzword will undoubtedly involve reference to all three concerns, the distinction between the three may not be self-evident. The first recommendation of this chapter is to be cognisant of these different concerns, and to make sure to first work out that there are different sorts of claims at play.

Step B – Prioritisation

The second step is to sort the three concerns into an order of priority, thereby allowing us to drill down to the crux of the issue. That is, to achieve the end goal of implementing something in practice (requiring the addressing of financial and other political factors), it is a prerequisite for a clear sense of the purpose served; for instance, having a blueprint for an institution taking such action, and having determined what laws or norms that institution would uphold. However, one must first have a sense of the nature and demarcation of the subject itself, involving an empirical knowledge base determining its size and scope, and also a sense of what, conceptually, the subject in fact is.

This process, therefore, is to 'drill down' from a concern for implementation (practice), through normative concerns (purpose), to the ontological basis (subject) informing both purpose and practice. The second recommendation of this chapter is the need to proceed through the structure in the right order.

Step C – Examination

Having oriented ourselves vis-à-vis a buzzword, and having established the order or prioritisation, it is then necessary to examine its subject for signs of intrinsic incoherence. That is, not evidence of falsehood, but indications of equivocation in meaning and real-world referent that may indicate that there is no certainty on which to speak meaningfully of purpose or implementation.

Here, six signs are proposed, indicative of something perhaps being rotten in the state of the art, that we should look for in any claim about what is, or what ought to be. These are:

(i) Universalist claims (U-claims): i.e., the use of categories that, while being collectively exhaustive and mutually exclusive – given that they cover all possible axes of analysis and all poles of those axes – and as such constituting a self-referential conceptual framework, are nevertheless of little utility, given that it remains wholly abstract. While superficially appealing, sketching as they do 'indistinct utopias' in 'language . . . so unexceptional and morally appealing that it [proves] . . . captivating', due to the fact that the definitions of the component categories are relational to each other, 'these phrases and categories [can] be arranged in every conceivable sequence and relationship to justify whatever [the actor was] claiming to do at the time' (Stewart 2011: 38). Universalist claims are unclear, in short, as to what they refer to in the real world, as distinct from what they mean in relation (and negation) of each other. They are truisms of language, rather than representations of real-world truths.

(ii) Particular claims (P-claims): i.e., statements in which no clear justification is given as to why a particular variable or theme has been foregrounded for analysis, ahead of other variables. While they may be defined in relation to a particular case, they do not indicate how one might generalise from that case to others more generally. As such, these cases are, at best, true in only an arbitrary sense.

(iii) Contradictory claims (C-claims): i.e., the existence of statements that contain opposing or conflicting claims.

(iv) Tautological claims (T-claims): i.e., statements in which the meaning of a claim is dependent upon a contrast with its binary negation, regardless of whether that binary is present or not.

(v) Laundering categories (L-claims): i.e., the tendency either to move on from the failure of, or to circumvent a critique of, a previous category or buzzword. Notably, such new categories maintain the intrinsic structural problems of their precursors, and have the effect of simply returning a critic to square one.

(vi) Equivocation (E-claims): i.e., statements to the effect that 'more research is needed', without identifying what precisely was wrong with existing research that requires more research, nor how future research could avoid the problems intrinsic to that existing research.

The third recommendation of this chapter is to be vigilant for the presence of any or all of these six tendencies. Being so is crucial if we are to see past well-meaning rhetoric, and have the capacity to interrogate the integrity of the buzzwords underpinning, informing and giving impetus to particular attempts to conduct societal interventions of one form or another.

Section II: Case Study

These three steps – orientation; prioritisation; examination – constitute a framework with which one may ensure we do our due diligence when we encounter a new buzzword or category that is held to have the capacity to render the world we live in more intelligible, and thereby the problems of the world more tractable. However, the presentation of these steps in Section I was highly abstract. It will now be applied to a case study of the Nansen Initiative's work on 'Disaster Displacement', a case study that has direct relevance to DRR including CCA.

The Nansen Initiative was launched in October 2012 by a consortium of governments, principally those of Norway and Switzerland. It claims to be a

> state-led, bottom-up consultative process intended to identify effective practices and build consensus on key principles and elements to address the protection and assistance needs of persons displaced across borders in the context of disasters, including the adverse effects of climate change.
>
> *(Nansen Initiative 2015: 8)*

In October 2015, the Initiative achieved the endorsement, by 109 governmental delegations, of 'The Agenda for the Protection of Cross-Border Displaced Persons in the Context of Disasters and Climate Change', abbreviated to the 'Protection Agenda'. Note how 'state-led' might contradict 'bottom-up', already providing a sign of the need to explore further.

This agenda identified 'three priority areas for action': 'collecting data and enhancing knowledge; enhancing the use of humanitarian protection measures for cross-border disaster-displaced persons; and strengthening the management of disaster displacement risk in the country of origin' (Nansen Initiative 2015: 60). Working from the three steps outlined in Section I, we can see that in order to implement anything (i.e., strengthen management of disaster displacement), one would have to have a clear sense of what purpose is being pursued (providing humanitarian protection for 'cross-border disaster-displaced persons'). In order to do that, one must first have a clear sense of who, precisely, are those in need of protection. In examining the work of the Nansen Initiative, the priority question must surely be what, exactly, is the state of knowledge with regard to 'cross-border disaster-displaced persons'?

A close reading of Nansen Initiative documents reveals evidence that implies a knowledge-base that is, at the very least, ill at ease with itself. This can be discovered if one reads the documents with the six tendencies, outlined in Section I – so-called U, P, C, T, L and E-claims – in mind.

First of all, the Nansen Initiative formulates a discussion of climate change, disasters, and migration by hedging its scope with ambiguous language, before acknowledging all possible axes of discussion. Although 'forced displacement' is framed as 'related to disasters', the 'effects of climate change' are held to be 'adverse' (Nansen Initiative 2015: 24), and in discussion of migration/displacement/mobility in the 'context of disasters' (Nansen Initiative 2015: 9), and displacement due to 'the effects of climate change' (Nansen Initiative 2015: 12), no specificity is given with regard to what 'related', 'adverse' or 'in the context of' or 'effect' do or do not refer to. Despite this, it is confidently stated that such forced displacement is 'a reality and among the biggest humanitarian challenges . . . in the 21st century' (Nansen Initiative 2015: 24).

Perhaps due to the discomfort of sitting with such ambiguity, the discussion then goes on to tack away from the Initiative's implied emphasis on climate change and environmental causes, noting that 'population growth, underdevelopment, weak governance, armed conflict and violence, as well as poor urban planning in rapidly expanding cities, are important drivers of displacement and migration as they further weaken resilience and increase vulnerability' (Nansen Initiative 2015: 25). The result is that it becomes unclear as to why climate change and environmental factors were privileged by the Initiative in the first place. In short, is its concern specifically migration/displacement from climate change and environmental degradation, or simply forced migration in general?

This tendency to make a statement privileging one thematic concern, and then to quickly caveat and contextualise to the point where the original statement becomes an arbitrary part of a largely all-inclusive whole, occurs across several other axes. For instance, the Initiative observes that 'cross-border disaster displacement could potentially be avoided or reduced if IDPs received adequate protection and assistance following disasters' (Nansen Initiative 2015: 19), a statement implying that cross-border displacement is a problem. Yet, on the previous page, the same document argues that 'managed properly, migration has the potential to be an adequate measure to cope with the effects of climate change, other environmental degradation and natural hazards' (Nansen Initiative 2015: 18), a statement implying that migration is something that can be a solution. Similarly, although the Initiative's explicit concern is 'cross border disaster displacement', it nevertheless goes on to concern itself with assisting internally displaced persons, too (Nansen Initiative 2015: 18–19). While this is a laudable sentiment, it raises issues of demarcation. Is the Initiative's focus on international or internal migration? Or is it simply migration in general? Perhaps not even that, as the Nansen Initiative also concerns itself with 'planned relocation' for those who are unable to migrate away from 'risks and impacts of disasters, climate change, and environmental degradation' (Nansen Initiative 2015: 18).

The Initiative's documents are also marked by contradictions. It states that, looking to the future,

> migration increasingly [becomes] an important response to both extreme weather events and longer-term climate variability and change. Sea level rise, in particular, is expected to force tens or hundreds of millions of people to move away from low-lying coastal areas, deltas and islands that cannot be protected.
>
> *(Nansen Initiative 2015: 25)*

It then goes on to state that 'a multitude of demographic, political, social, economic and other developmental factors also determines to a large extent whether people can withstand the impacts of [natural] hazard[s] or will have to leave their homes' (Nansen Initiative 2015: 28). Indeed, it also states that 'due to . . . multicausality and uncertainty regarding the extent to which States will be successful in their attempts to mitigate and adapt to climate change, accurate global quantitative projections are difficult to make' (Nansen Initiative 2015: 25). Similarly, the document states that 'while comprehensive and systematic data collection and analysis on cross-border disaster-displacement is lacking, based on available data, Africa along with Central and South America, in particular have seen the largest number of cross-border disaster-displacement' (Nansen Initiative 2015: 15). The contradictions here are between implying projections of expectations and also acknowledging the absence of projective capacity. The Initiative is here having its cake, and eating it, too.

Further evidence of a discussion that is ill-at-ease with itself is the multiplicity of terms used to discuss the central theme of 'disaster displacement'. While 'forced displacement related to disasters, including the adverse effects of climate change (disaster displacement)' (Nansen Initiative 2015: 15) appears to be the predominant formulation, at other points, phrases and terms such as 'human mobility in the context of disasters and climate change' (Nansen Initiative 2015: 9); 'climate change-related displacement' (Nansen Initiative 2015: 9); 'human mobility aspects of climate change' (Nansen Initiative 2015: 12); 'climate and disaster induced migration' (Nansen Initiative 2015: 13); 'disaster-displaced persons' (Nansen Initiative 2015: 16) appear.

Finally, there are indications that, despite the volume of existing research and the admonition throughout Nansen's documents, it remains a priority to '[collect] data and [enhance] knowledge on cross-border disaster-displacement' (Nansen Initiative 2015: 19). As the document notes (Nansen Initiative 2015: 31):

> While understanding of the causes, dynamics and magnitude of disaster displacement has been growing in recent years, these phenomena are still not fully understood and conceptualized. Therefore better data, concepts and evidence are needed to develop adequate policies. The development of tools and systems that allow for the systematic gathering and analysis of reliable data on displacement, and human mobility more generally, in the context of disasters and the effects of climate change is particularly needed.

The value of stating that 'better data, concepts and evidence are needed' is, in essence, by negation: all it reveals is that existing data, concepts and evidence are inadequate, and not fit for informing any normative framework or institutional intervention. To state the opposite of what is wrong is perhaps necessary, but not sufficient, if that wrong is to be righted. What is needed is an understanding of what was wrong with the failed data, concepts and evidence. Why was it not enough? But to ask this would require us not to strive forward, pushing for something new or something more, but to retreat into more reflexive, epistemological questions about what we take to be our threshold for meaningful knowledge claims.

In summary, the effect of axes of analysis being acknowledged (causal/multicausal displacement; migration as a problem/solution; international/internal migration; migration/non-migration) is that one is left with a constellation of well-meaning ideas, concepts, and concerns that are mutually exclusive and collectively comprehensive, but which remain ultimately meaningless with regard to how they relate to reality. Stating one's agenda as inclusive of everything has more in common with having one inclusive of nothing than with one that is inclusive of some things, and exclusive of others; elaborating essentially contradictory positions presents more questions than answers. Laundering categories and engaging in equivocation by suggesting more research is merely to tacitly admit the failure of the exercise thus far. Once the dust has settled, one thing is clear: the subject (and thus the purpose) of the Nansen Initiative, remains a mystery. The road to nowhere is, however, paved with good intentions.

Section III: Implications for Academics and Policy-makers

Inevitably, the foregoing framework, outlined in Section I and illustrated in Section II, has implications for how academics engage with buzzwords that circulate in the context of discussions of DRR including CCA, and thus in the context of how governments and institutions formulate policy. How might either group proceed so that they themselves avoid lapsing into using buzzwords to engage with the world as it is, and as it ought to be?

For academics, it is perhaps important to begin from an awareness, long since established by post-structural scholars, that knowledge-claims inevitably perform politically. Foucault argued that 'each society has its regime of truth, its "general politics" of truth'. That is,

> the types of discourse which it accepts and makes function as true; the mechanisms and instances which enable one to distinguish true and false statements, the means by which each is sanctioned; the techniques and procedures accorded value in the acquisition of truth; the status of those who are charged with saying what counts as true.
>
> *(Rabinow 1984: 131)*

In short, Foucault held that power and knowledge were mutually constitutive. From a post-structural perspective, the truth content of the claim is therefore not the issue, but rather its use in whatever is its context. Thus, buzzwords about climate change, 'climate migrants' or CCA may be equivocal in terms of to what, exactly, they refer, but once such claims are in circulation, they are open to expedient use and abuse by actors with their own agendas. This is the quiet danger of allowing buzzwords to gain currency by not doing due diligence. Their use and abuse occurs not in spite of their equivocal nature, but precisely because of it. If a claim is equivocal, it becomes almost impossible to define either the obligations or responsibilities of those occupying privileged positions in the political structure. If there is no certainty as to what, precisely, the buzzword refers to, those who use it cannot be held to account. In this sense, knowledge is not power – obfuscation is power.

On the one hand, equivocal claims allow one to talk about what is, or what ought to be, without creating obligations for specific action, thereby allowing the speaker to perform 'engagement' without consequence. On the other hand, equivocal claims allow one to avoid responsibilities for the consequences of any action taken. This is perhaps what Sen (2009: 26) meant when he wrote that 'the theory of justice, as formulated under the currently dominant transcendental institution-alism, reduces many of the most relevant issues of justice into empty – even if acknowledged to be "well-meaning" – rhetoric', with 'empty rhetoric' referring to what is here termed 'equivocal claim'. In short, the equivocal nature of our categories and buzzwords not only forecloses on the success of any venture they inform, but simultaneously makes it impossible to identify their failure.

Given that the equivocal nature of buzzwords thus acts, in effect, to disaggregate power from accountability, it stands to reason that, in an open democratic society in which keeping power accountable is one central organising tenet, academic engagement should take a particular form. Instead of conducting research seeking to produce certainty, academics would serve an important function if they worked to process spurious claims to have attained certain knowledge that can be of use in fashioning policy. This is not simply to exercise 'critical thinking'. Rather, it is more akin to doing 'due diligence'. While this may take various forms, the framework outlined in Section I is an example of a systematic process affording us the capacity to do our due diligence, and thereby inoculate ourselves against language that, while well-meaning in its sentiment, is nevertheless meaningless with regard to how it can be put to use to remake the world around us in a manner where the outcomes are what we intend. We would do better to see constellations of buzzwords as more theological than scientific. Indeed, the questions implied in such buzzwords (who is a climate migrant? What ought one to do about 'them'? How might we take action to achieve our goal of protecting 'them'?) arguably have more in common with ritualised religious catechism than scientific questions that can be met with objective answers.

By contrast, when it comes to avoiding the use of buzzwords and the equivocal claims they involve, policy-makers inevitably face a limitation that academics do not. This limitation is, of course, that policy-makers' work cannot be extracted from a political context, with its constitu-tive agendas. For them, doing something is a raison d'être. This then raises the question of what, precisely, could be attempted, in a manner avoiding the use of buzzwords and neologism?

To answer this, it is important to first note that buzzwords imply a claim to a sort of objective universality. In the context of concerns with DRR and/or CCA, to talk of 'climate change', 'environment', 'resilience' or 'migration' is to imply that there is something common to all cases, and thus an applicability of the buzzword in question across those cases in a manner maintaining its utility. Buzzwords attempt to capture 'essential' characteristics about the world, transcending the specificity of a specific context. Implied in any buzzword is therefore a 'billiard ball' model of society: that it is not only made up of discrete, discernible elements, each neatly captured by a buzzword, but that all these elements interact in a way that is, in principle at least, causally deter-minable. This reflects the presumptions of what Stewart calls the 'planning school' of thought, in which there is held to be 'a universal formula for intervention, specifying the exact quantity of resources required for each hypothetical country' (Stewart 2011: 77). Where buzzwords are equivocal, to talk about their causal relationships – say, between climate change and migration, or environment and resilience – is pointless. How can one talk of causal relations when one has not determined what the buzzwords involved even refer to?

What could be the alternative to an approach preoccupied with buzzwords, a reductive billiard ball model, and assumed causality? One might follow two approaches, one direct and one indirect.

The direct approach is as follows: rather than concerning ourselves with broad, apolitical general themes, such as 'disasters', 'environment', 'climate change', 'development' and 'migration', we could study the context-specific political contingencies conditioning the degree to which problems manifest. In this, we would be wise to follow the admonition of Stewart (2011), when he wrote that in addressing societal problems, and in particular prosecuting interventions, we should draw on mountain rescue for inspiration.

To Stewart, solving societal problems doesn't require 'theoretical knowledge', but 'practical wisdom: an activity in which there is no substitute for experience' (Stewart 2011: 77). Achieving societal change does not depend on a perfect formulation of concepts and buzzwords. It requires experience of and familiarity with the context in question, with 'the ideal education' being 'through an ever more detailed study of the history, the geography, and the anthropology of a particular place, on the one hand, and of the limitations and manias of the West, on the other' (Stewart 2011: 77). Indeed, Stewart exhorts us, when thinking about formulating some sort of societal intervention, to change our questions from 'what ought you to do?' to 'where are you and who are you?' (Stewart 2011: 77), as 'ought' does not necessarily imply 'can'. The suitability of the mountain rescue analogy is the caution inherent in that practice – proceeding only where the terrain is known, and being ready to turn back if the mission itself puts the rescuers at risk.

The mountain rescue approach, which appreciates contingency, is particularly suited in light of re-emerging recognition of 'multiple exposure' (Bankoff 2003; Kelman *et al.* 2015; Shaw *et al.* 2010; Wisner *et al.* 2012). That is, where:

> Climate change, globalization, poverty, earthquakes, injustice, tropical cyclones, lack of livelihood opportunities, inequity, landslides, overexploitation of natural resources, epidemics, and lack of water supply – amongst many other ongoing challenges – often converge to most affect those who have the fewest options and resources for dealing with those challenges.
>
> *(Kelman* et al. *2015: 23)*

As Kelman *et al.* (2015: 23) note, in many contexts, 'the most prominent or fundamental . . . challenges are neither climate change nor globalisation', but something more specific to the region. It would therefore, perhaps, be most appropriate to subsume a thematic concern for climate change and migration, or other elements of CCA, under DRR (Kelman *et al.* 2015: 24). This is because this umbrella theme has a long history of 'dealing with climate-related changes at all time and space scales and from multiple causes' (Kelman *et al.* 2015: 24). Indeed, 'whether climate change is a more . . . or . . . less significant contributor than other factors – such as relying on structural approaches for floods or increasing the social oppression that creates and perpetuates food-related vulnerabilities – depends on specific context' (Kelman *et al.* 2015: 25).

While more indirect, the second approach is no less important, albeit arguably more ambitious. That is, it would require policy-makers in liberal democracies to concern themselves not with the causal relationships between different component parts of society and the world, but with the relationship between people – each of us – and the increasingly intimate and connected society and world in which we all find ourselves. Policy-makers would have to recognise the existence of our 'whatist' culture, and the possibility of changing that culture.

The 'whatist' culture captures the idea that there is a strong societal norm to categorise and label everything and everyone. We often assume that this tendency is natural and inevitable, and that our options are only with regard to what categories we use, rather than with regard to whether we use any at all. That we could – and arguably should – downplay the importance of categorisation in our culture can be illustrated. In light of the media coverage of increasing numbers of Middle Eastern refugees seeking safety in Europe in 2015–16, many people have tended to see them through the prism of categories, be it as foreigners, migrants, Muslims, refugees, or potential terrorists.

The alternative, however, would be to simply view them without any prism, and see them as people. In our 'whatist' culture, we privilege abstractions (what and how the prevailing fashion compels us to think) over basic empathy (our emotional response to how someone else feels).

While a somewhat vague goal, foregrounding empathy as a cultural value, in whatever way, in conjunction with encouraging scepticism of categories, and thereby the dissolution of the strong 'whatist' tendencies of our prevailing culture, would surely be a step towards a culture where understanding forecloses on judgement, and where one knows better than to presume to know better about how others could or should organise their lives and communities. In the words of Kelman and Gaillard (2010: 31), our society would benefit from:

> The honesty to admit the root cause as being values, the honesty to admit that superficial approaches not tackling root causes can cause more harm than good, and the honesty to accept that many disaster-related concerns exist aside from climate change that need to be tackled with as much vigor as climate change. Focusing on a single climate change challenge is dishonest in failing to acknowledge other equally important concerns. If dishonesty is accepted in order to convince people and governments regarding appropriate behaviour, where does that dishonesty stop?

Put another way, we should consider the possibility that we are too generous (or optimistic) about the capacity of our institutions, governments and policy-makers to solve what we perceive to be categorical problems. Perhaps, then, instead of having faith in our categories and buzzwords, and the political and bureaucratic systems inextricably intertwined with them, we could try having faith in something else, which is the only thing, in reality, we have: community; people; each other; and the bonds of empathy intrinsic to all three.

Conclusion

This chapter has presented a framework for 'processing' buzzwords in order to dissolve them, and thereby to inoculate ourselves against both their capacity to bewitch our imaginations, and distort our perspective and thinking when prosecuting societal interventions of one form or another. The impetus for this critique is not some specialist understanding of the themes in question such as climate change, migration, vulnerability, resilience, and DRR including CCA. Rather, the critique derives from years closely observing the people making such claims, the patterns of their claims and their responses to those who dare to challenge the claims. Thematic specialisation is not the means by which we will solve our societal problems, but is largely constitutive of the societal problem itself. (It is worth noting that one meaning of the word 'special' is 'limited as to function, operation, or purpose'.) Simply because an endeavour is well-meant – such as the Nansen Initiative on 'disaster displacement' – does not ensure that it is meaningful. And where the former exists without the latter, the resulting statements are ripe for expedient use and abuse in the service of political agendas.

Whether such statements can ever be meaningful is beyond the scope of this chapter, as it is a reflexive epistemological question, as well as an ethical one (see Nicholson 2014 for a discussion of epistemology and ethics in this context). But at the very least, when presented with the claims of such endeavours, and with their constitutive 'buzzwords', it is incumbent upon us to do our due diligence to identify equivocal statements and assertions for what they are.

First, we must properly orient ourselves, ensuring we understand the different sorts of claims at play. Second, we must understand which claims are to be prioritised, and address them in the right order. Third, we must examine the priority claims for evidence of incoherence. We should not be surprised when we uncover incoherence.

Constellations of abstractions are not explanatory truths. Rather, they are an exercise in an activity that is most characteristic of being human: the construction of narrative. All myth-making is interpretative, and the idea that we, in our culture, have some special insight on the objective explanatory 'truth' is itself simply an overarching narrative we tell ourselves, and from within which we view the world, and our place in it. This chapter has presented a narrative, in a three-section structure, that allows us to gain perspective on, and thereby puncture, the nimbus of the narratives we are told, about what the world is like, and what we think it ought to be like.

The reality is that societal problems have existed and always will exist. Human life is fragile, and fraught with risk. Uncertainty is ubiquitous and inevitable. Given that uncertainty, the one thing we can certainly do is take pains to avoid obfuscation, dissemblance, and incoherent argument, which do nothing but muddy the waters, particularly (rather than despite) when such arguments are couched in agreeable sentiment. This chapter is not claiming that climate change does not exist, nor that migration does not occur. Neither is it suggesting that populations are not subject to varying degrees of vulnerability and resilience. Rather, the chapter questions our capacity to capture these processes using a constellation of categories, labels or buzzwords. Do any of these provide us any real utility, free from politics, and myriad unintended consequences?

The chapter has suggested that well-meaning but meaningless buzzwords, and by extension any actions they inform, not only fail to help us solve problems, but are constitutive of the problems themselves. Furthermore, they give us false hope, and in the meantime they exhaust our finite collective resources, both of finance and sentiment. We must remember that, as Stewart notes, 'ought' does not imply 'can'. This chapter has provided a framework to inoculate against buzzwords implying such equivalences, and which, by doing so, are primed for expedient use and abuse. This attempt will only have been successful if it has answered Stengers' (2005: 994) challenge, quoted at the outset, and aroused in the reader 'a slightly different awareness of the problems and situations mobilizing us'.

Bibliography

Bankoff, G. (2003) *Cultures of Disaster: Society and Natural Hazard in the Philippines*, London: Routledge.

Jeffares, N.A. and Gray, M. (eds) (1995) *Dictionary of Quotations*, London, UK: Harper Collins Publishers.

Kelman, I. and Gaillard, JC (2010) 'Embedding Climate Change Adaptation Within Disaster Risk Reduction', in R. Shaw, J.M. Pulhin and J.J. Pereira (eds), *Climate Change Adaptation and Disaster Risk Reduction: Issues and Challenges (Community, Environment and Disaster Risk Management Volume 4)*, Bingley, UK: Emerald Group Publishing, pp. 23–46.

Kelman, I., Gaillard, JC and Mercer, J. (2015) 'Climate Change's Role in Disaster Risk Reduction's Future: Beyond Vulnerability and Resilience', *International Journal of Disaster Risk Science* 6: 21–27.

Nansen Initiative (2015) 'The Nansen Initiative Global Consultation Conference Report, Geneva, 12–13 October 2015'. Online https://www.nanseninitiative.org/wp-content/uploads/2015/02/GLOBAL-CONSULTATION-REPORT.pdf (accessed 4 February 2016).

Nicholson, C.T.M. (2014) 'The Politics of Causal Reasoning: The Case of Climate Change and Migration', *The Geographical Journal* 180, 2: 151–160.

Rabinow, P. (ed.) (1984) *The Foucault Reader: An Introduction to Foucault's Thought*, New York: Random House.

Sen, A. (2009) *The Idea of Justice*, London: Allen Lane.

Shaw, R., Pulhin, J.M. and Pereira, J.J. (2010) *Climate Change Adaptation and Disaster Risk Reduction: An Asian Perspective*, Bingley, UK: Emerald.

Stengers, I. (2005) 'The Cosmopolitan Proposal', in B. Latour and P. Weibel (eds), *Making Things Public: Atmospheres of Democracy*, Cambridge, MA: MIT Press, pp. 994–1003.

Stewart, R. (2011) 'Can Intervention Work?', in R. Stewart and G. Knaus (2011), *Can Intervention Work?*, New York: W.W. Norton and Company.

Wisner, B., Gaillard, JC and Kelman, I. (eds) (2012) *Handbook of Hazards and Disaster Risk Reduction*, Abingdon, UK: Routledge.

8

CONCEPTS, CONNECTIONS, AND DISRUPTIONS

Disaster Risk Reduction and Climate Change Adaptation

Andrea Lampis

Introduction

Not only are concepts the building blocks for understanding, relating to and guiding practical actions on the ground, they are also the product and representations of contended domains of power, political interests, and epistemic communities who are often clustered around specific policy-driven beliefs. Concepts are institutionalised and become embedded within cultural practices of individuals and organisations. They are adopted in order to illustrate and clarify scientific debates, improve policies, strengthen networks and challenge related knowledge produced by others.

To deny wider power and interest-driven changes in the way knowledge is produced and uncritically adopt concepts that originated in the natural sciences, such as adaptation and resilience for instance, produces short-sighted and most often tautological explanations. However, the fact that the concepts of disaster risk reduction (DRR) and climate change adaptation (CCA) might have been perfectly and consistently elaborated within a number of original scientific disciplines that first proposed them (e.g., geography, engineering, sociology, ecology, and anthropology) or by a number of more interdisciplinary area studies (i.e., disaster or development studies) is not the issue at stake. One of the key issues discussed by this chapter is rooted in concerns regarding the shortfalls related to and caused by the translation of either specific or broad concepts to other scientific fields or areas. With regard to DRR and CCA, translation has only been partially successful; however, it has had a tremendous institutional and political impact, despite insufficient scientific evidence or agreement (Cannon and Müller-Mahn 2010). This becomes clearer when analysing the concepts of DRR and CCA inclusive of all their sub-sets of more specific concepts such as exposure, sensitivity, vulnerability, and resilience, amongst others. This chapter takes on board the challenge of connecting the concepts of DRR and CCA, analysing these two broad areas of scientific, political and cultural tension using a critical lens capable of opening up a dialogue both with a more traditional and positivist, and a more alternative and constructivist perspective.

Comparing DRR and CCA

DRR and CCA are not only concepts but also highly institutionalised areas of applied policy, relying in turn on complex apparatuses of production of meaning and its legitimisation throughout society. Nonetheless, within the determinate limits of human perception marked by time, space and culture, DRR and CCA have a material dimension that has enormous relevance for our individual and societal lives in terms of impact; i.e., the potential for determining our future risk or put simply, human well-being. Put another way, although we may know that a table is made up of molecules that are kept together only under certain conditions, for the material dimension of our lives and most often for our perception, what counts is that the table is solid and serves certain practical uses. As such, the objectivisation of risk plays a relevant role in the conceptualisation of DRR and CCA, reflecting what is possibly one of the most visible, valued and culturally relevant dimensions in our lives, the material one.

Hence, from a perspective that privileges the material dimension of risk, disaster risk and the social practices and policies geared to its reduction, as well as climate change with its respective social practices and policies aimed to improve adaptation, are also objectivated. Within the specialised literature on DRR and CCA, the objectivisation of disaster risk and climate change has produced a central debate regarding the reducibility of one area of concern to the other around the articulation of two objective categories, 'reduction' and 'adaptation'.

Objectivating Risk: Different Concepts in DRR and CCA

Almost the totality if not all institutions working on DRR and CCA in the international and national arenas tend to objectivate risk. That is, to tailor their definitions of these concepts, as well as all related policies and recommendations to the biophysical dimension attached to the risks that can be scientifically appraised regardless of differences and overlap between DRR and CCA.

The concept of DRR has evolved over the last six decades. This transformation is intimately bound to the evolution of both the prevailing policy frameworks and the changing conceptualisations of disasters brought about by epistemological debates.

The first social elaborations on disasters date back to at least the 1920s (Oliver-Smith and Hoffman 1999). However, in many national and local political environments the social dimension of disasters and its incorporation into DRR policies and intervention often remained limited or non-existent. In the 1940s, disasters were still firmly enclosed in a narrow conceptualisation linked to the implementation of emergency assistance in the face of critical events. This closely followed military approaches to crises and emergencies. Hence, this was very much a reactive approach based on an unquestioned 'naturalness' of the disaster events. That is, disasters were seen as bio-physical processes caused by 'normal' Earth dynamics. Disasters were also conceived as 'abnormal events' with an obvious need to mobilise further resources for material reconstruction as and when necessary.

From the 1950s until the 1970s, the 'disaster cycle' approach developed, which conceptualised disaster on a lineal sequence of phases, from the event causing the emergency to stages of humanitarian assistance and reconstruction. That sequence was conceptualised with little historical perspective and a scant interest for sociocultural patterns (Oliver-Smith and Hoffman 1999). At the beginning of the 1980s, this already more articulated albeit still essentially technical approach to disasters started to feel the influence of contributions coming from the social sciences. It was then that the concepts of 'preparation' and 'prevention' started to appear in the specialised literature, generally elaborated by and framed within ad hoc systems of disaster risk response and prevention.

Throughout the 1980s and the 1990s, the concepts of 'hazard' and 'vulnerability' found their way within a milieu still dominated by engineering and natural sciences. This placed on the table a number of critical issues, such as the role played by unequal access to assets, resources and power in the distribution of hazards, therefore moving the debate from a focus on physical geography to being inclusive of human geographies. The central role played by access to and control over assets and resources is thus conceptualised as a key factor in the creation of a stronger capacity to prevent, foresee, cope with and recuperate from damage and harm. Capacity (Blaikie *et al.* 1994; Chambers 1989) is ultimately dependent on socio-economic, political and cultural aspects and as such, is a concept re-taken by those scholars and practitioners working in DRR at least a decade and a half before the CCA community. International non-governmental organisations (INGOs) working on CCA started to use those same fundamental conceptual elements to build an apparently new approach, the Community-Based Adaptation approach (Lampis and Méndez 2013).

Over the last decade and a half, the concept of DRR has developed further to describe relationships between disasters and development. Knowledge of this relationship is a key element for understanding not only disasters per se but also their relationship with other phenomena such as climate change.

Henceforth, disaster risk evolved into an increasingly complex research field including at least five established communities. Representing the first two, scholars from the technical and natural sciences work to consolidate a scientific body of knowledge regarding hazards, whilst social scientists link this with the idea of disasters as socially constructed processes. On the operative side, governmental agencies act as a third community, elaborating policy frameworks aimed at institutionalising DRR, while the multi-layered community-based organisation constellation, the fourth community, has played a key operational role, especially in the areas of humanitarian aid, reconstruction and psycho-social support. The growing complexity of contemporary risks has strengthened even more the role of the fifth community, which is represented by international development actors including multilateral agencies like the World Bank, INGOs and the different agencies of the United Nations system.

The concept of 'adaptation' has its origins in the theory of evolution, specifically indicating biological transformations or acquired behaviours in the face of changing conditions, resource availability and environments (Simonet 2010). In the second half of the twentieth century, psychology and cybernetics take up the concept to signify processes by which human psyche or complex systems (among which are living beings) elaborate, come to terms with or master change. The concept also has important influences from within anthropology and sociology; e.g., in the processes throughout which culture is modelled by the environment and vice versa or by which individuals socialise or interiorise norms and values.

Within climate change science, the concept of adaptation has two fundamental roots. On the one hand the fact that the climate has changed in the past and that human societies adapted to or perished by those transformations has meant there is a strong chance they would potentially have options to adapt. On the other hand, the failures of international climate diplomacy, since the missing signature of the Kyoto Protocol by the US, up to the catastrophic Conference of Parties in Copenhagen in 2009 and Canada's withdrawal from the Kyoto Protocol, have given legitimacy to the already strong claim of those who advocate for a re-elaboration of the concept of adaptation on the basis of lessons from development. This last strand of the literature produces within the climate change community a similar effect to that of anthropologists in the domain of DRR in the 1970s–1980s (Oliver-Smith and Hoffman 1999); i.e., it brings the social into the broader frame and with it all the political, economic, environmental and cultural debates highlighting how adaptation within development is a process of power bargaining aimed at the reduction of inequalities and guarantee of human and social rights (Burton 1996).

In one of the earliest and still most complete analyses of the concept and policies related to adaptation, McGray *et al.* (2007) convincingly stressed the relationship between the way the concept of adaptation applies to climate change and the way it applies to the development agenda, pointing out three main categorisations based on a vast review of concrete experiences:

1   'Serendipitous' Adaptation: activities undertaken to achieve development objectives incidentally achieve adaptation objectives. The adaptation components of a given activity may even be noticed or emphasised only after the fact.
2   Climate-Proofing of Ongoing Development Efforts: activities added to an ongoing development initiative to ensure its success under a changing climate. Adaptation thus serves as means to achieve development ends – although the term 'climate-proofing' is dangerous as no element or system could ever be truly climate-proofed.
3   Discrete Adaptation: activities undertaken specifically to achieve CCA objectives. Development activities may be used as means to achieve adaptation ends.

Both definitions of DRR and CCA reflect the objectification of risk. According to UNISDR (2009, online):

> Disaster risk reduction is the concept and practice of reducing disaster risks through systematic efforts to analyse and reduce the causal factors of disasters. Reducing exposure to hazards, lessening vulnerability of people and property, wise management of land and the environment, and improving preparedness and early warning for adverse events are all examples of disaster risk reduction.

According to IPCC (2014, p. 116), CCA is:

> The process of adjustment to actual or expected climate and its effects. In human systems, adaptation seeks to moderate or avoid harm or exploit beneficial opportunities. In some natural systems, human intervention may facilitate adjustment to expected climate and its effects.

The institutional domains of DRR and CCA provide just one among many perspectives on disaster risk and CCA, albeit very influential and powerful ones.

Convergences and Divergences between Policy Communities

Following the publication of the highly criticised IPCC (2012) report, recent debates seem to have generated a number of interesting connections and convergences between the DRR and CCA policy communities. For instance, IPCC (2014) highlights how both climate change mitigation and CCA raise issues of equity, justice, and fairness – concepts that were long embedded in DRR, pollution prevention, and other sustainable development activities. Also remembering that many of those most vulnerable to climate change have contributed and contribute little to greenhouse gas emissions.

IPCC (2012) also touches upon a number of issues that clearly put into a close relation CCA, disaster risk and development, such as when it argues that CCA can reduce the risks of climate change impacts – one long-standing fundamental point in DRR. Nonetheless, IPCC (2012) suggests limits to the effectiveness of CCA, especially with greater magnitudes and rates of climate change. But, adds the report, once we take a longer-term perspective in the context of sustainable development more immediate CCA actions will also enhance future options and preparedness.

One major divergence within the international policy communities of DRR and CCA is articulated by two clearly opposite conceptualisations (Dilling *et al.* 2015). A first position, to which the majority of scholars in the DRR community adhere, is the 'low/no regret' approach. It responds to the need to find an operative answer to existing climate variability as a means of initiating the necessary actions to respond to climate change. From this standpoint, it is argued that reducing the vulnerability of the poor, specifically with respect to disasters and extreme hazards, will improve outcomes in the long-run for CCA.

The second position is the 'sceptical planetary' one, arguing that adapting to climate variability may not guarantee CCA in the long-run. With a clear origin in system science, it argues that since societal contexts in which climate experienced change over time can dramatically affect vulnerability we cannot assume that response to a particular physical stress at a particular point in time will still be effective as the societal landscape changes. Yet, despite the apparent divergence in modern DRR and CCA literature this point is not so different from much DRR work through the decades, simply being rearticulated for a CCA context.

In spite of some such policy convergences and those above on the more scientific level, it is undeniable that DRR and CCA are still often presumed to operate at different natural, bio-physical, geographical, and social scales. That is despite strong evidence for overlap and scientific convergence, as demonstrated throughout this handbook.

A Critical Genealogy of the Concepts

Rather than further indulging in a taxonomic exercise of tracking, comparing and displaying the differences between concept definitions, this section interrogates the reasons and the political motivation behind the main epistemic and political communities producing, working with and using concepts of DRR and CCA.

Assuming Modernity as an Unchanging Backstage or Challenging It

On the whole, both DRR and CCA as domains of knowledge and policy production at the level of multi-lateral agencies and international institutions tend to disregard key sociological transformations. In particular, the debate on the relationship between development and the project of modernity is not taken on board because it is considered not relevant for the advancement of scientific knowledge. That is, the role and relative power of countries within a number of changing anthropological, social, economic, and cultural geographies, is left aside as non-relevant. Indeed, both DRR and CCA policy domains pay far greater attention to environmental (bio-physical mostly) transformations than to social, political, economic, and cultural ones. In other words, institutional actors – e.g., UNISDR or the IPCC – see risk as a given, natural fact. One of the reasons why is that in DRR risk tends to be objectivated. DRR is a field where research concentrates on the role of specific organisations, people, or communities that affect the policy landscape (Hollis 2014).

However, as Beck (1992) pointed out, as traditional forms of life tended to fade away with the affirmation of industrial society at the beginning of the nineteenth century, the end of the twentieth century witnessed a similarly epochal transformation with the loss of centrality of social classes and established forms of social stratification, the nuclear family, gender roles and regulations, as well as the expectations of people regarding their work and professional lives. Beck's work is scantly acknowledged in IPCC (2012). The types of sentences one may find relating to a more sociological reading of DRR are like the one cited below, which does not say anything particularly complex; Beck, even if one were to criticise him, certainly said more important things than the fact that risk is unevenly distributed spatially: 'The capacity to manage risks and adapt to change is unevenly distributed within and across nations, regions, communities, and households' (IPCC 2012, p. 454).

As much as the concepts of 'risk', 'adaptation' and 'sustainability', the concepts of 'Planet Earth' and 'Nature' are also the outcome of a rationalisation of nature, which is not only the fruit of modernity but also the legacy of a colonial use of concepts that are part of a complex cultural elaboration that goes hand in hand with the exploitation of people and places (as resources and with their resources). No climate change mitigation, adaptation or sustainability would be possible without a powerful convincing discourse based on the sacred essence of our planet as a 'common house' that has to be protected. This rhetorical image is employed to a much lesser extent in the case of environmental justice, access to food, disaster recovery, and observation of building codes in the peripheries of many urban areas of both the global north and south.

Saving Planet Earth: System(s) and Social Vulnerability-Centred Approaches

Classic works from an earlier period in critical policy studies by Bernstein (1976) and Rein (1976) had already identified the incapacity of objective policy analysis to go beyond facts and incorporate motivations, and values to offer solutions to policy problems. The latter insight mirrors a limitation shared by many social and political science approaches, including in the areas of DRR and CCA.

In more recent elaborations within critical policy studies, as pointed out by Fischer *et al.* (2015) in their effective synthesis of recent debates, the analysis of what may be called 'fact-policy' is put under scrutiny due to its oversimplification of the politics of expertise. For instance, international institutions such as UNISDR and IPCC tend to promote consensus by means of ecumenical, all-embracing reports, overshadowing the fact that experts often disagree among each other or the fact that policy and science are often driven by vested interests.

Following these lines of thought, this section highlights how the concepts of DRR and CCA are value-laden by unveiling some of the normative constructs on which they ultimately rest, as well as their political implications, particularly for what concerns the directionality of applied policies.

System approaches, focusing on the coupled relationship and multiple interactions between systems labelled as social and ecological, dominate the epistemology of the IPCC as well as the institutional literature on sustainability and CCA. They do so by placing at the centre of the debate two key concepts, 'Planet Earth' and the 'Humans' as one specie (Rockström *et al.* 2009; Scoones *et al.* 2007). In one of the most well-known and devastatingly criticised planetary science manifestos where the word 'safe' appears 27 times, Rockström *et al.* (2009) spell out their conceptualisation of the Earth as a combination of space to be saved by the co-operation of all humans. From the critical perspective of Science and Technology Studies (STS), it is paramount to remember that in our societies nature is not only socially, culturally and politically constructed but also plays, and is made to play, a key role in science and politics (Latour 2004). As much as a national park or protected area, it is a rural ecosystem made up also of post offices, gas stations, roads, etc. Mining, food processing, climate change, other hazard influencers and disasters become non-human actors that shape not only society but also nature itself in their interaction with humans.

From a 'global south seat', 'the Planet' and 'the Humans' as pristine and universally recognised concepts and values, make much less sense and are much less legitimate as the best concepts/values than in the dominant scientific arena of the natural sciences. On the contrary, they can be definitely seen as concepts having a history embedded in power relations (see also the discussion on the relationship between modernity and colonial domination in the previous section), natural resources exploitation, reiterated human rights violations, and human suffering, on the basis of which implications of the utmost importance should be derived by anyone who wants to discuss sustainability issues. In fact, as remembered by critical political ecology scholars, as much as the 'Blue Planet vision' emerged in the global north in the 1970s, the unsustainable use of the 'Earth' (and its resources seen from an economic perspective) are the responsibility of industrial capitalism, with its more contemporary shift to predatory forms (Escobar 2008), rather than a shared responsibility or common goal.

At the other end of the spectrum, the political economy of disasters has demonstrated how disasters occur at the interface of society, technology, and the environment, and are fundamentally the outcomes of these features (Jones and Murphy 2009; Wisner *et al.* 2004). In particular, the concept of vulnerability, which is a complementary notion for both DRR and CCA, plays a different role (Oliver-Smith 2009, p. 13):

> The concept of vulnerability is, inherently of a political-economic nature, although not all vulnerability theorists use it as such. However, the vulnerability/risk approach has changed the way we think about disasters and human environment relations in fundamental ways.

Already in the strand of the political economy of disasters the supposedly 'new' resilience paradigm was criticised, indicating how it makes vulnerability to be seen as a negative feature of society. Vulnerability is a concept that, at least in the social sciences, underlines the pitfalls of development, whereas it has historically mobilised the intellectual energy of those who criticise development itself as a societal goal (Lampis 2015). Resilience, in turn, allows emphasising the positive aspects of certain definitions of vulnerability such as Blaikie *et al.*'s (1994, p. 14):

> The processes that generate 'vulnerability' are countered by people's capacities to resist, avoid, adapt to those processes, and to use their abilities for creating security, either before a disaster occurs or during its aftermath.

More recently, Cannon and Müller-Mahn (2010) have pointed out that the DRR community has historically recognised the social and political nature of vulnerability and risk, while the CCA community tends to de-politicise the debate emphasising the use of terms such as 'adaptation' and 'resilience' from an allegedly scientific position.

The Lost-in-translation Syndrome of DRR and CCA

Giddens (1984) underlined how to import concepts from natural to social sciences without questioning the origins of those concepts or how their different implications for different disciplines produce generalisations of little empirical use. This is still recognised nowadays even by those who work on and with systemic concepts, even in the global north, like 'resilience' (Bené *et al.* 2012, p. 49):

> Being a term that is used (loosely) in a large number of disciplines, resilience can be a very powerful integrating concept that brings different communities of practice together.

Indeed, many scholars, including some of the lead social science authors involved in the drafting of the last two IPCC assessment reports, reflect in their writing a clear preoccupation with the incorporation of a systems perspective capable of bridging social and ecological perspectives, or perhaps of legitimising the concepts elaborated by social scientists in the eyes of natural scientists dominating funding institutions. Within the CCA field, the 'lost in translation' syndrome tends to produce real taxonomies that, at best, work as normative prescriptions such as Romero-Lankao's (2012, p. 13) convoluted explanation of 'adaptive capacity':

> Adaptive capacity, which is different from and influences actual adaptation, is also dynamic and is determined by factors operating at different scales and ranging from age, gender and health conditions, to such assets and entitlements as individual/household resources (e.g., income, health services, air conditioning and/or heating appliances); the extent and quality of infrastructure and public services; and the quality and inclusiveness of governance structures, planning instruments and community organizations that provide or manage safety nets and other short- and longer-term responses.

The above reflect a self-predictive way of doing science, using concepts that everyone accepts but very few challenge and critically analyse. It produces a sort of self-fulfilling prophecy rather than robust science. The scholarship working on global environmental change has lived up to these contradictions with a certain ease.

The 'Double Agenda' Issue

Debates on DRR and CCA can also be read in terms of a number of 'double' agendas representing multiple tensions amongst contrasting political and epistemological positions. These 'double agendas' are in reality multiple agendas, insofar as they are the expression of a wide range of contributions and political positioning of institutional and individual actors from a wide range of areas: governmental, academic, and non-governmental. The way conceptual issues are proposed from these different positions determines a sort of crossroads where the power to influence policy and action is disputed (Lampis 2016). Four central agendas can be identified, which are relevant for the analysis of both DRR and CCA (without pretending to have with this proposed list an exhaustive classification as agendas tend to often be re-shaped, merged and re-elaborated under changing conditions).

The Double-agenda of Governance

This agenda is shaped by power and discourses that determine the way DRR and CCA are embedded within policies, and eventually territorialised (Lampis 2016). Although CCA is intrinsically a political issue, it appears within global and natural discourses as an apolitical one (Tanner and Allouche 2011). On the side of DRR, Hollis (2014) looking at its global dimensions has pointed out the need to analyse the structures that articulate this policy field, highlighting how at least three competing communities can be identified within the disaster risk management field (where differentiation between disaster risk management and DRR presents yet another double-agenda, but one which is not discussed here). These are the ethical community of care, the scientific community behind the rationality of interventions, and national and international political actors, each one claiming sovereignty over the field.

The Double-agenda of Climate Change and Development

One of the main debates presented within this chapter is the confrontation between the natural and the social science communities on concepts and principles about what should guide the implementation of adaptation policies at the local level (McGray *et al.* 2007). In particular, a range of potential strategies is found, marked at each end of the ideal double-tailed distribution by two rather opposite takes on the issue. At one extreme, those who suggest actions aimed at tackling the direct effect of climate change and at the other extreme, those who argue that CCA is basically a matter of unfinished development tasks. A specific position within the natural sciences has identified the problem with the need to reduce the emissions of greenhouse gases and to increase their uptake – collectively termed 'climate change mitigation' – thus also creating another important debate between those favouring climate change mitigation and those arguing that, especially in less affluent countries, the issue to be looked at is CCA.

The Double-agenda of Climate Change Epistemology

Several tensions and power struggles amongst different groups are fought on epistemological grounds. The (climate change) mitigation versus (climate change) adaptation debate is a clear example. Also, the social development versus risk management tension is notable, as much as the tension between the same risk management community and the meteorological-focused scientists, as became clear from the internal political struggle within the IPCC that led to the publication of the 'Managing the Risks of Extreme Events and Disasters to Advance Climate Change Adaptation (SREX)' report (IPCC 2012). The way risk is defined by the two communities differs, as discussed above, both in terms of time horizon, scientific interpretation, and proposed policy solutions.

The Double-agenda of Culture and Techno-Scientific Approaches to Risk Management

Krüger *et al.* (2015) have identified culture as the missing dimension in disaster and risk studies, an issue that applies to both DRR and CCA. How have people operationalised culture in everyday life to establish DRR practice? What constitutes a resilient culture and what role do cultures play in a society's decision making? As Krüger *et al.* (2015) point out, there is a disconnect between how institutions see risk as an area for rational action and policy-making, and the way people from local communities tend to see it. Two clashing visions, where techno-sciences tend to impose a sort of Foucauldian governmentality of the place, while choices related to livelihood sustainability and cultural beliefs inform everyday action as well as longer-term planning.

Conceptual Disruptions

Concepts with their definitions and uses talk to us about how reality should ideally be, what aspects of its dynamics are relevant and at the very end, what counts more and what counts less. A profound need exists for DRR and CCA to innovate and go beyond simply a risk-centred approach. Issues of socio-environmental justice, of the relationship between risk and adaptation and rights need to be incorporated, fully accepting the perspectives of the most critical scholars of DRR and CCA. History, especially of DRR, is also essential for CCA to learn.

In a landmark paper Anderson (1985) made the case for a greater interaction between what almost thirty years after, IPCC (2012) would re-invent and call DRR and development. Anderson (1985, p. 46) asked whether it would not be feasible 'to restructure both disaster emergency aid and long term development aid so that vulnerability is genuinely reduced?' This relationship was later defined in terms of (i) disasters as the indicators of the failure of development and (ii) development as the process of reducing vulnerability to disasters.

This chapter illustrated that some profound divergences show little change over a thirty year period, once we compare the DRR debates of the mid-1980s and the contemporary ones. In spite of extensive literature written on the topic and the calls for inter- and transdisciplinarity, the major disruption to the political and epistemological project of reducing CCA to a branch of DRR (or, bizarrely, even vice versa), is that marked differences remain among scholars, practitioners, and institutions belonging to those two broad areas. With little regard to the hundreds of thousands and often millions of dollars spent on international conferences, top-down science and politics cannot get hold of nor bring under their control the myriad of local instances, epistemic and cultural communities and territories claiming for their own vision, their own elaboration of conceptual frameworks and positioning of their local experiences. This irreducibility of local and often subaltern visions to the dominance of a few powerful discourses remains a source of hope for conceptual independence and the freedom of thought in general, both components that are paramount in a process of elaboration of concept that is not and should not be the monopoly of a few.

This statement should not be mistaken for an ingenious defence of the voice of the powerless or disempowered community at the court of a powerful king – which, nonetheless, is a position worth defending and fighting for. Conversely, it is a broader call in favour of the democratisation of concepts and policy production at diverse scales and levels of complexity, which seems to be the major disruption through which climate science, including on CCA, is still nowadays mainly directed by and for a very restricted group of elite policy-makers and politicians.

References

Anderson, M.B. (1985) 'A reconceptualization of the linkages between disasters and development', *Disasters* 9, s1: 46–51.

Beck, U. (1992) *The Risk Society*, London: Pluto Press.

Bené, C., Godfrey-Wood, R., Newsham, A. and Davies, M. (2012) *Resilience: New Utopia or New Tyranny? Reflection about the Potentials and Limits of the Concept of Resilience in Relation to Vulnerability Reduction Programmes*, Institute of Development Studies (IDS) Working Paper, Brighton, UK: IDS.

Bernstein, R.J. (1976) *The Restructuring of Social and Political Theory*, New York: Harcourt Brace Jovanovich.

Blaikie, P., Davis, I., Cannon, T. and Wisner, B. (1994) *At Risk: Natural Hazards, People's Vulnerability and Disasters*, London and New York: Routledge.

Burton, I. (1996) 'The growth of adaptation capacity: practice and policy', in J.B. Smith, N. Bhatti, G.V. Menzhulin, R. Benioff, M. Campos, B. Jallow, F. Rijsberman, M.I. Budyk and R.K. Dixon (eds), *Adapting to Climate Change: An International Perspective*, New York: Springer-Verlag, pp. 55–67.

Cannon, T. and Müller-Mahn, D. (2010) 'Vulnerability, resilience and development discourses in context of climate change', *Natural Hazards* 55, 3: 621–635.

Chambers, R. (1989) 'Vulnerability, coping and policy', *Institute of Development Studies (IDS) Bulletin* 20, 2: 2–17.

Dilling, L., Daly, M.E., Travis, W.R., Wilhelmi, O.V. and Klein, R.A. (2015) 'The dynamics of vulnerability: Why adapting to climate change will not always prepare us for climate change', *WIREs Climate Change* 6: 413–425.

Escobar, A. (2008) *Territories of Difference: Place, Movement, Life, Redes*, Durham, NC: Duke University Press.

Fischer, F., Torgerson, D., Durnová, A. and Orsini, M. (2015) 'Introduction to critical policy studies', in F. Fischer, D. Torgerson, A. Durnová and M. Orsini (eds), *Handbook of Critical Policy Studies*, Croydon, UK: Edward Elgar Publishing, pp. 1–24.

Giddens, A. (1984) *The Constitution of Society: Outline of the Theory of Structuration*, Berkeley and Los Angeles: University of California Press.

Hollis, S. (2014) 'Competing and complementary discourses in global disaster risk management', *Risk, Hazards and Crisis in Public Policy* 5, 3: 342–363.

IPCC (Intergovernmental Panel on Climate Change). (2012) *Managing the Risks of Extreme Events and Disasters to Advance Climate Change Adaptation*, Cambridge, UK and New York, USA: Cambridge University Press.

IPCC (Intergovernmental Panel on Climate Change). (2014) *Climate Change 2014: Synthesis Report. Contribution of Working Groups I, II and III to the Fifth Assessment Report of the Intergovernmental Panel on Climate Change*, Cambridge, UK and New York, USA: Cambridge University Press.

Jones, E. and Murphy, A.D. (2009) *The Political Economy of Disasters*, Lanham, New York, Toronto and Plymouth, UK: Altamira Press.

Krüger, F., Bankoff, G., Cannon, T. and Schipper, L. (eds) (2015) *Cultures and Disasters: Understanding Cultural Framings in Disaster Risk Reduction*, Abingdon, UK: Routledge.

Lampis, A. (2015) *UGEC Viewpoints*, Obtenido de Urbanization and Global Environmental Change Blogpost. Online https://ugecviewpoints.wordpress.com/2015/10/08/resilience-cities-critical-thoughts-on-an-emerging-paradigm (accessed 22 September 2016).

Lampis, A. (2016) 'Adaptation to climate change in Colombian cities. Which road ahead?', *Cambio Climático*, Bogotá D.C.: Universidad Externado de Colombia.

Lampis, A. and Méndez, O.L. (2013) 'Economía verde y enfoques centrados en las comunidades: Perspectivas frente al cambio ambiental global', in M. Flórez (ed.), *Medio Ambiente. Deterioro o Solución Río + 20*, Bogotá, D.C.: Aurora.

Latour, B. (2004) *Politics of Nature: How to Bring the Sciences into Democracy*, New York and London: Harvard University Press.

McGray, H., Hammil, A. and Bradley, R. (2007) *Weathering the Storm: Options for Framing Adaptation and Development*, Washington D.C.: World Resources Institute.

Oliver-Smith, A. (2009) 'Anthropology and the political economy of disasters', in E.C. Jones and A.D. Murphy (eds), *The Political Economy of Hazards and Disasters*, Lanham, New York, Toronto and Plymouth, UK: Altamira Press, pp. 11–28.

Oliver-Smith, A. and Hoffman, S.M. (1999) *The Angry Earth: Disaster in Anthropological Perspective*, New York: Routledge.

Rein, M. (1976) *Social Science and Public Policy*, The University of Michigan: Penguin.

Rockström, J., Steffen, W., Noone, K., Persson, A., Chapin, F.S., Lambin, III, E., Lenton, T.M., Scheffer, M., Folke, C., Schellnhuber, H., Nykvist, B., De Wit, C.A., Hughes, T., van der Leeuw, S., Rodhe, H., Sörlin, S., Snyder, P.K., Costanza, R., Svedin, U., Falkenmark, M., Karlberg, L., Corell, R.W., Fabry, V.J., Hansen, J., Walker, B., Liverman, D., Richardson, K., Crutzen, P. and Foley, J. (2009) 'Planetary boundaries: exploring the safe operating space for humanity', *Ecology and Society* 14, 2: 32.

Romero-Lankao, P. (2012) 'Governing carbon and climate in the cities: An overview of policy challenges and options', *European Planning Studies* 20, 1: 7-26.

Scoones, I., Leach, M., Smith, M., Stagi, S., Stirling, A. and Thompson, J. (2007) *Dynamic Systems and the Challenge of Sustainability*, Brighton: STEPS Centre.

Simonet, G. (2010) 'The concept of adaptation: interdisciplinary scope and involvement in climate change', *SAPIENS* 3, 1: 10.

Tanner, T. and Allouche, J. (2011) 'Towards a new political economy of climate change and development', *Institute of Development Studies (IDS) Bulletin* 43, 3: 1–14.

UNISDR (United Nations International Strategy for Disaster Reduction. (2009) *Terminology*, Geneva: UNISDR. Online http://www.unisdr.org/we/inform/terminology (accessed 22 September 2016).

Wisner, B., Blaikie, P., Cannon, T. and Davis, I. (2004) *At Risk: Natural Hazards, People's Vulnerability and Disasters (2nd ed.)*, London and New York: Routledge.

PART II

Cross-overs and Connections

9

EDITORIAL INTRODUCTION TO PART II: CROSS-OVERS AND CONNECTIONS

Less Alienation, More Inclusion

Ilan Kelman, Jessica Mercer, and JC Gaillard

Many researchers, practitioners, and policy makers seek silos. For instance, perhaps they research climate change adaptation (CCA). They frame their problem and collect their data according to CCA paradigms. They use only CCA theories and cite only publications covering CCA. They publish in a journal focusing principally on CCA with the readership being primarily those who do only CCA. If material, interests, thoughts, or items do not involve CCA, then they will not be relevant.

Is Specialism Appropriate?

No one can know everything. No one can do everything. We naturally specialise because we have to. Even if everyone worked all 28 hours of each day and all 9 days of each week, then we still could not keep up with all the material available.

Therefore, we create disciplines and we narrow down the topics to which we pay attention. In Isaac Asimov's short story 'Sucker Bait', disciplinarism has led to high partitioning, with trust amongst scientists to represent their discipline correctly being necessary to maintain the silos. The story is a brilliant rendition of consequences when disciplinarians and generalists do not find a balance to collaborate.

It is also a warning for what is often seen in development research, policy, and practice today. Many pick a sector and stick to that sector. Cross-overs and connections are not always of interest.

Such select siloisation appears within disaster risk reduction (DRR) including CCA. Seismologists and volcanologists do not always read each other's papers. Hydrology journals might not be interested in papers on riverine values, landscapes, and livelihoods. Meteorologists and climatologists can delineate rigid boundaries between their fields. CCA has deliberately and methodically aimed to separate itself from DRR as a distinct and new field, with its own publications, conferences, policies, and funding streams. Even CCA and climate change mitigation remain firmly separated, despite repeated calls to link them from the beginning of these fields' formal founding.

Specialism is not detrimental in and of itself. It is, in fact, essential to achieve the desired and required depth for a topic. Unparalleled richness can emerge by diving deep and knowing everything – the full history, all the nuances, and all the critiques – of a highly specific and narrow topic. This wealth is then enhanced by connecting with others who have sought breadth, covering a wide range of topics without dipping much beneath the surface. A marriage between breadth and depth, with mutual exchange and respect, will unleash the best of human knowledge and wisdom, cooperating to seek positive change.

Unfortunately, as with many topic areas, some groups within DRR, particularly those focusing on CCA, do not take this approach, Rather than learning and building on what exists already, CCA has been presented as an entirely new and innovative field, with its own different vocabularies, theories, and framings. While words and phrases might be new, or presented as such, Part I of this handbook demonstrates the infrequency with which genuine newness is seen. Even DRR more widely, while being a comparatively young term and field, presents little which was not already seen in earlier development, environment, and sustainability work.

This is not to say that re-presentation is counterproductive, provided that the representation solidly acknowledges what has gone before. The danger comes when not making cross-overs, when not acknowledging overlaps, and when not accepting connections. The danger is in enforced separation and the construction of boundaries, rather than using specialism and depth to complement generalism and breadth – and vice versa. The danger is particularly acute in not learning from the past while making present-day suggestions less robust than previous work, which includes those in DRR who often do not acknowledge the rich history of ground-breaking disaster research dating back to at least the 1910s with contributions from previous centuries.

An example is a suggestion emerging from CCA literature that global environmental change and economic globalisation experienced together tend to increase vulnerability due to the dual challenges in tandem, referred to as 'double exposure'. Notwithstanding opportunities which emerge from global environmental change and economic globalisation, separately and together, this statement is largely a truism.

It had a much wider and deeper form in development studies through the decades as 'multiple exposure'. The point of tackling poverty, oppression, discrimination, and inequity is that those who have the fewest options, resources, and opportunities in life and livelihoods tend to be most harmed by most hazards (e.g., storms, earthquakes, and volcanoes) and hazard influencers (including climate change). They also tend to have the fewest possibilities for grasping any available advantages, such as business opportunities due to climate change, stockpiling emergency supplies in a personal cache, or retrofitting properties to deal with tornadoes. Global environmental change and economic globalisation are part of this set of challenges and opportunities, at times being the most prominent and at times being almost irrelevant.

In summary and accepting many exceptions, where people, social groups and societies are identified as being highly vulnerable to one challenge, they then tend to be highly vulnerable across multiple challenges simultaneously. This notion of 'multiple exposure' was one underlying ethos of defining and understanding vulnerability in the context of development as DRR developed. It remained a tenet of much DRR work, meaning that CCA could easily have embraced it and deepened it for developing CCA approaches.

The chapters in this part, though, do not simply criticise. Instead, they explain how to move forward creating and supporting positive cross-overs and connections. In themselves, they are sectoralised. After all, a chapter in this handbook could not be titled 'Everything'. Is it helpful or hypocritical to separate 'Ecosystems' from 'Development and Livelihoods'? Why would 'Gender' not be incorporated as part of 'Human Rights' which in turn would be part of 'Ethics'? How could 'Sustainability' be achieved without 'Humanitarian Protection' and other forms of protection? 'Violent Conflict' directly impacts all other cross-overs and connections.

The divisions could have been produced in many ways. The key is whether or not the topics permit an adequate level of depth to be achieved without sacrificing the links amongst them. Many of the chapters display cross-cutting themes which could, perhaps should, have been their own chapters. Examples would be separate chapters on 'Identity', 'Culture', 'Ethnicity and Race', 'Disability', 'Religion and Spirituality', 'Sexuality', 'Caste', and 'Language', with other possibilities apparent.

A good proportion of such titles, topics, and separations are arbitrary. Rather than parsing the titles and refining the boundaries, the importance is recognising and accepting wider scopes and engaging with those, while nonetheless going forth with one's own expertise and focus. Specialism can include, rather than alienate – especially when combined with generalism for DRR including CCA.

10

DEVELOPMENT AND LIVELIHOODS FOR DISASTER RISK REDUCTION INCLUDING CLIMATE CHANGE ADAPTATION

Bob Alexander

Introduction

People's livelihoods play important roles in enabling self-protection toward addressing their everyday, seasonal, and extreme risks (Alexander *et al.* 2006; Cannon 2008; Twigg 2015). In the mid-1990s, a framework for sustainable livelihoods (SL) was developed to address these risks. Concepts, development, and contributions of SL approaches are explained in Carney (2002), Sanderson (2012), Solesbury (2003), and Twigg (2015). Likewise, the relationship between development processes and the reduction, creation, or exacerbation of the vulnerability context underlying these risks is explained in Cardona (2004), DFID (2004), Hewitt (1995), IFRC (2002), UNDP (2004), and Wisner *et al.* (2012). Recent discussion has focused on how creeping environmental problems such as climate change are contributing to changes in hazards and vulnerability with increasing future uncertainty (Kelman *et al.* 2015; Kelman and Gaillard 2010; Mercer 2010; Schipper 2009). And discussions of the need for an adaptive approach (effectively revisiting much earlier work on adjusting socially to reduce vulnerability) resulted from this focus on the potential effects of development and environmental change on risk and opportunities (DKKV 2011; Jones *et al.* 2010).

Dominant approaches to reducing socio-economic vulnerability to environmental hazards in the past few decades include disaster risk reduction (DRR), environmental management, poverty reduction, and climate change adaptation (CCA). SL effectively bridges the first three of these by focusing on poverty reduction in a way that manages environmental and other resources for long-term sustainable use and manages risks to reduce disaster losses. But the original framework did not explicitly bridge these approaches with rapid changes in everyday, seasonal, and extreme vulnerability due to rapidly changing conditions from development and environmental change. The interaction of changing hazards with changing vulnerability contributes to disaster risks that hinder local development, necessitating further evolution to prospective risk reduction approaches that focus on anticipating risks not yet realised and incorporating their management into future development planning (Lavell 2008). Doing so further enables evolution to a development-based prospective risk reduction SL approach that improves sustainability and performance by incorporating risk into development. By modifying it to incorporate CCA along with other approaches for dealing with rapid environmental changes, SL can serve as a framework for better understanding the connections of all of these approaches and related implications for utilisation.

This chapter builds upon these concepts to explain how livelihood strategies that are part of an adaptive integrated system approach can address such changes in hazards and vulnerability contexts that are driven by development and environmental processes. It begins with an explanation of the basic elements of the initial progressive corrective SL framework iterations. Modification of this framework to enable focus on prospective risk reduction is then explicated. Examples highlight how relevant additional considerations regarding exogenous and endogenous changes can be incorporated.

The SL Framework for Progressive Corrective Risk Reduction

The SL framework was designed to enable specific programming strategies based on holistic analysis of local system factors affecting vulnerability. Figure 10.1 depicts a synthesis of the most common variations of this model in which there are:

- An overall box containing components representing endogenous development that is internal to livelihood decision-making. Such internal decision-making has primarily been at the household level but has also been used as relevant to livelihood strategy determination at the community or local level (see Lavell 2004 and Maskrey 2011 for relevant differences between community and local).
- Exposure and vulnerability to everyday, seasonal, and extreme trends and events that comprise the vulnerability context.
- An endowment of natural (N), physical (P), financial (F), social (S), and human (H) assets to which their access is influenced by processes including marginalisation and previous unsustainable uses. These asset categories often overlap or are otherwise difficult to analyse individually in practice. But they serve to represent sufficiently different characteristics among some types of assets so that the focus can be on how overall asset endowments can be combined in strategies to improve and stabilise livelihoods.
- Local policies, institutions, and practices (PIPs) that can transform these assets as resources in this vulnerability context in sustainable livelihood strategies. For these purposes, livelihoods are generally defined as comprising 'the capabilities, assets (including both material and social resources) and activities for a means of living' (Scoones 2009, p. 175). A livelihood is defined as being sustainable 'when it can cope with and recover from stresses and shocks, maintain or

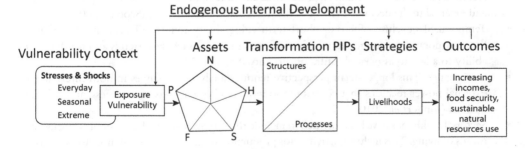

Figure 10.1 Synthesis of Common Sustainable Livelihoods Frameworks

(By Author)

enhance its capabilities and assets, while not undermining the natural resource base' (quoted by Scoones 2009, p. 175).

- Positive and negative changes in outcomes in income, food security, and asset conditions then contribute to the vulnerability context, asset base, and PIPs that can be considered in continuous strategy decision-making.

Designed to be flexibly adapted (Carney 2002), SL has previously been modified to address specific purposes or perceived limitations. Such modifications allow for incorporating tools for market analysis, for bridging to higher-level policies, for local understanding of how assets are used and how people substitute between them, and for incorporating SL into other aspects of decision-making such as Poverty Reduction Strategy Papers, implementing community-based planning processes linked to local government planning, and for using SL in post-disaster and post-conflict situations (Carney 2002). SL has also been incorporated into recent resilience frameworks (Twigg 2015).

Since original modifications of the SL approach, differences amongst different types of risk reduction have been explicated as follows (UNISDR 2015):

- Residual risk reduction enhances capacity to respond and cope with what has been deemed as acceptable risk.
- Conservative corrective risk reduction enhances capacity to reduce what has been deemed undesirable exposure and vulnerability of assets that suffered damages and losses in previous events.
- Progressive corrective risk reduction enhances capacity to address underlying factors that contributed to this undesirable exposure and vulnerability.
- Prospective risk reduction enhances capacity to anticipate and adapt to changes in the vulnerability context that would result in undesirable risks of damages and losses that are different from those previously experienced.

SL's focus on livelihood and poverty-related factors underlying vulnerability was a breakthrough in the evolution from conservative to progressive corrective risk reduction approaches (Lavell 2008). Addressing risk exacerbation from system changes caused by development activities and changing environmental conditions through prospective risk reduction requires further evolution to address the following SL limitations:

- Spatial: Limiting analysis to endogenous internal development processes neglects effects on assets and the vulnerability context of environmental and development trends with origins exogenous and external to the decision-making of the household or community (Scoones 2009).
- Temporal: SL's exclusive focus on short-term vulnerability context effects neglects the need for incorporation of longer-term effects of such exogenous external trends and for dynamic ability to adapt to associated future uncertainty (Scoones 2009).
- Governance: This longer-term perspective requires analysis of changes in both internal and external power relations that contribute to vulnerability contexts, asset access, and strategy options (Jones *et al.* 2010).
- Sectors: Livelihoods have been defined to include the activities in which households engage as a means of living. This includes activities for production of income, food, and non-food items to sustain and enhance their lives. Some aspects of enabling household access to water, health care, shelter, power, and education and training are included in these activities. But some aspects of providing these and other basic social services (BSS) are organised in activities considered to be

external to decision-making regarding livelihoods. Limiting SL analysis to how the asset types can be transformed into adaptive livelihood strategies neglects important synergy opportunities for developing such BSS provision strategies that strengthen engagement in such livelihoods while contributing toward overall current and future well-being outcomes.

A SL Framework for Prospective Risk Reduction

Figure 10.2 depicts a conceptual framework for overcoming these limitations.

Prospective Endogenous Internal Development and Exogenous External Trends

The original SL framework focused on livelihood strategy decision-making at the household and community level to reduce poverty while achieving other improvements in vulnerability reduction and well-being (Carney 2002). Although subsequent SL framework modifications recognised the importance of processes at the macro and meso levels, they continued to focus on internal influences on the vulnerability context and asset conditions. Prospective risk management must address this spatial limitation by recognising potential effects on internal assets, trends, and events from external development and environmental change processes at all levels.

The prospective SL framework depicts the exogenous external trends outside of the original endogenous internal development box. Climate change is one of many environmental change processes such as soil erosion, land desertification, land subsidence, and deforestation. These environmental processes mix with development processes external to local decision-making to change natural and physical capital assets and everyday, seasonal, and extreme trends and events. These external effects combine with the impacts of internal development to determine overall potential changes in assets, hazards, and vulnerability. Although climate change itself can influence

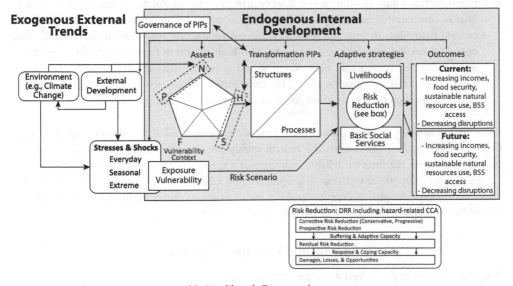

Figure 10.2   The Prospective Sustainable Livelihoods Framework

(By Author)

frequency and intensity of natural hazards and everyday and seasonal climate conditions, vulnerability and resulting disaster risk will depend on the combination of all such change factors.

The temporal limitation is addressed by analysing these trends and potential impacts based on predictions of potential future changes. As appropriate to decision-making needs, predictions can be translated into indicative risk scenarios that can be utilised to prioritise potential issues and to inform the adaptive livelihood and BSS strategy decision-making process. But especially in contexts where immediate needs tend to favour short-term interventions, long-term climate information need not be considered in every decision. When projections of future impacts are considered helpful, timescales for climate information should be matched with time frames of other relevant available data and decision-making needs.

Examples of climate risk scenario creation include the following:

- In the preliminary step towards determining improved livelihood options in the Mainstreaming Climate Change Adaptation in Irrigated Agriculture project in China's Huang-Huai-Hai River basin, an external scientific model created climate change scenarios based on potential impacts of climate change on agriculture and water resources, agriculture economics changes over spatial and temporal scales, and related changes in farmer behaviour.
- In USAID's Adaptation and Resilience to Climate Change Program in the Lower Mekong Basin, implementing partners developed scientific climate stories (SCS) from local vulnerability assessments and downscaled scientific information on past, present, and future climate change effects on agricultural livelihoods. Community members then created community climate stories (CCS) based on a climate awareness survey of what community members know about climate change and how specified weather changes in their area are impacting livelihoods. The SCSs and CCSs were then compared to merge knowledge into a shared scenario from which a visioning process of scenarios of what they desire in selected future years was used for backcasting to develop adaptation plans.
- CARE International's Participatory Scenario Planning (PSP) in Kenya brings together national and local meteorological services; local forecasters; research institutions; government departments such as water, agriculture and disaster management; information intermediaries such as NGOs, community leaders and media; and representatives of information users such as farmers and pastoralists to interpret seasonal climate forecasts into implications for hazards, risks, opportunities and impacts in the local area. This leads to effective communication of resulting information for community understanding and action to manage climate risk, uncertainty and opportunities in a given season. The development of this new forum of non-traditional relationships also enables improved governance for risk-informed community CCA planning.

Regardless of improvements in climate prediction accuracy and global trend downscaling for local decision-making, these predictions will remain highly uncertain and imprecise. Where available information about the future is reliably indicative of a particular impact such as less water access, communities can use these scenarios to prepare for and adapt to those indicative changes. Where scenarios are less indicative of expected impacts, more emphasis should be placed on overall ability to rapidly modify livelihoods as changes are identified (Pasteur 2011).

Internal and External Governance

Adaptive capacity in this longer-term perspective requires individual and institutional capacity to identify and address potentially complex chains of changes across sectors of the holistic biocultural system. Although many progressive corrective sustainable livelihoods projects have focused on natural, physical, and financial assets or on general institutional capacity building, prospective risk reduction requires emphasis on human and social assets to enable such adaptive capacity and longer-term forward-looking strategy decision-making. In Figure 10.2, while all assets continue to be transformed through PIPs in adaptive strategies, human and social assets have a mutually reinforcing relationship with PIPs that strengthens internal governance. Overcoming power relations in governance requires such internal micro-level mechanisms and PIPs to ensure that vulnerable households have access to assets and strategy options.

It also requires that endogenous community leaders have sufficient input about issues affecting constituents and power to influence macro-level PIPs. Communication and participation in micro-level processes should improve governance by enabling macro-level strategies to be informed directly by people's concerns (Carney 2002). These leaders then must be able to raise these issues at higher levels of governance so that any external processes potentially adversely affecting assets, risks and strategies can be addressed. In Figure 10.2, strengthening micro-level PIPs has a mutually reinforcing relationship with the governance of macro-level PIPs to influence external development and all processes that it affects.

An example of such micro and macro governance improvements is in Nepal's Karnali River Basin where the communities were caught in a flood–poverty trap. Intensity and magnitude of precipitation events have increased significantly in the past few decades. Larger floods and landslides have become more frequent. With the beds of rivers rising, some communities lower than rivers have flooded frequently during rainy seasons. Rapid population growth, weak land use management, slow economic development, deforestation, poor building practices, floodplain encroachment, and poor flood preparedness and management capacity further increased flood vulnerability and destabilised natural assets. In addition to some direct livelihood-related interventions, residents benefited from local government and NGO collaboration to help them manage flood risk. Human and social assets were strengthened within and between communities and across sectors through establishment and improvement of community disaster management committee capacities.

Village savings and loan association (VSLA) development is an example of financial investment that can contribute sufficiently to human and social assets and participatory local governance. In villages in the Karamoja region of northern Uganda, both livelihoods and BSS provision were severely lacking because of recurrent drought and years of armed cattle raiding. Financial capital and training to develop management capacity were made available to groups. Resulting improved participation in decision-making led to initiatives to improve natural and physical assets which contributed to adaptive strategies for livelihoods, BSS, and risk reduction. Similarly, local governance and social assets were strengthened in the Chars Livelihoods Programme in northwest Bangladesh by establishing VSLAs and also requiring beneficiaries to attend group meetings and training on community and socio-economic development topics and disaster preparedness.

As poverty and marginalisation increase people's vulnerability to both climate variability and change, amongst other hazard drivers and hazards, CCA needs to address such root causes of this vulnerability. To do so, it must engage with enhancing the human and social assets of formally marginalised groups such that their actions and voices can overcome entrenched inequitable processes. One such example is Operation Mworio in a drought-prone aid-dependent village in Kenya. Participatory assessment of vulnerabilities and reduction options first resulted in pooling resources to build water pans for subsistence farming. In this process, participation and leadership developed sufficiently to garner support to build water reservoirs for expansion into commercial farming. Such location-based livelihood initiatives that enhance human and social assets can be strengthened in addressing current vulnerabilities in a way that creates adaptive capacities for continuing to overcome challenges (see also Box 10.1).

Box 10.1   Integrating Current and Future Vulnerability Reduction in Southern Peru

Bob Alexander[1]

[1] Independent, World Citizen

Southern Peru's Andes Mountains highland rural communities face poverty, low human development, fragile mountain ecosystems, and environmental deterioration. Although high climate variability has always existed, climate change involves increased temperatures, glacial melting, rainfall and drought timing and distribution variations, frost, and storm frequency and severity. Changes in everyday, seasonal, extreme trends and events result from such climatic changes combining with vulnerability-modifying externally driven environmental and development processes such as soil erosion, changing markets, water use conflicts, and internally driven resource degradation through overgrazing, grassland fires, land use changes, and native forest loss.

   Many programmes in these communities aim to prospectively improve people's everyday access to livelihoods and BSS and reduce seasonal and extreme event access disruptions. The Peru Environment Ministry and Swiss Agency for Development Cooperation's Programa de Adaptacion al Cambio Climatico (PACC) utilises development strategies to address vulnerability's root causes and how they and disaster risks are changing. For example, its 'Growing Water' project targets areas with future risk scenarios emphasising insufficient water source levels. Resulting asset-enhancing strategies for risk-informed livelihoods incorporating water-fed agriculture and family gardens and BSS focusing on nutrition and early childhood development (ECD) include:

- Natural: Swampland restoration, forestation, pasture rotation, and terrace soil conservation.
- Physical: Reservoirs, dykes, infiltration ditches, pasture irrigation, traditional aquifer and lake channels, fences protecting springs, greenhouses, animal pens, and warmer housing with kitchens.
- Financial: Microfinance.
- Human: Training in irrigation, water distribution, resilient agriculture techniques, agroforestry, value-added production, nutrition, gardening, greenhouses, animal raising, and ECD.

Most importantly, adaptive capacity development through social assets and PIPs includes community association leader training for relevant change identification, community adaptation and preparedness training, related participatory decision-making and elevating external issues to higher government levels; related local government training; water management statute strengthening; microfinance communal banks; and women's group strengthening for nutrition, ECD, and microenterprise activities.

Prospective Risk-informed Livelihood and BSS Adaptive Strategies

Livelihoods are drivers of a sustainable self-protection system (Alexander *et al.* 2006; Cannon 2008). Sustainable sources of food and income allow people to avoid needing social protection for these items and enable them to expend earnings toward improving BSS access. But bottlenecks in BSS provision can prevent such access and also cause reductions and disruptions in livelihoods. For example, disruptions in nutrition, health care, and education services can all cause adults to miss work. And disruptions in water supply can adversely affect household, crop, livestock, and service facility access. The progressive corrective SL framework focused exclusively on livelihood strategies that could have beneficial food, income, and asset strengthening outcomes.

A prospective SL framework overcomes this sector's limitation by additionally considering the important dynamic synergies involved in implementing strategies that address strengthening both livelihoods and BSS provision in the integrated local system.

Examples of such integration include Box 10.2 and:

- OneWASH initiatives in which water supply project strategies aim to jointly improve water supply access for households, agricultural, livestock, and service facilities.
- OneHealth initiatives in which people in areas with insufficient access to separate doctors for human and livestock health are trained in basic competencies to care for both.
- India's Western Orissa Rural Livelihoods 'Watershed Plus' Project in which livelihood and service strategies strengthen human and social assets to enhance physical, natural, and financial asset effectiveness for corrective risk reduction in a way that enables prospectively better identifying and adapting to changes.

Box 10.2   Location-based Convergence of Ethiopia's National Flagship Programmes to Enhance Adaptive Capacity

Bob Alexander[1]

[1] Independent, World Citizen

Ethiopia's most development-restricting hazard is drought. Root causes of drought vulnerability factors such as poverty and food insecurity include population pressure, degraded land, low-input and low-technology smallholder agriculture, livestock and crop pests and diseases, and asset disposal during drought recurrence. Vulnerability is exacerbated by reduced utilisation of temporary migration and other coping strategies. Climate change involving more frequent heavy rainfall events worsens flooding and soil erosion while higher temperatures and rainfall variability increase drought intensity and frequency. Increased pollution threatens water quality and livelihood-supporting ecosystems.

The Productive Safety Net Programme (PSNP) was designed in 2004 to reduce household vulnerability to seasonal drought-related food insecurity, improve household and community resilience, and break the food aid dependence cycle. Vulnerable households avoid productive asset depletion through cash or food payment for public works employment on natural resource management and BSS infrastructure construction and maintenance. Subsequent modifications include credit and technical assistance access for asset stabilisation and contingency fund risk financing mechanism support to people temporarily needing post-event assistance.

Recent approach modifications link PSNP's asset-building aspects with other national platforms. This enables location-based risk-informed livelihood and BSS strategies that address vulnerability's root causes. Linking to the Sustainable Land Management Programme helps focus natural and physical assets public works on improving natural resource and farm management. Links to the National Nutrition Programme involve strategies addressing nutrition insecurity. Related public works on structures and roads contribute to human assets through improved education, water, and Health Extension Programme training that contributes to reduced illness and undernutrition. Linking to the Disaster Risk Management framework helps ensure that risk-informed strategies and asset strengthening include public works and preparedness for disaster reduction. Overall adaptive capacity to identify and respond to changes in natural resources, health, nutrition, and risks increases through ever-evolving community-based health extension, women's groups, and nutrition governance systems.

In Figure 10.2, this relationship is depicted as livelihood and BSS adaptive strategies reinforcing each other as influenced by risk reduction decisions in the expanded box at the bottom of Figure 10.2. In previous frameworks, risk information for SL decision-making was limited to conservative corrective estimates of loss and damage to such items as assets, food, income and BSS access in previous events and progressive corrective information about underlying factors contributing to these losses and damages. The current prospective risk scenarios also incorporate how these losses and damages are expected to be affected by climate change effects on hazards and other relevant environmental and development influenced changes over time.

Some beneficial opportunities may also arise as a result of such disaster events. And there are many potential everyday and seasonal benefits from development and environmental changes that decision-makers may wish not to address with adaptive measures. Such opportunities, any positive outcomes associated with risks, and limited budgets for risk reduction encourage decisions regarding which levels of which potential damages and losses are prioritised as unacceptable and which are otherwise considered acceptable. For those considered acceptable, the decision-makers hope that negative event impacts are not realised. They may choose to engage in residual risk reduction and preparedness activities to increase response and coping capacities to help minimise such negative impacts. For those considered unacceptable, decision-making regarding livelihood and BSS provision strategies needs to incorporate buffering and adaptive capacity strengthening so that associated negative impacts can be resisted. In all such considerations, optimisation may require choosing between potential short-term and long-term effects from different strategy alternatives but may best focus on strategies that simultaneously address both. The distinction between current well-being outcomes and future well-being outcomes in Figure 10.2 can be modified to whatever different times are relevant for consideration in decision-making. In this manner, decision-making on adaptive strategies aims to dynamically optimise well-being by balancing positive income, food, asset, and service access outcomes with minimised disruptions to them dynamically.

Developing these strategies requires an integrated risk identification, analysis, assessment, and reduction process (Alexander and Mercer 2012). As the livelihoods approach is location-based, many SL approaches focusing on development through DRR have utilised a Community-Based DRR (CBDRR) methodology. CBDRR is a process of community participation, empowerment, and problem solving to better enable them to prepare for, withstand, and cope with potential disasters. Community-Based Adaptation (CBA) is a similar process that was designed for CCA and used in many projects to focus exclusively on adaptation to climate change. But different current proponents of CBA approach its use differently. Those espousing use of adaptation as part of wider development planning argue that CBA should incorporate participatory climate change impact scenarios to inform strategies to address all factors contributing to natural, social, and economic problems that people face rather than just climate change (Forsyth 2013; Sabates-Wheeler *et al.* 2008; Tanner *et al.* 2015). By this definition, CBA contributes to a development-focused CBDRR methodology. All manners of adaptation could be subsumed under CBA as part of a CBDRR process for determining adaptive livelihood and BSS strategies that prospectively reduce risks toward desired development well-being outcomes. In Gaibandha in northern Bangladesh where flash flooding and erosion are increasing, CBA resulted in villagers adapting a freshwater weed into floating rafts that act as platforms for soil and compost for food gardens that allow farmers to maintain agricultural productivity despite flooded lands.

Other examples of utilising community-based risk identification, analysis, assessment, and reduction methodologies to determine adaptation strategies include the following:

•   Practical Action's 'Vulnerability to Resilience' methodology was used with communities in the Chitwan District of Nepal seeking to address adverse human life, livelihood, and BSS access impacts from weather-related hazards that are increasing in frequency and intensity due to climate change. Community-led analysis and prioritisation resulted in strategies such as strengthening livelihoods, preparedness, and governance toward continuous assessment and adaptive action. These strategies integrated CCA within development-oriented DRR by aiming to strengthen overall community abilities to address climate change impacts while also addressing everyday issues, regular climatic variability, other factors affecting future disaster vulnerability, and other trends.

- The 'Livelihoods Adaptation to Climate Change' methodology of FAO and the Department of Agricultural Extension in various parts of Bangladesh bridged local and scientific knowledge by engaging stakeholders in assessing current vulnerability and future climate risks. Steps included training in methods to downscale the global climate outlook into local impact scenarios, determining local perceptions, evaluating livelihood asset portfolios of different access groups, and assessing levels of potential vulnerability impacts; synthesising, validating, and prioritising among indigenous, local, and scientific options potentially suitable to location-specific conditions for field testing; and, due to uncertainty in downscaled information, designing and upscaling appropriate 'no regrets' adaptation strategies based on results, learning, and capacity building. By determining how climate change might interact with other changes to vulnerability, these strategies address CCA in a manner consistent with a broader development-oriented DRR approach.

- In southwest Bangladesh's Subarnabad area, the 'Reducing Vulnerability to Climate Change' project of the then-existing Canadian International Development Agency and CARE started with a bottom-up participatory risk identification approach. Significant livelihood and living condition vulnerability was attributed to changes in production systems from land-based crops to shrimp farming due to increasing soil salinity. This salinity was caused by environmental changes associated with saltwater intrusion from both external infrastructure projects and climate change-related sea-level rise and increased storm surge magnitude. Although climate change projections are not precise, projections for all of South Asia were deemed sufficiently indicative that Subarnabad will further experience development challenges exacerbated by climate change through increased temperatures and precipitation, sea-level rise, cyclone and storm surge intensity, saltwater intrusion and associated flooding, coastal soil erosion, freshwater resource stress, human health problems and migration, and agricultural production disruption. As both current and future vulnerabilities are related to power relations and economic conditions as well as climate change, strategies mainstreaming CCA within development-oriented DRR focused on human assets enhancing training and technical support and financial assets through loans and a savings bank. Rather than a formal prioritisation process, this enabled trial-and-error based on sharing through adoption and diffusion of any successful new livelihood or BSS strategies. Resulting social assets strengthening helped enable current climate, economic, and political sources of vulnerability to be addressed in ways that would be easily adapted as conditions continue to change.

Conclusions

Risk-informed development planning must consider people's access to what enhances their lives. Such planning must integrate often-pervasive daily and seasonal vulnerability to lack of such access with how this vulnerability is modified during and after hazards. In the context of rapid development, inherent uncertainty in projections of continuing climate change and other environmental processes, and the complex relationship between these changes and local impacts, it must also incorporate how both such aspects of vulnerability are changing over time. Any climate change approach that considers only the future and only the effects of climate change insufficiently considers current and extreme hazards vulnerability and the effects of other change processes. Any disaster risk approach that considers only current and past vulnerability neglects the potential effects of changes on what will happen in the future. In this perspective, CCA helps DRR fulfil its full mandate and vice versa.

Some reviews of DRR consider it to frequently focus operationally on current risk and previous experience (Mercer 2010; Twigg 2015). But broader views of DRR acknowledge the relevance of forward-looking decision-making and incorporate a future perspective that temporally integrates all climatic, development, and intervention trends across all time scales (Alexander and Mercer 2012; Mercer 2010).

Integrated CCA approaches inform how risks are changing so that development-mainstreamed DRR approaches can then incorporate this information into decision-making that sufficiently addresses all aspects of current and future vulnerability. Integrated livelihoods strategies play an important role towards risk-informed development utilising a DRR approach that includes CCA. Sanderson (2012) and others have concluded that an SL focus naturally facilitates, bridging development, DRR, and CCA across scales and sectors. The revised framework and accompanying discussion in this chapter aim to guide that facilitation.

The objective of the prospective SL framework is to better enable such DRR supported by CCA through an adaptive prospective risk reduction system in which local governments, households, markets, and providers of services are equipped to identify and adapt to potential changes that they determine are likely to adversely affect livelihood and service access and vulnerability. To do so effectively, they would also need to employ corrective risk reduction to adapt for vulnerabilities and root causes already identified and residual risk reduction to be prepared to respond to situations for which they have not adapted. Many places have only recently embraced the transition from reactive to proactive DRR. So enabling this system is likely to require a location-specific process that continues to evolve from building on local residual preparedness and response strengths towards developing corrective buffering and ex-post adaptive capacities while gradually building towards prospective identification and corresponding optimisation over ex-ante adaptation, ex-post adaptation, and preparedness for response for what remains.

Spatial critiques such as those in Scoones (2009) of SL being too holistic and micro-scale to be relevant for development decision-making have diminished recently because of renewed interest in location-based approaches. Failure of SL approaches to link to external actors and higher-level decision-making processes has been addressed through the addition of connections to external governance and to potential effects on assets and risks from climate change and other environmental and development changes.

This link to higher-level decision-making only helps overcome governance limitations if power relations are such that concerns that people have about the external processes affecting their vulnerability contexts, asset access, and strategy options will be heard and addressed. In situations in which additional participation and communication will enable this to be achieved, the modified framework better reveals potential interactions amongst asset types, identification of internal and external causes of potential problems, internal PIP development, and influence on external PIPs. This same modification and set of processes also applies to how political or economic transformation can be achieved in situations in which embedded power structures at local or higher levels prevent marginalised voices from effectively influencing internal or external PIPs.

Beyond addressing external governance issues, greater focus on human and social capacity also helps internally address temporal constraints to embedding dynamic change and CCA within

DRR. This includes capacity to integrate local and external information in development and utilisation of scenarios of long-term changes and their potential impacts when indicative information is available. It also includes capacity to identify and adapt to short-term changes that were previously unaddressed due to future uncertainty or other factors.

In contrast to previous perceived limitations that SL was too inter-sectorial for implementation, the need for holistic location-based approaches encourages further understanding of how livelihoods, BSS, and DRR including CCA interact in short-term and long-term strategies. The revised framework encourages consideration of these interactions and utilising a mix of strategies that can optimise over current, short-term, and longer-term well-being outcomes. As in the Southern Peru and Southwest Bangladesh examples, the focus should be on choosing livelihood and BSS strategies that consider current everyday, seasonal, and extreme problems and how to develop capacities in current strategies to address these problems and how they may change in the future.

To do so, the chosen strategy identification, analysis, assessment, and choice methodology will need to consider CCA as part of DRR which in turn is part of development planning. CBA approaches that consider adaptation beyond only climate change (in effect, being how 'adjustment' was frequently used in earlier disaster-related studies) can contribute to CBDRR as such a comprehensive methodology. The resulting strategy decision-making for prospective SL toward DRR including CCA can broadly incorporate climate and other change projections as appropriate for choices of risk-informed livelihood and BSS provision strategies to dynamically optimise development well-being outcomes.

Acknowledgments

Helpful suggestions regarding explication of the prospective SL conceptual framework and comments on an early draft were provided by Wei Liu and Adriana Keating from the International Institute of Applied Systems Analysis.

References

Alexander, B., Chan-Halbrendt, C. and Salim, W. (2006) 'Sustainable livelihood considerations for disaster risk management: Implications for implementation of the government of Indonesia tsunami recovery plan', *Disaster Prevention and Management* 15, 1: 31–50.

Alexander, B. and Mercer, J. (2012) 'Eight components of integrated community based risk reduction: A risk identification application in the Maldives', *Asian Journal of Disaster and Environmental Management* 4, 1: 57–82.

Cannon, T. (2008) *Reducing People's Vulnerability to Natural Hazards: Communities and Resilience*, United Nations University World Institute for Development Economics Research (UNU-WIDER) Research Paper 34, Helsinki: UNU-WIDER.

Cardona, O.D. (2004) 'The need for rethinking the concepts of vulnerability and risk from a holistic perspective: A necessary review and criticism for effective risk management', in G. Bankoff, G. Frerks and D. Hilhorst (eds) *Mapping Vulnerability: Disasters, Development & People*, London: Earthscan, pp. 37–51.

Carney, D. (2002) *Sustainable Livelihoods Approaches: Progress and Possibilities for Change*, London: Department for International Development (DFID).

DFID (Department for International Development). (2004) *Disaster Risk Reduction: A Development Concern*, London: DFID.

DKKV (German Committee for Disaster Reduction). (ed.) (2011) *Adaptive Disaster Risk Reduction. Enhancing Methods and Tools of Disaster Risk Reduction in the light of Climate Change*, Bonn: DKKV.

Forsyth, T. (2013) 'Community-based adaptation: A review of past and future challenges', *WIREs Climate Change* 4, 5: 439–446.

Hewitt, K. (1995) 'Sustainable disasters? Perspectives and powers in the discourse of calamity', in J. Crush (ed.) *Power of Development*, London: Routledge, pp. 115–128.

IFRC (International Federation of Red Cross and Red Crescent Societies). (2002) *World Disasters Report: Focus on Reducing Risk*, Geneva: IFRC.

Jones, L., Jaspars, S., Pavanello, S., Ludi, E., Slater, R., Arnall, A., Grist, N. and Mtisi, S. (2010) *Responding to a Changing Climate: Exploring How Disaster Risk Reduction, Social Protection and Livelihoods Approaches Promote Features of Adaptive Capacity*, Overseas Development Institute (ODI) Working Paper #319, London: ODI.

Kelman, I. and Gaillard, JC (2010) 'Embedding climate change adaptation within disaster risk reduction', in R. Shaw, J.M. Pulhin and J.J. Pereira (eds) *Climate Change Adaptation and Disaster Risk Reduction: Issues and Challenges*, Bingley: Emerald Group Publishing Limited, pp. 23–46.

Kelman, I., Gaillard, JC and Mercer, J. (2015) 'Climate change's role in disaster risk reduction's future: Beyond vulnerability and resilience', *International Journal of Disaster Risk Science* 6, 1: 21–27.

Lavell, A. (2004) *Local Level Risk Management: From Concept to Practice*, Quito: CEPREDENAC-UNDP.

Lavell, A. (2008) *Relationships between Local and Community Disaster Risk Management & Poverty Reduction: A Preliminary Exploration*, A Contribution to the 2009 ISDR Global Assessment Report on Disaster Risk Reduction, Geneva: United Nations International Strategy for Disaster Reduction (UNISDR).

Maskrey, A. (2011) 'Revisiting community-based disaster risk management', *Environmental Hazards* 10, 1: 42–52.

Mercer, J. (2010) 'Disaster risk reduction or climate change adaptation: Are we reinventing the wheel?', *Journal of International Development* 22, 2: 247–264.

Pasteur, K. (2011) *From Vulnerability to Resilience: A Framework for Analysis and Action to Build Community Resilience*, Oxfordshire: Practical Action Publishing.

Sabates-Wheeler, R., Mitchell, T. and Ellis, F. (2008) 'Avoiding repetition: Time for CBA to engage with the livelihoods literature?', *Institute of Development Studies (IDS) Bulletin* 39: 53–59.

Sanderson, D. (2012) 'Livelihood protection and support for disaster', in B. Wisner, JC Gaillard and I. Kelman (eds) *Handbook of Hazards and Disaster Risk Reduction*, Abingdon: Routledge, pp. 697–710.

Schipper, L. (2009) 'Meeting at the crossroads?: Exploring the linkages between climate change adaptation and disaster risk reduction', *Climate and Development* 1: 16–30.

Scoones, I. (2009) 'Livelihoods perspectives and rural development', *The Journal of Peasant Studies* 36, 1: 171–196.

Solesbury, W. (2003) *Sustainable Livelihoods: A Case Study of the Evolution of DFID Policy*, Overseas Development Institute (ODI) Working Paper #217, London: ODI.

Tanner, T., Lewis, D., Wrathall, D., Bronen, R., Cradock-Henry, N., Huq, S., Lawless, C., Nawrotzki, R., Prasad, V., Rahman, Md., Alaniz, R., King, K., McNamara, K., Nadiruzzaman, Md., Henly-Shepard, S. and Thomalla, F. (2015) 'Livelihood resilience in the face of climate change', *Nature Climate Change* 5, 1: 23–26.

Twigg, J. (2015) *Disaster Risk Reduction: Good Practice Review 9*, London: Overseas Development Institute (ODI).

UNDP (United Nations Development Programme). (2004) *Reducing Disaster Risk: A Challenge for Development*, Geneva: UNDP.

UNISDR (United Nations International Strategy for Disaster Reduction). (2015) *Making Development Sustainable: The Future of Disaster Risk Management. Global Assessment Report on Disaster Risk Reduction*, Geneva: UNISDR.

Wisner, B., Gaillard, JC and Kelman, I. (2012) 'Framing disaster: Theories and stories seeking to understand hazards, vulnerability, and risk', in B. Wisner, JC Gaillard and I. Kelman (eds) *Handbook of Hazards and Disaster Risk Reduction*, Abingdon: Routledge, pp. 18–33.

11

A SUSTAINABLE DEVELOPMENT SYSTEMS PERSPECTIVE ON DISASTER RISK REDUCTION INCLUDING CLIMATE CHANGE ADAPTATION

Tom R. Burns and Nora Machado des Johansson

Sustainability and Sustainable Development

Sustainability: Background and Definition

The literature on the concepts 'sustainability' and 'sustainable development' is vast (Burns 2016). These influential concepts emerged out of political and administrative processes, not scientific ones. Like the concept of development itself, sustainable development has been a contentious and contested concept, not only with respect to controversies between advocates of capitalism and those of socialism and social democracy, but between industrialised and developed countries, or between modernisation advocates and their diverse opponents. In other words, environmental issues have been added to earlier contentious issues. These have been and continue to be divisive, for instance between those who, on the one hand, advocate limiting or blocking much socio-economic development in order to protect or reclaim the environment and those who, on the other hand, stress the need of socio-economic development to alleviate poverty and inequality, if necessary at the expense of the state of the environment.

Historically, the linkage of sustainability and development has been, in large part, the result of global political and administrative processes and the diverse interests driving these processes. The term 'sustainable development' was coined as a political-administrative term to bridge differences between developed and developing countries in the context of UN negotiations and resolutions. The UN World Commission on Environment and Development (hereafter, World Commission 1987), chaired by Gro Harlam Brundtland (former Norwegian Prime Minister), produced an influential report in 1987, *Our Common Future* (World Commission 1987). The Brundtland Commission had been established by the UN in 1983 in response to growing awareness and concerns about the deterioration of the human environment and natural resources at the same time that developing countries were pushing for higher levels of economic growth (with the likelihood of increased damage to the environment). The Commission was to address the environmental challenge as it was intertwined with economic and social issues.

The Commission concerned itself with environment and growth/development as well as a number of related issues. The term 'sustainability development' (SD) was intended to build bridges between the economic, ecological, and social areas of concern. Above all, the concept was meant to refer to development that meets the needs of the present generation without compromising (or jeopardising) the ability of future generations to meet their needs (numerous other definitions have been proposed (Burns 2016); among others see Drummond and Marsden (1999), Goodland (1995), Opschoor and van der Straaten (1993), and WWF (2002)). It was intended to build bridges between, on the one hand, developed countries particularly concerned about issues of sustainability and, on the other hand, developing countries determined to industrialise and develop themselves economically and socially (World Commission 1987).

A precise *definition of sustainable development*, based on entirely technical or ecological criteria is not feasible; concepts such as 'sustainable development' and 'sustainability' are normative and political ones (Opschoor and van der Straaten 1993), much like 'democracy', 'social justice', 'equality', 'liberty', etc. rather than precise, scientific concepts. As such, they are contested and part of the struggles over the direction and speed of social, economic, and political initiatives and developments in the global context (Baker 1996; Lafferty 1995). Similarly, DRR and CCA (defined and illustrated below) are also normative concepts. This means that – unlike purely scientific concepts such as force, electric charge, negative feedback, complex social network – they provide policy orientation and a normative or moral 'force'.

Sustainability, as a normative and political concept, is used, among other things, to refer to a fair distribution of natural resources among populations of the world today as well as among different generations over time. It concerns also values and 'rights' to the existence of other species as well as notions about how much environmental capital one generation should bequeath to the next (Opschoor and van der Straaten 1993: 2). In the language of policymaking, reference is often made to the three or more pillars or fields of sustainable development: effective and enduring normatively satisfying economic functioning and prosperity, social welfare and justice, political deliberation and decision-making, and environmental protection. The most difficult challenge is to determine how one combines and balances these in a sustainable manner, particularly since under many conditions they are contradictory: economic growth may be accomplished at the cost of environmental protection and conservation, or economic growth is sacrificed for the sake of increased public welfare and distributive justice.

The concept's power and also contentiousness relates to it bringing together these apparently contradictory environmental, economic, social, and political imperatives. Harris (2001: 3) emphasises that its contestation arises not only from the emphasis placed on diverse imperatives but from the difficulties encountered in their practical combination and realisation.

The sustainability perspective calls for a complex systems model of society and the latter's multiple interactions with its physical and social environment (Burns 2016). In part, this is because sustainable development focuses our attention on multi-functionality and multiple-dimensional interactions.

Social System Model

Drawing on earlier work (Burns 2006b; Burns 2016; Burns *et al.* 1985), our point of departure is a model of complex systems (ecological as well as social) with which to characterise and analyse DRR, CCA, and SD, all of which are instances of complex systems. The model is based on a social science theory of dynamic, complex systems (Baumgartner *et al.* 2014; Burns *et al.* 1985; Burns 2006a, 2006b; Burns and Hall 2012). After introducing the theory here, we apply it in formulating models applicable to complex, multi-faceted, dynamic systems: socio-technical systems such as adaptation subsystems and disaster reduction subsystems. Such systems are not only multi-dimensional but typically have multiple vulnerabilities to diverse hazards internally or externally generated (Burns and Machado 2009). The systems are subject to perturbation/ change – due to external forces or internal processes. Their responses are usually non-linear, not fully predictable (in part because there are unintended and unanticipated impacts), and also, therefore, not fully controllable.

Consider a complex social system with a population of social actors, rule regime(s) underlying social relationships and interaction patterns, resources (materials and technologies), and multiple production processes functioning in their social and ecological contexts:

1   Our perspective entails the following dynamics patterns: Complex 'social systems' have multiple subsystems with multiple interrelated functionalities (economic, socio-cultural, educational, political, etc). In the complex system a number of DRR subsystems are likely to be included (including CCA subsystems) designed to deal with particular stressors, for example earthquakes, floods, spread of dangerous diseases (or their vectors). DRR and CAA stressors are addressed by one or several selected subsystems (in many cases, they may be largely local in character).
2   Functioning systems are subject to external and internal stresses. Some *stressors* may result in performance failing, possibly disasters. That is, failure to avoid some hazardous stressors or to regulate/control/block them results in malfunctioning and even possibly substantial damage to the system, the people involved, the property, and essential resources, in short, system disaster.
3   Our·systems approach views agents as acting within (embedded in) and also upon systems (performing in them as well as adapting and transforming them through multi-level governance arrangements). Agents in any given system (and also possibly external agents) make judgments about adaptation and risk reduction measures as well as sustainability measures. They assess performances, risks, hazards, vulnerabilities, adaptation, and sustainability needs. This entails normatively oriented judgment and risk logic(s).
4   Social systems are subject to external and/or internal stressors of varying degrees of hazard. Responses to these are in the form of built-in algorithms and procedures that are activated as well as agent-responses (through monitoring, innovating, taking actions to regulate, ban, or avoid the stressors). Multiple agents are typically involved, managers, experts, operatives who make use of appropriate materials and technologies, and operate according to a design.

The systemic model enables a robust conception of sustainability and sustainable development. As we have seen, it identifies hazards and opportunities for DRR including CCA as well as risk reduction for other challenges. DRR (including CCA as well as other systems or subsystems) entails constructions (or re-constructions) of subsystems for the purpose of response to stressors and involves such operations as 'avoiding', regulating/blocking, reducing vulnerability to them, and transforming the stressors and/or the social system (see Table 11.1).

Table 11.1 Agential and Systemic Response to Stressors

I. Powers and capabilities are mobilised to reduce exposure or contact; withdrawal, migration, temporary or seasonal retreat.

II. Powers and capabilities mobilised to partially block, regulate, countervail (for instance, building dams, walls or other barriers, or canals, or eliminating a disease through vaccination).

III. Powers and capabilities are mobilised to adapt and transform the system or its subsystems: construction of waterways and canals combined with irrigation, elimination of a population of parasites or threatening animals, or eliminating disease-vectors; i.e., drainage of mosquito breeding areas; establishing or reforming educational and research subsystems, etc.

The concerns here include restructuring or transforming a given system in order to adapt to or to overcome stressors such as climate change or other global environmental change (GEC), reducing exposure to hazards, reducing vulnerability, and reducing or repairing damages to the system or system components effectively. System adaptation/transition may then accomplish a reduction in a stressor or vulnerability to it. A systems perspective enables us to identify social structural, political, and agential factors relating to system stressors such as destruction of infrastructure, weather changes, new disease vectors, and other vulnerabilities to them – and how major agents in a system respond to these with strategies to overcome, or perhaps fail in their efforts. In particular:

- Multiple powerful agents – those controlling essential resources and expertise in the system – must be mobilised and coordinated to address the hazards and vulnerabilities (and to overcome barriers and opposition to adaptation and systemic transition).
- That is, mobilising and coordinating agents gain access to essential resources, authorities' experts, stakeholders, etc. to address system stressors.
- There are typically multiple adaptations to one or more stressors: for instance, in the case of climate change, flood control response may be combined with retrofitting buildings to be more robust or less vulnerable to moisture and mildew.

Thus, we emphasise not only the importance of agents – and their culture, capabilities, and access to resources – but also the complex social systems in which agents interact and upon which they operate. A complex social system has populations of social actors, rule regimes of social and cultural structures, resources (materials and technologies), and multiple production processes in a social and ecological context (see the representation in Figure 11.1).

The systems model enables us to systemically describe and analyse such constructed subsystems as those of DRR including CCA. These subsystem constructions were launched and developed by agents in order to deal with climate change and other system challengers. But many system functions do not relate directly to climate change or DRR. Sustainable development entails a global perspective that encompasses multiple subsystems, which are differentiated and have diverse functionalities.

Case Studies of Response to Stressors

For our purposes here we consider two types of stressors (which may overlap): disaster with the action of DRR and the hazard influencer of climate change with the action of CCA – and actors' responses to these.

DRR

Two significant instances of DRR, which we have investigated, relate to EU policymaking and regulation regarding food and chemical safety. Although these are both large sectors with multiple stressors and disaster risks, they cover only narrow fields of societal stressors and vulnerabilities.

Table 11.2 explores food safety governance as a DRR topic, through examining reduction or control of disease or disease vectors (humans, animals) related to food ('Safe Food as a Public Good') (Carson *et al.* 2009). Table 11.3 presents chemical security governance as a DRR topic, through reduction of levels and exposures to hazardous chemicals that threaten human health and the environment (Carson *et al.* 2009).

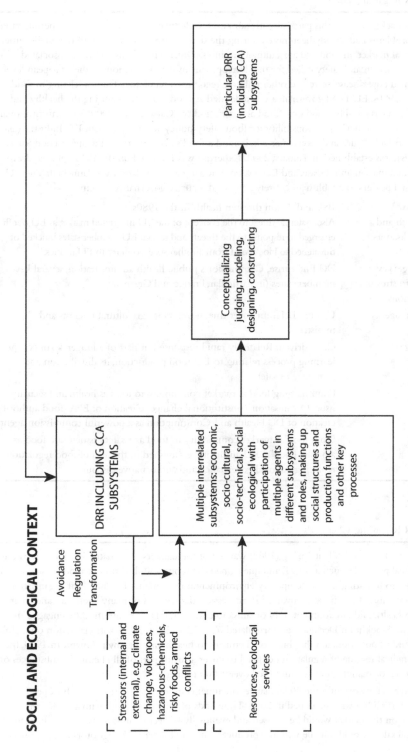

Figure 11.1 Multi-factor Model of Social System Response to Stressors

(By Authors)

Table 11.2 DRR Regarding Food

OVERVIEW: As background to this particular stressor and transformation in governance, numerous 'risks' and 'technical' problems had already been noted during the deregulation associated with the establishment of the EU 'internal market' in food and agricultural products. Diverse problems and issues associated with food contamination and safety called for the development of new regulations at the European level. Following the 'mad cow disease' scare and other publicly recognised regulatory failures relating to food in the 1980s and 1990s, EU Food governance was redefined as food safety (prioritising the healthy and safety aspects of food as a public good over 'food as a commodity' (Carson *et al.* 2009). New institutional arrangements were implemented: responsibility for food safety policy was moved from DG Industry (later DG Enterprise) to DG Health and Consumer Affairs, and a food safety authority, European Food Safety Authority (EFSA), was established in January, 2002. Expertise was recruited on the EU level to address problems of contamination and disease, and EU level monitoring and regulation was launched. The EU used its centralised powers to establish an entirely new food security governance system.

Issue or problem threatening health and environmental disasters	BSE and dioxin threaten health in the 1980s. Also, systemic threat to the concept of the EU integrated market as EU conflicts emerged in response to the threats and several EU member states blocked or threatened to block food from another state, contrary to EU treaties.
Drivers of change: key agents respond to stressors with reform actions	DG Enterprise, Commission's public health administration; several key member states (in particular, France and Germany).
Main opposition agents	UK and Belgium governments, relevant agricultural interests and ministries.
Modes of change	Crisis driven. Relatively rapid response but part of a longer-term complex learning process relating to EU food production, its distribution and guarantee of safety.
Paradigm shift	From framing food as market commodity to food as health and security issue. Organisational/institutional change: Creation of EFA (food authority), creation of DG Health and Consumption as a powerful commission agent.
Discourse shift	From food as market commodity to food as a safety issue ('safe food as a public good'). 'EU authority' established in the area of food regulation replacing national governance and authoritative discourses.

Table 11.3 DRR Regarding Chemicals

OVERVIEW: Rachel Carson's Silent Spring (1962) warned of the dangers to all natural systems from the misuse of chemical pesticides such as DDT, and questioned the scope and direction of modern science; the work contributed to initiating the contemporary environmental movement. Thousands of organisations were over time established and became involved in addressing the dangers of many chemicals and their threat to human health and environment. UN organisations and OECD were significantly engaged in development. The 'Stockholm Declaration', formulated in 1972, called for international action on a global scale against chemical pollution and the threat of chemicals to health and the environment. In 1990, the EU launched a radical regime of regulation designed to cover a major part of all chemicals (substances of either high concern or manufactured or imported over 1,000 tonnes).

The law replaced a patchwork of some 40 legislative instruments in force at the time. The European Chemical Agency (ECHA) was established in 2007. Thousands of substances (approximately 30,000 expected) already on the market would be assessed and eventually subject to authorisation. The burden of proof of chemical safety would now lie with the producers, and no longer with regulators or consumers.

Dangerous chemicals could be readily banned – or if their value to society were judged to be very high they could be allowed but under very tight controls, and a safer alternative would be expected in the future. The REACH law (Registration, Evaluation, Authorization, and restriction of CHemicals) came into effect in June, 2007 following a process involving almost a decade of debate, mobilisation, struggle, and negotiations between the Commission, the Council of the European Union, the EU Parliament, industry, labour unions, environmental, consumer and animal rights NGOs, and others (see below).

Issue or problem threatening health and environmental disasters including the 'silent hazards' such as in the case of asbestos	Many chemical threats to human health and environmental health not sufficiently recognised and regulated.
Drivers of change: key agents respond to stressors with reform actions	Change in international context with the emergence of influential NGOs and environmental movements (even 'political' such as the Green parties). Several new EU member states such as Sweden, Austria and Finland prioritised a stringent EU regulation of chemicals. DG environment led by a major Swedish politician (Margot Wallström) and the environmental movement (Green political parties and NGOs) pressed for far-reaching legislation.
Main opposition agents	European chemical industry (and the American and Japanese industries). UK, French, and German top political leaderships. The USA also lobbied against, in particular, EU member states themselves.
Modes of change	Attention increasingly focused on widespread risks (chemicals in the environment, in human bodies, in pregnant women, in children) and on European inconsistencies in regulating some chemicals and materials such as asbestos. Scientific indications and growing public awareness that only a few of the most dangerous chemicals were recognised and adequately recognised. Scientific and technical research along with legal preparations were launched for far-reaching legislation, policy shifts, and institutional innovation and development.
Paradigm shift	From relatively low and sporadic regulation to a system where all significant chemicals would be assessed for safety. Creation of ECHA to monitor and sanction compliance. The burden of proof of safety or low risk is the responsibility of the producers, not the consumers or government. Costs of determining low risk fall upon the producers and suppliers. Chemicals deemed unsafe for humans and the environment are subject to ban. Some exceptions may be made temporarily.
Discourse shift	From chemicals as all 'good' to chemicals as potentially health and environmental hazards and requiring 'systematic regulation' – as opposed to 'market freedom'. European integrated authority to replace member-state authority.

Key agents in the cases mobilised sufficient powers to impose safety controls, to regulate, and even to ban and thus successfully reduce, disaster risks. Note the multi-level character of these two cases. Global EU legislation – and monitoring; while implementation takes place at the member-state or national level.

CCA

Table 11.4 illustrates the pertinence of placing CCA within wider sustainability contexts.

Table 11.4 CCA Extended to Sustainability Issues in the Gothenburg Metropolitan Area

OVERVIEW (Lundqvist 2016): Climate change in the Gothenburg area threatens increased flooding, slides, erosion, and new diseases and disease vectors. Gothenburg Metropolitan area (GMA) – expected to be the part of Sweden most affected by the impacts of climate change – has been addressing and acting upon relevant issues of climate change adaptation. This has been done within the framework of Sweden's Planning and Building act which puts responsibility with local governments for dealing in the 'common interest' of climate adaptation

GMA has launched a variety of policies and institutional response to actual and potential climate change stressors. In the course of conducting investigations and launching projects GMA developed a perspective and strategy for an environmental – and a sustainable development – programme as opposed to purely climate programmes. GMA emphasises that climate work needs to be coordinated and guided toward long-term sustainability; climate needs to be considered in the context of multiple environmental, economic, and social challenges and diverse measures to deal with them.

Issue or problem threatening health and environmental disasters	Examples of climate change stressors are changed rain intensities, altered precipitation patterns such as clustering or longer rain duration, runoff following snowmelt in combination with rain, resulting in increased flooding, slides, and erosion. There have also emerged new diseases and disease vectors.
Drivers of change: key agents respond to stressors with reform actions	Municipal political leadership, public energy companies, insurance companies, research communities, Green Party, Green NGOs (Greenpeace and local environmental groups).
Main opposition agents	After considerable flooding experience, there was very little opposition in GMA. But typically municipalities in the region compete for resources. Issues of administrative jurisdiction and resource scarcities and competition for resources are key 'opposition' or constraining factors.
Modes of change	Disaster driven initially; but also reports and experiences of disasters from similar areas in Sweden and abroad. Examples of climate change adaptation are the introduction and development of infrastructures to reduce associated peak flows and flooding consequences in the local environment, development of new designs against moisture and flooding for built environments, development of more effective urban drainage systems, particularly since on the Swedish West Coast (Gothenburg and Malmo) higher sea levels are likely to be combined with heavier rainfalls resulting in flooding of coastal and low land areas, with probable damage to infrastructures and buildings and threat to human life.
	Among other things, this entailed changing standards of expected flood levels and developing plans and programmes to deal with expected impacts such as flooding, erosion, and slides as well as loss or contamination of freshwater supply. In restructuring waterways and flood barriers, there is both climate change adaptation and disaster risk reduction; regulation of water levels of the three major rivers is also both CCA and DRR.

Broad comprehensive programmes, integrated strategies, characterise Gothenburg's response to the disaster threats of climate change (DRR). In addition, there are pressures for regional 'socio-economic growth', protection of coastal areas for recreation and leisure, promotion of GMA's role as the logistic centre of the Nordic Region. There are also many voices calling for plans for the harbours, industrial and business activities, and shaping an 'attractive city environment'. Gothenburg is developing energy efficiency strategies and renewable energy production. Also, there are groups conducting research on issues, strategies, and consequences of having non-sustainable institutions and practices and determining alternative sustainable systems and their development.

To effectively avert the impact of flooding, slides and erosion and to secure GMA's freshwater supply, ultimately governance and coordination arrangements have been needed to cut across two nations, four regions and a substantial number of local governments. This compounds problems of gaining acceptance and legitimacy for large-scale plans to adapt appropriately to the impact of climate change (Lundqvist 2016).

Paradigm shift

Gothenburg has developed a perspective and strategy for an environmental and sustainable development programme as opposed to a purely climate change programme. The city is stressing that climate work needs to be coordinated and guided toward long-term sustainability; climate needs to be considered in the context of multiple environmental, economic, and socio-political challenges and the measures to deal with them. Issues of a sustainable economy for GMA, education (for instance, sustainable conscious consumption), relevant research, sustainable water supply regulation, and new transportation conceptions. The aim is to develop a conceptualisation and institutional order for sustainability relevant to GMA.

Discourse shift

Discourses emerged about climate change and its meaning for MGA in response to increasingly serious flooding and erosion problems. But over time the discourses have not only been about CCA but about sustainability issues: the future of the local economy, energy, transport, water supply regulation, building and infrastructure, education, consumption, social planning, among others.

'Sustainable Gothenburg' has become a key normative public concept.

Sustainable Development as a Societal Development System

CCA and DRR are Only Partial Systems

Sustainability judgments, designs, institutional arrangements, and developments would optimally include (or embed), among other things, appropriate and satisfactory CCA as well as appropriate and satisfactory DRR. System sustainability is multi-dimensional and with defined standards that surpass 'sufficient adaptation' or 'sufficient disaster risk reduction'. As suggested above, the latter are only partial. In contrast to sustainable development arrangements where multiple functionalities and the interrelationships of multiple 'pillars' are built in conceptually and practically, CCA and DRR models tend to leave out or leave implicit a number of the key functionalities, for instance, agential aspects (the training and education, the necessary professionalism, the reliable appropriate materials and technologies, the sustained over time budgeting (private and public)). These must be included in a broad or systemic perspective, as in sustainability models, in ways that are not inherent to DRR or CCA subsystems.

In the conceptualization of CCA and DRR, the economic, socio-cultural, and political bases for financing and operating the systems are typically taken for granted or simply ignored, while the main focus is on technical and cost minimization issues. Sustainability's multi-functionality approach, in contrast, typically takes explicitly into account sustainable budgeting levels, the political support and legitimation of policies and key societal actors, and the education of experts to manage and operate CCA and DRR subsystems.

The argument here suggests that 'necessary' CCA or DRR arrangements may or may not be sustainable over the medium to long run. This may be for economic reasons, ecological/natural resource constraints, or socio-cultural forces (refusal of the population to sufficiently change life-style or make necessary institutional and economic changes; e.g., replacing fossil fuels with renewable energy sources and controlling pollution and environmental degradation, among others).

Although CCA and DRR are only partial systems, they relate to one another in systemic terms, with CCA embedded within DRR. But does that assist in connecting them with sustainable development?

Possible Incompatibility of CCA and DRR Systems with
Overall Social System Sustainability

As suggested above, DRR and CCA may not be compatible with sustainable development, in that their scope may be far too narrow, ignoring a number of key dimensions of sustainable development in the social systemic context such as increased reliance on renewable energy, biodiversity support, and chemical pollution reduction. At the same time, policymaking agents may accept disaster risks (and reject adapting to climate change or other threats because of lack of resources, or poor judgment of cost/benefit imbalances or priorities elsewhere).

Or, as pointed out below, it may be a matter of prioritisation or practices of risk-taking. In responding to stressor(s), agents may prioritise broad sustainability over particular disaster risks including those influenced by climate change, like sacrificing a leg for survival of the rest of the body. Or vice versa, one may risk the social system as a whole to save an important part of the system; for instance, the system's elite or its particular social structure (e.g., an apartheid system). Or, for the sake of technological advancement, one takes unknown risks with Genetically Modified Organisms in the context of 100 per cent sustainable food supply. In these cases, CCA could be said to trump sustainability from the perspective of a high risk-taker. Thus, one may succeed in accomplishing DRR including CCA without achieving or maintaining medium- to long-run sustainability. Consequently, successful DRR (possibly involving CCA, or not) may accomplish (immediate objective(s)) but may be unsustainable over time or over an extended space. Unsustainability may be caused by (1) insufficient resources (including technologies); (2) agents' insufficient knowledge and/or capabilities; (3) insufficient coordination and effective governance (organisational); and (4) insufficient socio-political legitimacy.

DRR Including CCA Should Be Subjected to Sustainability
Standards, Discourses, and Judgments

The DRR subsystem, including the CCA subsystem, is judged to be sustainable if and only if (1) essential inputs into these subsystems are sustainable (or replaceable with sustainable factors when used up); (2) subsystem outputs do not undermine, block, or interfere with the sustainability of other subsystems or with the social system as a whole. That is, a given subsystem in its interactions with other subsystems may produce disaster risks (or increases in disaster risks) or developments that undermine, block, or interfere with sustainability conditions and mechanisms.

In general, from a holistic, societal perspective, DRR including CCA can be viewed and assessed in relation to the societal multi-functionality and long time scale of sustainable development. Sustainability incorporates capacities to deal with such potential systemic disturbances as climate change (as well as other hazardous global changes), either through prevention or avoidance and/or adaptation.

Several key points in this part are: (1) One may accomplish risk reduction which in the medium to long run is not sustainable in terms of ecological, economic, or socio-political dimensions; (2) One may also have adaptation with respect to a stressor X, for instance climate change, without system or even subsystem sustainability; (3) DRR including CCA may be pursued separately or pursued as part of an overall initiative to adapt to particular climate change impacts in such a way as to reduce disaster risk; (4) DRR including CCA – while important sources of conceptual and methodological tools – may lead to non-sustainable judgments and developments. In other words, improvements in DRR including CCA may not be sustainable over the long run because of their unintended or unanticipated impacts on key sustainability dimensions; (5) Note that the systems model alerts us and enables us to analytically differentiate DRR, CCA, and sustainability development. It is argued here that DRR including CCA should and can be related to the medium- and long-term goals of sustainable development; and (6) In general, the factor of sustainability trumps DRR including CCA in terms of impact assessment and design, but there are other considerations as discussed below.

Distinctions between Sustainable and Non-Sustainable DRR Including CCA

DRR for a particular hazard (e.g., volcanic eruption, earthquake, soil erosion, salination of the soil) may or may not be sustainable in the long run. It may be unsustainable because, for instance:

- The DRR action or socio-technical system makes use of non-renewable, non-replaceable resources.
- The DRR system blocks necessary adaptation of the more encompassing social system, for instance by draining away resources essential to the sustainability of the larger system. As a result of this systemic non-sustainability, the DRR subsystem will fail to be maintained over the medium to long run.
- In the prevailing socio-cultural and economic perspective, people are unwilling to make the sacrifices for DRR, for instance they are willing to accept the risk of a volcano or earthquake, or the rising seas.

In sum, sustainability developments, judgments, designs, and institutional arrangements may include (or embed), among other things, appropriate and satisfactory CCA placed within appropriate and satisfactory DRR. System sustainability is multi-dimensional and with defined limits that surpass 'sufficient adaptation' or sufficient DRR. The latter are only partial.

On the other hand, a 'necessary' CCA or DRR infrastructure may or may not be sustainable over the medium to long run. This may be for economic reasons, ecological/natural resource constraints, or socio-cultural forces (refusal of the population to sufficiently change life-style or to make necessary institutional and economic changes (for instance, shifting from fossil fuels to renewable energy sources)).

Concluding Remarks

The sustainability or sustainable development perspective provides a systemic view, indicating the complex interdependencies among essential subsystems to human welfare and survival. Societal sectors or specialised subsystems like DRR or the CCA subsystem within it are insufficiently comprehensive when viewed from the perspective of a sustainability model; they do not explicitly take into account the systemic economic, socio-cultural, and political factors in societal survival and even the question of the availability of long-term resources for CCA and DRR subsystems themselves (resource availability and sufficient prioritisation of the resources in competition with other demands).

The conception of sustainable development orients us not only to ecological factors and the interaction of the ecological, and the social. It also orients us to the economic conditions and processes, the essential conditions for financing and maintaining subsystems or programmes of DRR including CCA; and it orients us to consider socio-cultural conditions essential to policy legitimation and the maintenance of societal support and a readiness to aim for wider sustainability. Sustainable development is concerned about social systems as a whole and the long timescale. A complex systems perspective encompasses then not just functionality and simple cost–benefit issues but multi-functional and multi-dimensional sustainability: it considers that a subsystem may contribute not only to ecological sustainability but to the economy, to cooperation and coordination, and other governance mechanisms as well as socio-cultural integration, legitimacy, and political order.

And, as we have seen, a system may be sustainable and include its climate change adaptability (as well as perhaps other environmental change adaptability) within its DRR capacity as well as DRR capacity beyond CCA. But, as indicated earlier, the designs, realisations and practices of CCA and DRR subsystems should be incorporated into a sustainable development model, its standards/ values, discourses, plans, and institutional arrangements. Failure to do this would mean, in general, that DRR subsystems including CCA – and the social system in which they are embedded – may not be sustainable or only precariously sustainable (vulnerable to particular internal and/or external stressors). Thus, as discussed earlier in the chapter, a well-functioning DRR subsystem including a well-functioning CCA subsystem may be non-sustainable for long-term economic reasons, or for socio-cultural or political reasons; or indirectly, the DRR subsystem (or CCA subsystem) drains away essential resources or interferes with key systemic mechanisms essential to global system sustainability.

Part of sustainable development discourses entails exploring alternative DRR subsystems, within which could sit alternative CCA subsystems, namely finding or developing DRR including CCA subsystems that are more likely to be sustainable. They would be designed and assessed so as to take into account economic, socio-cultural, and political dimensions, increasing systemic chances of medium- to long-term sustainability.

References

Baker, S. (1996) 'The evolution of EU environmental Policy: From growth to sustainable development', in S. Baker, M. Kousis, D. Richardson and S.C. Young (eds), *The Politics of Sustainable Development: Theory, Policy and Practice within the European Union*, London: Routledge, pp. 91–106.

Baumgartner, T., Burns, T.R. and DeVille, P. (2014) *The Shaping of Socio-economic Systems*, New York and Oxford: Routledge.

Burns, T.R. (2006a) 'System theories', in *The Blackwell Encyclopedia of Sociology*, Oxford: Blackwell Publishing, pp. 4922–4928.

Burns, T.R. (2006b) 'The sociology of complex systems: An overview of actor-systems-dynamics', *World Futures: The Journal of General Evolution* 62: 411–460.

Burns, T.R. (2016) 'Sustainable development: Agents, systems and the environment', *Current Sociology* 64(6): 875–906.

Burns, T.R., Baumgartner, T. and DeVille, P. (1985) *Man, Decision and Society*, London: Gordon and Breach.

Burns, T.R. and Hall, P. (2012) *The Meta-power Paradigm: Causalities, Mechanisms, & Constructions*, Frankfurt, Berlin and Oxford: Peter Lang.

Burns, T.R. and Machado, N. (2009) 'Technology, complexity, and risk: Part I: Social systems analysis of risky sociotechnical systems and the likelihood of accidents', *Sociologia, Problemas e Práticas* 61: 11–40.

Carson, R. (1962) *Silent Spring*, New York: Houghton Mifflin Company.

Carson, M., Burns, T.R. and Calvo, D. (eds) (2009) *Public Policy Paradigms: Theory and Practice of Paradigms Shifts in the EU*, Frankfurt, Berlin and Oxford: Peter Lang.

Drummond, I. and Marsden, T. (1999) *Sustainable Development: The Impasse and beyond from the Condition of Sustainability*, London: Routledge.

Goodland, R. (1995) 'The concept of sustainable development', *Annual Review of Ecological Systems* 26: 1–24.

Harris, J.M. (2001) 'Basic principles of sustainable development', *The Encyclopedia of Life Support Systems*, Paris: UNESCO. Online http://ase.tufts.edu/gdae (accessed 3 September 2016)

Lafferty, W.M. (1995) 'The implementation of sustainable development in the European Union', in J. Lovenduski and J. Stanyer (eds), *Contemporary Political Studies*, Proceedings of the Political Studies Association (PSA), Vol. 1, Belfast: PSA, pp. 223–232.

Lundqvist, L. (2016) 'Planning for climate change adaptation in a multi-level context: The Gothenburg metropolitan area', *European Planning Studies* 24, 1: 1–20.

Opschoor, J. and Van der Straaten, J. (1993) 'Sustainable Development: An Institutionalist Approach'. Online: www.geocities.ws/saxifraga_2000/costanza.doc (accessed 3 September 2016).

World Commission on Environment and Development. (1987) *Our Common Future*, New York: United Nations.

WWF (World Wide Fund for Nature). (2002) *Living Planet Report 2002*, Gland, Switzerland: WWF International.

12

ECOSYSTEMS' ROLE IN BRIDGING DISASTER RISK REDUCTION AND CLIMATE CHANGE ADAPTATION

Nathalie Doswald, Marisol Estrella, and Karen Sudmeier-Rieux

Introduction and Rationale

Ecosystem-based disaster risk reduction (Eco-DRR) and ecosystem-based adaptation (EbA) are increasingly recognized as effective approaches to reducing disaster risk, including supporting adaptation, through the sustainable use, conservation and restoration of ecosystems as natural buffers to hazards. This recognition is evidenced by increased numbers of field projects, policies at the national and global levels and publications on Eco-DRR and EbA. One of the reasons for the increased interest and mainstreaming of ecosystem-based approaches is that they are often applied to both disaster risk reduction (DRR) and climate change adaptation (CCA). Ecosystems and their services are increasingly recognised as important to both DRR and CCA, and environmental degradation is clearly a leading driver in increasing disaster risk (IPCC 2014; UNISDR 2015). This chapter recognises that climate change should be considered as a contemporary 'creeping environmental change' alongside other important environmental drivers of risk (Kelman and Gaillard 2010). It also highlights that at the community level, distinctions between DRR, CCA, Eco-DRR and EbA are not relevant; rather, it is the sustainable utilisation of resources to sustain livelihoods and maximize human security which is important. However, at the global and national policy levels because DRR and CCA have been operating in different institutional and policy spheres, so have Eco-DRR and EbA practices evolved in parallel tracks, despite their common ecosystem-based approaches. Whilst there are strong arguments for embedding CCA within DRR as outlined in other chapters of this volume, Eco-DRR and EbA remain two strong parallel tracks, with different constituents, funding sources and spheres of influence. Yet ecosystems are the basis for DRR and CCA, and as we outline in this chapter, the similarities are much greater than the differences, providing a strong case for bridging the divide between the two.

Eco-DRR and EbA are essentially about promoting sustainable development in hazard-prone areas. Because ecosystem degradation is accompanied by increasing risks, costs, and poverty for many people, the sustainable management of ecosystems is essential to supporting livelihoods for present and future generations (Renaud *et al.* 2013). Against this background, Eco-DRR and EbA approaches comprise much more than just preserving or restoring ecosystems, such as mangroves or dune systems. Rather, Eco-DRR and EbA approaches are essential components of integrated land and water management with the overall goal of reducing disaster risk and supporting sustainable development. However, all approaches have their limits, and ecosystems, especially when degraded, may not be able to protect against all hazards (Sudmeier-Rieux *et al.* 2013). Ecosystem-based approaches need to be based on rigorous understanding of local ecological conditions, socio-cultural and economic circumstances, existing hazards, and technical requirements of the intervention (Estrella and Saalismaa 2013). This chapter is based on the 2015 study, 'Promoting Ecosystems for Disaster Reduction and Climate Change Adaptation: Opportunities for Integration' (Doswald and Estrella 2015), which provides a review of Eco-DRR and EbA projects. The chapter reflects on how Eco-DRR and EbA have become mainstreamed into global frameworks, and will offer perspectives on further integrating DRR and CCA policies and practices.

The Role of Ecosystems in DRR and CCA

Ecosystems and ecosystem services are central, though not primary in the policy sphere, to the discussion of DRR and CCA. Ecosystem services are the benefits people obtain from ecosystems, which have been classified by the Millennium Ecosystem Assessment (2005) as: supporting services, such as seed dispersal and soil formation; regulating services, such as carbon sequestration, climate regulation, water regulation and filtration, and pest control; provisioning services, such as supply of food, fibre, timber and water; and cultural services, such as recreational experiences, education and spiritual enrichment.

It has been demonstrated that ecosystem services can be used for DRR and CCA (CBD 2009; IPCC 2014; Munang *et al.* 2013; Renaud *et al.* 2013). For example, forests provide flood and landslide regulation services, a phenomenon that is harnessed in watershed management programmes (Doswald and Osti 2011; Renaud *et al.* 2013). Coastal mangroves have been shown to protect adjacent areas from storm surges, although not in all circumstances (Badola and Hussain 2005; Spalding *et al.* 2014). Nevertheless, ecosystems are vulnerable to current anthropogenic pressures and are being degraded (Millennium Ecosystem Assessment 2005). The capacity of ecosystems to provide these services may be further undermined by climate change or hazard impacts, as well as by potentially unsustainable DRR or CCA measures. Strategic management of ecosystems is necessary to ensure provision of services that are important to society in the face of climate change and other hazards.

Ecosystem-based Disaster Risk Reduction

Eco-DRR emanated from a DRR perspective, with the recognition of ecosystems as 'natural infrastructure' that can assist in protecting people and communities against hazard impacts. The operational principle of Eco-DRR is relatively simple – it entails combining sustainable ecosystem management approaches with DRR methods, such as early warning systems, disaster preparedness and emergency planning to reduce disaster impact upon people and communities, and support disaster recovery. Eco-DRR aims to manage the environment in such a way that risk to communities is reduced (Estrella and Saalismaa 2013). Hence, Eco-DRR is 'the sustainable management, conservation and restoration of ecosystems to reduce disaster risk, with the aim to achieve sustainable and resilient development' (Estrella and Saalismaa 2013: 30). Well-managed ecosystems, such as wetlands, forests and coastal systems, act as natural infrastructure, reducing physical exposure to many hazards and increasing socio-economic resilience of people and communities by sustaining local livelihoods and providing essential natural resources, such as food, water and building materials (Sudmeier-Rieux *et al.* 2013; IPCC 2014). In many cases, Eco-DRR focuses more on the immediate threats caused by hazard events and encourages use of tools such as risk, vulnerability and hazard assessments and mapping which include an ecosystem component. In parallel, it also encourages long-term planning tools, such as risk-sensitive land use planning, which consider the combination of environmental planning and ecosystem management together with DRR (Sudmeier-Rieux *et al.* 2014). Eco-DRR awareness was elevated by several international frameworks in 2014 and 2015, which either passed resolutions on Eco-DRR and/or EbA, or directly recognised their importance (see section on global policies below). Until recently, and in contrast to EbA, Eco-DRR had not yet received significant attention in DRR policy contexts.

Ecosystem-based Adaptation

EbA emerged simultaneously in international climate policy platforms as a 'new' approach, although nothing is substantively different from the long-standing ecosystem-based management discourse. According to the Convention on Biological Diversity (CBD), EbA involves the use of biodiversity and ecosystem services through sustainable management, conservation and restoration of ecosystems, to help people adapt to the adverse effects of climate change (CBD 2009, 2010). EbA was introduced to the international policy arena by the conservation community in the context of CCA negotiations during the United Nations Framework Convention on Climate Change (UNFCCC) processes. EbA is currently gaining more widespread interest in the climate change policy arena, with inclusion in the Intergovernmental Panel on Climate Change Fifth Assessment Report, and with the production of case studies, research, guidelines and tools including under the Nairobi Work Programme on Adaptation. EbA is not mentioned directly in any agreement under the UNFCCC aside from a decision to hold a technical workshop on EbA. The UNFCCC also has a database on projects, which complement an information paper (FCCC/SBSTA/2011/INF.8). However, EbA is defined and outlined within decision X/33 of the Convention on Biological Diversity.

Similarities and Differences Between EbA and Eco-DRR

EbA is an outgrowth of a long history of using environmental management to adapt to climatic variations and reduce risks from natural hazards (Doswald *et al.* 2014). EbA projects primarily focus on maintaining and increasing the resilience of biodiversity and ecosystem services as a way to help people adapt. It takes into account the notion of uncertainty and long-term climate variability more so than Eco-DRR. EbA projects also have a greater focus on water management, and generally combine livelihoods and agricultural concerns with natural resources management (Doswald and Estrella 2015), while Eco-DRR projects include DRR components mentioned above, on which EbA does not usually focus. Increasingly however, EbA projects incorporate DRR components, particularly early warning and disaster preparedness (Doswald and Estrella 2015).

In many cases, there is a focus on ecosystems in relation to addressing climate-related hazards as well as climate change. This is so because ecosystem-based approaches are not widely applied for non-climatic hazards or geohazards, such as earthquakes or volcanic eruptions, although several studies have shown how re-vegetation and forest management can sometimes reduce risk of rock falls or landslides triggered by earthquakes (e.g., in the case of protection forests in Switzerland; see also Peduzzi 2010). Tsunamis are another non-climatic hazard that has sparked a range of research surrounding the effectiveness of coastal vegetation in absorbing such waves. A general conclusion from the literature is that this will depend on the width and health of the coastal vegetation as compared to the strength of the tsunami wave (Tanaka *et al.* 2013; Vo-Luong and Massel 2008).

In general, Eco-DRR and EbA projects tend to follow the main similarities and differences between DRR and CCA practices. Table 12.1 shows a summary of the key similarities and differences found between Eco-DRR and EbA in policy and practice.

Table 12.1 Key Differences and Similarities Between Eco-DRR and EbA (Doswald and Estrella 2015)

Differences and similarities		Points of convergence
Eco-DRR	EbA	
Usually adopts UNISDR terminology in defining disaster risk (as a function of hazard, exposure, and vulnerability).	Usually adopts UNFCCC terminology in defining vulnerability (as a function of sensitivity, exposure, and adaptive capacity).	Greater convergence towards adopting common terminologies.
Deals with climate-related hazards including climate change, but also non-climate hazards such as tsunamis, earthquakes, avalanches and rockfall.	Deals with climate-related hazards, but also deals with climate change impacts, including sea level rise, glacial lake outbursts, and broad changes to temperature and rainfall patterns.	Most Eco-DRR and EbA projects deal with water- and climate-related hazards; Eco-DRR increasingly factoring in climate change impacts.
Aims to 'reduce disaster risk', 'increase protection and resilience against hazards'.	Aims to 'reduce vulnerability', 'increase resilience to climate change', 'undertake appropriate adaptation'.	Key differences in stated aims are purely semantics in how terminology is being used. Both Eco-DRR and EbA emphasize the multiple benefits of ecosystem services, including for sustainable livelihoods.
Conducts disaster risk assessments (DRA), usually starting with a focus on hazards, exposure and vulnerabilities as core elements to understanding disaster risk, but also assessing linkages to environmental conditions and natural resource management.	Conducts vulnerability assessments (VA), usually starting with an ecosystem focus (e.g., impact of climate change on biodiversity loss and ecosystem integrity), and developing future change scenarios.	Both seek to incorporate ecosystems and environmental factors within their assessment frameworks; with growing appreciation in Eco-DRR to incorporate future climate trends. But given difficulties in determining future climate change projections, especially at a field/local level, both Eco-DRR and EbA projects tend to rely on examining past and current risks, a key characteristic of DRR practice.
Implementation approach – Less focus on biodiversity conservation and protection as a primary aim; focus is on optimizing ecosystem services for increasing resilience of people or reducing exposure and vulnerability to hazard impacts.	Implementation approach – Greater emphasis (but not always) on the health status of ecosystems and their services, and on biodiversity conservation; focus on maintaining and increasing resilience of biodiversity and ecosystem services to enable people to adapt to climate change impacts.	Both apply sustainable ecosystem management principles and utilize a common set of tools and approaches, such as: integrated water resource management (IWRM), integrated coastal zone management (ICZM), protected area management, drylands management, among others.

(Continued)

Table 12.1 Continued

Differences and similarities		Points of convergence
Eco-DRR	*EbA*	
Typically incorporates other key aspects of disaster risk management, such as establishing early warning systems and undertaking disaster preparedness.	Emphasis is on strengthening 'adaptive management' due to uncertainty of climate change impacts.	Both incorporate disaster preparedness/mitigation measures, including early warning systems.
Less attention given to monitoring and evaluation, apart from standard project reporting requirements.	Active discussions on developing monitoring and evaluation frameworks and guidelines for EbA/CCA projects.	Both face challenges of attribution in evaluating effectiveness and impacts through an ecosystem-based approach. Little attention overall given to developing indicators for EbA and Eco-DRR projects.
Actors involved – Typically involve environmental agencies/ ministries, conservation NGOs but also humanitarian and disaster management actors at local and national levels, as well as climate change focal points.	Actors involved – Typically involve environmental agencies/ministries, conservation NGOs, climate change national focal points; usually does not engage with humanitarian or disaster management actors.	Both increasingly recognize the importance of bringing together different communities and sectors, including from disaster management, climate change, environment and other key sectors (e.g., water, agriculture).
Policy advocacy can target a broad range of policies, including climate change adaptation strategies, environmental policies, and other sectoral policies (e.g., water and agriculture).	Policy advocacy generally focuses on the national adaptation strategy as well as other development policy sectors affected by climate change (e.g., water); rarely works on DRR-related policies.	Both typically engage with the environmental ministries/ agencies and the conservation community, but still with a tendency to operate in separate policy tracks, depending on whether the project is more oriented towards DRR or CCA.

One of the strongest rationales for using ecosystem-based approaches within DRR and CCA is that, in addition to reducing hazard impacts, ecosystem-based approaches provide multiple social, economic and cultural benefits for communities (Doswald *et al.* 2014). For example, agroforestry mitigates erosion, improves water balance and regulates the micro-climate, whilst providing communities with essential as well as cash resources (Vignola *et al.* 2015). Protecting areas for floodplains or forests also provides recreation areas for human well-being and tourism. Multiple benefits derived from ecosystem services are especially effective in terms of CCA, as successful adaptation needs to be undertaken in a multi-faceted and integrated manner, thus strengthening long-term resilience of communities (Doswald *et al.* 2014).

Resilience refers to the capacity of people to bounce forward after a shock (Manyena *et al.* 2011). A number of case studies exist that outline the benefits of ecosystem-based approaches, especially with respect to CCA (i.e., EbA) (Andrade Pérez *et al.* 2010; Doswald and Osti 2011; Doswald *et al.* 2014). Furthermore, studies show that EbA is mainstreamed within many sectors (e.g., in coastal protection, agriculture and forestry, urban areas) albeit the term EbA is not used (Doswald and Osti 2011). It is worth pointing out, however, that there is often a cross-over between the use of EbA and Eco-DRR terms in the case studies (Doswald and Osti 2011; Renaud *et al.* 2013). Interest from the climate change arena is one of the reasons why case studies have been subsequently 'labelled' as EbA rather than Eco-DRR. In summary, it is clear that both types of ecosystem-based approach are strong contributions to more sustainable, long-term reduction in disaster risk, including climate change as an additional aggravating factor.

Ecosystem-based Approaches to DRR and CCA at the Project Level

Projects involving some form of EbA, Eco-DRR or a mixture of both have been undertaken or are currently underway in many countries worldwide. Using ecosystem management, restoration and conservation to reduce vulnerability to climatic hazards is not new (Doswald *et al.* 2014). In recent years, projects dedicated to applying ecosystem-based approaches specifically for adaptation or DRR (or for both purposes) have mushroomed in a disjointed way, stemming from either Eco-DRR proponents within DRR communities or EbA proponents within CCA communities with little, if any, collaboration or integration. This has led to confusion and overlaps between the two fields of practice, missing out on the benefits which could be gained from a more informed, and integrated practice. In some cases, the lack of DRR/CCA integration could potentially lead to mal-adaptation or increased disaster risk by not taking into account the linkages between development, climate change and multiple hazards.

Boxes 12.1 and 12.2 provide concrete examples of Eco-DRR projects and highlight that both Eco-DRR/EbA projects often seek to address multiple objectives, usually including a livelihood component. In many cases, such projects are based on combining natural resource management approaches, such as IWRM, with DRR actions such as early warning systems and CCA actions such as anticipating and planning for climate uncertainty.

Box 12.1   Environment and Natural Resources Thematic Programme (ENRTP) Project in Haiti

Marisol Estrella[1]

[1] United Nations Environment Programme, Geneva, Switzerland

An example of an Eco-DRR project is the Environment and Natural Resources Thematic Programme (ENRTP) project in Haiti, where the United Nations Environment Programme (UNEP) is working with the Government of Haiti on strengthening the marine protected area management to maintain healthy, coastal, and marine ecosystems for two main objectives: disaster resilience and sustainable livelihoods and diversification. The project established a coastal nursery with 137,000 seedlings (of fruit, forest, mangrove (Figure 12.1), and seagrape tree species), and carried out re-vegetation and reforestation activities with community residents, covering a total area of 141 hectares. The project also trained the local fishers association on early warning and safety procedures at sea, in the event of hurricanes or tropical storms.

Figure 12.1 Mangroves in Port Salut, Haiti

(Photo by UNEP)

Box 12.2 Ecosystems Protecting Infrastructure and Communities (EPIC) Project in Nepal

Karen Sudmeier-Rieux[1]

[1] University of Lausanne, Lausanne, Switzerland

The Ecosystems Protecting Infrastructure and Communities (EPIC) project in Nepal is being implemented by the International Union for Conservation of Nature (IUCN) and the University of Lausanne, Switzerland and focuses on reducing economic and physical damages from rural road construction in mountainous areas (Figure 12.2). Three sites were established in the Panchase area of Western Nepal to demonstrate the effectiveness of low-cost and community-based, soil bioengineering (e.g., use of deep-rooted plants together with small drainage works) for 'eco-safe roads'. Community-supported research is being conducted to determine the most drought-resistant species, considering climate change impacts, while providing the best root systems and livelihood opportunities. At the national level, the project is working with the Government of Nepal to support their integrated water resource management (IWRM) approaches, which are of special interest in light of the 2015 Nepal earthquake.

Figure 12.2 Bio-engineering, Ecosystems Protecting Infrastructure and Communities (EPIC) Project, Nepal

(Photo by Karen Sudmeier-Rieux)

Bridging the gap between DRR and CCA at the local level, understanding how EbA, Eco-DRR and hybrid Eco-DRR/EbA projects have been and are being undertaken, can facilitate improved DRR-CCA integration in practice, as well as in policy and institutional contexts. The Doswald and Estrella (2015) study undertook a global review of 38 projects (15 EbA projects, 12 Eco-DRR projects and 11 Eco-DRR/CCA hybrid projects). Eco-DRR and EbA projects were analysed according to each phase of the project implementation cycle:

1   Defining aims of the project;
2   Conducting risk and vulnerability assessments;
3   Project implementation: methods, approaches, tools; and
4   Monitoring and evaluation.

The project review showed that, in practice, Eco-DRR and EbA have much more in common than they are different, primarily because of the sustainable ecosystem management approach that is central to both. Furthermore, there exist ecosystem-based 'hybrid projects' that integrate DRR and CCA. Yet, due to the largely different policy and institutional contexts of CCA and DRR, EbA and Eco-DRR also tend to operate in separate silos. Moreover, hybrid Eco-DRR/EbA projects tend to have either an EbA or Eco-DRR 'slant' depending on the experts involved in the project. Nevertheless, some key differences exist between Eco-DRR and EbA at the project level, some of which are due to their aims, such as disaster preparedness for Eco-DRR and long-term change within ecosystems or agri-ecosystems for EbA, while other differences are incidental due to the backgrounds of the development of EbA and Eco-DRR. For example, there is currently a greater emphasis on biodiversity conservation in EbA projects due to links to the Convention on Biological Diversity.

At the same time, there are practical similarities and overlaps between Eco-DRR and EbA projects which could be capitalized on through collaboration between DRR and CCA communities. For example, the development of Monitoring and Evaluation assessments (M&E) methodologies in DRR and CCA, and by extension in Eco-DRR and EbA, are currently undertaken separately with little coordination. These differences and similarities will be discussed below along with potential key integration points.

Phase 1: Defining Aims of the Project

Within the projects reviewed (see Doswald and Estrella 2015), the differences in stated aims of projects are more the result of how terminology is used than real differences in stated objectives. For example, Eco-DRR projects generally aim to 'reduce vulnerability or disaster risk', while EbA projects typically 'increase adaptive capacities to climate change'. In practice, Eco-DRR and EbA projects seek to harness the multiple benefits of ecosystem services for DRR and CCA, including for sustainable and resilient livelihoods.

In spite of seemingly small or superficial differences in stated aims, there can be large differences in the approach taken in Eco-DRR and EbA projects in terms of project assessment and implementation, depending on the project's orientation towards either DRR or CCA, and on the nature of the implementing institutions. Moreover, whether a project focuses on reducing the impacts of long-term climatic change or reducing the risks posed by certain hazards makes a difference in the articulated project aims. To formulate an integrated Eco-DRR/EbA project, it would be essential to understand current and future changes with the help of future scenarios that take into account climate change, multiple hazards and environment and development trends. This would then help determine who should be involved and how best to implement the project.

Phase 2: Conducting Risk and Vulnerability Assessments

As Eco-DRR and EbA are emerging fields in their own right, each are developing assessment methods and tools, in which data availability plays a large role. There is sometimes cross-over in assessment needs, which may result in duplication or missed opportunities due to lack of coordination or consultation. For example, within EbA projects, existing future climate change models cannot be reliably downscaled to the local level. Assessments then have to fall back on existing hazard data or simply use broad indicators of future change. Modelling and analyses commonly used in Eco-DRR assessments could be better utilized within EbA project vulnerability assessments in these cases. Conversely, Eco-DRR could learn from EbA assessment developments. Through fostering collaboration between fields, knowledge and practice would be strengthened.

Phase 3: Project Implementation: Methods, Approaches, Tools

Implementation approaches and activities are broadly similar between Eco-DRR and EbA, which is expected since both seek to apply sustainable management, restoration and conservation of ecosystems. These ecosystem-based or environmental management approaches include:

- Land use planning and zoning;
- Sustainable (natural resource) management within forestry, agriculture and pastureland;
- Integrated coastal zone management (ICZM);

- Integrated water resource management (IWRM);
- Integrated watershed or river basin management (IWM);
- Integrated land management (ILM) or Sustainable land management (SLM);
- Protected Areas Management;
- Drylands management;
- Community-based natural resource management; and
- Stewardship systems.

A key difference, however, is that many Eco-DRR projects incorporate early warning systems (and other disaster preparedness activities) and sometimes involve post-disaster activities in contrast to EbA. EbA and hybrid Eco-DRR/CCA projects reviewed, on the other hand, often include the establishment or improved management of protected areas and protected area networks, including corridor establishments, whereas such approaches are less common in the Eco-DRR projects surveyed.

EbA and Eco-DRR both involve some form of re-vegetation and reforestation: for example, for land rehabilitation, to improve ecosystem functions and services, to prevent or mitigate hazard impacts, such as soil erosion, landslides and floods, to increase water security and to act as windbreaks and for storm surge protection. EbA and Eco-DRR projects sometimes involve removal or control of invasive/alien species, sand dune re-establishment, agro-forestry, river re-naturalization, and soil conservation techniques.

However, there is often more of an emphasis in some EbA projects on conservation and enabling ecosystems themselves to adapt, and selecting species suitable to future climatic conditions. Adaptive management, which is strongly promoted in the EbA community, is an approach that recognises uncertain future conditions and therefore embeds learning-oriented, flexible decision-making processes. In contrast, EbA could learn from Eco-DRR's multi-hazard, integrated disaster management approach.

Mainstreaming and Scaling-up Eco-DRR and EbA

Mainstreaming ecosystem-based approaches into DRR and CCA policy and decision-making processes helps to support project sustainability. That is as well as the potential scaling-up of Eco-DRR and EbA approaches through replication in other locations or implementation at a larger geographic scale. A key to scaling-up Eco-DRR and EbA is through mainstreaming Eco-DRR and EbA across development sectors and embedding Eco-DRR/EbA practice in development policies and planning.

Fostering Collaborative, Cross-sectoral Partnerships

Eco-DRR and EbA projects work mostly with environmental ministries to influence policy. However, DRR and CCA are broader policy objectives than the reach of environmental policies. Increasingly, Eco-DRR and EbA projects work through other development sectors, such as agriculture, water and urban development, to influence national and local development policies, programmes and plans. Eco-DRR, EbA and hybrid Eco-DRR/EbA projects recognise the importance of bringing together and working with different government ministries and other stakeholders, including civil society, universities and the private sector. For instance, UNEP's Eco-DRR project in the Democratic Republic of the Congo involved the Ministries of Environment (including the national climate change focal point), Interior, and Social and Humanitarian Affairs.

One major difference between EbA and Eco-DRR projects is that EbA projects rarely involve working on DRR-related policies or with humanitarian agencies and disaster management-related NGOs. EbA projects often focus on the national adaptation strategy or may work within specific development policies, for example in the agriculture or water sectors. On the other hand, Eco-DRR projects seek to influence both DRR and environmental policies, and in many cases, also work to mainstream Eco-DRR into national adaptation policies and programmes, and engage with national CCA focal points.

Although there are many cases within Eco-DRR, EbA and hybrid Eco-DRR/EbA projects of bringing together stakeholders from different sectors, there is still a tendency to work in separate DRR and CCA policy tracks at the national level. Environmental government ministries and environmental organisations play an important role in mainstreaming ecosystem-based DRR and CCA approaches.

'Incentivising' Eco-DRR and EbA

One key challenge in promoting wider applications of Eco-DRR and EbA is that the expected DRR and CCA benefits are not always tangible or may take time to be fully demonstrated. It is critically important to provide incentives that obtain individual or community support (i.e., 'buy-in') for undertaking Eco-DRR or EbA interventions. Incentives may be in the form of potentially increasing livelihood incomes (e.g., cash-for-work schemes, increasing farm yields), or creating new livelihoods (e.g., eco-tourism, selling of new products obtained from the project such as honey), or improving access to and quality of food, water or other natural resources on which livelihoods depend. Other types of incentives include creating opportunities for individuals or communities to articulate their needs (i.e., giving people a 'public voice', especially amongst women). It may include establishing new or strengthening existing institutions that facilitate collective actions.

Eco-DRR and EbA projects potentially offer a range of incentives because of the multiple and direct benefits of ecosystem services that could be derived by local communities. Designing and articulating incentives that provide immediate benefits to people (e.g., improved access to water) helps enhance community participation and support to Eco-DRR and EbA over the long-term. The challenge is ensuring the right mix of incentives is in place to ensure sustained support for Eco-DRR and EbA, beyond the project's lifespan.

Translating Global Policies into Action

Another way of scaling-up Eco-DRR and EbA is to create an enabling policy environment that catalyzes action at national and local levels. Over the past two years (2014 and 2015), there has been significant progress in global policy agendas that lend strong impetus for Eco-DRR and EbA implementation.

In October 2014, at the 12th Conference of the Parties of the Convention on Biological Diversity (CBD), countries adopted a decision explicitly linking biodiversity, disasters and climate change, and recognized the critical role of ecosystems in climate change mitigation and adaptation and in DRR. In March 2015, the 3rd UN World Conference on DRR adopted the post-2015 global DRR framework known as the Sendai Framework for Disaster Risk Reduction (SFDRR) (2015–2030), subsequently endorsed by the UN General Assembly in June, 2015. The SFDRR explicitly recognises sustainable ecosystem management as a key DRR measure in building resilience and calls for greater coherence between DRR and climate change agendas. In June 2015, the Ramsar Convention on Wetlands held its 12th Conference of the Parties, adopting a Resolution calling for the integration of DRR in wetland management plans in countries. In September 2015, the UN General Assembly endorsed the post–2015 Sustainable Development Goals, which make explicit references to mitigating and adapting to climate change and disaster risks. Finally, in December 2015, UNFCCC met in Paris to elaborate on a post–2015 climate change agenda.

All of these frameworks and agreements include aspects of Eco-DRR and EbA to various degrees, signifying a readiness from countries to integrate environment, DRR and CCA and promote Eco-DRR and EbA globally. A key challenge will be to translate these global agreements into national and local contexts for which key activities are raising the awareness of national and local decision-makers regarding the implications of these global agreements, developing implementation guidelines, and establishing innovative financial mechanisms, including public–private sector partnerships.

Conclusion: Limitations and Opportunities for Eco-DRR and EbA

The analysis of practice in EbA, Eco-DRR and hybrid projects showed that often it is more a question of differences in discourse (and use of terminologies) than a real difference in practice at the local level. Nevertheless, EbA and Eco-DRR are generally carried out by separate communities due to differences in policy and funding tracks. Fostering collaboration at the project level would provide good lessons for future practice and facilitate integration of EbA and Eco-DRR, and by extension DRR and CCA. This would then facilitate the development of much needed integrated DRR/CCA tools. Gaps in knowledge in both communities should be filled through inter-disciplinary research and practice, appropriate M&E frameworks that support learning and knowledge exchange platforms.

Investing in ecosystems is not a single solution to DRR or CCA, but should be used in combination with other risk management and adaptation measures. For instance, ecosystem-based solutions often require a lot of land which may not be available (Doswald and Osti 2011; Temmerman *et al.* 2013), or may not provide sufficient protection against certain types and magnitude of hazards (Renaud *et al.* 2013). In order to be effective, ecosystem-based approaches need to be based on rigorous understanding of local, ecological conditions, socio-cultural and economic circumstances and livelihoods, existing hazards, and technical requirements of the intervention. In some cases, ecosystem thresholds may be surpassed depending on the type and intensity of the hazard event and/or health status of the ecosystem, which may therefore be insufficient to provide adequate protection against hazard impacts. For instance, mangroves may not provide as much protection against tsunamis as they would for storm surges. However, ecosystems provide other benefits to support livelihoods and buffer the long-term effects of climate change, such as regulating clean water supplies, regardless of the occurrence of hazard impacts. At the project level, communities do not distinguish between Eco-DRR and EbA activities, and rather are concerned about the actual benefits they could derive. As ecosystem-based approaches are about taking a holistic approach to DRR and CCA, they offer solutions on how we can continue to forge bridges between the two.

References

Andrade Pérez, A., Herrera Fernandez, B. and Cazzolla Gatti, R. (eds) (2010) *Building Resilience to Climate Change: Ecosystem-based adaptation and lessons from the field*, Gland, Switzerland: International Union for Conservation of Nature.

Badola, R. and Hussain, S.A. (2005) 'Valuing ecosystem functions: An empirical study on the storm protection function of Bhitarkanika mangrove ecosystem, India', *Environmental Conservation* 32: 85–92.

CBD (Convention on Biological Diversity) (2009) *Connecting Biodiversity and Climate Change Mitigation and Adaptation: Report of the Second Ad Hoc Technical Expert Group on Biodiversity and Climate Change, Technical Series No. 41*, Montreal, Canada: Secretariat of the CBD.

CBD (Convention on Biological Diversity) (2010) *Decision Adopted by the Conference of the Parties to the Convention on Biological Diversity at its 10th Meeting*, New York: United Nations Development Programme (UNDP).

Doswald, N. and Estrella, M. (2015) *Promoting Ecosystems for Disaster Risk Reduction and Climate Change Adaptation: Opportunities for Integration – Discussion Paper*, Geneva, Switzerland: United Nations Environment Programme Post-conflict and Disaster Management Branch.

Doswald, N., Munroe, R., Roe, D., Guiliani, A., Castelli, I., Stephens, J., Moller, I., Spencer, T., Vira, B. and Reid, H. (2014) 'Effectiveness of ecosystem-based approaches for adaptation: Review of the evidence-base', *Climate and Development* 6: 185–201.

Doswald, N. and Osti, M. (2011) 'Ecosystem-based approaches to adaptation and mitigation – good practice examples and lessons learned in Europe', *BfN-Skripten 306*, Bonn, Germany: Federal Ministry of Environment, Nature Conservation and Nuclear Safety.

Estrella, M. and Saalismaa, N. (2013) 'Ecosystem-based disaster risk reduction (Eco-DRR): An overview', in F.G. Renaud, K. Sudmeier-Rieux and M. Estrella (eds), *The Role of Ecosystems in Disaster Risk Reduction*, Tokyo: United Nations University Press, pp. 26–54.

IPCC (Intergovernmental Panel on Climate Change) (2014) *Climate Change 2014: Synthesis Report. Contribution of Working Groups I, II and III to the Fifth Assessment Report of the Intergovernmental Panel on Climate Change*, Geneva, Switzerland: IPCC.

Kelman, I. and Gaillard, JC (2010) 'Embedding climate change adaptation within disaster risk reduction', in R. Shaw, J.M. Pulhin and J.J. Pereira (eds), *Climate Change Adaptation and Disaster Risk Reduction: Issues and Challenges Community, Environment and Disaster Risk Management*, Bingley, UK: Emerald, pp. 23–46.

Manyena, S.B., O'Brien, G., O'Keefe, P. and Rose, J. (2011) 'Disaster resilience: A bounce back or bounce forward ability?', *Local Environment* 16, 5: 417–424.

Millennium Ecosystem Assessment (2005) *Ecosystems and Human Well-being*. Cambridge, UK: Synthesis Island Press.

Munang, R., Thiaw, I., Averson, K., Liu, J. and Han, Z. (2013) 'The role of ecosystems services in climate change adaptation and disaster risk reduction', *Current Opinion in Environmental Sustainability* 5: 47–52.

Peduzzi, P. (2010) 'Landslides and Vegetation Cover in the 2005 North Pakistan Earthquake', *Natural Hazards Earth Systems Science* 10: 623–640.

Renaud, F.G., Sudmeier-Rieux, K. and Estrella, M. (eds) (2013) *The Role of Ecosystems in Disaster Risk Reduction*, Tokyo: United Nations University Press.

Spalding, M.D., Ruffo, S., Lacambra, C., Meliane, I., Zeitlin Hale, L., Shepard, C.C. and Beck, M.W. (2014) 'The role of ecosystems in coastal protection: Adapting to climate change and coastal hazards', *Ocean and Coastal Management* 90: 50–57.

Sudmeier-Rieux, K., Ash, N. and Murti, R. (2013) *Environmental Guidance Note to Disaster Risk Reduction, Revised edition, Ecosystem Management Series No. 8*, Gland, Switzerland: International Union for Conservation of Nature.

Sudmeier-Rieux, K., Fra Paleo, U., Garschagen, M., Estrella, M., Renaud, F.G. and Jaboyedoff, M. (2014) 'Opportunities, incentives and challenges to risk sensitive land use planning. Lessons from Nepal, Spain and Vietnam', *International Journal for Disaster Risk Reduction* 14: 205–224.

Tanaka, N., Yagisawa, J. and Yasuda, S. (2013) 'Breaking pattern and critical breaking condition of Japanese pine trees on coastal sand dunes in huge tsunami caused by Great East Japan Earthquake', *Natural Hazards* 65: 423–442.

Temmerman, S., Meire, P., Bouma, T.J., Herman, P.M.J., Ysebaert, T. and De Vriend, H.J. (2013) 'Ecosystem-based coastal defence in the face of global change', *Nature* 504: 79–83.

UNISDR (United Nations International Strategy for Disaster Reduction) (2015) *Global Assessment Report on Disaster Risk Reduction 2015,* Geneva, Switzerland: UNISDR.

Vignola, R., Harvey, C.A., Bautista-Solis, P., Avelino, J., Donatti, C. and Martinez, R. (2015) 'Ecosystem-based adaptation for smallholder farmers: definitions, opportunities and constraints', *Agriculture, Ecosystems and Environment* 211: 126–132.

Vo-Luong, P. and Massel, S. (2008) 'Energy dissipation in non-uniform mangrove forests of arbitrary depth', *Journal of Marine Systems* 74: 603–622.

13

THE GENDERED TERRAIN OF DISASTER RISK REDUCTION INCLUDING CLIMATE CHANGE ADAPTATION

Sarah Bradshaw and Brian Linneker

Introduction

There have been a number of calls to integrate Disaster Risk Reduction (DRR) and Climate Change Adaptation (CCA) strategies within international development policy to reduce people's livelihood risks. However, from a gender perspective this presents a major challenge. While women may be involved in DRR including CCA at the grass roots level, the national and international policy discourses remain largely ungendered and unrelated. This chapter re-visits what Enarson and Morrow (1998) termed the 'Gendered Terrain of Disasters', to examine the ways in which gender has been incorporated into DRR and also into CCA (see Bradshaw and Linneker 2014) to explore what this means for a gendered DRR including CCA approach. It begins by addressing the fundamental question of why consider gender. It highlights that, to explain the position and situation of women and men, gender must be understood as intersecting with other sites of oppression. It then explores how gender has been conceptualised within the wider development and environment discourses and how in turn this has influenced policy debates around DRR and CCA. It seeks to problematise the 'engendering' of the two discourses, seeing a commonality as being a move toward a 'feminisation' of responsibility in policy and practice which needs to be addressed in any DRR including CCA approach.

Why Gender?

While now in common usage, gender is a relatively new term and has only been widely used in relation to differences between men and women since the 1970s. Up until then the term 'sex' had been used to define difference based on biological characteristics (see Bradshaw 2013). To draw a parallel with disasters: whilst a hazard is generally seen to be a natural event, a disaster is only what society understands or constructs it to be. So, too, while biological differences are natural, the meanings given to them are social, and it is this social construction of what it means to be a man or a woman (ideas of masculinity and femininity) that the term 'gender' seeks to capture. To study gender is to study the relationships between men and women, and why and how inequalities based on gender are produced and reproduced and how they can be changed. More recently gender scholars, including within the fields of DRR and within climate change, have been stressing the need also to recognise differences between women by applying an 'intersectional' lens (see Bradshaw and Fordham 2013). That is, recognising that gender intersects with other characteristics such as age, ethnicity and class to construct individual and collective identities and related inequalities.

Gendered inequalities are actively produced and reproduced across time and space, within the household, community and society but also via wider global systems of capitalism and patriarchy. Patriarchal power relations, whereby male-dominated systems of social relations place women in a relative subordinate position and reinforce gender inequality, shape day to day lives of women and men and also processes of capitalist development (Walby 1990). Economic relations based on processes of environmental resource exploitation and capital accumulation have historically shaped an intensely gendered, racialised and increasingly urbanised global development (Harvey 2014). Consequently, men and women may experience the same event differently, and this may also be the case for hazards, including those related to climate change. However, what evidence do we have on which to base such claims?

The assertion that women will be more adversely affected than men by hazards including those related to climate change appears to be based on understandings that women may have fewer resources (including assets, education, income and political voice) to respond. There is now a limited but growing body of work challenging the assumed gender neutrality of climate change (Alston and Whittenbury 2013; Cela, Dankelman and Stern 2012; Dankelman 2010). However, much of the evidence on whether climate crises are gendered has been focused on rural areas and agricultural practices. While evidence around the gendered experience of disasters is more readily available, even basic data such as reliable disaster fatalities data disaggregated by gender and age is still largely missing.

A number of studies do suggest a gendered impact of disasters in terms of higher deaths among women but the most often cited source is a study that constructed indicators of disaster magnitude and of women's socio-economic status and explored how these relate to the size of the gender gap in life expectancy (Neumayer and Plümper 2007). They concluded that in countries where a disaster had occurred and the socio-economic status of women is low, more women than men die or die at a younger age. This supports the idea that it is not gender alone that predicts vulnerability but gender inequalities, and these may be related to how gender intersects with class and caste, for example. Other studies have noted that women's supposed greater vulnerability is related to socialised gender roles rather than any biological 'weakness' of women. For example, girls are biologically as able to swim as boys, but in many societies it is not seen to be appropriate so they are not taught to swim, putting them at greater risk in floods and tsunamis. This highlights that vulnerability, like gender roles and relations, is socially constructed, and gender alone cannot predict who will be most vulnerable as it is the intersection of gender with characteristics such as ethnicity and cultural norms that determines vulnerability.

Risk is another concept that may help to explain any differential impact of hazards including climate change. Risk is a subjective notion and there may be systematic gender differences in the perception of risk. For example, research suggests white males are more likely to rank a variety of risks significantly lower than women, and also lower than men of colour or other 'minority' ethnic groups in the same society. This male risk taking is highlighted during bushfires (Pacholok 2013); not only is fire-fighting a masculinised profession, but whilst most female and child fatalities occur while sheltering in the house or when fleeing, usually too late, men are most often killed outside while attempting to protect the home and other assets. Lack of information – often transmitted by men to men – and lack of education and engagement with preparedness activities means women when faced by a perceived risk often do not know when to act or how to act on warnings. All this means that socially constructed gender roles and relations may result in disasters having a feminised impact. If loss of life is to be reduced this would imply the need to focus DRR efforts on women, or at least to recognise that risk is a gendered concept.

In climate change literature the gendered nature of risk has also been noted (Alston and Whittenbury 2013). While men and women both relate negative change in agricultural production to wider climate change, their understandings of what this will mean differ in line with gender roles – with men's concerns focussed on production while women are more concerned with wider impacts on the family. A number of studies highlight that women are more likely than men to acknowledge ecological problems and risks, express higher levels of concern and engage in activities beneficial to the environment (Alston and Whittenbury 2013). Similarly, research suggests that women are more likely to act upon early warnings and when early warning is in the hands of women, fatalities from disasters appear to be lower.

However, while a tendency toward feminised fatalities suggests the need to reduce women's vulnerability and risk prior to an event, and to target adaptation and mitigation activities at women, where women have been most included in disaster-related initiatives is in relief and reconstruction. This is interesting as there is little data to demonstrate a feminised impact in terms of loss of material goods. This may be due to what is measured and how it is measured. Losses recorded tend to be land and large 'productive' resources such as mechanised farming equipment, large livestock such as cattle, as well as housing. Women's livelihood losses, such as small livestock and chickens for example, are less often recorded; and items used for housework, such as ovens and cooking utensils etc., tend to be unrecorded (Bradshaw 2013).

The impact of a hazard may be gendered and the secondary impacts on women mean that women and girls often face a 'double disaster' (Bradshaw and Fordham 2013). Perhaps the biggest impact on women and girls is the escalation of hours in the working day as the event has an impact on each of women's 'triple roles'. Women may move into income generating activities (so called 'productive' work) to help support the household while at the same time their activities focussed on the home such as cooking, cleaning and washing ('reproductive' work) become more burdensome and time consuming. At the same time the hours engaged in reconstruction activities ('community management' role) may be large, not least given women are often targeted by donors and agencies. Age adds to this, and girls may have to juggle four roles – their productive, reproductive and community management roles and their role – often promoted and supported by non-governmental organisations and other agencies such as the United Nations anxious to get children back into education – as 'school girls', blurring further their dual identity as adult/child and with potentially important consequences for the future (Bradshaw and Fordham 2013).

Other 'secondary' impacts felt by women include deterioration in reproductive and sexual health, increased early and forced marriage, increased poverty, insecure employment, forced and transactional sex, trafficking, and migration. Sexual violence in particular is highlighted as increasing post-event and a change from private to public violence, or from the hands of knowns to unknowns, has been noted (Bradshaw and Fordham 2013). There is also some limited but growing evidence that there may be climate change-related gender-based violence (Alston and Whittenbury 2013). This is related to tensions over access to and control over resources designed to mitigate disasters, such as irrigation resources, rather than a 'disaster' event itself. While an increase in violence is now assumed to occur post-event, it is important to note there is still a lack of systematic studies to support this and more reliable and longitudinal data is needed to fully understand the processes involved.

While studies might still be limited, the evidence is growing to support a gendered impact of hazards including climate change, which supports the need to integrate gender into related policies and practice. How best this can be done is an interesting question.

The Evolution of Thinking: Gender, Environment, and Development

Although a relatively new but growing field of academic enquiry, gendered DRR has historical antecedents in debates concerning the role of gender in environment, and gender in development (see Bradshaw 2013). Gender and development emerged as a field of academic enquiry and as a policy practice in the 1970s, beginning with the Women in Development (WID) approach. WID sought to better integrate women into what was constructed as a benign development process. The WID approach focused on women's 'practical gender needs' – such as providing better access for women and girls to water. The WID approach was critiqued for its focus on women only and its rather narrow understanding that 'empowerment' could be achieved via women's education. WID also focussed on increasing women's involvement in 'productive' work such as in agriculture and paid employment as a means to 'economic empowerment', which in turn would lead to wider empowerment. It did not recognise that women's entry into income generating activities may leave them 'time poor' as they juggle their domestic work and a 'double day' of work outside and inside the home. The Gender and Development (GAD) tradition emerged from critiques of WID and focuses on gender roles and relations that are at the basis of women's exclusion from development, and problematises the nature of development. GAD projects are more holistic and address women's 'strategic gender interests'. So, instead of seeing training women in modern farming techniques as a solution to their low participation in agriculture, for example, it seeks to eliminate the institutionalised forms of discrimination around land rights that deny women access to land ownership (see Bradshaw 2013).

The 1980s saw the emergence of a Women, Environment and Development (WED) approach that drew on the WID tradition and also 'ecofeminist' thinking that highlighted how women's dominant role as mothers and the resulting sexual division of labour meant women were closer to nature, with some even going as far as constructing women 'as' nature through giving birth. A major critique was that they were 'essentialising' – that they prioritised biology as an overarching explanatory variable. This led to a number of subsequent approaches that recognised the material basis of the gender division of labour and differences between women (see Bradshaw and Linneker 2014). In the mid-1990s, in parallel with development toward a GAD approach, the Gender, Environment and Development (GED) approach emerged and sought to apply gender analysis tools to the environment. This and other approaches, such as 'feminist environmentalism', and 'feminist political ecology', all share a common idea of gender–environment relations as embedded in dynamic social and political relations (Leach 2007).

In the contemporary context, gender issues are a fixed item on the international development agenda, as witnessed by a stand-alone gender goal in both the Millennium Development Goals (MDGs) and the Sustainable Development Goals (SDGs). While attempts to 'engender' development have helped shape gendered understandings of disaster, it is suggested that the dominant disaster discourse appears to remain 'a number of decades' behind the development discourse (Bradshaw 2014). Similarly, despite the rich literature on gender and environmental concerns, after promising beginnings there has been a 'strange silence' on the gender dimensions of climate change, particularly in global policy discourse (MacGregor 2010). This may in part be due to the fact that while much of the environmental literature is focussed on the conceptualisation of earth as nurturing mother, this stands in contrast to the historical conceptualisation in both the disaster and climate change discourses of nature as wild, uncontrollable and something to be dominated – via science and technology. This has meant that at the international and national level both the DRR and particularly the CCA agendas are more 'scientific' than social agendas (Seager 2009), and given they often privilege knowledge constructed by men over more qualitative and local knowledges of women they are largely masculinised fields of enquiry (Alston and Whittenbury 2013).

While the lack of inclusion of gender in mainstream disaster and climate change discourse may be seen as a problem, inclusion of gender in the development and environmental discourses has also been critiqued (Jackson 1993; 1996). Both WID and WED were attractive to policymakers. Actors such as the World Bank combined the desire for greater gender equality with the desire for wider policy aims such as poverty reduction and economic growth. They suggest gender equality to be 'smart economics' with women constructed as an 'untapped' economic resource. Playing on the presumed 'natural' altruistic tendency of women, they have been targeted with resources designed to bring benefits to children. This efficiency approach was replicated in environmental policy, where the presumed closeness of women to nature has led to women being constructed as 'chief victims-and caretakers' (Resurreccion 2012). In what they saw to be a 'win–win' approach to environment and gender actors such as the World Bank played on the 'naturalness' of what are ultimately socially constructed gender roles to appropriate women's unpaid labour in activities to protect the environment. This led to the suggestion that to celebrate women's responses to environmental problems, particularly when connected to their gender specific responsibilities for social reproduction, might be 'somewhat dangerous' (MacGregor 2010).

While there are similarities in the thinking underpinning gendered DRR and CCA, the literatures remain largely separate and only a small number of authors have sought to bridge the divide (see Enarson 2013). Before exploring how gender has been incorporated into DRR and CCA policy, and attempts to embed CCA within DRR, it is important first to examine the coping strategies adopted by women, men and households to respond to hazards including climate change.

Gendered Response to Hazards

The DRR and CCA literature suggest that social relations of power bear upon women's abilities to act on their knowledge in the face of climatic uncertainty (Bradshaw and Linneker 2014). While women may have good strategies to reduce risk and adapt to change, they often have less access to the resources needed to put these in place. In particular, their lack of decision-making power in the household and beyond may mean the strategies they propose will not be heard and adopted. Women's voice and agency are not just related to gender but other characteristics such as age and stage in life course; i.e., if they are seen as a daughter, a wife or an elder, and cultural norms, religion, race and ethnicity (see Box 13.1).

Box 13.1 The Intersection of Gender, Generation, Ethnicity and Education

Sarah Bradshaw[1] and Brian Linneker[2]

[1] Middlesex University, London, UK
[2] Independent scholar, London, UK

Young women of a 'minority' ethnic group may face specific issues as they try to navigate the expectations of their family and ethnic community and those of friends and the wider society in which they live. Their self-identity will be determined by the intersection of the characteristics that define them: gender, age, ethnicity and religion.

Research that focussed on people born in England but of South Asian descent explored how flood risk is understood, communicated and acted upon through an intersectional lens (Riyait 2016). The research found first generation migrants with flood experience abroad perceived flooding in England as a 'bit of water' and did not take flood risk or response seriously. Although the interviews

---

were about floods and risk, what emerged as a key 'risk' for many older people was the westernisation of women and this led to strict patriarchal control of young women by elders, including how they dressed and what jobs they aspired to. It also limited women's voice in the home.

Inter-generational communication about risk was also guided by intersecting characteristics including education. The study found that while first generation, uneducated women were dependent on men in a flood event, those women born in England and with more education had the power to advise their family on flood risk and what to do. However, although educated young women have knowledge on flood risk response, they are still bound by their gender and age, and conflict with elders over other elements of their lives, such as appropriate dress. This limits their desire to communicate with grandparents and the desire of grandparents to listen and act on the advice.

Applying an intersectional lens – migration, gender and generation – this study provides new insights into the complexity of how flood risk and knowledge may be constructed and transmitted.

Environmental hazards may impact older women through increasing the time burden to fulfil reproductive tasks. Their impact on girls may have short- and long-term consequences, limiting access to schooling now, and access to productive land and employment in the future. Yet, while differences exist between women, the existing discourse often focusses on how women's response to environmental issues differs from men's – highlighting, for example, male response as institutionalised compared to a more individual/collective female response. Studies across a range of regions, countries and agricultural practices (see Alston and Whittenbury 2013) show that climate impacts do affect men and women differently and that women tend to suffer more negatively in terms of their assets and well-being, including in urban settings (see Moser 2016). There may be exceptions to the pattern; where men may be more negatively impacted by environmental hazards including drivers such as climate change because they own land, or because women are able to invoke cultural norms that make men responsible for household food security, for example.

In terms of the longer-term changes to hazards, including from climate change, it has been suggested that women farmers may be most affected, given their lower asset base, and while having knowledge of how to adapt, they may be least able to adopt appropriate adaptation strategies. For example, they may already be farming the least productive land, have less access to scarce resources such as water and inputs such as drought resilient seeds, and less access to technical knowledge and support. However, because of this lack of access to resources on a daily basis, they may have learned to be more adaptable to crisis than their male counterparts. The strategies women assume to reduce risk in this and other contexts may include taking on non-traditional roles. For example, they may move into off-farm employment taking them away from the home. While this potentially adds to their time burden, earning an income also has the potential to change their status and position in the home.

Positive change can occur out of crisis, but once again an intersectional lens highlights this may not be for all women. Women with very poor levels of education may not be able to find employment, for example, or the stigma associated with women working outside the home might be too great to allow women of some cultural or religious groups to take up existing opportunities. Men, too, are not a homogeneous group and different men may react differently to crisis and the changing gender roles this may bring (see Box 13.2). While some men may take on more reproductive work in the home others may feel threatened by women's strategies and seek to re-establish patriarchal control (Bradshaw 2016). How strategies to reduce risk and adapt to change are viewed may vary between men and women, and there may be divergent views on whether a strategy is adding to coping or adding to vulnerability.

Box 13.2 Involving Men

Sarah Bradshaw[1] and Brian Linneker[2]

[1] Middlesex University, London, UK
[2] Independent scholar, London, UK

To study gender implies studying gender roles and relations – and this means considering both women and men. However, much gender literature focusses on women only and 'gender roles and relations' is often shorthand for inherently oppositional relations (see Bradshaw 2014).

Men's behaviour is often presented as irresponsible as compared with women's altruistic behaviour, whereby women use all their money on children and the home, while men often spend a proportion of their money on themselves and socialising with other men (Chant 2008). To challenge this behaviour demands a focus specifically on men, and by the end of the 1990s there was a new interest in the notion of masculinity and the rise of masculinity projects. Often, these projects focussed on men's personal constructions and understandings of 'maleness' and the implications of this for relationships with others.

In particular, the role of men in reducing violence against women was recognised, including post-disaster. The Nicaraguan feminist NGO Puntos de Encuentro designed a campaign to reduce violence against women post-hurricane Mitch which, for the first time, targeted men. It used the slogan 'Violence against women, a disaster men CAN prevent' on hats and T-shirts, while radio slots, billboards by the roads, and also calendars distributed to bars explained seven steps men could take to avoid becoming violent, such as talking to friends or taking a walk.

The evaluation of the campaign suggests this focus on men may have been successful in helping prevent the escalation in violence often seen post-disaster. It suggests if they are to be successful, DRR including CCA projects need to include both women and men.

For some women then the event may not be a 'disaster' in that it may bring positive changes as well as negative (Cupples 2007). This may mean women feel able to leave an abusive partner (Enarson and Morrow 1998) or find a new collective voice (David and Enarson 2012). Some people may find positive changes through struggling with the aftermath of trauma, for example greater intimacy and compassion for others, feeling personally stronger, and a deeper appreciation of life. This can have a profound impact on their lives, for example influencing the decision of young women to 'come out' and live openly as lesbians (see Box 13.3).

Box 13.3 Gender and Sexuality

Sarah Bradshaw[1] and Brian Linneker[2]

[1] Middlesex University, London, UK
[2] Independent scholar, London, UK

Adolescent girls are only just beginning to be understood as having unique interests and needs that are specific to their cohort and that do not always coincide with those of children or with adults. Very often, youth is subsumed into adulthood or young people are constructed as 'genderless' children and this is problematic for understanding the impacts of disasters and indeed for understanding the notion of 'disaster'.

While recognising young, childless women as having unique concerns and knowledges, they still cannot be seen as a homogeneous group. Youth is a time to explore who we are, including sexual exploration. Research post-Hurricane Katrina in New Orleans explored how youth, gender and female sexuality intersect after a disaster to not only determine the gendered experience of the event but the extent to which it was a 'disaster' for those involved (Overton 2017). The research focussed on young women growing up and coming out, finding that the changes Katrina brought to some young women's lives, such as their leaving their communities, were positive in the sense they allowed them, or caused them, to more openly express their sexualities. Through engaging in creative acts they also created strong bonds with other young women and the formation of an informal network that proved able to reach out in solidarity to other young women after a subsequent hurricane. Thus, while the 'disaster' was recognised as having led to loss, hardship and pain, it was also seen as bringing positive change to this group of women highlighting the need to apply an intersectional lens to better understand what a 'disaster' means to those that experience it.

However, while hazards may have a profound impact on the lives of young women, they are often provided very little support. Young women across the globe are seldom targeted in policies and projects in times of crisis, as their construction as daughters rather than mothers, and dependants rather than income earners or care-givers, makes them largely invisible in the policy context.

Policy Exclusion and Inclusion

Within the DRR including CCA discourse, women tend to be assumed to be the ones responsible for addressing the effects of decreasing supplies of clean water, and decreasing access to crop residues and biomass for energy, due to their traditional gender 'reproductive' roles. The discourse seems to be based on existing gender roles and related assumptions around women's vulnerability and their assumed characteristics and desire to nurture and care for the people and environment that surrounds them (Resurreccion 2012). They are then targeted not as 'women' but based on the assumed characteristics related to being a woman. Women's role as mother and assumed desire to care for family and community also explains why they have become a target for resources and beneficiaries of relief and reconstruction post-disaster. In this construction some women are made invisible – the very old for example, and young women who are not yet mothers are instead conceptualised as daughters or 'children' rather than 'women'.

However, while the exclusion of some women from DRR initiatives can be seen as problematic, the benefits to be gained from their inclusion in projects can also be questioned. An in-depth study of four communities in Nicaragua impacted by Hurricane Mitch suggested that while half the women interviewed perceived that they participated most in the projects for reconstruction, only a quarter felt that it was women who benefited from this 'participation' (Bradshaw 2013). The majority stated that it was the family that benefited from their participation in reconstruction. Such perceptions are supported by interviews with representatives from a number of the organisations involved that targeted resources at women, including 'non-traditional' resources such as giving women collective ownership of cows. For example, when asked how men had reacted to this focus on women, one representative commented that there had not been any major problems since 'the women have their cows and the men are drinking the milk'. Thus, while portrayed as a project focused on women and promoting gender equality, in fact the outcome might bring more work for women with few personal benefits. While disasters may bring changes to gender

136

roles, they may not bring changes in gender relations, or if they do bring changes, they may not be uncontested or positive changes for women (Bradshaw and Linneker 2016).

Women's existing gendered role in social reproduction, as mother and carer, has often meant that women are expected to take on new roles as the 'beneficiary' of projects that provide economic resources for the household and in particular children especially in times of crisis. Writing on poverty alleviation projects, Molyneux (2007) suggests that, as they do little to change the situation and position of women, women are at the service of these new policy agendas, rather than being served by them. The way governments and non-governmental organisations have increasingly incorporated gender frameworks into their planning for disaster relief is also of concern and mirrors what has been seen to be a 'feminisation of obligation and responsibility' within development policy (Chant 2008).

Gender guidelines often conceptualise women in limited, essentialised terms as mothers charged with protecting others, or as 'weak' women needing protection. This double identity of women as both virtuous and vulnerable (Arora-Jonsson 2011) is at the root of the focus on women by policymakers in the environmental field. In the disaster context women 'beneficiaries' of aid may similarly be targeted as 'virtuous-victims' (Bradshaw 2014). The post-2015 international framework for DRR is a good example of this as it constructs women as both vulnerable and as a resource that can be utilised for more effective DRR, yet makes no mention of gender inequality as a root cause of vulnerability (Bradshaw 2015).

In the 1990s, critics suggested that the proclaimed 'success' of Women Environment and Development projects had often been gained at the expense of women – adding unremunerated environmental protection roles to women's existing burdens to bring positive collective benefits but few personal gains. Research suggests patriarchal relations are highly resilient to disasters (Bradshaw 2016). This raises the question of how best to ensure the inclusion of women within ongoing initiatives around DRR including CCA in a way that ensures they are served by these programmes, rather than being at the service of them.

Conclusions

While the body of gender knowledge in DRR including CCA is growing, there remain many gaps, and what is not known is much greater than what is known. Fundamental gaps such as a lack of routine collection of sex-disaggregated data post-disaster on deaths, injuries and losses remain. Where evidence does exist, its micro level, qualitative nature means it is not seen to be 'robust' and thus is often ignored. This means policymakers largely base policy decisions on stereotypical assumptions about the role of men and women. The gendered division of labour constructs women as better protectors of the environment, better service providers, and better able to responds to hazards. This would not be an issue if women were 'rewarded' for this efficiency in economic terms or via society valuing these characteristics more highly. However, as the nurturing characteristics of women are assumed to be 'natural' they are little valued, and further exploited by policymakers. If this is to be addressed there is a need to adopt a focus on the unpaid care economy that acknowledges its value and leads to a reconceptualisation of its worth within society.

Both women and men are still constructed by policymakers as homogeneous groups and differences based on age, ethnicity, and sexuality among others, are often ignored and little understood. There is a need for further studies at the micro level, particularly household studies, to show how men and women experience environmental change differently, and how these differences, and those that arise from other characteristics such as age, income and ethnicity, make disasters a gendered issue.

The extent to which integrating gender into DRR including CCA can lead to improved livelihoods and gender strategic developments for women is open to question, and how best to bridge the existing gender policy divide within and between DRR and CCA remains an important policy challenge. Studies that consider hazards, including climate change, and that look at short- and long-term risks are a necessity if the issues raised are to be tackled in a way that improves, rather than harms, the position and situation of women.

References

Alston, M. and Whittenbury, K. (2013) *Research, Action and Policy: Addressing the Gendered Impacts of Climate Change*, Netherlands: Springer.

Arora-Jonsson, S. (2011) 'Virtue and vulnerability: Discourses on women, gender and climate change', *Global Environmental Change* 21, 2: 744–751.

Bradshaw, S. (2013) *Gender, Development and Disasters*, Northampton, UK: Edward Elgar.

Bradshaw, S. (2014) 'Engendering development and disasters', *Disasters, Special Issue: Building resilience to disasters post-2015* 39, 1: 54–75.

Bradshaw, S. (2015) 'Gendered rights in the Post-2015 development and disasters agendas', *IDS Bulletin Special Edition '20th anniversary of the Beijing Platform for Action'*, 59–65.

Bradshaw, S. (2016) 'Re-reading gender and patriarchy through a "Lens of Masculinity": The "known" story and new narratives from post-Mitch Nicaragua', in E. Enarson and B. Pease (eds), *Men, Masculinities and Disaster: Revisiting the Gendered Terrain of Disaster*, London: Routledge, pp. 56–66.

Bradshaw, S. and Fordham, M. (2013) *Women, Girls and Disasters: A Review for Department for International Development (DFID)*, London: DFID. Online: www.gov.uk/government/uploads/system/uploads/attachment_data/file/236656/women-girls-disasters.pdf (accessed 8 September 2015).

Bradshaw, S. and Linneker, B. (2014) *Gender and Environmental Change in the Developing World*, London: International Institute for Environment and Development (IIED), Human Settlements Group. Online http://pubs.iied.org/pdfs/10716IIED.pdf (accessed 8 September 2015).

Bradshaw, S. and Linneker, B. (2016) 'The gendered destruction and reconstruction of assets and the transformative potential of "disasters"', in C. Moser (ed.), *Gender, Asset Accumulation and Just Cities: Pathways to Transformation*, London: Routledge, pp. 164–180.

Cela, B., Dankelman, I. and Stern, J. (2012) *Powerful Synergies: Gender Equality, Economic Development and Environmental Sustainability*, New York: United Nations Development Programme (UNDP). Online: www.undp.org/content/dam/undp/library/gender/Gender%20and%20Environment/Powerful-Synergies.pdf (accessed 8 September 2015).

Chant, S. (2008) 'The "feminisation of poverty" and the "feminisation" of anti-poverty programmes: Room for revision?', *Journal of Development Studies* 44, 2: 165–197.

Cupples, J. (2007) 'Gender and Hurricane Mitch: Reconstructing subjectivities after disaster', *The Journal of Disaster Studies, Policy and Management* 31, 2: 155–175.

Dankelman, I. (ed.) (2010) *Gender and Climate Change: An Introduction*, London: Earthscan Publications Ltd.

David, E. and Enarson, E. (eds) (2012) *The Women of Katrina: How Gender, Race and Class Matter in an American Disaster*, Nashville: Vanderbilt University Press.

Enarson, E. (2013) 'Two solitudes, many bridges, big tent: Women's leadership in climate and disaster risk reduction', in M. Alston and K. Whittenbury (eds), *Research, Action and Policy: Addressing the Gendered Impacts of Climate Change*, Netherlands: Springer, pp. 6–74.

Enarson, E. and Morrow, B. (eds) (1998) *The Gendered Terrain of Disasters*, Connecticut and London: Praeger.

Harvey, D. (2014) *Seventeen Contradictions and the End of Capitalism*, London: Profile Books.

Jackson, C. (1993) 'Doing what comes naturally? Women and environment in development', *World Development* 21, 12: 1947–1963.

Jackson, C. (1996) 'Rescuing gender from the poverty trap', *World Development* 24, 3: 489–504.

Leach, M. (2007) 'Earth mother myths and other ecofeminist fables: How a strategic notion rose and fell', *Development and Change* 38, 1: 67–85.

MacGregor, S. (2010) 'A stranger silence still: The need for feminist social research on climate change', *Sociological Review* 57, Issue Supplement S2, Special Issue: Sociological Review Monograph Series, B. Carter and N. Charles (eds), *Nature, Society and Environmental Crisis*, 124–140.

Molyneux, M. (2007) 'Two cheers for conditional cash transfers', *IDS Bulletin* 38, 3: 69–75.

Moser, C. (ed.) (2016) *Gender, Asset Accumulation and Just Cities: Pathways to Transformation*, London: Routledge.

Neumayer, E. and Plümper, T. (2007) 'The gendered nature of natural disasters: The impact of catastrophic events on the gender gap in life expectancy 1981–2002', *Annals of the Association of American Geographers* 97, 3: 551–566.

Overton, L. (2017) *Girls Interrupted: Young Women's Life Histories after Hurricane Katrina in New Orleans.* Unpublished PhD thesis, forthcoming, London: Middlesex University.

Pacholok, S. (2013) *Into the Fire: Disaster and the Remaking of Gender*, Canada: University of Toronto Press.

Resurreccion, B.P. (2012) 'The gender and climate debate: More of the same or new ways of doing and thinking?', in L. Elliot and M. Caballero-Anthony (eds), *Human Security and Climate Change in Southeast Asia*, London and New York: Routledge, pp. 95–111.

Riyait, S. (2016) *'I'm big you're small, I'm right you're wrong'. The influence of gender and generation on migrant response to flood risk in England*, Unpublished PhD thesis, London: Middlesex University.

Seager, J. (2009) 'Death by degrees: Taking a feminist hard look at the 2°C climate policy', *Kvinder, Køn & Forskning* 3: 11–21. Online https://tidsskrift.dk/index.php/KKF/article/viewFile/44305/84084 (accessed 3 September 2016).

Walby, S. (1990) *Theorizing Patriarchy*, Oxford: Basil Blackwell.

14

HUMAN RIGHTS FOR DISASTER RISK REDUCTION INCLUDING CLIMATE CHANGE ADAPTATION

Jean Connolly Carmalt

This chapter will discuss the ways in which human rights must be respected, protected, and fulfilled in the context of adaptation to climate change that is part of disaster risk reduction (DRR). As Kelman and Gaillard (2010) illustrate, embedding climate change adaptation (CCA) within broader discussions about DRR is important because it avoids the pitfalls of falsely separating interrelated issues of sustainability and development. It therefore raises questions about how international development policy can integrate short- and long-term goals at local, national, and international scales. Human rights contributes to this conversation by setting out legal and ethical responsibilities that range from specific and immediate actions by local actors (e.g., non-discriminatory evacuation of a flooded area) to structural and long-term issues addressed by national or international actors (e.g., the impact on certain groups of an eroding coastline).

International human rights law provides a comprehensive set of standards that address the creation of vulnerability to disasters and the way in which that vulnerability intersects with hazards in general and the negative effects of climate change specifically. Existing literature has focused primarily on the legal obligation to respect, or not directly violate, human rights as part of humanitarian disasters arising from physical processes. In addition, human rights are relevant in terms of the legal obligations to protect (i.e., ensure others do not violate) and fulfil (i.e., enact requisite measures for realization). Moreover, while international law primarily refers to state-based responsibility, human rights provide universal standards with possibilities for defining transnational obligations. Therefore, the obligations to respect, protect, and fulfil provide a way to prioritize and streamline human rights within an integrated discussion of CCA within DRR.

This chapter will begin with a short introduction of the international legal obligations associated with human rights, paying close attention to the way those obligations relate to different scales of time and space. It will then examine which human rights are most likely to be violated in the context of DRR, including CCA, drawing on existing conversations about how human rights relate to both. Finally, it will analyse the obligations associated with the human right to health as they relate to CCA within DRR. The chapter will conclude with recommendations about how human rights can contribute to future discussions.

International Legal Obligations

International human rights law is rooted in the basic requirement that human beings be treated with dignity throughout their lives, as articulated by the 1948 Universal Declaration of Human Rights. Human rights are defined through multiple treaties, customary norms, authoritative commentary, and jurisprudence, which encompass binding (hard) legal norms and persuasive (soft) law. In addition, human rights law overlaps with, and draws from, humanitarian law and international criminal law. However, human rights law also differs from these areas because it applies in both peace and wartime, and because it is framed in terms of state-based obligations instead of individual criminal responsibility. That said, humanitarian law is also relevant to responsibilities related to emergency situations, displacement, and humanitarian operations in the aftermath of disasters.

Human rights encompass civil, political, economic, social and cultural rights, all of which entail three levels of legal obligation: the requirement to respect, to protect, and to fulfil (Eide *et al.* 2001). Civil and political rights like non-discrimination, the right to life, and the right to movement are interdependent with economic, social, and cultural rights, including the rights to health, housing, and food. Interdependency means that these rights cannot be considered separate from one another because they depend on each other for realization. The prohibition against discrimination is the best example of this, since non-discriminatory implementation is required of all human rights.

All human rights also have procedural requirements for implementation, including the requirements for participation, information, non-discrimination, non-retrogression, and remedy. In the context of DRR including CCA, the question of remedy is particularly important, since effective remedies must address the intersecting root causes at issue. Although there is some allowance under international law for derogations from these responsibilities during times of emergency, there are certain fundamental rights, including the prohibition against discrimination and the right to life, from which derogation is never allowed (ICCPR 1966, article 4, and see Box 14.1).

Box 14.1   Human Rights and Corporate Social Responsibility (CSR)

Jean Connolly Carmalt[1]

[1] City University of New York, New York, USA

As non-state actors, multinational and transnational corporations (MNCs and TNCs) cannot be signatories to the primary international human rights treaties. Nonetheless, private corporations are subject to direct and indirect human rights obligations. Direct obligations arise from a broad interpretation of human rights that has the basic and universal requirement that all actors must respect human dignity. Soft law instruments, including the consent-based UN Global Compact, also provide for direct obligations, albeit nonbinding ones. Finally, some domestic courts have ruled that corporations are obligated to refrain from serious violations of human rights; in the United States before the 2013 *Koibel v. Royal Dutch Petroleum* decision, for example, corporations were held financially liable for gross violations of human rights associated with their production and development activities (Holzmeyer 2009).

Indirect legal obligations arise from the state's responsibility to protect human rights. States may be held accountable for failing to protect from the activities of corporations when corporate behaviour clearly violates human rights standards. In 2001, for example, the African Commission ruled that Nigeria had failed to protect the people of Oganiland from human rights violations perpetrated by oil companies in the region (*SERAC and ESCR v. Nigeria* 2001). The violations included short-term, specific instances (e.g., burning of particular villages), but they also included a violation of the right to a healthy environment, which involves a broader spatial and temporal scale.

The obligation to respect human rights requires states to refrain from violating the right directly through state action or policy. A failure to respect often equates to a particular time and place in which a practice or policy violates international norms. It is therefore more enforceable than the obligations to protect and fulfil, which may involve prospective requirements instead of completed violations. For example, a state that gave out humanitarian assistance on the basis of race would violate its obligation to respect the prohibition against discrimination (CEDAW 1979). Under international law, the discrimination can be direct, or intentional, but it can also be indirect, or resulting in disparate impact for particular groups. For example, in the aftermath of Hurricane Katrina in the United States, a federal housing programme meant to rebuild damaged homes in New Orleans awarded money based on pre-Katrina home values. Because New Orleans, like other American cities, is characterized by residential racial segregation that correlates to property values, the result of this post-Katrina policy was that African Americans were given less money to rebuild the same homes as whites, whose houses were in more valuable neighbourhoods before the floods (GNOFHAC 2008). The effects of the policy therefore violated the international legal responsibility to refrain from enacting discriminatory policies. When the federal government settled the case in 2011, it also changed the award policy in order to address the disparate impact, thus bringing it in line with international requirements that the state respect the prohibition against direct or indirect forms of discrimination. While the outcome in this case was satisfactory, the question of how to respect non-discrimination across broader scales of time and scale remains more complex, since post-Katrina urbanization processes continue to reinforce discriminatory patterns of residential development.

The obligation to protect human rights requires states to ensure that rights are not violated by third parties or other outside forces. Courts and experts have interpreted this obligation to apply in cases of indirect threats, including disasters. In cases of indirect threats, the foreseeability of the harm is part of any determination of governmental responsibility. For example, in the *Budayeva* case, the European Court of Human Rights found that Russia had violated its duty to protect the lives of people living in the village of Tyrnauz because it had failed to warn or evacuate citizens in the face of predictable mudslides (*Budayeva v. Russia* 2008). Despite Russia's arguments that mudslides were acts of nature that could not be prevented by the government, the Court found that a state's duty to protect life encompasses threats arising from physical hazards. Russia's duty in this case was not to prevent the mudslides themselves, but to warn and evacuate its citizens, which illustrates the importance of defining a remedy clearly in order to enforce a human rights claim. Because the remedy sought in this case was compensation for individual survivors, the scope of the obligation in question was defined to be narrowly limited to a specific instance of failed governmental action. However, the obligation to protect against such mudslides could also be interpreted to include systemic shifts to governance (e.g., as related to water management or transportation networks) that would widen the spatial and temporal scope of the obligation.

Finally, the obligation to fulfil requires states to take the steps required to ensure a right, including the enactment of new policies or laws that prioritize human rights over competing concerns. Fulfilling human rights related to climate change and disasters, for example, requires states to engage in data collection and analysis related to the root causes of vulnerability to disaster, and then to enact policies and laws prioritizing risk reduction as a mechanism to ensure particular rights. For example, in Colombia's Cauca River Basin, one programme has performed a vulnerability assessment that includes the participation of local agricultural workers, as well as the social and political dimensions of challenges to the changing climate (IDRC/CRDI 2013; Tran *et al.* 2015). This attention to how and why vulnerability to physical hazards occurred, combined with an approach that prioritized the participation of the affected population, provides one example of how governance approaches to DRR, including CCA, can fulfil the human rights of people in the region. Nonetheless, the success of this programme will also depend on the transnational economic and political relationships that influence agricultural practices in the region; in other words, while national implementation of policies that prioritize human rights obligations are a step in the right direction, if such efforts are to be successful, they must also address the long-term, transnational relationships that contribute to vulnerability in the region.

Framing CCA in terms of DRR highlights interrelated issues of sustainability and development. However, at the heart of these discussions is a looming question of responsibility. Human rights law frames responsibility in terms of human dignity, which means that all other considerations – including economic and political ones – must be justified in terms of what best respects, protects, and fulfils the rights of human beings.

Violations of Human Rights Arising from Disasters and Climate Change

Human rights are important before, during, and after disasters. The conversation about human rights and disasters has focused largely on how humanitarian assistance, displacement, relocation, and rebuilding efforts should be informed by international human rights obligations. At a minimum, all humanitarian efforts must be carried out in a non-discriminatory way that respects the dignity of all human beings. Humanitarian assistance must be equally accessible to all, and it must be carried out in a participatory, culturally appropriate way. When evacuations are necessary, it is crucial that they be carried out in a non-discriminatory fashion. Indirect discrimination is particularly relevant in disaster contexts since those most affected by disasters are typically those who had less access to resources before the disaster occurred (IASC 2011; Wisner *et al.* 2004).

In addition to approaching the aftermath of a disaster from a rights-based perspective, there is a significant body of work around the rights of displaced populations, including a comprehensive analysis following the Indian Ocean tsunami of how the Guiding Principles on the Human Rights of Internally Displaced Persons (IDPs) relate to situations of disaster (Kälin 2005). Internal displacement is often one of the most severe and lasting human tolls of a disaster; the Indian Ocean tsunami, for example, displaced more than a million people within national borders (Kälin 2005). Since these displaced persons do not cross international boundaries, their human rights remain the responsibility of their home state; however, the displacement itself prompts a range of human rights concerns, including family reunification, housing, and participation in relevant political processes. The time frame of events prompting the displacement is a crucial consideration in the rights of IDPs; when shifts in climate result in rising sea levels, altered water tables, or otherwise long-term physical changes at the local level, displacement may become confused with economic migration.

Displaced populations also highlight the way in which particular rights shift when they are applied to a disaster situation. For example, the right to freedom of movement applies to everyone, but for a person who is displaced, that right translates into a right to choose whether to return home or resettle elsewhere (IASC 2011, p. 3). Similarly, the prohibition against forced evacuations may be suspended for situations where a serious, imminent threat to human life is present. Such evacuations must still adhere to other human rights principles, however, which means that evacuees are at all times treated with dignity and respect.

In addition to displaced populations, groups who are more vulnerable to disasters because of their role or status within society require special consideration in the context of rights-based approaches to DRR including CCA. Indigenous populations must be consulted about traditional practices, and any pre-existing vulnerabilities arising from indirect discrimination against minorities must be considered in DRR including CCA measures. The role of women also requires consideration of how women's roles in society relate to economic and political processes, since those roles frequently intersect with the construction of vulnerability. For example, in Papua New Guinea (PNG), women have traditionally produced staple food crops such as the sweet potato (Bayliss-Smith 1991). Since many families rely on subsistence crops, the role women play in producing those crops is directly related to sustainable food access, and by extension, to the resilience of the community to disasters. PNG faces a range of climate-related disaster risks that compound existing concerns about land degradation, demographic shifts, and negative effects of globalization (Mercer 2010). As Mercer illustrates, one of the crucial steps for CCA as part of DRR in PNG is the participation of the local population in defining potential threats and implementing adaptive measures. A human-rights-based approach would articulate that basis in terms of the obligation to implement the right to food through participatory measures. Since the right to food requires food to be accessible in a sustainable way, a human rights approach would view the steps needed to adapt to climate change and reduce risk as responsibilities of the PNG government to respect, protect, and fulfil the rights to life and food (CESCR 1999).

In 2009, the UN Office of the High Commissioner for Human Rights (OHCHR) published a report detailing the links between climate change and human rights (HRC 2009). While all human rights are impacted by the effects of climate change, the report highlighted two difficulties related to making legal claims that rights have been violated by the actions of governments that contribute to climate change. First, in order to make a valid claim, there must be a clear causal relationship between governmental actions and negative outcomes. The causal link is made more complicated by the construction of vulnerability through myriad interacting processes; nonetheless, causality does not mean complete causality. In other words, although the report rightly identifies the causal link as a hurdle for making human rights claims in relation to climate change, it is also possible to show how governmental actions (and inactions) contribute to, rather than exclusively cause, specific negative outcomes associated with climate change.

The second difficulty highlighted by the report pertains to DRR more broadly. The report points out that claims of harm are typically made after a violation occurs, whereas climate change processes involve ongoing processes with future predictions of harm (HRC 2009: 104). Legally, when no harm has yet occurred, it is not possible to claim that a remedy is needed. In this respect, the legal obligation of states to protect against indirect threats is particularly important; the harm of creating a more vulnerable population can be seen as an ongoing form of harm that might open the door to legal action. In addition, however, human rights standards can articulate the responsibilities of different actors in relation to DRR before a disaster occurs. In other words, using human rights law does not need to be limited to court actions that claim specific violations. Despite the legal challenge of bringing court-based claims related to climate change, it is clear that CCA is required in order to realize all human rights, and in particular, the rights to life, food, water, health, adequate housing, and self-determination.

Impacts of climate change tend to be more acute in poorer locations, where pre-existing vulnerabilities exacerbate the severity of weather-related events, outbreaks of disease, and fires or floods. Moreover, those same pre-existing vulnerabilities can make it harder to protect the right to life through adaptation measures. For example, poor housing conditions can increase a person's vulnerability to natural hazards, including extreme temperature fluctuations, some of which may be exacerbated by climate change. Heat waves can result in high fatality levels, as evidenced by the 2003 heat wave in Europe. One measure for both DRR and CCA might involve installing more air conditioning units or strengthening available care networks for the elderly. However, the condition of housing structures may present technical barriers to installing air conditioning, while the poverty giving rise to poor housing conditions may also interfere with the ability to pay the electricity costs of air conditioning as well as the ability of community members to provide increased care to the elderly during a heat wave. In either case, the same pre-existing vulnerabilities that contribute to disasters may make it difficult to adapt to climate change in ways that protect life, so climate change adds little new to the human rights challenges.

Traditionally, human rights analysis has focused on claimants who can demonstrate that a violation has already occurred, which then leads to a legal remedy (CCPR 2006). However, given the ongoing nature of the processes and relationships which create vulnerability to disasters, human rights can also be used to frame legal and ethical responsibilities of state and non-state actors before violations take place. In particular, the rights of these groups must be incorporated throughout any DRR including CCA approach, and policies affecting these populations should be implemented through participatory processes. Both offer an opportunity to incorporate human rights before a disaster occurs, which is to say that they offer mechanisms by which human rights can be incorporated into ongoing processes of DRR including CCA.

Health as a Human Right

DRR strategies conceive of disasters as constructed through multiple socio-economic and political processes (Hewitt 1983; Lewis 1999; Wisner *et al.* 2004). Vulnerability to disasters arises from the way in which those processes intersect with specific situations and risks of physical hazard. In other words, reducing risk and vulnerability to disaster requires policies and practices that empower marginalized groups most likely to suffer the adverse effects of natural hazards (Mercer 2010). Although many DRR strategies focus on community-level mechanisms by which to reduce vulnerability, the root causes of vulnerability also arise from transnational globalization processes that affect issues of land use, socio-economic conditions, and resource availability. These issues are particularly salient for disasters related to changing water availability, rising sea levels, or increased desertification, all of which are linked to climate change.

While all human rights are relevant to discussions of CCA and other DRR topics, the rights to life, health, movement, food, water, and non-discrimination are of particular importance. The remainder of this section uses the human right to health to consider how state obligations relate to CCA as part of DRR. There has already been a significant amount of discussion about the relationship between DRR and health, as well as discussion of health rights and climate change (see, e.g., UN 2014; McMichael *et al.* 2006). However, framing health as a human right links traditional concerns about health and disasters (e.g., accessibility to hospitals) to broader questions of governance and health systems.

The human right to health is not the right to be healthy, but is instead the right to the highest attainable standard of health (ICESCR 1966: art. 12). The right to health therefore draws from the World Health Organization's definition of health as 'a complete state of physical, mental and social well-being and not merely the absence of disease or infirmity' (WHO Constitution 1948). Legally, governments must respect, protect, and fulfil the availability, accessibility, acceptability, and quality of health services, along with the underlying preconditions to health, such as clean drinking water and a clean and healthy environment (CESCR 2000). These elements must be implemented in a participatory and non-discriminatory way; obligations are also non-retrogressive, meaning a country cannot undo positive changes.

The obligation to respect health means that states must refrain from directly creating unhealthy conditions, which include the availability of the underlying preconditions to health. Clean water and air, for example, must exist in sufficient amounts for human beings to attain the highest possible standard of health. This means that a governmental policy that degrades the physical environment in a way that contributes to decreased quantities of clean water (so that there is insufficient water for drinking and sanitation purposes) would violate the obligation to respect available preconditions to health. For example, when the Soviet Union began diverting water from the Aral Sea in the 1950s for agricultural purposes, it enacted a policy that significantly altered the local environment and decreased clean water supplies in the region. The Soviet policy therefore violated the right to health of residents in the region because it degraded available water supplies needed for the health of the population. More recent processes of urban and economic development, coupled with climate-change-related temperature fluctuations, have complicated the physical processes already underway in the region, which means that the initial violation of health rights has been compounded by subsequent local and non-local governance (Glantz 1999).

While availability of clean water is one prerequisite to the highest attainable standard of health, nondiscriminatory physical and economic access to health goods and services is also required under international law. Accessibility includes being able to reach a particular health facility, which is of crucial importance for displaced populations following a disaster, but it also includes non-discrimination and economic access to the entire health system. Health systems, like justice systems, are core institutions of any society (Hunt and Backman 2008). Therefore, systemic issues of discrimination or barriers to access constitute violations of the right to health. In many instances, these systemic violations are also relevant to DRR, including that connected to climate change. Like the mudslides in the *Budayeva* case, governments must protect people against known hazards, including through the provision of a functioning health system that decreases pre-disaster vulnerability and provides post-disaster care. Given the long-term nature of climate change, efforts to protect the population must also be long-term, which means that the way in which the health system is organized is of crucial importance.

Health as a human right means that availability and accessibility are coupled with the requirement that systems, services, and goods are acceptable in terms of basic standards and cultural norms. The requirement for acceptability can be particularly important for minority or indigenous communities that may have different understandings of what it means to be healthy. The Māori in New Zealand, for example, conceptualize health in a holistic way that incorporates the health of the environment in which they live. Their negative experiences with climate-related changes to the immediate physical environment are therefore part of an analysis of acceptable health. One potential concern for those who live on the coast might be the increased likelihood that mosquito vectors will increase the spread of vector-borne diseases (Jones *et al.* 2014). From a right to health perspective, in other words, acceptability in DRR including CCA measures requires a carefully defined, bottom-up risk assessment of the type used by the Red Cross and others. The Māori are also a good example of how the right to health may prompt transnational obligations, since there is a basic transnational human rights obligation to do no harm (Hunt and Backman 2008).

Finally, the right to health requires that all health goods and services, as well as the underlying preconditions to health, are of high quality and reflect best practices. Health systems must prioritize best practices, including best management practices before, during, and after disaster situations. The requirement that health goods and services be of high quality has transnational components linked to the obligation to do no harm; for example, if the United States bans the use of expired medicines, thus meeting its requirement to provide quality health goods, it cannot then approve the export of those same medicines to a poorer country (Hunt and Backman 2008). Similarly, countries engaged in lowering their carbon emissions in order to decrease the impacts of climate change globally are fulfilling an obligation to take measures that will increase the quality of air, water, and other determinants of health.

Analysing climate change as part of DRR means that physical phenomena must be understood as continuing processes, rather than single events, and it means that the construction of vulnerability is deeply entwined with globalized trade and development processes. In other words, from a human rights perspective, analysing CCA as part of DRR means that both the spatial and temporal scope of analysis must be expanded. Using human rights to frame the analysis provides an opportunity to specifically articulate which responsibility must be assumed by which actor, and to ensure that the dignity and wellbeing of the population remains the utmost priority for all actors.

Recommendations

As Kelman and Gaillard note, 'climate change resulting from human activity is a long-term, global disaster' which, like many other hazard drivers, has root causes that include human actions, decisions, and policies (Kelman and Gaillard 2010: 30). Human rights provide a mechanism by which to understand the legal and ethical responsibilities associated with those root causes. However, future research and action is needed in order to incorporate international human rights standards into the CCA/DRR conversation. In particular:

- Governments can craft policies that prioritize non-discrimination, participation, and protection of particular rights (especially life, health, and food) and implement such policies as part of larger DRR including CCA initiatives.
- Private corporations can prioritize basic human rights principles as a matter of corporate policy.
- Human rights lawyers and experts can expand on the legal responsibilities involved with long-term, extra-territorial, and ongoing processes that construct vulnerability, and where appropriate, bring suits to challenge the validity of policies and practices supporting such processes (see Box 14.2).
- Policy experts can prioritize non-discrimination (direct and indirect) and participation of affected populations throughout policy-creation processes.
- Scholars can analyse and clarify how the violation of human rights with regard to actions, decisions, and policies contributes to vulnerability and risk. Given the complex nature of how vulnerability comes about, scholars have a key role to play in explaining how physical processes at multiple geographic scales are connected to economic, political, and social relationships.

Box 14.2   Legal Responsibilities Associated with Long-term Processes and Extraterritorial Harm

Jean Connolly Carmalt[1]

[1] City University of New York, New York, USA

There are two distinct challenges for using human rights law to define responsibilities for DRR: first, how human rights violations might be prevented by shaping policies that change long-term

---

processes associated with vulnerability, and second, how obligations apply extra-territorially. In both cases, the way in which violations occur (which for DRR including CCA means understanding the complex creation of vulnerability) is key to understanding the legal responsibilities involved.

Implementing rights-based strategies to decrease the risk of disasters, including those associated with climate change, speaks to the need for a legal link between indirect discrimination, the creation of vulnerability, and the violations of rights associated with disasters. The example of heat waves and elderly populations without care networks mentioned above provides one such link, as does the racial and income segregation of pre-Katrina New Orleans. Human rights lawyers who wish to expand upon current legal arguments for rights-based policy practices as part of DRR might, therefore, look to these links in order to argue for new policies that prioritize non-discrimination (direct and indirect) as a matter of DRR including CCA. Such preventive policies could be framed as remedies, since indirect discrimination violates human rights, or they could be framed in terms of protecting against likely threats to life. Either way, a rights-based approach would require that all policy decisions prioritize the importance of human dignity.

Extraterritorial responsibilities similarly require a complex understanding of how vulnerability is created, together with a clear legal link between specific processes associated with that vulnerability and actions taken by extraterritorial actors. The Responsibility to Protect (R2P) is often cited as one way in which extraterritorial obligations require outside states to intervene when a government is unable or unwilling to respond to a disaster (Allan and O'Donnell 2013); however, R2P has largely been discussed in terms of actions taken instead of in terms of curtailments of existing activities that have an extraterritorial effect. The latter warrants consideration in the context of DRR including CCA, particularly in light of the way in which extraterritorial jurisdiction is readily exercised in other issue areas (e.g., the US application of the Foreign Corrupt Practices Act (Koehler 2012)).

References

Allan, C. and O'Donnell, T. (2013) 'An offer you cannot refuse? Natural disasters, the politics of aid refusal and potential legal implications', *Amsterdam Law Forum* 5, 1: 36–63.

Bayliss-Smith, T. (1991) 'Food security and agricultural sustainability in the New Guinea Highlands: Vulnerable people, vulnerable places', *IDS Bulletin* 22, 3: 5–11.

Budayeva and others v. Russia. (2008) *European Court of Human Rights 15339/02 & Ors (20 March 2008)*, European Court of Human Rights.

CCPR (Centre for Civil and Political Rights). (2006) *Aalbersberg and 2,084 other Dutch citizens v. Netherlands*, Communication No. 1440/2005, United Nations document No. CCPR/C/87/D/1440/2005, New York: United Nations.

CEDAW (Convention on the Elimination of All Forms of Discrimination Against Women). (1979) *United Nations General Assembly, Convention on the Elimination of All Forms of Discrimination Against Women, 18 December 1979*, New York: United Nations, Treaty Series 1249: 13.

CESCR (Committee on Economic, Social and Cultural Rights). (1999) *General Comment No. 12: The Right to Adequate Food (Art. 11 of the Covenant), 12 May 1999*. Online http://www.refworld.org/docid/4538838c11.html (accessed 7 August 2015).

CESCR (Committee on Economic, Social and Cultural Rights). (2000) *General Comment No. 14: The Right to the Highest Attainable Standard of Health (Art. 12 of the Covenant), 11 August 2000*. Online http://www.refworld.org/docid/4538838d0.html (accessed 5 September 2016).

Eide, A., Krause, C., and Rosas, A. (eds) (2001) *Economic, Social, and Cultural Rights: A Textbook, Second Revised Edition*, UK: Martinus Nijhoff Publishers and Brill Academic Publishers.

Glantz, M.H. (ed.) (1999) *Creeping Environmental Problems and Sustainable Development in the Aral Sea Basin*, Cambridge, UK: Cambridge University Press.

GNOFHAC (Greater New Orleans Fair Housing Action Center). (2008) *Greater New Orleans Fair Housing Action Center v. United States Department of Housing and Urban Development, Complaint for Declaratory*

and Injunctive Relief. Online http://www.gnofairhousing.org/wp-content/uploads/2011/12/11-12-08_
RoadHomeComplaint.pdf (accessed 16 July 2015).

Hewitt, K. (ed.) (1983) *Interpretations of Calamity from the Viewpoint of Human Ecology*, Boston: Allen & Unwin.

Holzmeyer, C. (2009) 'Human rights in an era of neoliberal globalization: The Alien Tort Claims Act and Grassroots Mobilization in Doe v. Unocal', *Law and Society Review* 43, 2: 271–304.

HRC (Human Rights Council). (2009) *Report of the Office of the United Nations High Commissioner for Human Rights on the Relationship between Climate Change and Human Rights.* UN Doc. A/HRC/10/61, New York: United Nations.

Hunt, P. and Backman, G. (2008) 'Health systems and the right to the highest attainable standard of health', *Health and Human Rights* 10, 1: 81–92.

IASC (Inter-Agency Standing Committee). (2011) 'IASC operational guidelines on the protection of persons in situations of natural disasters', *Brookings-Bern Project on Internal Displacement.* Online http://www.ohchr.org/Documents/Issues/IDPersons/OperationalGuidelines_IDP.pdf (accessed 17 July 2015).

ICCPR (International Covenant on Civil and Political Rights). (1966) *International Covenant on Civil and Political Rights, 16 December 1966*, New York: United Nations, Treaty Series 999: 171.

ICESCR (International Covenant on Economic, Social and Cultural Rights). (1966) *International Covenant on Economic, Social and Cultural Rights, 16 December 1966,* New York: United Nations, Article 12.

IDRC/CRDI (International Development Research Centre/Centre de recherches pour le développement international). (2013) 'Climate change, vulnerability, and health in Colombia and Bolivia', *IDRC/CRDI.* Online http://www.idrc.ca/EN/Themes/Natural_Resources/Pages/ProjectDetails.aspx?ProjectNumber= 106914 (accessed 17 July 2015).

Jones, R., Bennett, H., Keating, G. and Blaiklock, A. (2014) 'Climate change and the right to health for Māori in Aotearoa/New Zealand', *Health and Human Rights* 16, 1: 54–68.

Kälin, W. (2005) 'Protection of internally displaced persons in situations of natural disaster: A working visit to Asia by the representative of the United Nations Secretary-General on the human rights of internally displaced persons', *United Nations 27 February to 5 March 2005.* Online http://www.ohchr.org/Documents/Issues/IDPersons/Tsunami.pdf (accessed 6 August 2015).

Kelman, I. and Gaillard, JC (2010) 'Embedding climate change adaptation within disaster risk reduction', in R. Shaw, J.M. Pulhin and J.J. Pereira (eds), *Climate Change Adaptation and Disaster Risk Reduction: Issues and Challenges*, Bingley: Emerald, pp. 23–46.

Koehler, M. (2012) 'The story of the Foreign Corrupt Practices Act', *Ohio State Law Journal* 73: 929–1004.

Lewis, J. (1999) *Development in Disaster-prone Places: Studies of Vulnerability*, London: Intermediate Technology Publications.

McMichael, A. J., Woodruff, R.E. and Hales, S. (2006) 'Climate change and human health: Present and future risks', *The Lancet* 367, 9513: 859–869.

Mercer, J. (2010) 'Disaster risk reduction or climate change adaptation: Are we reinventing the wheel?', *Journal of International Development* 22, 2: 247–264.

Tran, T., Mrad, S. and Mantilla, G. (2015) 'Climate change and health in Colombia', in W. Filho (ed.), *Handbook of Climate Change Adaptation*, Berlin: Springer, pp. 195–213.

UN (United Nations). (2014) *Health and Disaster Risk: A Contribution by the United Nations to the Consultation Leading to the Third World Conference on Disaster Risk Reduction (WCDRR)*, Geneva: United Nations. Online http://www.wcdrr.org/documents/wcdrr/prepcom1/UN/ATTR8FWA.pdf (accessed 4 August, 2015).

Wisner, B., Blaikie, P., Cannon, T. and Davis, I. (2004) *At Risk. Natural Hazards, People's Vulnerability and Disasters*, London: Routledge.

World Health Organization (WHO) Constitution. (1948) *Preamble to the Constitution of the World Health Organization as adopted by the International Health Conference, New York, 19–22 June, 1946; signed on 22 July 1946 by the representatives of 61 States (Official Records of the World Health Organization, no. 2, p. 100) and entered into force on 7 April 1948,* New York: WHO.

15

VIOLENT CONFLICT AND DISASTER RISK REDUCTION INCLUDING CLIMATE CHANGE ADAPTATION

Richard S. Olson and Vincent T. Gawronski

The Question

As argued throughout this volume, climate change adaptation (CCA) is an essential component of a larger set of disaster risk reduction (DRR) strategies and programmes. While both may be considered loss reduction values in and of themselves, they may also be considered *instrumental* values to achieving another, and in some ways a higher order value: avoiding or at least reducing violent conflict within or between communities, societies, or nation-states. That is, from this latter perspective DRR, including CCA, constitutes conflict mitigation potentialities, and may even contribute to resolving ongoing conflicts. The obvious challenge, however, is to address the prior question: Is there a relationship between (a) environmental hazard events or processes and their drivers, including climate change, and (b) violent conflict, and (c) if so, how strong are those respective relationships? If there are no such connections, then DRR including CCA as an instrumental value is moot, and it can be valued more simply as a loss reduction measure including for climate change events and disasters.

Before exploring that question, however, we should first frame this chapter's challenge in the larger context of disaster research, where disaster risk has long been understood as an outcome of an equation based on hazards, exposures, and vulnerabilities, mitigated by capacities for response and resilience. Because climate change generally influences one set of hazards – hydro-meteorological – it can have a ripple effect on disaster (and in extreme combinations, catastrophe) risk. CCA should therefore be seen as a subset of DRR, because the latter comprises measures to reduce risk from a range of hazards broader than just the hydro-meteorological, including earthquakes, tsunamis, volcanic eruptions, lahars, and others. In that sense, CCA is 'nested' within a broader set of risk reduction measures, including prevention, mitigation, preparedness, and improved community resilience.

Climate Change and Violent Conflict: Everything Is Tentative

Uncertainty dominates much of the academic literature on climate change, climate change effects, and their longer-term outcomes, and the breadth and depth of these uncertainties manifest in interesting ways. For example, it is worth noting the prevalence of so many words that connote tentativeness in the literature on potential climate change–conflict linkages: *may*, *could*, *might*, *uncertain*, *possibly*, and *perhaps* pepper academic works especially, and it seems that the more deliberative the work, the more caution is expressed (a 2013 US National Research Council report is a good example).

The logical place to begin exploring possible relationships between climate change effects and violent conflict is the literature on disasters and their complex and usually indirect relationships with socioeconomic and political processes and outcomes. In this literature disasters are predominantly seen as both cause and consequence of economic underdevelopment, crises, violence, crime, and migration.

Hazards, including those influenced by climate change, occur in contexts ranging from developed, wealthy, and stable nation-states to less-developed, poor, and fragile nation-states. Moreover, extreme events and disasters in rural areas are very different from those experienced in urban areas. Impact effects may be immediate and severe or slow, creeping, and highly localised. Small island developing states, for example, are already feeling direct climate change consequences, namely sea-level rise and changing precipitation regimes.

Central America is a good example of how vulnerabilities interact indirectly and directly with violence, crime, and migration. Hundreds of thousands of Central Americans have migrated to the United States since 1980, including tens of thousands of unaccompanied children, and tens of thousands have been deported back, only to exacerbate existing conditions in Central America. Hurricanes, tropical storms and droughts (and criminality and violent conflict) affect everyone in the region, and they have contributed to the push-pull factors of migration.

Disasters and Violent Conflict: Not Conclusive Except at the Case Level

At the international level, there is little to suggest that 'disasters' broadly defined are related in any way to interstate aggression or war. Nelson (2010), however, argues that a large-scale disaster may contribute to an increased likelihood of conflict initiation, but there is evidence that in certain situations, a damaging natural event may also lead to 'disaster diplomacy' and improved relations between previously adversarial states (Kelman 2012); at the very least, disaster diplomacy can catalyse public compassion and positive diplomatic action between previously rival states (Akcinaroglu *et al.* 2011). In their more recent review of how disasters might contribute to conflict *or* cooperation, Streich and Mislan (2014: 70, our emphasis) arrived at three conclusions: '(1) Disasters generally do not lead to conflict initiation, (2) Disasters generally do not lead to *new* cooperative processes, but (3) Disasters can catalyse or reinforce *existing* rapprochement processes between conflict-prone dyads.'

At the intrastate or internal level, and in the case-study tradition, research over many years has built up an impressive array of stories where a disaster, especially if a government has handled the response and/or reconstruction poorly, has catalysed, triggered, reignited, or accelerated violence or armed conflict, usually anti-governmental and at times reaching the level of guerrilla or civil war (Brancati 2007; Gawronski and Olson 2013; Nel and Righarts 2008; Quarantelli and Dynes 1976). The linkages are complex, however, and can even be reciprocal, as disasters can impact ongoing conflicts, and ongoing conflicts can affect disaster impacts. Disasters occurring in conflict zones are especially problematic, and prevailing negative forces tend to compound into complex humanitarian crises.

A particularly striking example of complexity occurred in the aftermath of the 2004 Asian tsunami, where the government of Indonesia and separatists in the heavily impacted province of Aceh negotiated an end to a long-running conflict, while the same hazard event contributed to vastly increased violence in another long-running conflict, in Sri Lanka. In neither case did the earthquake and tsunami cause the conflict-related outcome, but the hazard event did influence those outcomes.

Overall, while the great majority of case examples of a disaster–violent conflict intrastate nexus point to violence, the entire set may simply be the result of selection bias where authors (to be transparent, including us) found post-impact violence and worked causally backwards. This cautionary observation explains why attempts at more systematic and quantitative analyses of the putative relationship between disasters and violent conflict are so important.

Expanding and improving upon some early quantitative work by one of the authors here (Drury and Olson 1998), Nel and Righarts (2008) built a multi-hazard disaster event and violent civil conflict dataset with 8,203 observations covering 187 political units from 1950 to 2000 inclusive. They demonstrated a significant positive relationship between 'natural disasters' and risk of violent civil conflict. Brancati (2007) had previously found earthquake disasters specifically to be important factors in explaining social conflict based on a statistical analysis of 185 countries over the period 1975–2002. Brancati's results indicated that earthquakes increase the likelihood of internal conflict, but their effects are greater for higher magnitude events striking densely populated areas in countries with a lower gross domestic product and pre-existing conflicts, similar to what Omelicheva (2011) later found: Disasters generally become catalysts of political instability primarily in conflict-ridden or conflict-prone states.

Specifically focusing on climate events, however, Bergholt and Lujala (2012) did not find a connection between climate-related disasters and increased risk of armed conflict. Slettebak (2012: 174) tested whether storms, floods, and droughts could add explanatory power to an established model of civil conflict – and found that climate-related disasters can sometimes contribute to reducing conflict. More recently, Nardulli *et al.* (2015: 330) analysed the destabilising effects of storms and floods and cautiously concluded that they 'have a highly variable effect on civil unrest, particularly violent unrest. These variable effects differ across space and time'.

In sum, there is some case evidence (albeit selected on the dependent variable) for a relationship between (a) disasters involving environmental hazards, and (b) intrastate or internal violence, but it is highly dependent on specific national histories and contexts. In-depth qualitative case-study research may reveal more disaster–violent conflict linkages. However, such research is time consuming, requiring an historical perspective and significant country expertise (Box 15.1).

Box 15.1   Monitoring Myanmar and Thailand

Richard S. Olson[1] and Vincent T. Gawronski[2]

[1] Florida International University, Miami, Florida, USA
[2] Birmingham-Southern College, Birmingham, Alabama, USA

It often takes decades to determine if, and how, major hazard events lead to violent conflict, and Myanmar and Thailand are two countries that merit close monitoring in coming years for two reasons. First, climate change might already be affecting droughts, storms, floods, sea-level rise, decreasing agricultural and fishery yields, and health-related problems. Second, just since 2008, disasters have killed nearly 140,000 (Myanmar) and 1,400 (Thailand) people, with economic losses totalling more than USD 4 billion in Myanmar and more than USD 41 billion in the more developed Thailand. More specifically, Cyclone Nargis hit the Ayeyawady Delta in May 2008, causing nearly all of Myanmar's losses, and the 2011 floods constituted Thailand's worst disaster in recent memory.

In Myanmar, Cyclone Nargis hit during efforts to hold a constitutional referendum, and Taylor

(2015: 915) observed that 'Cyclone Nargis struck against a background of twenty years of instant but usually uninformed politicisation of any event, including now a natural disaster with huge humanitarian implications'. Myanmar's military junta initially denied entry of international relief organisations, apparently seeing foreign aid workers as 'uncontrollable' and therefore a potential threat. The regime even circulated stories that the United Nations tenet of Responsibility to Protect (R2P) was a cover for American/Western military intervention. In a country rife with political rumouring, refusing international assistance may have been a domestic political manoeuvre to discredit certain government officials. Other rumours circulated that the disaster provided an opportunity to punish the opposition Kayin ethnic minority, and it certainly exacerbated the plight of the persecuted Rohingya people.

By 2010, political conflict in Thailand had turned violent, with populist business tycoon Thaksin Shinawatra's party having won elections in 2001 and 2005. Thaksin served as Prime Minister until he was overthrown in a coup in 2006, but his earlier successes had threatened established elites. Coincident with the 2010 drought, which contributed to rural-to-urban migration, political tensions led to violent clashes between mostly rural, pro-Thaksin 'red shirts' and pro-royalist 'yellow shirts', primarily in Bangkok.

When the 2011 floods hit the country, they seemed to have temporarily diverted central government attention and resources. Local conflicts, sometimes armed or violent, emerged over flood barriers and diversion efforts. While conflicts over water management, flood control, and resources are nothing new in Thailand, a lack of political will and institutional weaknesses had prevented actual implementation of DRR strategies, and the existing flood control infrastructure was not well maintained/managed due to unclear or competing agency mandates. Indeed, the long-term political fallout of the 2011 floods in a context of climate change requires an equally long-term assessment.

Results from the quantitative analyses are mixed and simply not strong enough for conclusive comfort in either direction. In the end, we agree with Nardulli *et al.* (2015) that because disaster impacts are so often highly localised, higher resolution temporal and spatial data on any ensuing violence or civil unrest are needed for anything beyond tentative conclusions. Disasters over the next few decades and the application of ever-improving technologies and sophisticated methodologies (GIS mapping, 'big data', and supercomputing) should produce more reliable research results.

For this chapter we have grouped the more specific literature on potential climate change–conflict linkages around three nodes or poles: (1) Popular or 'Gloom and Doom', (2) National Security, which is quite contentious and requires sidebar treatment of climate change and 'human' security, and (3) Scholarly/Academic.

Pole One: Climate Change, Violence, and Popular 'Gloom and Doom' Treatments

Homer-Dixon's (2006, for example) series of articles and books brought the issue of 'resource scarcity', which was quickly dubbed 'neo-Malthusian' for its bleak outlook, to wide attention. In agenda-setting terms, Homer-Dixon's works elevated the possibility of resource-based conflicts onto public, governmental, institutional (including national security), and academic agendas.

A set of publications then followed Homer-Dixon (e.g., Welzer 2012). Often bestsellers, books in this genre tend to be fatalist, alarmist, and/or sensational (and seemed to have inspired movie scripts). The academic research community has consistently challenged these various works and their underlying thematic, arguing that apocalyptic scenarios may sell books but vastly simplify cause-and-effect relationships and underestimate conflict resolution mechanisms from the individual to the international levels. Indeed, the imagined gloom-and-doom scenarios found in some of this literature are so dark and pessimistic (with even the human species ending) as to make the implementation of DRR (including CCA) efforts seem hopeless endeavours.

Pole Two: The (Contentious) 'Securitisation' of Climate Change

What might be called the 'national security pole' of the climate change effects–violent conflict outcomes literature derives principally from the defence, intelligence, and related consulting communities, which together comprise the U.S. military/security–industrial complex. This pole centres on how climate change may affect U.S. national interests and on how climate change could exacerbate hunger, poverty, instability, and conflict. In fact, U.S. defence strategy now refers to climate change as a 'threat multiplier', and suggests possible indirect climate change–terrorism linkages. Military planners also think of climate change in terms of increased mission loads.

Of course, the U.S. military and intelligence services had been paying attention to climate change-related security issues since the end of the Cold War (around the time of Homer-Dixon's first publications), but those concerns are now much more institutionalised, and the threat multiplier theme pervades global security and development discourses. Various intergovernmental organisations and U.N. agencies now understand how climate change, extreme weather, and poor policy responses can exacerbate humanitarian crises and ongoing conflicts.

The problem remains, nonetheless, that while the climate change discourse and national security interests (broadly defined) seem to remain inextricably linked, little consensus exists on precisely – and reliably – *how*, and then on what to do about it. Even in the one area where scholarly research indicated reasonable consensus (that climate change-induced interstate violence was highly *un*likely), the melting of Arctic ice and the opening of the Northwest Passage to commercial shipping and resource exploitation may contribute to friction among the Arctic states of Canada, Denmark, Finland, Iceland, Norway, Russia, Sweden, and the United States.

As noted above, the securitisation of the climate change discourse alarmed a substantial portion of the academic community, who were and remained concerned about conclusions (and more worrying, policies) based on weak, distorted, or non-existent empirical evidence. As Gleditsch (2012: 3) argued, '[F]raming the climate issue as a security problem could possibly influence the perceptions of the actors and contribute to a self-fulfilling prophecy'.

Another word of caution here, however: This back and forth between the national security community and the academic research community (parts of it) on climate change effects and possibly violent outcomes may appear to be a collision of worldviews, but it is actually more like two worldviews sliding past each other, because they start from very different premises. The academic research community values the 'Search for General Patterns' and takes note of, but does not particularly emphasise, exceptions or outliers – to which disaster researchers are inexorably attracted. The national security community, particularly in the United States, however, lives in a high stress post-9/11 world of 'We Can't Be Wrong Even Once on a Big Thing', even if that Big Thing is a statistical outlier or exceptional case in the academic frame. That is, the national security community cannot afford a false negative, to have 'missed' a crucial instance of climate change triggering or fuelling a major national security threat. Viewed that way, the national security and academic research communities are actually more complementary than they often appear, even to each other.

The challenge is conceptualising solutions, advocating policies, and applying strategies to mitigate climate change threat-multiplier effects, which is where the folding of CCA into DRR comes back into play. The U.S. military and many foreign national militaries are increasingly moving beyond mere disaster and humanitarian response. Many are now applying DRR principles, concepts, and strategies, which has also been changing public perceptions of the military in countries with unsavoury histories of military rule and/or repression.

Climate Change and Human Security

Before moving to our Pole Three ('scholarly/academic') of the climate change effects literature, an important variant in the climate change security literature derives from those concerned with *human* security and humanitarian crises, and it comfortably bridges researchers and practitioners. Climate change effects will very likely influence localised shortages of arable land, living space, food, and water. Competition for scarce resources and the movements of people may result in an increasing number of complex humanitarian crises (as of this writing an ongoing example being the migration flows out of the Middle East into Europe). People moving to places where others already reside tends to generate a host of unexpected positive and negative consequences.

The human security pole also tends to adopt a human rights perspective on climate change, and sometimes on other hazards, so that human rights are linked to all disasters and embrace a much broader understanding of what constitutes violence. The core argument is that all types of disasters are physically violent, but they can also generate other forms of violence – direct, indirect or structural, and/or symbolic or cultural – and they can occur within larger contexts of conflict, violence, and humanitarian crises.

A particular human security dimension revolves around migration, which is a common but sometimes underappreciated response or adaptation to disasters. In 2015, the number of refugees in the world surpassed 60 million, but 'climate/environmental refugee' per se is not yet an official classification in international law, so that number may be low in a practical sense. Because of the already-existing tensions over the legally recognised status of 'political' refugee, however, it seems unlikely that 'climate' or 'crisis' refugees will acquire the same protections as 'political' refugees, at least anytime soon.

Indeed, the migration of individuals, families, and communities has been a climate change coping or adaptation strategy since people began moving about the planet. The mass migration of populations may sometimes be a significant conflict driver, but the empirical evidence is thus far uneven and inconclusive. That is, the complexities of push-pull migration factors make establishing simple climate change-migration–conflict relationships impossible. Nonetheless, a special note should be made of climate-induced migration to congested urban areas specifically. The argument is that population pressures combined with higher than normal migration will put additional strain on urban resources, institutions, and infrastructures, and social disorder may result, especially when conflicts are already present and effective political institutions and good governance are lacking. High levels of poverty and economic inequality simply exacerbate the social stressors.

The human security research and practitioner communities, including nongovernmental and intergovernmental organisations, have already been nesting CCA into DRR with their ethos of sustainable development and promotion of community resilience. Moreover, providing basic needs, improving quality of life, and ensuring a better future are often understood as conflict mitigation strategies.

Pole Three: Scholarly/Academic Analyses of (Possible) Climate Change–Conflict Connections

The academic research community is cautious by inclination and training. As noted at the outset of this chapter, conclusions in peer-reviewed journal articles, reports, books, and book chapters about a possible nexus between climate change and violence are rife with caveats, cautions, and tentativeness. In addition, a quantitative-qualitative methodological tension is evident among scholars, where the latter argue that in-depth case studies are able to draw out linkages not evidenced in cross-national dataset analyses. The counterargument, of course, is that quantitative treatments more than case studies will reveal the general rule or relationship between the variables of interest.

Possible climate change–conflict connections and pathways have been analysed, and no clear, direct connections have been empirically demonstrated. This might be due to methodological problems and imprecise data collection. Conceptually, it has been difficult to measure precisely climate change effects and various types and scales of conflict. Climatic changes, however, have contributed to the historical rise or demise of civilisations such as the Maya (Box 15.2) and to wars and dynastic cycles in China (Zhang *et al.* 2006).

Box 15.2   Climate Change and the Collapse of the Maya

Richard S. Olson[1] and Vincent T. Gawronski[2]

[1] Florida International University, Miami, Florida, USA
[2] Birmingham-Southern College, Birmingham, Alabama, USA

Archaeologists and historians have attributed the decline of several ancient civilisations to multiple factors, including environmental degradation and climate change. Climates influence agricultural production, and when they change, food shortages can follow. This is particularly true of drought. While the pathways are complex, droughts and conflict likely contributed to the demise of Mesopotamia's Akkadian Empire (BC 2334–2154), the Moche (AD 100–800) in coastal Peru, the Tiwanaku (AD 200–1000) in the Peru–Bolivia *altiplano*, and the Anasazi (AD 100–1600) in the American Southwest, among others.

Perhaps most clearly in Mesoamerica, climate change played a role in the collapse of the Classic Maya (AD 250–900) civilisation (Peterson and Haug 2005). Permanent surface water is rare in the Yucatán Peninsula because it tends to dissolve into the limestone, forming caverns and *cenotes*. An unusually wet 200+ year period (AD 440–660) enabled the Maya to flourish. Agricultural production expanded and the population surpassed 13 million. During this period, the Maya built approximately 60 cities, many with massive pyramids and temples. Tikal, Caracol, Calakmul, Coba, and Copán were the largest urban areas, with populations ranging from 20,000 to more than 125,000 inhabitants. Celestial observatories, such as the one at Chichen Itzá, and smaller ceremonial centres also emerged, with the Maya practising a complex cosmology and developing an accurate calendar, a base-20 number system with a concept of zero, and hieroglyphic writing.

At least three boom-and-bust, wet–dry cycles contributed to the decline of the Classic Maya civilisation. Control of water resources was connected to centralised political authority, but drought, exacerbated by rainforest clear cutting, undermined ruling elite legitimacy, as indicated by increased conflicts among the numerous city-states, heightened rivalries, strategic alliances, and wars during one mega-drought period (AD 600–1000). As the Mayan population reached its environmental limits and competition for poorly managed scarce resources increased, dynastic rule withered and people eventually began to abandon the (unsustainably) large cities for smaller and more subsistence-oriented settlements. Of course, the Maya did not contribute to climate change, but rulers failed (or were unable) to adapt. As Kennett and Beach (2013: 95) concluded:

> The Classic Maya collapse was first and foremost a political failure with initial effects on the elite sectors (kings and their courts) that ultimately undermined the economy and stimulated the decentralisation of Maya civic-ceremonial centres and the reorganisation of regional populations.

It was this pattern of remnant pieces of a once great civilisation that the Europeans encountered in the early sixteenth century.

At a sub-regional level, however, qualitative case arguments (e.g., Davis 2002) focused on the dire consequences of particularly strong ENSOs (El Niño-Southern Oscillations), and Hsiang *et al.* (2011) found a positive and robust statistical association between strong ENSOs and onsets of civil violence. Subsequently, Burke *et al.* (2015) found in their meta-analysis of 55 quantitative studies that the median effect of one standard deviation change in climate variables is associated with a 14 per cent change in the risk of intergroup conflict and a 4 per cent change in interpersonal violence.

Several academic publishing houses have dedicated entire volumes to potential climate change–conflict relationships. The weight of scientific evidence, however, has not yet confirmed a causal relationship between climate change and violent conflict – but neither has it been dismissed. There are at least three perspectives: (1) climate change will directly cause resource wars and violence, (2) climate change will indirectly result in violent conflict and national security challenges, and (3) the evidence is lacking to demonstrate any linkages.

An interesting strand of the Pole Three literature is wound around food, or more precisely its price volatility and access to it – and their possible connections to civil unrest or other types of violence. United Nations experts are anticipating negative climate change effects on food security. Again, however, there is little agreement in the literature on how the mechanisms might interact or not interact, but a pattern of resource scarcities, economic shocks, and food price volatility and violent responses is emerging. Indeed, access to food is the crucial underlying concept and in most places price determines access. So, with any real or perceived food supply problem affecting price, climate change impacts will be a factor in (or blamed for) food price spikes and food security crises, which may then lead to civil unrest.

Another more specific Pole Three strand focuses on water. From one perspective, climate change is all about water, what state it is in, and where it is. Historically, there seems to be as much cooperation as conflict over water resources, but the uncertainties that climate change generates mean that the past is an unreliable guide to future outcomes. Indeed, water abundance might even be linked to interstate war: 'Political violence is more prevalent when basic needs are met and when the tactical environment is more conducive to attacks – conditions that hold when water is comparatively abundant' (Salehyan and Hendrix 2014: 239).

Similarly, Selby and Hoffman (2014: 360) analysed Sudan and (now) South Sudan and found 'plentiful evidence that water abundance, and state-directed processes of economic development and internal colonisation relating to water, have had violent consequences'. Reinforcing these conclusions from the opposite direction, Devlin and Hendrix (2014) found a conflict-dampening effect when neighbouring countries experience water scarcity.

As usual, however, there are cautions, with von Uexkull (2014: 16) arguing that the literature on water scarcity or abundance is interesting but inconclusive because 'existing research has not sufficiently taken into account the local vulnerability and coping capacity that condition the effect of drought'. Along that line, the question of drought effects – and drought outcomes – has recently achieved both high profile and immediacy because of its apparent contributory effects to the complex of interrelated factors that led to the ongoing civil/proxy war in Syria (Box 15.3).

Box 15.3   Climate Change and the Syrian Conflict

Richard S. Olson[1] and Vincent T. Gawronski[2]

[1] Florida International University, Miami, Florida, USA
[2] Birmingham-Southern College, Birmingham, Alabama, USA

It seems unlikely that war or even violent conflict will be attributed – directly at least – to climate change. Nonetheless, strong circumstantial evidence suggests that a climatic event was a significant driver contributing to the outbreak of the Syrian conflict (Gleick 2014; Kelley *et al.* 2015). More specifically, an extreme and extended drought (2006–2011) led to the internal migration of nearly two million people from rural farming areas to city peripheries, which then generated conflict in and around those cities. That particular drought, however, is part of a much larger pattern.

Scientists have been tracking and documenting a century-long warming and drying trend in the Eastern Mediterranean, where from 1900 to 2005 what is now Syria experienced six significant droughts. According to Kelley *et al.* (2015), climate change made the 2006–2011 drought more than twice as likely to occur and where, as in the United States during the 1930s 'Dust Bowl', the dry winds further depleted already poor-quality topsoil.

As a hazard event, the drought problem in Syria was exacerbated by increasing human exposure. In 1961 the country counted 4.5 million inhabitants and 1.3 hectares of arable land per capita. By 2011, however, the population had grown to 22.5 million with arable hectares per capita falling to 0.21. Refugee inflows from Lebanon, Palestine, and Iraq compounded the stresses – keeping in mind the Syrian government's role in those conflicts. As if that were not enough, President Hafez al-Assad (1971–2000) had pursued water-intensive and fundamentally unsustainable agricultural policies, which his successor son Bashar al-Assad maintained, deepening the rural crisis.

In the end, the war in Syria, which as of 2015 has generated more than 7.5 million internally displaced persons and 4 million refugees, was triggered by a number of factors in addition to the 2006–2011 drought, including increasing poverty and unemployment, corruption, lack of basic political freedoms, and a growing rural–urban divide. These factors are neither independent nor isolated, but reciprocally affect each other.

If there is an academic consensus, it might be that climate change effects, including water scarcity or abundance, sea-level rise, influences on hazards, and resource scarcities, interact with so many other socioeconomic and political phenomena, such as corruption, poor governance, poverty, inequality, migration, conflict, and instability, that establishing direct causality is impossible.

In sum, we arrive at essentially the same conclusions as so much of the literature: There are no clear, clean, and direct linkages between climate change effects, climate change outcomes, and violent conflict, with the past providing unsure guidance to forecast the future. However, we do believe the academic community that focuses on climate change effects would intuitively understand how CCA fits into DRR and how such a merger would generate value-added benefits at both the conceptual and operational levels, including conflict mitigation potentialities.

Conclusion: DRR Including CCA as Conflict Avoidance, Reduction, and Management

This chapter began by asking a pivotal question: Is there a relationship between (a) environmental hazards and their drivers, including climate change, and (b) violent conflict, and (c) if so, how strong are those respective relationships? We conclude that the relationship between climate change and violent conflict depends upon specific contexts and a multitude of complex interactive forces. Our intuition is that there *is* a relationship, but it is highly dependent on variable selection, and not adequately addressed is how disasters, including those with hazards influenced by climate change, impact circumstances in ongoing hot-conflict zones. The consequences are generally terrible for local residents, but an extreme climatic or other hazard might also put the violent conflict on hold, if only temporarily. More frequently than not, a complex humanitarian crisis emerges.

In the end, assessing the strength of the relationships among hazards (including climate change), vulnerabilities, and conflict variables is an evolving, if not burgeoning, research endeavour that requires better qualitative and quantitative evidence. It is a key research challenge for the next twenty years if we intend to get a reasonable handle on the second half of the twenty-first century.

It bears emphasising that climate change in and of itself is not a 'disaster', but if vulnerability is not addressed, then disasters – and even catastrophes – will become more frequent irrespective of how climate change affects hydro-meteorological hazards. For that reason alone, embedding CCA within a larger DRR framework is logical. It makes further sense because such an integration or nesting will facilitate the much-needed improved analyses of the disaster–conflict nexus and inform actions to deal with both together, rather than focusing only on disasters or only on conflict. That is, CCA should fold into DRR, which will ideally then be subsumed into the even larger discourse on sustainable development and its strategies. However, CCA and DRR researchers and practitioners have created, for the most part, distinct communities with different institutions, stakeholders, methodologies, policy frameworks, and funding sources.

In addition, climate scientists and disaster experts have adopted somewhat different world-views and conceptualisations of time. Certain climate change effects are understood to be more permanent while many disasters tend to recur or be episodic. Disaster experts, usually social scientists, and climate scientists also adopt somewhat different discourses and do not always speak the same professional language. For example, 'mitigation' and 'adaptation' are understood and used somewhat differently while 'coping' is sometimes confused with 'adaptation' by professionals in both camps. For climate scientists, mitigation refers to strategies and efforts to reduce or absorb greenhouse gas emissions in order to slow global warming. Disaster experts see mitigation as the application of specific strategies to diminish or limit the adverse impacts of specific types of hazard events or processes.

Coping generally refers to the mechanisms or strategies whereby individuals, families, and/or communities deal with emotional stress, trauma, or sudden loss, or how they struggle to overcome challenges or difficulties. For disaster professionals, coping capacities can contribute to reducing disaster risks and mitigating disaster impacts. Coping, however, is not adaptation, which is an essential component of building long-term community resilience. Adaptation in the CCA frame tends to be more evolutionary, top-down, and permanent while adaptation for DRR tends to be more grassroots or community-based and intended to mitigate recurring disaster impacts.

In conclusion, while folding CCA into DRR will encounter some resistance, it is necessary for a coherent epistemic community to emerge. Furthermore, promoting and implementing DRR, including CCA, may very well ameliorate/manage conflicts, promote cooperation, and contribute to peace. In the end, even with a nesting or integration of CCA into DRR, the sustainable development and urban/community planning communities must also embrace the DRR–CCA combination.

References

Akcinaroglu, S., DiCicco, J.M. and Radziszewski, R. (2011) 'Avalanches and olive branches: A multimethod analysis of disasters and peacemaking in interstate rivalries', *Political Research Quarterly* 64, 2: 260–275.

Bergholt, D. and Lujala, P. (2012) 'Climate-related natural disasters, economic growth, and armed civil conflict', *Journal of Peace Research* 49, 1: 147–162.

Brancati, D. (2007) 'Political aftershocks: The impact of earthquakes on intrastate conflict', *Journal of Conflict Resolution* 51, 5: 715–743.

Burke, M., Hsiang, S.M. and Miguel, E. (2015) 'Climate and conflict', *Annual Review of Economics* 7: 577–617.

Davis, M. (2002) *Late Victorian Holocausts: El Niño Famines and the Making of the Third World*, New York: Verso.

Devlin, C. and Hendrix, C.S. (2014) 'Trends and triggers redux: Climate change, rainfall, and interstate conflict', *Political Geography* 43: 27–39.

Drury, A.C. and Olson, R.S. (1998) 'Disasters and political unrest: An empirical investigation', *Journal of Contingencies and Crisis Management* 6, 3: 153–161.

Gawronski, V.T. and Olson, R.S. (2013) 'Disasters as crisis triggers for critical junctures? The 1976 Guatemala Case', *Latin American Politics and Society* 55, 2: 133–149.

Gleditsch, P.N. (2012) 'Whither the weather? Climate change and conflict', *Journal of Peace Research* 49, 1: 3–9.

Gleick, P.H. (2014) 'Water, drought, climate change, and conflict in Syria,' *Weather, Climate, and Society* 6: 331–340.

Homer-Dixon, T. (2006) *The Upside of Down: Catastrophe, Creativity, and the Renewal of Civilization*, Washington, DC: Island Press.

Hsiang, S.M., Meng, K.C. and Cane, M.A. (2011) 'Civil conflicts are associated with global climate', *Nature* 438–441.

Kelley, C.P., Mohtadi, S., Cane, M.A., Seager, R. and Kushnir, Y. (2015) 'Climate change in the Fertile Crescent and implications of the recent Syrian drought', *Proceedings of the National Academy of Sciences (PNAS)* 112, 11: 3241–3246.

Kelman, I. (2012) *Disaster Diplomacy: How Disasters Affect Peace and Conflict*, New York and London: Routledge.

Kennett, D.J. and Beach, T.P. (2013) 'Archeological and environmental lessons for the Anthropocene from the Classic Maya collapse', *Anthropocene* 4: 88–100.

Nardulli, P.F., Peyton, B. and Bajjalieh, J. (2015) 'Climate change and civil unrest: The impact of rapid-onset disasters', *Journal of Conflict Resolution* 59, 2: 310–335.

Nel, P. and Righarts, M. (2008) 'Natural disasters and the risk of violent conflict', *International Studies Quarterly* 52, 1: 159–85.

Nelson, T. (2010) 'When disaster strikes: On the relationship between natural disaster and interstate conflict', *Global Change, Peace and Security* 22, 2: 155–174.

Omelicheva, M.Y. (2011) 'Natural disasters: Triggers of political instability?', *International Interactions* 37: 441–465.

Peterson, L.C. and Haug, G.H. (2005) 'Climate and the collapse of Maya civilization', *American Scientist* 93: 322–329.

Quarantelli, E.L. and Dynes, R.R. (1976) 'Community conflict: Its absence and its presence in natural disasters', *Mass Emergencies* 1: 139–152.

Salehyan, I. and Hendrix, C.S. (2014) 'Climate shocks and political violence', *Global Environmental Change* 28: 239–250.

Selby, J. and Hoffman, C. (2014) 'Beyond scarcity: Rethinking water, climate change and conflict in the Sudans', *Global Environmental Change* 29: 360–370.

Slettebak, R.T. (2012) 'Don't blame the weather! Climate-related natural disasters and civil conflict', *Journal of Peace Research* 49, 1: 163–176.

Streich, P.A. and Mislan, D.B. (2014) 'What follows the storm? Research on the effects of disasters on conflict and cooperation', *Global Change, Peace and Security* 26, 1: 55–70.

Taylor, R.H. (2015) 'Responding to Nargis: Political storm or humanitarian rage?', *Journal of Social Issues in Southeast Asia* 30, 3: 911–932.

von Uexkull, N. (2014) 'Sustained drought, vulnerability and civil conflict in Sub-Saharan Africa', *Political Geography* 43: 16–26.

Welzer, H. (2012) *Climate Wars, Why People Will Be Killed in the Twenty-First Century*, Malden, Massachusetts: Polity.

Zhang, D.D., Jim, C.Y., Lin, G.C.S., He, Y-Q., Wang, J.J. and Lee, H.F. (2006) 'Climatic change, wars and dynastic cycles in China over the last millennium', *Climatic Change* 76: 459–477.

16

HUMANITARIAN PROTECTION PERSPECTIVES FOR DISASTER RISK REDUCTION INCLUDING CLIMATE CHANGE ADAPTATION

Elizabeth Ferris

Remarkable global efforts to address climate change, reduce the risk of disasters, bring about sustainable development, and strengthen humanitarian response are currently in motion. Each of these efforts emerged from a different group of actors at the national and international levels. Each uses different terminology and is based on different conceptual understandings and normative frameworks. The convergence of developments in these four areas offers exciting opportunities for cross-fertilisation and even synergies in actions and policies to address present and future human needs. But this cross-fertilisation does not happen automatically. Scholars, practitioners and policy-makers alike are more comfortable talking with those in their own field and those who use their own jargon.

This chapter explores ways in which the concept of humanitarian protection can enrich actions in disaster risk reduction (DRR), including climate change adaptation (CCA) and conversely the ways in which DRR including CCA can contribute to more effective protection in humanitarian settings, including in situations other than disasters. These observations are also relevant to the 2030 Agenda of the Sustainable Development Goals (SDGs), although space precludes a fuller discussion of all of these interlinkages. In many ways, DRR including CCA lies at the nexus of development and humanitarian approaches. While humanitarian actors focus on responding to people who have experienced disasters and often call for more DRR, it is development actors who are charged with incorporating DRR into development plans as recognised by the SDGs.

The concept of protection is used in many fields as is evident from expressions as diverse as consumer protection, protection of civilians, protection under the law and environmental protection. But protection has a particular meaning and resonance in humanitarian work. Introducing the concept of protection into DRR including CCA not only enriches the understanding and approaches of DRR, but also offers the possibility of increasing engagement of humanitarians in the field of DRR by recognising the unique contribution they can make. To be blunt, humanitarians do not have a lot to contribute to DRR policies (beyond, of course, making sure that their own programmes reduce the risk of disasters to the extent possible). For the most part, they are not engaged with communities before disasters occur. Nor do humanitarian actors have much to contribute to designing DRR policies, including CCA, in the context of development plans or even working with national authorities responsible for development plans. But they do have something unique to contribute: an understanding and experience with protection and a set of tools which may be helpful in DRR activities. Protection has been central to humanitarian work for 150 years and is grounded in international human rights law, international humanitarian law and refugee law.

This chapter begins by looking at conceptual issues around the term 'protection', offers some examples of humanitarian protection concerns in disasters and suggests ways in which protection can be included in DRR, also encompassing CCA. The chapter concludes with some reflections on humanitarian protection in CCA specifically and suggests that a focus on protection may serve to overcome some of the silos which characterise much of the work in the field.

Protection: A Loaded Term

The concept of protection in the humanitarian world has a long history, beginning with international humanitarian law (IHL) where agreement was needed on how to protect soldiers who were wounded, captured or otherwise *hors de combat*. IHL was expanded in 1949 to include protection of civilians. In 1951, the Refugee Convention (UNHCR 1951/1967) was developed precisely to provide international protection to individuals fleeing persecution who no longer enjoyed the protection of their governments. Protection has also been a central concept in international human rights law: the 1948 Universal Declaration of Human Rights uses the word 'protection' ten times. Over the last twenty years, political actors have increasingly incorporated protection into their action as evidenced by the explicit reference to protection which has emerged in Security Council resolutions and mandates of peacekeeping operations and in the General Assembly with adoption of the Responsibility to Protect doctrine in 2005.

The generally accepted definition of protection in the humanitarian world was developed by the International Committee of the Red Cross and later endorsed by the Inter-Agency Standing Committee as:

> [A]ll activities aimed at obtaining full respect for the rights of the individual in accordance with the letter and spirit of the relevant bodies of law (i.e., human rights law, international humanitarian law, refugee law.
>
> *(ICRC 2001)*

Although this definition has been critiqued for being so broad that virtually all activities are included (Ferris 2011), it remains the central organising concept in humanitarian response. The basic rationale is clear: people's need to be kept safe is as vital as their need for food, shelter, and medical care. Over the years, the concept of protection as used by humanitarians has become both more specialised (e.g., child protection) and more broad (e.g., social protection).

Protection is much less grounded in the development context where concepts such as poverty alleviation, resilience and vulnerability have been central – and heavily critiqued. However, a clear trend has emerged in recent decades to adopt rights-based approaches to development which call for protecting human rights and protection of persons (Weerelt 2001).

While protection is a well-established central concept in humanitarian work, it has only been rigorously applied to disasters in the last decade – and widely incorporated into DRR measures, including CCA, even more recently.

Protection in Disasters

In many respects, the 2004 Indian Ocean tsunami marked a turning point in the international practitioners' perception of disasters. Although scholars and practitioners, such as Cuny (1983) and Anderson (1999), had earlier raised the political dimension of humanitarianism, disasters occurring before the 2004 tsunami were primarily seen by the humanitarian system in terms of the need to mobilise rapid humanitarian aid – an area in which logistical expertise and often military capacity was prioritised. After the tsunami, awareness grew that human rights had to be built into all phases of disaster management – prevention and risk reduction, response and recovery (Kälin 2005; Lewis 2006).

While the Convention on the Protection of Persons with Disabilities is the only human rights treaty to explicitly reference disasters, the applicability of human rights law to disasters is receiving greater attention from both the scholarly community and intergovernmental bodies at the regional and international levels (Harper 2009; Kälin 2012). The International Law Commission has finalised Draft Articles on the Protection of Persons in the event of disasters and affirms that '[p]ersons affected by disasters are entitled to respect for their human rights' (ILC 2010, article 8). UN treaty bodies are increasingly taking up issues related to disasters in carrying out their monitoring duties (Valencia-Ospina 2013). The UN Human Rights Council, for the first time, devoted a special session to human rights issues arising from a disaster: the Haitian earthquake of 2010 (UN Human Rights Council 2010). Presently, the Human Rights Council is engaged in further work on the relationship between the promotion and protection of human rights in post-disaster and post-conflict situations (UN Human Rights Council 2013).

All international human rights conventions include the right to life and the subsequent obligation of the state to protect life. OHCHR's message on DRR sums up the linkages and provides a useful context to this discussion:

> All states have positive human rights obligations to protect human rights. Natural hazards are not disasters, in and of themselves. They become disasters depending on the elements of exposure, vulnerability and resilience, all factors that can be addressed by human (including state) action. A failure (by governments and others) to take reasonable preventive action to reduce exposure and vulnerability and to enhance resilience, as well as to provide effective mitigation, is therefore a human rights question.
>
> *(UNOHCHR 2011)*

In particular, states have a responsibility to reduce the risks of disasters and to protect those at imminent risk of disasters through timely warnings and evacuations. When they fail to exercise this responsibility, they face domestic and international criticism and potential legal action. This has occurred in a number of cases, from Hurricane Katrina in the United States in 2005 (Kromm and Sturgis 2008) to Japan's 2011 earthquake, tsunami and nuclear power plant meltdown (Mosneaga 2015).

Protection in Disaster Risk Reduction (DRR)

It is the responsibility of governments to protect their populations from disasters and central to that effort is DRR. While governments cannot prevent cyclones or earthquakes, they can take measures to reduce their impacts on their people. DRR is the concrete expression of a government's responsibility to protect people living within its jurisdiction.

There have been three United Nations conferences on risk reduction and one sees the evolution of thinking about protection in the texts resulting from those three conferences. The first, held in Yokohama, Japan, in 1994, mentions protection in the tenth of its ten principles: 'Each country bears the primary responsibility for protecting its people, infrastructure, and other national assets from the impact of natural disasters' (UNISDR 1994: 8).

The second, held in Kobe, Japan in 2005, led to a Hyogo Declaration and a Hyogo Framework for Action 2005–2015 (UNISDR 2005), which draws the relationship between protection and DRR. Principle 4 of the Hyogo Declaration states 'We affirm that States have the primary responsibility to protect the people and property on their territory from hazards, and thus, it is vital to give high priority to disaster risk reduction in national policy'. The Hyogo Framework set out three strategic goals: the integration of DRR into sustainable development policies and planning; the development and strengthening of institutions, mechanisms and capabilities to build the resilience of communities to hazards; and the systematic incorporation of risk reduction approaches into emergency preparedness, response and recovery programmes.

In March 2015, in Sendai, Japan, at the Third World Conference on Disaster Risk Reduction, UN Member States signed the Sendai Framework for Disaster Risk Reduction 2015–2030 (SFDRR) (UNISDR 2015). Later in June, the document was formally adopted by the UN General Assembly. The SFDRR, intended as a follow-up to the Hyogo Framework for Action, includes many more references to protection than the 2005 Framework. Article 5 of the SFDRR states: 'it is urgent and critical to anticipate, plan for and reduce disaster risk in order to more effectively protect persons, countries and communities, their livelihoods, health, culture, heritage, socioeconomic assets and ecosystems, and thus strengthen their resilience' (UNISDR 2015). Article 19d explicitly links DRR with 'protecting persons and their property' and includes the clause 'while protecting and promoting all human rights'. Other sections refer to financial protection (30b), protection of culture (30d), and protection of economic assets and livelihoods (30p). Early warning is seen as a protective measure (36d).

This represents a sea change in emphasis on protection from the Hyogo Declaration of a decade before which refers to taking 'effective measures to reduce disaster risk, including for the protection of people on its territory, infrastructure and other national assets from the impact of disasters' (13b); 'risk assessment and early warning systems are essential investments that protect and save lives, property and livelihoods' (13i); 'protect and strengthen critical public facilities and physical infrastructure' (19f) and protection of 'past development gains' (27) (UNISDR 2005).

Protection, Disasters and Humanitarian Response

At the same time that practitioners working in DRR were integrating into their work long-standing ideas from the literature regarding the need to include protection in their approaches, humanitarian actors who were not typically engaged with disasters were looking at protection in these settings. Again, it was the 2004 Indian Ocean tsunami which marked a turning point for practitioners as organisations such as UNHCR and ICRC were heavily involved in disaster response – a field relatively new to them, although both organisations had responded to disasters in the past when they had existing operations in an affected area.

A review of UNHCR efforts to respond to disasters (UNHCR-PDES 2010) was followed in 2013 by an analysis of protection risks in disasters (UNHCR-PDES-DIP 2013). This study identified four main categories of protection risks: risks related to the disaster itself; those stemming from the humanitarian delivery of assistance; the creation of vulnerabilities; and protection risks in early recovery and reconstruction (UNHCR-PDES-DIP 2013: 2). Most disaster responders, the study found, focus on immediate life-saving priorities and may simply not be aware of vulnerabilities and protection needs. Unlike most conflict situations, where fragile states are the norm, in disasters, governments often play strong national roles on the basis of institutions and policies established before the disaster occurs.

Addressing protection risks in disasters thus means supporting the government; while not a new role for UNHCR, it is more prominent in disasters than when UNHCR responds to refugees fleeing conflict. Overall, the case studies in the 2013 study highlight the tenuous acceptance of the concept of protection within disaster response efforts; neither governments nor agencies thought in terms of protection, but rather used different terms, such as affected populations and vulnerabilities. The UNHCR-PDES-DIP study (2013) found that protection needs were particularly acute and particularly unmet in the case of those displaced by disasters. The study concluded with some observations of how UNHCR could contribute its operational expertise in protection to disasters, including by responding to gender-based violence, displacement, legal aid, documentation, access, identification of vulnerable groups and participatory assessments (2013: 4). We will return to this theme of displacement as a possible unifying concept between DRR and protection. But first we look at the development of the key operational tool for protection in disasters.

The *IASC Operational Guidelines on the Protection of Persons in Situations of Natural Disasters*

Following the 2004 Indian Ocean tsunami, the Representative of the Secretary-General on the Human Rights of Internally Displaced Persons initiated a process to provide guidance to humanitarian organisations working in the aftermath of disasters. This was based on a working visit in February 2005, where he found myriad human rights issues in his assessments of internal displacement following the disaster (Kälin 2005). He found, for example, that in some cases internally displaced persons (IDPs) faced discrimination and that within the IDP population, different responses were implemented for those displaced by conflict and by natural hazards and sometimes between IDPs living inside and outside of camps. He recommended that the *Guiding Principles on Internal Displacement* (United Nations 2004) serve as a basis for development of policies by states as well as by international and national humanitarian actors to protect and assist those displaced by disasters. He also called for the development of more general guidance on human rights in humanitarian settings, recognising that while IDPs have specific needs related to their displacement, others affected by the disaster also face protection concerns. In 2006, he presented draft guidelines to the Inter-Agency Standing Committee on Protection of Persons in Natural Disasters, which after field-testing, were revised and adopted by the IASC in 2010 (Brookings 2011).

These *Operational Guidelines* emphasise that persons affected by disasters should enjoy the same rights and freedoms under human rights law as others in their country and not be discriminated against. They reassert the principle that states have the primary duty and responsibility to provide assistance to persons affected by disasters and to protect their human rights. They also state that all communities affected by the disaster should be entitled to easily accessible information concerning the nature of the disaster they face, possible mitigation measures that can be taken, early warning information, and information about ongoing humanitarian assistance.

The *Operational Guidelines* point out that people affected by disasters may face four kinds of human rights violations that require humanitarian action:

1   Harm (past, present and future) caused by the action or neglect of those with responsibilities for upholding their rights.

2   Access to goods and services protected by human rights, such as food, water, sanitation, shelter, health care, and education.

3   Incapacity and obstacles which prevent them from claiming their rights, including lack of: information, documentation, effective remedies and accountability for violations.

4   Discrimination on the basis of their race, colour, sex, language, religion, political or other opinion, national or social origin, property, disability, birth, age or other status.

(Brookings 2011: 6–7)

The *Operational Guidelines* are based on the realisation that while all human rights are important, priorities must be set in the aftermath of a disaster. The first priority is to protect life, personal security, and the physical integrity and dignity of affected populations while a second priority relates to providing the basic necessities of life including adequate food, water and sanitation, shelter, clothing and essential health services. Other economic, social and cultural rights (e.g., education, housing, livelihoods) are a third priority, followed by other civil and political rights (such as freedom of movement, assembly, electoral rights, etc.) as a fourth priority.

The *Operational Guidelines* offer concrete guidance to those responding to specific types of disasters, with a particular focus on humanitarian actors. For example, in the immediate aftermath of a flood, governments are usually not able to provide necessary educational facilities for affected children. This can (and must) come later, once the children are protected against violence and have access to the basic necessities of life. Similarly, the right to documentation is a crucial issue for many affected by emergencies, but affected communities have a more urgent need for sufficient food and water. The Guidelines offer further suggestions on incorporating a human rights approach to DRR and to recovery efforts.

While the *Operational Guidelines* are fairly comprehensive in drawing the connections between protection and response and recovery in disasters, they do not systematically deal with preparedness and risk reduction. However, it should be possible to use the basic outline of the *Operational Guidelines* to develop guidance for DRR including CCA. For example, if lack of documentation is a protection challenge, how can this be addressed in DRR activities, including CCA, at the national level?

Social Protection

Coming from a different perspective from a focus on protection as part of humanitarian response are those working on social protection (Box 16.1). As Midgley (2013) traces the evolution of the field, scholarly work in the field dates back to the early decades of the twentieth century and is associated with the introduction of statutory social assistance and social insurance programs. Scholars studying social security from a social policy perspective have traditionally focused on state-run social assistance programs and primarily in developed countries, initially devoting little attention to social protection activities at the household or community levels. Such social insurance and social security schemes can be seen as institutionalised mechanisms for addressing risk through collective means. Midgley (2013) explains that there are debates within the social protection field about whether such programs should primarily help households manage risks or whether they should focus on poverty reduction or promotion of social rights and equality (2013: 10). For example, the World Bank's risk management framework has shifted over the years away from collective provision through the state to households, communities and NGOs. Some examples of social protection policies are microfinance and microinsurance – both of which may be useful in the context of DRR including CCA.

**Box 16.1   Aligning Social Protection and Disaster Risk Reduction
Including Climate Change Adaptation**

Lindsey Jones[1]

[1] Overseas Development Institute, London, UK

While many aspects of the humanitarian and development divide remain frustratingly siloed, links between social protection, disaster risk reduction (DRR), and climate change adaptation (CCA) are increasingly being forged. Indeed, despite differing entry points, all three approaches share a common objective: supporting the capacity of vulnerable groups to prepare for and respond to a range of current and future risks. Social protection is notable for the breadth of activities that it encompasses, ranging from protective (providing relief from deprivation) and preventative (seeking to avert deprivation) measures through to promotive (enhancing incomes and capacities) and transformative (addressing issues of social justice and exclusion) measures (Devereux and Sabates-Wheeler 2004).

In tackling the root causes of social and economic vulnerability, social protection offers a helpful medium to extend the remit of traditional DRR including CCA. A further advantage that social protection provides is a focus on multiple and overlapping risk rather than on a singular shock or stress – supporting households to deal with not only the impacts of hazards, but also consequences such as food price shocks or knock-on effects such as pest outbreaks. Additionally, many of the central tenets of social protection – such as the provision of social safety nets and cash transfer schemes – can contribute directly to enhancing the core characteristics needed by households to implement DRR including CCA (Levine *et al.* 2011). Interventions such as weather-indexed insurance that offer valuable opportunities for risk transfer and mitigation (and traditionally associated with social protection) are increasingly being re-framed as, and integrated into, DRR including CCA, further blurring the lines between the various approaches.

Another recent development is the support of many funders and development agencies for 'adaptive social protection' which combines elements of social protection and DRR including CCA, placing a stronger emphasis on productive livelihoods and ensuring that social protection schemes help to buffer the negative impacts of hazards and hazard drivers (Davies *et al.* 2008). Such sectoral alignments are inherently positive, though they still face a considerable challenge in promoting meaningful cross-sectoral integration. For instance, despite considerable efforts to promote greater alignment, most countries still assign responsibility for the coordination of social protection, DRR, and CCA to separate ministries and sectoral coordination groups at the national level (World Bank 2011). Key to maximising opportunities, however, will be ensuring that cross-sectoral alignment moves past issues of frame bridging and financial mobilisation towards more meaningful sharing of ideas, co-production and collaboration.

Holzmann and Jorgensen (2000: 1–2) contrast the traditional definition of social protection as 'public measures to provide income security for individuals with its focus on labour market intervention, social insurance and social safety nets' into a broader framework with different strategies to deal with risk (prevention, mitigation and coping), different levels of formality of risk management (informal, market-based and public) and different actors – ranging from individuals to international organisations. Social protection is defined by the UK's Department for International Development (DfID) as 'a sub-set of public actions that help address risk, vulnerability and chronic poverty' (DfID 2006: 1) and, like Midgley (2013), notes its expansion beyond the original idea of the state as provider and protector of citizens. In reviewing the literature on social protection, Scott (2012) notes that social protection traditionally focused on short-term protective safety nets to protect people from the impact of 'shocks', but it has expanded to include measures to address the structural causes of chronic poverty. She notes for example that social protection now includes such programs as cash and in-kind transfers, food assistance, and public work programs (2012: 12–15).

Jones *et al.* (2010: 10) base their analysis of the relationship among DRR, CCA and protection on the understanding that social protection programs 'aim to protect poor and vulnerable households from the shocks and stresses that have negative impacts on their wellbeing'. They argue that social protection can be considered to be a component of adaptation interventions as it addresses both vulnerability and response capacity (Jones *et al.* 2010: 12–13). They also compare the assets, policies and institutions targeted in three different approaches: DRR, social protection and livelihoods, concluding that what is 'important to note is that each of the various approaches typically uses differing labels for many of the same community responses and actions' (Jones *et al.* 2010: 18).

While humanitarian agencies generally focus on protecting people once they have been affected by disasters, a social protection approach seems more relevant to DRR, including CCA, as it focuses on longer-term actions before a disaster occurs. While humanitarian agencies tend to see protection as something to be incorporated into their own programmatic work, social protection focuses more on actions by the state or by communities and households themselves.

Climate Change Adaptation and Protection

Climate change influences many hazards in complex ways. CCA measures are thus also ways of decreasing the risk of disasters, although it is telling that often those working on CCA and DRR are pursuing their work independently of each other. Adapting to climate change must include measures to reduce vulnerability and build capacity as well as to manage climate risk and to address the impacts of climate change (Jones *et al.* 2010: 5).

The Paris Agreement on climate change (UNFCCC 2015) makes only two references to protection – to protecting biodiversity in the preamble (UNFCCC 2015: 20) and in chapter 7 where it refers to adaptation as a

> key component [that] makes a contribution to the long-term global response to climate change to protect people, livelihoods and ecosystems, taking into account the urgent and immediate needs of those developing country Parties that are particularly vulnerable to the adverse effects of climate change.
>
> *(UNFCCC 2015: 24)*

It seems that at least some of the work on protection and CCA focuses on social protection as it interacts with environmental change. For example, Chaudhury *et al.* (2011) look at social protection as CCA strategies to changing environmental conditions in Honduras and Zambia, finding that in both cases communities were able to use or adapt existing social protection mechanisms to minimise the impact of drought.

Displacement, Migration and Planned Relocations in DRR

The Conference of Parties to the UNFCCC meeting in Cancun in 2010 adopted the following resolution under the general framework of CCA, calling for '[m]easures to enhance understanding, coordination and cooperation with regard to climate change induced displacement, migration and planned relocation, where appropriate, at the national, regional and international levels' (UNFCCC 2010, article 14[f]). In other words, displacement, migration and planned relocations may all be appropriate measures to adapt to – and protect people from – the effects of climate change.

There are important differences between these three terms. Displacement always conveys a sense of coercion or involuntariness. People are forced from their homes by rising flood waters, for example, or the ground on which they live simply disappears because of coastal or riverbank erosion. They do not have a choice if they are to survive, so they must leave. Although displacement is to be avoided wherever possible, it is clearly a protection strategy. Migration, under international law, is always considered to be voluntary. People choose to leave their homes, communities or countries for many reasons, including economic opportunities, family reunification, educational pursuits, and adventure. In practice, people also migrate as a protection strategy to avoid difficult or dangerous situations, including to avoid current and future risks. In the case of climate change, it is likely that many people will 'see the handwriting on the wall' and move before they are forced to do so. In some cases, such as Kiribati, the government has pursued efforts to enable people to 'migrate with dignity' by providing them with skills to leave their country before they are forced out (McNamara 2015). Migration is thus a DRR measure, which can also be framed in CCA terms, as well as a form of protection. Planned relocation has received much less attention from academics and policy-makers alike, but is also a measure in which governments relocate people to other areas in order to protect them from disasters or the effects of environmental change, including climate change (Brookings-LSE Project 2015). DRR measures including for CCA thus should include measures to avoid displacement, while supporting migration and planned relocation efforts.

Although there are as yet no global estimates on the number of people migrating or who are relocated because of disasters, including those with hazards influenced or potentially influenced by climate change, displacement from disasters tends to be increasing in absolute numbers and will certainly increase in the future in both absolute and proportional metrics if appropriate measures are not taken. One of the perhaps unheralded success stories of the past few decades has been the dramatic decrease in fatalities associated with sudden-onset hazards. While more people are being affected by disasters, fewer are dying. When people survive, but their homes are damaged or destroyed, they can be displaced. For the past few years, the Internal Displacement Monitoring Centre has produced annual global estimates of the numbers of people displaced by sudden-onset hazards, finding that some 200 million people were displaced by sudden-onset hazards in the last five years (IDMC 2015a). Most displacement occurs within national borders and disaster-induced displacement can be protracted, as in the Philippines where reports are that half a million people are still displaced over two years after Typhoon Haiyan (IDMC 2015b).

The protection needs of IDPs have received considerable attention over the years and, in fact, led to the reform of the humanitarian system in 2005 (Ferris 2014). DRR has been much slower to look at the relationship between displacement or mobility and DRR, since DRR also means – or should mean – reducing the risk of displacement and preparing for it when it does occur.

The only reference to population movements included in the Hyogo Framework (UNISDR 2005) was paragraph 19.i., which called on States to '[e]ndeavour to ensure, as appropriate, that programmes for displaced persons do not increase risk and vulnerability to hazards'. It recognised that forced population movements (whether induced by disasters or conflicts) – and the efforts that go into assisting those who are forced to move and addressing their situation – contribute to determining exposure and vulnerability. It also called for the management of their risk outcomes.

The final text of the Sendai Framework (UNISDR 2015) includes the three following references on the need to include migrants in DRR and disaster management work at all levels:

7:     governments should engage with relevant stakeholders, including [. . .] migrants [. . .] in the design and implementation of policies, plans and standards.

27.h: empower local authorities, as appropriate, through regulatory and financial means to work and coordinate with [. . .] migrants in disaster risk management at local level.

36.a.vi: Migrants contribute to the resilience of communities and societies and their knowledge, skills and capacities can be useful in the design and implementation of disaster risk reduction.

Analysis of the relationship between DRR, including CCA, and human mobility – migration, displacement and planned relocation – would seem to be a promising area for further research, policies and actions. Although the literature on the scale of mobility likely to result from the effects of climate change is presently quite speculative, it is generally assumed that larger numbers of people will move in the future. In Alaska, Bronen (2012) demonstrated that many small indigenous communities can no longer remain where they are because of the effects of climate change – including melting permafrost, stronger storms, coastal erosion. Additionally, well-intentioned mitigation efforts to keep climate change in check by investing in renewable energies, protecting forests and conserving habitats are sometimes causing further displacement – compelling people to leave their homes and livelihoods, often with little safeguards or compensation for their losses.

Concluding Thoughts

A focus on protection in DRR offers a number of advantages. By looking at how the protection risks from disasters can be reduced, it focuses on a somewhat different set of actions than just vulnerabilities. It allows humanitarian actors to be involved in DRR by using the particular expertise they have acquired over decades in responding to – and being prepared to respond to – displacement, gender-based violence, child protection and other needs and consequences within humanitarian situations. This expertise is needed to develop comprehensive DRR including CCA strategies. By becoming more involved in DRR including CCA initiatives, humanitarian agencies can be nudged to work more closely with national authorities and national development plans. It has long been recognised that one of the shortcomings of traditional humanitarian operations has been their tendency to bypass the state, at times substituting for the state. While this may be necessary to save lives when governments are fragile and needs are urgent, it is inappropriate and counter-productive in DRR. By working on DRR, humanitarian agencies can develop different ways of working with national and subnational authorities, which could well have concrete payoffs if they are called to respond to movements of refugees or IDPs at some future point.

A focus on protection in DRR also has the advantage of bringing humanitarian and development actors closer together. For humanitarians, it is a stretch to work on DRR. For development actors, working on protection, particularly with its strong human rights basis, means venturing out of their comfort zone. A specific recommendation might be to put together a small group of experts from both the humanitarian and the DRR fields to work together to develop operational guidance on reducing the protection risk of disasters, using the *Operational Guidelines* as an outline. Humanitarians should feel comfortable as they would be working on the basis of a document formulated and adopted by humanitarian actors while DRR experts should be comfortable as it is their expertise that will be needed to develop such guidance. Such approaches would overcome separations, better connecting the various protection approaches and ensuring that humanitarianism and DRR including CCA are better linked.

References

Anderson, M.B. (1999) *Do No Harm: How Aid Can Support Peace or War*, Boulder, CO: Lynne Rienner Publishers.

Bronen, R. (2012) 'Climate induced community relocations: Creating an adaptive framework based in human rights doctrine', *New York University Review of Law and Social Change* 35: 357–407.

Brookings-Bern Project on Internal Displacement. (2011) *Operational Guidelines on the Protection of Persons in Situations of Natural Disasters*, Washington DC: Inter-Agency Standing Committee, The Brookings Institution. Online http://www.brookings.edu/research/reports/2011/01/06-operational-guidelines-nd (accessed 6 September 2016).

Brookings-London School of Economics (LSE) Project on Internal Displacement. (2015) *Planned Relocations in the Context of Disasters and Environmental Change, including Climate Change*, Washington DC: The Brookings Institution. Online http://www.brookings.edu/about/projects/idp/planned-relocations (accessed 6 September 2016).

Chaudhury, M., Ajayi, O.C., Hellin, J. and Neufeldt, H. (2011) *Climate Change Adaptation and Social Protection in Agroforestry Systems: Enhancing Adaptive Capacity and Minimizing Risk of Drought in Zambia and Honduras*, Working Paper No. 17, Nairobi: World Agroforestry Centre.

Cuny, F. (1983) *Disasters and Development*, Oxford, UK: Oxford University Press.

Davies, M., Guenther, B., Leavy, J., Mitchell, T., and Tanner, T. (2008). 'Adaptive social protection: Synergies for poverty reduction', *IDS Bulletin*, 39(4): 105–112.

Devereux, S. and Sabates-Wheeler, R. (2004). *Transformative Social Protection*, IDS Working Papers 232, University of Sussex: Institute of Development Studies. Brighton: UK.

DfID (Department for International Development). (2006) *Social Protection in Poor Countries*, Social Protection Briefing Note Series, Number 1, London: DfID.

Ferris, E. (2011) *The Politics of Protection: The Limits of Humanitarian Action*, Washington DC: Brookings Institution Press.

Ferris, E. (2014) *Ten Years After Humanitarian Reform: How Have Internally Displaced Persons Fared?* Brookings-LSE Project on Internal Displacement, The Brookings Institution, Washington DC. Online http://www.refworld.org/pdfid/54cb8e484.pdf (accessed 6 September 2016).

Harper, E. (2009) *International Law and Standards Applicable in Natural Disaster Situations*, International Development Law Organization. Online http://eird.org/publicaciones/natural-disasters-manual-idlo-2009.pdf (accessed 6 September 2016).

Holzmann, R. and Jorgensen, S. (2000) *Social Risk Management: A New Conceptual Framework for Social Protection and Beyond*, Social Protection Discussion Paper Series No. 0006, Social Protection Unit, Human Development Network, the World Bank. Online http://siteresources.worldbank.org/SOCIALPROTECTION/Resources/SP-Discussion-papers/Social-Risk-Management-DP/0006.pdf (accessed 6 September 2016).

ICRC (International Committee of the Red Cross). (2001) *Strengthening Protection in War: A Search for Professional Standards*, ICRC. Online https://www.icrc.org/eng/resources/documents/publication/p0783.htm (accessed 6 September 2016).

IDMC (Internal Displacement Monitoring Centre). (2015a) *Global Estimates 2015: People Displaced by Disasters*, IDMC. Online http://www.internal-displacement.org/assets/library/Media/201507-globalEstimates-2015/20150713-global-estimates-2015-en-v1.pdf (accessed 6 September 2016).

IDMC (Internal Displacement Monitoring Centre). (2015b) *Long-term Recovery Challenges Remain in the Wake of Massive Displacement*, IDMC. Online http://www.internal-displacement.org/assets/library/Asia/Philippines/pdf/201502-ap-philippines-overview-en.pdf (accessed 6 September 2016).

ILC (International Law Commission). (2010) *Draft Articles on the Protection of Persons in the Event of Disasters*, United Nations Document A/CN.4/L.776, New York: United Nations. Online http://legal.un.org/ilc/guide/6_3.shtml (accessed 6 September 2016).

Jones, L., Jaspars, S., Pavanello, S., Ludi, E., Slater, R., Arnall, A., Grist, N. and Mtisi, S. (2010) *Responding to a Changing Climate: Exploring How Disaster Risk Reduction, Social Protection and Livelihoods Approaches Promote Features of Adaptive Capacity*, London: Overseas Development Institute. Online https://www.odi.org/sites/odi.org.uk/files/odi-assets/publications-opinion-files/5860.pdf (accessed 6 September 2016).

Kälin, W. (2005) *Protection of Internally Displaced Persons in Situations of Natural Disaster: A Working Visit to Asia*, Washington DC: The Brookings Institution. Online www.brookings.edu/research/reports/2005/04/tsunami (accessed 6 September 2016).

Kälin, W. (2012) 'The human rights dimension of natural or human-made disasters', *German Yearbook of International Law* 55, 2012: 119-147.

Kromm, C. and Sturgis, S. (2008) 'Hurricane Katrina the guiding principles on internal displacement', *Southern Exposure* Special Report xxxvi, 1 & 2, Institute for Southern Studies.

Levine, S., Ludi, E. and Jones, L. (2011). *Rethinking Support to Adaptive Capacity to Climate Change: The Role of Development Interventions*, London: Overseas Development Institute (ODI).

Lewis, H. (2006) *Human Rights and Natural Disaster: The Indian Ocean Tsunami*, Boston: School of Law Faculty Publications, Northeastern University. Online http://hdl.handle.net/2047/d20001074 (accessed 6 September 2016).

McNamara, K. (2015) 'Crossborder migration with dignity in Kiribati', *Forced Migration Review*, 49. Online http://www.fmreview.org/climatechange-disasters/mcnamara.html (accessed 6 September 2016).

Midgley, J. (2013) 'Social development and social protection: New opportunities and challenges', *Development Southern Africa*, 30, 1: 2–12.

Mosneaga, A. (2015) *The Sendai Framework and Lessons from Fukushima – Our World*, Tokyo: United Nations University. Online http://ourworld.unu.edu/en/the-sendai-framework-and-lessons-from-fukushima (accessed 6 September 2016).

Scott, Z. (2012) *Topic Guide on Social Protection*, Birmingham, UK: Governance and Social Development Resource Centre, University of Birmingham. Online http://www.gsdrc.org/docs/open/sp1.pdf (accessed 6 September 2016).

United Nations. (2004) *Guiding Principles on Internal Displacement, 2nd ed.*, New York: United Nations. Online http://www.brookings.edu/~/media/Projects/idp/GPEnglish.pdf (accessed 6 September 2016).

UNFCCC (United Nations Framework Convention on Climate Change). (2010) *Decision1/CP.16 The Cancun Agreements: Outcome of the Work of the Ad Hoc Working Group on Long-term Cooperative Action under the Convention, FCCC/CP/2010/7/Add.1*, Bonn: UNFCCC. Online http://unfccc.int/resource/docs/2010/cop16/eng/07a01.pdf (accessed 6 September 2016).

UNFCCC (United Nations Framework Convention on Climate Change). (2015) *Conference of the Parties. Twenty-first Session Paris, 30 November to 11 December 2015. Adoption of the Paris Agreement.* FCCC/CP/2015/L.9. Bonn: UNFCCC. Online http://unfccc.int/resource/docs/2015/cop21/eng/l09.pdf (accessed 6 September 2016).

UNHCR (United Nations High Commissioner for Refugees). (1951/1967) *Convention and Protocol Relating to the Status of Refugees*, Geneva: UNHCR.

UNHCR-PDES (United Nations High Commissioner for Refugees Policy Development and Evaluation Services). (2010) *Earth, Wind, and Fire: A review of UNHCR's role in recent natural disasters*, Geneva: UNHCR-PDES. Online http://www.unhcr.org/4c1228e19.html (accessed 6 September 2016).

UNHCR-PDES-DIP (United Nations High Commissioner for Refugees Policy Development and Evaluation Service (PDES) and Division of International Protection (DIP)). (2013) *A Review of Protection Risks and UNHCR's Role in Natural Disasters*, Geneva: UNHCR-PDES-DIP. Online http://www.unhcr.org/51408d589.pdf (accessed 6 September 2016).

UN (United Nations) Human Rights Council. (2010) *Human Rights Council in Solidarity with Haiti*, New York: United Nations. Online http://www.ohchr.org/EN/NewsEvents/Pages/HRCouncilinsolidarity withHaiti.aspx (accessed 6 September 2016).

UN (United Nations) Human Rights Council. (2013) *Promotion and Protection of Human Rights in Post-disaster and Post-conflict Situations A/HRC/RES/22/16*, New York: United Nations. Online www.ohchr.org/Documents/HRBodies/HRCouncil/AdvisoryCom/A-HRC-RES-22-16_en.pdf (accessed 6 September 2016).

UNISDR (United Nations International Strategy for Disaster Reduction). (1994) *Yokohama Strategy and Plan of Action, Guidelines for Natural Disaster Prevention, Preparedness and Mitigation.* Japan: UNISDR. Online http://www.unisdr.org/files/8241_doc6841contenido1.pdf (accessed 6 September 2016).

UNISDR (United Nations International Strategy for Disaster Reduction). (2005) *Report of the World Conference on Disaster Reduction, Kobe, Hyogo, Japan, 18–22 January 2005*, Geneva: United Nations. Online https://www.unisdr.org/2005/wcdr/intergover/official-doc/L-docs/Final-report-conference.pdf (accessed 6 September 2016).

UNISDR (United Nations International Strategy for Disaster Reduction). (2015) *Sendai Framework for Disaster Risk Reduction 2015 – 2030*, Geneva: UNISDR. Online http://www.preventionweb.net/files/43291_sendaiframeworkfordrren.pdf (accessed 6 September 2016).

UNOHCHR (United Nations Office of the High Commissioner for Human Rights). (2011) *Organization Profile: Policies and Programmes in DRR*, New York: UNOHCHR. Online www.preventionweb.net/english/professional/contacts/profile.php?id=1370 (accessed 6 September 2016).

Valencia-Ospina, E. (2013) *Sixth Report on the Protection of Persons in the Event of Disasters, A/CN.4/662*, New York: International Law Commission, United Nations.

Weerelt, P.V. (2001) *A Human Rights-based Approach to Development Programming in UNDP – Adding the Missing Link*, New York: United Nations Development Programme (UNDP). Online http://www.undp.org/content/dam/aplaws/publication/en/publications/democratic-governance/dg-publications-for-website/a-human-rights-based-approach-to-development-programming-in-undp/HR_Pub_Missinglink.pdf (accessed 6 September 2016).

World Bank. (2011) 'Social protection and climate resilience', report from the international workshop *'Making Social Protection Work for Pro-poor Disaster Risk Reduction and Climate Change Adaptation', Addis Ababa, 14–17 March 2011*, Washington DC: USA.

17

ETHICS AND DISASTER RISK REDUCTION INCLUDING CLIMATE CHANGE ADAPTATION

Christopher Gomez

Introduction

In recent years, both the scientific literature (e.g., Kelman and Gaillard 2010; Shaw *et al.* 2010) and policy documents have emphasised the importance of integrating Climate Change Adaptation (CCA) into the Disaster Risk Reduction (DRR) framework. The Commission on Climate Change and Development has also expressed it as a necessity for sustainable development (CCCD 2009), and similarly the Sendai Framework for Disaster Risk Reduction (SFDRR; UNISDR 2015) has pointed out that DRR policies need to be coherent with CCA frameworks (UNISDR 2015, article 19-h). SFDRR also noted up front that these frameworks need to be supported by international collaborations and by the assistance of 'developed countries' to poorer nations, in order to be effective, putting to the fore inequalities in the face of the consequences of climate change. Indeed, there is a growing amount of evidence that the least wealthy communities and the poorest in communities will be more vulnerable to the impacts of climate change, and that such inequalities will also strongly affect the existing differential of wealth concentration (UNDP 2013), reinforcing disequilibrium of centres and peripheries.

Emerging from this picture are the issues of fairness and responsibility on the national and international scenes. Who should and can pay? How do we deal with the dilemma of future generations in making choices and in considering what is right or wrong? These different themes articulate and inform some of the ethical paradigm of DRR including CCA.

Overview of the Common Underpinning Ethical Theories

As a whole, ethics can be defined as a branch of philosophy that investigates what is right or wrong, good or bad for the individual and for the community. Within this field exists a cloud of research directions, from which we can distinguish two major clusters: metaethics (interested in the nature of ethics), and normative ethics that attempts to answer the questions of 'how we should live our lives' or 'what kind of person we should be', etc. In this chapter, we will not discuss the nature of ethics but we will focus on a small portion of 'normative ethics', for it can help us discuss questions on DRR including CCA. Such questions can take the shape of: are you right or wrong when you keep your car engine running while waiting for your favourite coffee brew, and can anyone blame you for it, and arbitrating some of these discussions are a series of ethical theories, of which one of the most important is utilitarian ethics.

Consequentialism: Utilitarian Ethics

One of the most important theories that feed our attitude towards human activities having an impact on the global environment is consequentialism. It describes a group of ethical theories that considers an action or a rule as right or wrong based on its sole consequences. For instance, if we consider greenhouse gas emissions, it is because those emissions have negative impacts on our societies and because it may lead to a global disaster that the Conference of the Parties 21 (COP21) met in Paris at the end of 2015 to take international action against climate change. It is not because there is something ethically wrong with releasing greenhouse gases into the atmosphere. The consequence is what matters, and what drives ethical considerations in this case. Such behaviour is typical of consequentialist views and ethics, and they are prevalent in our contemporary societies.

There are several schools of consequentialist ethics, out of which the most prominent is utilitarianism. At the beginning of the nineteenth century, Jeremy Bentham articulated first his theory of utilitarianism as being the quest for the greatest pleasure for the greatest number. As pleasure could hardly be defended as a source of right and wrong, John Stuart Mill rapidly modified the theory, which became the greater good for the greater number. From this theory utilitarian ethics is therefore the maximisation of utility. The theory has evolved during the last 200 years and many different variations such as Act Utilitarianism and Rule Utilitarianism have emerged, and it has become prominent on the scene of DRR (Zack 2009).

Such conceptualisation is not without opposition. If one lies to you, should you only assess whether it is right or wrong based on the consequences? Do you expect your friends and your family to be honest with you? The second issue utilitarianism meets concerns the rights of minorities when only the greater good of the majority is considered. Indeed, utilitarians can argue that it is right to keep the car air conditioning on for the majority's well-being and avoid heat stroke even when the very same gas you emit pushes some small island states into disarray in the face of sea-level rise.

A Counterpoint to Utilitarian Ethics: Kantian Ethics

Popular movements such as *Green America* propose alternatives to fossil fuel by 'divesting'; i.e., investing out of fossil fuel energy, because it is the right thing to do for the planet. On their webpage (http://app.greenamerica.org/fossil-free/), they do not refer to benefits or to the greater good for the greater number, but they argue that being fossil free is the right thing to do. Such an ethical stand is certainly at odds with consequentialist forms of ethics, as it is interested in the 'good will' behind an action. This form of ethics is often at the root of various forms of activisms and it is informed by Kantian ethics, which is part of a different group of ethical theories named deontological moral theories. These are not interested in the results of actions but the rightness of the action itself. Referring back to the example of greenhouse gas emissions, deontological ethics would state that we should stop polluting the planet not because greater harm than greater good may arise, but because it is intrinsically wrong to do so, because one cannot consciously take the decision to action anything that is wrong. The moral issue stands with the motivation of the action, not the result.

Based on this line of thought, Kantian ethics, postulated by the German philosopher Immanuel Kant (1724–1804), is based on what Kant named the 'categorical imperative' (CI). CI is an unconditional command that you have to obey, even if you do not want to do so. Even if you would like to cheat to win a card game and win a lot of money, to be morally right the CI commands you not to. Cheating cannot be right, even if the outcome is positive for a majority (concept prevailing in Utilitarianism). To support this idea, Kant explains that you should not act in a way that you are not willing to allow others to do as well. You should not cheat at the card game, because cheating only gives you an advantage if you do not allow the others to do so. Consequently, Kant's ethics is interested in the 'will' behind an action and not the results.

Another element of the CI, which Kant saw as integrated to the 'I shall allow myself what I would allow others to do as well', is the necessity to inspect whether an action respects the interests and goals of other human beings and is not only serving one's own interests solely. If the action benefits my own purpose over the respect of other human beings, this action cannot be seen as ethical either. Applying this ethical principle to DRR research, Gaillard and Gomez (2015) have put forth the importance of considering the action itself (what is right or wrong) and not only the outcome. They argue that the plundering of a disaster impacted area by a tsunami of researchers and practitioners cannot be justified by the potential good it would bring for future DRR – which would be a typical utilitarian view. They argue that the action itself needs to be informed by a sound ethical judgment and not the sole result.

Another Position: Rawls' Ethics

John Rawls is another opponent of Utilitarianism, but he interrogates ethics from a different perspective. Rawls, in his book *A Theory of Justice* (1971), outlines that Justice should be the first virtue of social institutions, and it is Justice that provides individuals with rights that are inviolable and that those cannot be taken away even for the greater good of the community as a whole. For Rawls (1971), the greater good for the greater number cannot justify the loss of freedom and liberty, because those cannot be separated from the physical individual as they are an integral part of the being. His approach also differs from other philosophers as he bases his ethics on the concept of justice. Although he does not believe that inequalities in the face of justice can be totally erased, he postulates that decisions on what is right or wrong need to be taken from an egalitarian position, an original position that minimises the influence of one's position and characteristics in society.

If we consider the example of Kiribati, which might be threatened by sea-level rise, and imagine that the world is one single vast community with international agreement (and we are not totally there yet), Rawls (1971) would argue that it is unacceptable for the international community to deprive the I-Kiribati people of the freedom to live on their islands, even if the rest of the planet benefits from greenhouse gas emissions, because the gain of freedom of some, even a majority, does not make it right if it is to the detriment of others. According to Rawls (1971), such decisions should be taken in the circumstance of not knowing whether one will be one of the I-Kiribati islanders or not. He stipulated that one should choose the rules by which a community should live 'under a veil of ignorance' (p. 136). In other words, when choosing the rules by which a community wants to live, one should not know where one will be on the social ladder, in such a way that the rules are being chosen without being dictated by one's personal status in a community. For Rawls (1971), this ensures that the ethical rules are good for everybody and fair.

If there are other popular ethical theories, such as Aristotelian ethics, this chapter mostly concentrates on the interplay of Utilitarianism, Kantian Ethics and Rawls' conception of Justice, which are most relevant to DRR including CCA most certainly because they are the pillars of the ethics behind our contemporary institutions.

Responsibility and Blame

Responsibility and blame in DRR inclusive of CCA can be considered from the point of view of (1) who is responsible/blameworthy for climate change and (2) who is responsible/blameworthy for not taking any action when anthropogenic climate change modifies the realms of DRR. Both have an impact on DRR, but the responsibility can be placed at two different levels.

Responsibility and Blame for Climate Change

One of the favourite media witch-hunts consists in finding individuals, companies, or countries, responsible or blameworthy for anthropogenic climate change, because it has partly (but not entirely) a human trigger. The human trigger is not necessarily different for many other hazards, such as floods caused by dams or earthquake shaking amplified locally by building foundation design, even if some see it as being different. An example of a law case that proved a company could be condemned for harm caused to others through the environment is the Exxon Valdez case. On 24 March 1989, the super-tanker spread oil over 7,700 square kilometres impacting 350 kilometres of shoreline in the Prince William Sound. The Exxon Corporation was sentenced to pay USD 900 million in damages, an extra USD 100 million in unforeseen damages, and also USD 25 million for criminal charges. In other words, the Corporation had to pay for the impacts of their actions on people, through the medium of the environment, even if there was no intent to cause harm. It also asserted the undeniable right of individuals and communities living on the coasts of the Prince William Sound to a clean water resource. But what about the water itself? See Box 17.1.

Box 17.1   The Rights of Water

Gustavo Wilches-Chaux[1]

[1] Universidad Externado de Colombia, Bogotá, Colombia
Translated by Ilan Kelman

The Internet has approximately 65 million hits for 'el derecho AL agua' in addition to approximately 897 million on the 'right TO water'. Yet there are almost no hits for 'los derechos DEL agua' or 'the rights of water', despite more than 92 million on 'los derechos de la naturaleza' and 1.3 billion on 'nature's rights'.

Ecuador's political constitution from 2008 enshrines the rights of nature, as does Bolivia's from 2009. Here, we are not going to discuss if this has had any practical effect on the way in which development relates to ecosystems in these two Andean countries.

There are also complicated discussions regarding whether or not it makes sense to consider non-human beings as having rights. Those who oppose are convinced that rights are cultural concessions which human beings have made in favour of ourselves and that a right exists only through the metric of recognition through social acceptance and legal order.

Water, however, is neither interested nor caught up in these discussions. Straightforwardly, when some of these rights are violated, water feels the pain and tries to recover.

When a water-related disaster occurs, we ask ourselves what rights have been violated as a cause of the disaster. The observation of these processes has convinced me that water has at least the following fundamental rights:

- The right to exist without pollution: If this is not respected, then human beings cannot exercise our right to water.
- The right to a riverbed: Water has a memory that persists beyond human memory. Often, when a disaster manifests, a popular idiom explains: 'The water returns to retake what was taken'.
- The right to flow freely: If this is not respected, then water sooner or later will overflow its natural channels and boundaries, break through obstructing barriers, and generate huge flows which produce disasters.

- The right to expand in precipitation-intensive storms: If the riverbanks and wetlands have been invaded by development and human activities (including livelihoods), and through dams and dikes water is deprived of these spaces, then water will try to break these walls and belligerently recover these spaces. If this fails, then floods will occur downstream.
- The right to be absorbed by the ground: If the ground is impermeable due to urbanisation and paving, water has no other option except to run off over the city cement, overwhelming the sewage system's capacity and yielding floods which generate disasters.

To be effective, managing disaster risk from hydrometeorological hazards, including climate change adaptation, has as one of its principal challenges to guarantee that development does not violate the rights of water.

What differs from the Exxon Valdez example is the global nature of the anthropogenic climate change regardless of the 'source of harm'. Therefore, blaming 'entities' delineated in time and space for the cause of it is particularly difficult. To resolve this issue a simple strategy has been to consider present-day emissions of a greenhouse gas such as carbon dioxide. If a country, for instance the USA, contributes 20 per cent of the carbon dioxide emissions worldwide, should they contribute 20 per cent of the world funds for CCA? A proportional distribution would seem fair and simple at first sight, especially because the same greenhouse gas that has allowed the development of one country is now allegedly hampering the development and resilience pathways of another.

This attractive simplicity, however, does not withstand close examination. Firstly, in the present example of the USA, the country would certainly argue that part of the carbon dioxide production in the country also benefits other countries, as it provides goods and services, funds and salaries used outside the USA.

Secondly, climate change is not a problem that can be measured from the last few years of emissions. It has historical roots, and during the nineteenth century, Western European nations were the biggest polluters, and their earlier heavy contributions should be accounted for in the present state of the planet. Effectively measuring those past productions can be extremely difficult as we cannot use the same tools scientists have in their toolbox to survey carbon production.

Thirdly, the 20 per cent refers to the causes of anthropogenic climate change and not the impacts, making the link between the polluters and the impacted tenuous and difficult to prove. There is also no linear regression between carbon increase and negative impacts on societies, and no linear relationship can thus be drawn between the amount of carbon production and the amount of harm caused.

Responsibility for Being 'DRR Including CCA Ready'

Because DRR is directly influenced by anthropogenic climate change influencing some hazards, several countries have started to integrate originally separated strategies within their DRR plans, giving rise to DRR inclusive of CCA (Kelman and Gaillard 2010). Even national agencies that traditionally have not directly dealt with anthropogenic climate change and CCA have integrated their realms in their framework, because anthropogenic climate change hinders existing DRR frameworks (FEMA 2011). Politics that have tried to hide behind the excuse of anthropogenic climate change for failing DRR plans have been openly criticised, even in the popular press. Hence, responsibilities and blameworthiness for anthropogenic climate change do not preclude responsibilities for DRR including CCA. There is a tacit expectation that DRR evolves to integrate CCA. For example in the USA, the Federal Emergency Management Agency (FEMA) has taken a proactive approach in integrating CCA into DRR (FEMA 2011). The 'FEMA Climate Change Adaptation Policy Statement' recognises that anthropogenic climate change will have impacts on 'mitigation, preparedness, response and recovery operations' (p. 1), on the 'resiliency of critical infrastructure and various emergency assets' (p. 1), and that anthropogenic climate change 'could trigger indirect impacts that increase mission risks' (p. 1). DRR including CCA is not modifying the responsibility framework of FEMA, and FEMA cannot be blamed for anthropogenic climate change, but it can be blamed for not providing appropriate DRR strategies, even in the face of anthropogenic climate change.

Considering DRR inclusive of CCA, responsibilities and blame can either fall on the causes of anthropogenic climate change, or on the fact of not acting on its consequences. In both cases, however, finding individuals or groups that are blameworthy or responsible is a difficult task, because even in the presence of linkages between one action and a consequence they are difficult to quantify, and because some of them may already belong to a previous generation. From an ethical standpoint, the issues of DRR including CCA belong to more than one generation, creating what is termed the 'intergenerational problem'.

The 'Intergenerational Problem'

Deciding What is Right or Wrong in the Present and for the Future

The intergenerational problem of climate change and DRR including CCA is the issue of the temporality of our actions that outlive ourselves and ripple into future generations. It is generated by the Earth phenomena operating over periods longer than the life expectancy of individuals. Past actions have present impacts on anthropogenic climate change, and present actions and decisions concerning anthropogenic climate change and CCA will ripple to future generations as well, even if we cannot foretell what those consequences will be. It is, however, a legacy that will outlive the twenty-first century and morph into an uncertain future.

The haphazard quality of the future renders any decision ethically difficult to assess, because we are not able to predict how it will impact our future and the future of other generations as well. As Karl Popper wrote, we are fundamentally unable to make long-term predictions, because scientific progress cannot be foreseen. Although the common stand is that the future will be better, with technological and societal improvement there is, however, no assurance that we aren't heading for a bleak future either. Such an issue has recently emerged in Japan, where the country has been covered with Sabo dams, built to stop sediments in mountain streams and minimise the impacts of sediment-laden flows. As the population is ageing and leaving the countryside, the question of structure maintenance is arising. Economic decline could even accelerate this effect, while anthropogenic climate change influence on rain patterns could also render the engineered structures inadequate.

Despite those emerging examples, the commonly progressivist view is deeply engrained in all of us – who is not hoping for a better future for her/his children? – and it has been mostly fuelled by two elements. First, scientists who look at those issues are usually progressivists and believe in a brighter tomorrow. Second, this same idea is at the root of the societal change of the Western and westernised world, which has seen the ascension of democratic systems and human rights over the last few centuries. One must not forget, however, that these advances were mostly allowed because the economic mechanisms of capitalism need an expanding market, and thus the creation of a consumer-class that is not controlled by coercion – as was seen in feudal systems for example – but by the need to always consume ever more. If this model were to fail in the face of anthropogenic climate change or if climate change cannot be turned into an opportunity to resolve the long-standing problems, then it is most likely that the 'ever-brighter' future might be compromised.

Pushing the Problems towards the Next Generation

Beyond examining if we are taking the right or wrong decisions for future generations, Tremmel and Robinson (2014) have also asserted that the conception of a brighter future is the source of highly immoral behaviour. It allows present individuals to push the problems towards the next generation. The authors stated that

> discounting the future because it is far away or not here yet is inadmissible. Temporal and spatial distances have a lot in common. Whether a murder occurs one mile or one hundred miles away, whether it happens in one year on in 100 years – it is a murder and inherently wrong. In the same way, discounting the damages due to climate change in the distant future is morally inadmissible.
>
> *(p. 103)*

The ethical issue of DRR including CCA is not so straightforward. In the short-term, the progress in addressing anthropogenic climate change and even integrating CCA into DRR can come to a halt when more pressing issues interrupt the process. For instance, when the subprime-mortgage crisis hit the USA and the rest of the world in 2008, individuals who lost their houses were not likely to be making consciously any economic effort to curb greenhouse gas emissions. The economic crisis naturally reduced the consumption and the emission of greenhouse gas emissions during the period 2008–2009 (Peters *et al.* 2012). What seemed to have derailed the plan is actually helping us reach objectives we thought hardly attainable. This situation is symptomatic of the ethical dilemma we are facing. Utilitarian ethics is not well-suited to take into account hypothetical future populations. It will not sacrifice the greater good of the present population for the benefit of hypothetical future populations. As the 2008 crisis shows, it might be a necessary evil. Kantian ethics is not of much help either, because it investigates the 'will' and not the consequences.

A potential solution to this issue may come from an extension of Rawls' (1971) idea of the 'veil of ignorance' to multiple generations. Instead of not knowing what one social status might be, we can imagine that one also would not know whether she or he is bound to live in the present or in the future. In such case, one is most likely to make a fair decision for all, both in terms of anthropogenic climate change and DRR including CCA. Such a stand could allow us to work around the difficulties that arose from the accountant approach of utilitarianism and the 'will' of Kantianism.

Winners and Losers, and Problems of Multiple-Memberships

Winners and Losers: Profit over People

Glantz (1995) wrote that countries could become 'climate-related world powers' (p. 42), emphasising the importance of the concept of 'winners and losers' (p. 41) of climate change. This idea underpins the concepts of equity and justice, which are inherent to CCA and DRR on the world stage. While some French wineries are buying land in the southern part of the United Kingdom to ensure their production when the average temperature will be too high to produce in France, the cow herders of the Cuvelai Basin in Namibia are facing greater threats of droughts and flash floods, and they certainly do not have the possibility to invest in a piece of land in the United Kingdom.

Anthropogenic climate change is obviously reinforcing existing inequalities and broadening the gap between the 'winners and the losers'. In terms of DRR, it also means that countries and communities that are already struggling with existing disaster risk will face greater challenges. An action to reduce anthropogenic climate change (like drastically reducing greenhouse gas emissions) would benefit the struggling countries, but cutting greenhouse gas emissions would reduce the economic leverage that a group of countries have over others. It is therefore most unlikely that any action that would benefit the whole international community will be taken to curb the progress of anthropogenic climate change. Countries will be left to work at the state scale to include CCA in DRR. At the same time this process lets the winners off the hook and divides a global problem at the state-scale. This will result in a divide between countries able to pay for adaptation and the others, increasing present inequalities and even further reinforcing power relations.

In the United Kingdom for instance, the Thames Barrier, which is a huge tidal gate system, protects about 200 billion pounds of assets in London, and originally cost 1.4 billion pounds (2015 value) – although most of those assets would not have been constructed without the Thames Barrier being in place. In 2002, the 'Thames Estuary 2100' project began the update of the structure to address the anthropogenic climate change impact of sea level rise, drawing even more resources to the project. At the other end of the spectrum, countries like Mauritania, Senegal, Gambia, Guinea Bissau, and Cape Verde are working through the United Nations Educational, Scientific and Cultural Organisation (UNESCO) programme 'Adaptation to Climate Change in Coastal Zones of West Africa'. Because the cost of adaptation has been estimated at 5–10 per cent of the GDP, too often this cost has outweighed the benefits of taking any action against anthropogenic climate change impact of coastal erosion.

Such obvious inequality in developing DRR including CCA strategies is part of what is named the problem of 'distributive justice'; and the necessity of a 'Robin Hood' redistribution model is certainly the biggest obstacle to fair adaptation at the world scale. It is indeed most unlikely that rich and powerful nations will renounce their benefits to help poorer and weaker nations; most especially because the former would eventually lose some of the leverage they have on the world stage. Shue (1992, p. 376) wrote 'if one is profiting from injustice, it is hardly going to be in one's interest to pursue justice'. Ethically, one could argue that Kantian ethics should preclude such unfair profit, but the ethic that prevails in most sovereign states and especially in democracies is Utilitarian ethics, which look after the interest of the majority of the members within the boundaries of the community. The interests of the wider, global community are thus more complicated to advocate and defend, because of the existing state structures.

Like the European countries during the colonial period, Noam Chomsky explains that the present day USA has developed international policies that assure their economic and military superiority, in order to preserve their advantages (Chomsky 1999), and it doesn't seem that such practice will decline any time soon, as we are now seeing China flexing its muscles to reclaim islands in territory which is also claimed by other sovereign states.

The Role of Multiple-Membership in Opening the Gap

The power relation between 'winners and losers' is exacerbated by the mechanism of multiple memberships, which operates from both a moral and a contractual stand. Multi-membership is the recognition that we all belong to different groups, communities that are all imbricated within each other, from the family, to the village, the country, a continent, the world.

Figure 17.1   Can We Really Care About What Happens on the Other Side of the Planet?

(By Author)

In one of his lectures at Harvard University, Professor of Justice M.J. Sandel asked his students whether they would keep quiet or whether they would report their room-mate if they caught him/her cheating for an exam. The fundamental question was actually asking where the student's loyalty lay. Similarly, my students invariably choose their own dog over a distant neighbour, when asked to make a choice between experiencing the death of one of the two. One would rather choose another species – i.e., an animal over a human – just based on the distance-relationship. This spatial gradient also applies to how we interact with remote communities (Boylan 2011). It is therefore most unlikely that we will stop using our car to help the inhabitants of the small island states of the Pacific who might rapidly experience the consequences of sea level rise (Figure 17.1). For this reason, humans are most unlikely to naturally try closing the gap between the 'winners and the losers' of anthropogenic climate change.

This issue of spatial moral judgment is seconded by the mechanics of memberships, in being an obstacle to distributive justice. The communities we are part of rule some of our duties – i.e., the military service some owe to their country, or the government tax we have to pay, duties that are bounded by the limits of the community we have a tacit contract with. This contract is reciprocal and allows us to benefit from the community we are part of. This philosophy has been championed by Rousseau and Rawls – philosophers sometimes named contrarian because their philosophy is based on the concept of a contract, which provides a framework to define what is right or wrong in a community (with its laws, rules, culture, etc.). The role of the contracts with the nation appears clearly during disasters, in the USA or New Zealand for instance. Respectively, the governmental agencies of the FEMA and Ministry of Civil Defence and Emergency Management (MCDEM or Te Rakau Whakamarumaru) organise relief in disaster-impacted areas, and they are expected to do so. It is part of the benefits of a contract financed by the taxpayers, and therefore there is no expectation that the FEMA or MCDEM will go and provide relief to any disaster-impacted area on the planet. Consequently, the delimitations of nations and nations' obligations through contracts limit any obligation of DRR including CCA from one country to another.

From a theoretical perspective, Boylan (2011, p. 26) has tried to provide solutions to the problem created by the 'Russian Dolls' community fences and hierarchy of obligations that has little space left for the remote community. He wrote that

> each agent must educate himself and others as much as he is able about the peoples of the world – their access to the basic goods of agency, their essential commonly held cultural values, and their governmental and institutional structures – in order that he might create a world view that includes those of other nations so that individually and collectively the agent might accept the duties that pursue from those people's legitimate rights claims and act accordingly within what is aspirationally possible.

Boylan therefore argues that knowing the diversity would help us become closer across cultures, regardless of the distance, and it is only then that DRR, including CCA, can be considered at a global scale, erasing the gap between 'winners and losers', because one will be ready to help individuals and groups in remote and very near places, in the very same way.

Conclusion and Other Directions

The ethics of DRR including CCA is ever-changing and often apparently contradictory, depending on the ethical theories one considers. Consequentialist theories such as Utilitarianism are certainly not in tune with Kantian ethics or even contractualism. This uneven cushion supports, in turn, a handful of other challenges that are intrinsic to DRR including CCA. In the eye of the storm are the spatial and temporal scales of the issue as well as the complexity of the physical and human chain of processes, which encompasses all the planet at different periods and through different generations. This situation presents numerous challenges to determine who is to blame, who is responsible, what should be done by whom, and in the end it may even challenge the present day socio-political structures in place. From this imbroglio of scales and issues, one will retain the following important points from this chapter (Figure 17.2):

- DRR including CCA involves three main schools of ethics: consequentialism, which looks at the outcome; deontology, which is interested in the intention; and the contrarians who investigate the 'ecology' of human relations and how we enter into contracts with each other.
- Practically, deciding what is right or wrong on the world stage is complex, because of the power-relations between countries, and it is hampering fair distributive justice. Difficulties are not only theoretical, but also practical.
- The imbalance is also most likely to remain because powerful nations do not want to lose their leverage by helping less favoured nations.
- Consequentialist and deontological ethics have difficulties dealing with the spatial imbricated scales and the multigenerational scales of the problem. Deciding what is right may need to rely on an extended contrarian ethics.
- Taking decisions for future generations is also an ethical dilemma. Utilitarianism has intrinsic difficulties in taking into account populations that are not born yet, and other forms of ethics are still denying future generations their liberty and freedom of choice, regardless of how virtuous one's choice is.

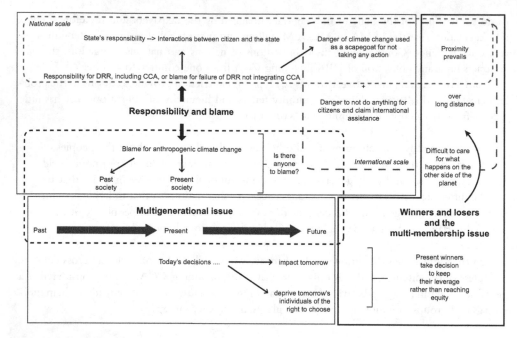

Figure 17.2 Interactions Between the 'Responsibility and Blame Issue', the 'Multigenerational Issue' and the 'Winners and Losers and the Multi-Membership Issue' of Anthropogenic Climate Change and DRR Including CCA

(By Author)

- The intergenerational issue and the common idea that tomorrow's technologies will be able to solve the problem can lead to pushing back the problems to future generations, which is highly immoral.

Further Outlook and Potential Solutions

Although DRR including CCA presents numerous challenges to traditional ethics, several propositions have been made to solve the issues encountered (see also Box 17.2). Ecofeminism, for instance, has proposed a different framework, which includes the environment and the ecology in the ethical framework. Hourdequin (2015, p. 87) wrote 'One important diagnosis of our problematic relationship to the natural world focuses on anthropocentrism and our associated failure to value nonhuman organisms and ecosystems'. The relation of domination between human beings and nature as explained by Drewermann (1981) and Merchant (2014), for instance, has fermented an ecofeminism that relates the domination of the environment as being similar to the domination of women (Warren 1996).

From an ethical perspective, ecofeminism proposes a solution to some of the challenges we have expressed earlier, because it redefines nature as being more than just a vector of human activity but an ethical end by and for itself. As Kaneko Misuzu wrote (1923, p. 15):'。。。海のお魚は、なんにも世話にならないし、いたずら一つしないのに、こうして私に食べられる。' ('... Fish of the sea, do not owe anything to anyone, they don't do even one thing wrong, and I who eat them . . .'). In other words, nature exists for itself, it is not here just to provide us with a service, especially because there is no reciprocal contract between humanity and nature.

Like ecofeminism does, Kaneko Misuzu explained a century earlier that nature cannot be seen as 'other' and needs to be integrated at an equal level and deserves respect as well.

Box 17.2   The Necessity of a Bio-ecocentric Ethic

Gustavo Wilches-Chaux[1]

[1] Universidad Externado de Colombia, Bogotá, Colombia
Translated by Ilan Kelman

At various times in history – and across all the continents – basic ethics have ruled in which only an elite had rights, inseparably wielding different powers (religious, political, economic, and military) and possessing the right to kill people from other sectors of society and other communities.

An ethnocentric ethic prevailed in the world, no different from before, in which a race has rights that assume a legitimate capacity to impose slavery on others, to dispose of their lives, to decide their destinies, and to seize their territories and goods.

Since the United Nations General Assembly adopted the Universal Declaration of Human Rights in 1948, an anthropocentric ethic has been consolidated, centred on the human species. It is strengthened with important instruments recognising the existence of economic, social, and cultural rights of all the people of the Earth. The Declaration of the Rights of the Child, and others, point to similar objectives.

These international declarations, and their equivalents in the majority of national constitutions, are based on efforts to avoid, in practice, those who today hold economic power (many times also connected with political, technological, and military power – and sometimes religion) trying to control the destinies, territories, and resources of those who lack those powers. Yet there exists both globally and inside countries a plutocratic ethic based on economic wealth. This lacks full legitimacy and would not have international recognition under human rights.

From the planet's point of view, however, the problem that still prevails is that the anthropocentric ethic legitimises whatever type of action humans do to ecosystems and to all the others who share the Earth with us. Ethics based on an equally anthropocentric worldview is constructed on the conviction that God created the Earth and all that the planet contains to be exclusively at the service of humanity without restrictions.

This worldview and this ethic ignore that human beings are one node within the complex web of the biosphere. Our opportunity to form part of the planet is inseparable from the existence of other beings, some living and some apparently not living, such as water, which are nonetheless indispensable for life. A consequence of this way of thinking is what we term climate change, a set of adjustments in which the Earth, in its capacity for resilience, is seeking a new stable state to compensate for the impact of human activity on its systems.

A jump from the anthropocentric ethic towards a biocentric ethic based on respect for all aspects of life and the recognition of our dependency on the Earth is not a philosophical challenge but a necessity for survival of our species.

References

Boylan, M. (2011) *Morality and Global Justice, Justifications and Applications*, Boulder, Colorado: West View Press.
CCCD (Commission on Climate Change and Development). (2009) *Climate Change: Impacts and Roles of Humans*. Online http://www.preventionweb.net/publications/view/11618 (accessed 2 October 2016).

Chomsky, N. (1999) *Profit over People: Neoliberalism and Global Order*, New York: Seven Stories Press.

Drewermann, E. (1981) *Der Tödliche Fortschritt*, Germany: Herder Verlag.

FEMA (Federal Emergency Management Agency). (2011) *FEMA Climate Change Adaptation Policy Statement*. Online https://www.fema.gov/media-library-data/20130726-1919-25045-3330/508_climate_change_policy_statement.pdf (accessed 2 October 2016).

Gaillard, JC and Gomez, C. (2015) 'Post–disaster research: Is there gold worth the rush?', *Jàmbá: Journal of Disaster Risk Studies* 7, 1: 1–6.

Glantz, M.H. (1995) 'Assessing the impacts of climate: The issue of winners and losers in a global climate change context', in S. Zwerver, R.S.A.R. van Rompaey, M.T.J. Kok and M.M. Berk (eds) *Climate Change Research: Evaluation and Policy Implications*, New York: Elsevier, pp. 41–54.

Hourdequin, M. (2015) *Environmental Ethics from Theory to Practice*, London and New York: Bloomsbury Academic.

Kaneko, M. (1923) お魚 (fish), in 私と小鳥とすずと (Me, a small bird, and a bell). Tokyo: JULA edition of 1984, pp. 14–15.

Kelman, I. and Gaillard, JC (2010) 'Embedding Climate Change Adaptation Within Disaster Risk Reduction', in R. Shaw, J.M. Pulhin, and J.J. Pereira (eds.), *Climate change adaptation and disaster risk reduction: Issues and challenges. Community* pp. 23–46.

Merchant, C. (2014) 'Ecofeminism and feminist theory', in M. Boylan (ed.) *Environmental Ethics*, Oxford: John Wiley & Sons, pp. 59–63.

Peters, G.P., Marland, G., Le Quere, C., Boden, T., Canadell, J.G. and Raupach, M.R. (2012) 'Rapid growth in CO2 emissions after the 2008–2009 global financial crisis', *Nature Climate Change* 2: 2–4.

Rawls, J. (1971) *A Theory of Justice*. Cambridge, Massachusetts: Harvard College.

Shaw, R., Puhlin, J.M. and Pereira, J.J. (2010) 'Climate change adaptation and disaster risk reduction: An Asian perspective', in R. Shaw, J.M. Puhlin and J.J. Pereira (eds) *Climate Change Adaptation and Disaster Risk Reduction: Issues and Challenges*, Bingley, UK: Emerald, pp. 1–18.

Shue, H. (1992) 'The unavoidability of justice', in A. Hurrell and B. Kingsbury (eds) *The International Politics of Environment: Actors, Interests and Institutions*, Oxford: Oxford University Press, pp. 373–397.

Tremmel, J. and Robinson, K. (2014) *Climate Ethics*, London: I.B. Tauris.

UNDP (United Nations Development Programme). (2013) *Humanity Divided: Confronting Inequality in Developing Countries*, New York: UNDP Bureau for Development Policy Press.

UNISDR (United Nations International Strategy for Disaster Reduction). (2015) *Sendai Framework for Disaster Risk Reduction*. Online http://www.unisdr.org/we/coordinate/sendai-framework (accessed 2 October 2016).

Warren, K. (1996) *Ecological Feminist Philosophies*, Bloomington, Indiana: Indiana University Press.

Zack, N. (2009) *Ethics for Disaster*, New York: Rowman & Littlefield Publishers.

18

FROM CONNECTIONS TOWARDS KNOWLEDGE CO-PRODUCTION FOR DISASTER RISK REDUCTION INCLUDING CLIMATE CHANGE ADAPTATION

Jake Rom D. Cadag

Introduction

Chapters in this part agree that implementing effective disaster risk reduction including climate change adaptation (DRR including CCA) requires knowledge and understanding of the various but interconnected social and environmental processes as well as the issues and problems that arise from such connections. In order to understand why marginalised sectors (e.g., poor people and ethnic minorities) choose to live in hazardous areas without proper DRR measures despite awareness of the risks they face, one should be able to link, for instance, people's lack of livelihood opportunities and some cultural attachment to places and the dynamics of natural hazards (e.g., frequency, magnitude, and scope) that affect them. The impact of the latter is largely influenced by the state of the environment and ecosystems, which to some extent is shaped by the actions and policies implemented within a framework for sustainable development.

To address DRR including CCA, collaboration with actors of development and environmental protection agendas is necessary. As argued by Wisner et al. (2012, p. 1), 'no single person can possess the knowledge and skill to map out and successfully implement DRR'. Yet, if CCA is to be embedded as one of the components of DRR, hence the phrase 'DRR including CCA', there is a need to rethink knowledge production and strategies for DRR to deal with the anticipated implications of climate change in shaping disaster risk (i.e., characteristics of climate-related hazards and their impacts upon people especially to the most vulnerable sectors) and eventually in action-planning and policy-making. It is equally important to formulate strategies to foster collaboration between different actors (e.g., academe, government and non-government organisations (NGOs), humanitarian groups, and local communities, amongst others) from the international down to the local levels in order to address and draw upon everyone's needs and strengths.

The purpose of this chapter is to reassert the arguments that DRR including CCA should be integrated into achieving development goals and addressing environmental issues and vice versa. The integration of these goals requires recognition of their connections (and disconnections), which should guide co-production of knowledge by different actors at different geographical and temporal scales. This chapter also argues that the integration of DRR including CCA within development goals should occur at all levels from practical through to methodological, theoretical, and policy in order to support local communities and to ensure that local plans and actions are consistent with those of government and non-government institutions. Thus, frameworks, methodologies, and the desired outcomes should be flexible and inclusive to address upward accountabilities (i.e., the accomplishment and compliance reports to donors and funding agencies) and, most importantly, downward accountabilities (i.e., the needs, rights and demands of local communities).

Connections: DRR Including CCA and Development Goals

In order to understand the connection between DRR including CCA and development goals, and later to justify their integration, it is important to understand the concepts that bind them including vulnerability, resilience, and hazards. Development goals, as used in this chapter, may refer to the specific Sustainable Development Goals (SDGs), or to development in general, to include other priority development agendas at the local or community level.

In the literature of disaster studies, it has been thought that disasters associated with natural hazards are partly consequences of non-inclusive, if not failed, development (e.g., Lewis 1999; Wisner *et al.* 2004). Such 'development', which is characterised by social injustice and oppression, results in many social problems and marginalises some sectors of society, making them vulnerable to disasters. Such connections between disaster, development, and environmental issues are not just based on theoretical abstractions. In the Philippines, Holden and Jacobson (2013) argued that mining-based development encouraged by the government could worsen the condition of poor people in the country by exposing them to several hazards, such as landslides, typhoons, earthquakes, and acid rain. Some of those hazards can be exacerbated directly and indirectly by the mining operations. The authors then examined whether such a development strategy is an example of 'digging to development or digging to disaster?' (Holden and Jacobson 2013, p. 2). Based on empirical research at a global scale, less wealthy countries have suffered more from the impacts of disasters than more affluent countries in terms of numbers of deaths and affected people, and economic impacts (Guha-Sapir and Hoyois 2012). Further, Kim (2012) claims that, poor people – defined as those earning under USD 2 per day – are twice as exposed to 'natural' disasters as people not in that poor classification.

The empirical findings above imply a direct correlation between a household's and individual's living conditions and social status, and susceptibility to suffer from hazards. Such a position has led to a shift in paradigm influencing academic research, humanitarianism, and government policies at all levels and geographical scales to address the underlying causes of people's vulnerabilities, while maintaining the focus on hazard mitigation (Lewis 1999; Wisner *et al.* 2004, 2012). It is thus difficult to understand such correlation or causality without enquiring about the concept of vulnerability. Vulnerability, as defined by Wisner *et al.* (2012, p. 22), refers to the 'degree to which one's social status (e.g., culturally and socially constructed in terms of roles, responsibilities, rights, duties and expectations concerning behaviour) influences differential impact by natural hazards and the social processes which led there and maintain that status'. Wisner (2001) also argued that while it is important to recognise the needs of vulnerable sectors (usually termed as being the poor, women, children, and elderly), vulnerability reduction should address the situations or conditions (i.e., landlessness, ethnic segregation, deprivation of human rights, etc.) that make people vulnerable. Since the 1970s, a great deal of research has emphasised the human dimension of disaster risk or the conditions of the society that make people vulnerable to disasters (e.g., O'Keefe *et al.* 1976).

For instance, health programmes may lower the number of malnourished children who are more susceptible to diseases in different phases of a disaster. Similarly, environmental protection and preservation may contribute to hazard mitigation by, for example, allowing mangroves to grow to serve as a natural buffer against some storm surges and, at the same time, as a breeding ground for fish and other marine animals. Such programmes are more effective in reducing people's vulnerabilities and thus reducing disaster risks especially when they are designed to serve the interests of the most marginalised sectors.

Thus, achieving development goals, though not intentionally meant to address disaster risk, could make people or communities more resilient. And since the 1990s, development goals (e.g., the SDGs and their predecessor, the Millennium Development Goals) and the respect of human rights are being mainstreamed in DRR including CCA (e.g., a human rights based approach) to lessen people's vulnerabilities that arise from social injustice and prejudice (e.g., Asian Disaster Preparedness Center 2006).

Very recently, CCA has emerged as a distinct but related goal to DRR. It has captured the attention of many concerned stakeholders of DRR and development (i.e., international and local government and non-government agencies, donor agencies, and scholars in related academic disciplines). The most recent synthesis report of the Intergovernmental Panel on Climate Change (IPCC 2014) claimed that human activities characterised by unprecedented economic and population growth since the pre-industrial era are responsible for the anthropogenic climate change. The report also indicates significant changes in weather and climate phenomena (e.g., temperature extremes, sea level rise and precipitation), which, according to the findings presented in the report, are expected to become more extreme (IPCC 2014). Climate change and its influence on the characteristics of natural hazards could reshape disaster risk as well as development and environmental issues, or at least our understanding of them, as well as the current and future actions and policies to address them.

Yet, some scholars who recognise the implication of climate change on DRR have reaffirmed their earlier arguments that efforts for DRR including CCA should invest in vulnerability reduction and sustainable development (e.g., Gaillard 2010; Kelman and Gaillard 2010). It is also argued that climate change should not be used as a 'scapegoat' but rather as a justification to pursue 'comprehensive disaster risk reduction, environmental management, and sustainability processes' (Kelman and Gaillard 2010, p. 41).

Recognising More Connections Through the Disconnections

The translations of connections from the theoretical level to actual practice (e.g., from the conceptual framework to the conceptualisation of methods and tools, actions and policies that aim to address inclusively the above mentioned goals) is challenged by some conceptual discrepancies; e.g., the multiple meanings of resilience (Gaillard 2010) and implementation barriers; e.g., technocratic institutional frameworks. In this chapter, those challenges are referred to as 'disconnections'.

The literature suggests that development, climate change, and disaster literatures utilise in different ways the concepts of vulnerability, capacity, and resilience, and, as a result, influence the formulation of strategies and policies as well as the implementation of programmes at different levels and scales (Gaillard 2010). On the one hand, a long history of research, dominated by social scientists and development advocates (e.g., humanitarian sector and NGOs), emphasises that vulnerabilities are products of historical socioeconomic processes characterised by injustice and prejudice. Thus, present and future vulnerability reduction measures should address such societal conditions (Kelman and Gaillard 2010). On the other hand, in the climate change literature (e.g., IPCC 2014), vulnerability is understood based on the projection of climate change models in the future and their potential impacts on exposed areas and people. In this sense, vulnerability misleadingly appears to be a sole function of hazards rather than the product of social processes as argued from a DRR perspective. Clearly, there is a discrepancy in the understanding of some concepts by DRR (e.g., UNISDR, the United Nations Office for Disaster Risk Reduction) and CCA (e.g., the IPCC) communities, which would have serious implications on the operationalisation of plans and actions to reduce disaster risk.

Addressing mismatches in scales is one of the main challenges for DRR including CCA (Gaillard and Mercer 2013). Scale mismatches refer to discrepancies in interactions characterised by shortcomings or failures of human organisations to properly grasp the management of the natural environment resulting in many forms of environmental problems (Lee 1993). According to Lee (1993), there are three ways in which 'scale mismatches' can be analysed: at the spatial, temporal, and functional scale. These mismatches, according to several recent studies, make the integration of DRR and CCA challenging and/or difficult (e.g., Gaillard 2010; Mercer 2010).

On the one hand, spatial scale mismatches in CCA often occur due to its heavy reliance on climate analysis (including climate forecasting) covering a larger geographical scope (e.g., region, large country, and continent). Localising CCA, like development goals being promoted at the international scale and implemented at the local level (i.e., communities), becomes a real challenge especially for outside actors (i.e., national government agencies, non-government organisations, and scientists). On the other hand, DRR has been emphasising community-level risk analysis to address specific vulnerabilities of the local communities and the hazards that affect them (Gaillard 2010; Mercer 2010). Climate change findings may not always be consistent with the experiences of the local authorities and people in the communities. If implemented separately, challenges may arise when translating DRR, including when integrating it into actions and policies. For example, in hazard mitigation, taking into account the implication of climate change at the regional level, CCA could mean measures covering a large geographical area and numerous communities. They are likely to undermine local DRR plans. In due course, CCA measures, if not validated locally, would adversely affect the livelihoods and practices in the communities (e.g., relocation of settlements along rivers or coasts identified as being hazard-prone areas).

In order for CCA to be integrated into DRR, it is necessary to downscale present global climate change models to the local level. Presently, there are several efforts to downscale climate change models in order to attribute local weather to global climate phenomena. On the one hand, some recent studies suggest the potential of different methods (i.e., statistical downscaling methods) in localising climate change projections (e.g., Gutiérrez *et al.* 2013). On the other hand, some scholars remain sceptical about the reliability of downscaling climate predictions at the regional level (e.g., Pielke and Wilby 2012). The doubts on the accuracy of regional climate downscaling complicate, for example, the use of climate-scenario-based maps that are often regional in scale in efforts for risk reduction at the community or city (or municipal) levels.

With regard to temporal scale mismatches, effective DRR including CCA means developing long-term strategies geared towards the practice of sustainable development. However, present strategies, shaped by political interests and upward accountabilities to humanitarian donors, focus heavily on short-term interventions. Though necessary at some point in the implementation of any DRR including CCA programmes, they do not encourage DRR including CCA.

DRR including CCA, as well as development goals, are also disconnected by their distinct political economies that influence their conceptualisation and realisation at different levels and geographical scales. Begum *et al.* (2014) highlight the difficulties posed by having two different institutions to manage DRR and CCA separately from the international (e.g., United Nations Framework Convention on Climate Change and United Nations International Strategy for Disaster Risk Reduction / United Nations Office for Disaster Risk Reduction) down to the national level (e.g., National Security Council and Ministry of Natural Resources and Environment in the case of Malaysia). Guided by different objectives and global frameworks, which require different levels of commitments and capacities from various actors, DRR and CCA remain institutionally disconnected from the international down to the local level. Presently, the Sendai Framework for Disaster Risk Reduction (UNISDR 2015) by the UNISDR and the Paris Agreement under UNFCCC appear to pursue common goals and principles such as the mainstreaming of DRR and CCA in achieving sustainable development goals, fostering a more inclusive and participatory process, and enhancement of the capacities of local communities, particularly within the vulnerable sectors. The challenge now is how to put together the different efforts of several institutions at different levels to achieve their common goals and to facilitate a coordinated and efficient inclusion of CCA within DRR.

The integration of DRR including CCA, and development goals both in theory and in practice can be justified because they are connected by common objectives (i.e., reduce risks and promote sustainable development and environmental protection) and pursuing them would have common benefits for the different actors involved. It is important to recognise that each goal (e.g., DRR including CCA and SDGs' seventeen goals among others) is by itself a difficult and long-term task. For example, achieving the first SDG articulated as 'no poverty' is a long-term social struggle and may require drastic economic reforms. Combining efforts to address some, if not all, development goals is another concern. It is also difficult to assess the mutual benefits that can be derived, for instance, from mainstreaming poverty reduction into the environmental protection goal and vice versa. The positive impacts of the former to the latter may not be immediately visible. And while achieving those goals may contribute to DRR, other factors (e.g., gender and ethnic identities, armed conflicts, and environmental degradation) may come into play and influence people's vulnerabilities to natural hazards.

The connections and disconnections between DRR including CCA and development goals discussed above are influenced by and, in return, are shaping theoretical debates and research methodologies. Such influences are also apparent in the actual practice in the field by the practitioners and policies institutionalised at different levels. The ever-changing perspectives on any social issues or goals (e.g., DRR including CCA, development and environmental issues) are driven by a much more complicated evolution of theoretical understanding in the academe characterised by endless debates, constructive criticism, and discovery and rediscovery of knowledge. The debates often originate in the different and sometimes conflicting definition of concepts, which usually serve as the basis of research frameworks and paradigms, and later methodologies and policies.

In DRR, Hilhorst and Heijmans (2012) identify four major phases in the academic understanding of disaster risk including (1) hazard or engineering orientated, (2) behavioural, (3) structural, and (4) complex and mutuality paradigms. Hazard and behavioural paradigms suggest that disaster risk is determined by the extremity of natural hazards (see Gaillard 2010 for a review of the history of the hazard and behavioural paradigm). Such an argument leads to risk reduction measures that aim to prevent or mitigate hazards through structural measures (e.g., flood dikes and seawalls) and improve people's risk perception and behavioural adjustment to hazards. Since the 1970s, the so-called vulnerability paradigm has challenged the deterministic hazard-centred approach to DRR and argued that increasing disasters and their impacts are mainly due to people's increasing vulnerability derived from poverty, gender discrimination, and ethnic segregation amongst others. As suggested by Wisner *et al.* (2004), these many forms of social injustice can be rooted in dominant social and economic structures (e.g., distribution of power, wealth, and resources), ideologies (e.g., neoliberalism), and history and culture (e.g., colonial history) (Wisner *et al.* 2004, 2012).

Moving away from the perspectives that are too hazard-centric (hazard and behavioural paradigms) or social-centric (vulnerability paradigm), the complex and mutuality paradigm recognises the influence of the relationship between society and the environment in shaping disaster risk. The human (society) and the environment are 'mutual' processes as they constantly shape each other. The 'complex' refers to the non-linear interactions between several social and environmental (or natural) factors that produce and reproduce disaster risks (i.e., hazard, people's vulnerabilities, and capacities). This complex mutuality of human–environment relationships and how they shape disaster risk is best exemplified in the pressure and release framework (PAR; also called progression of vulnerability) proposed in the two editions of 'At Risk' (Blaikie *et al.* 1994; Wisner *et al.* 2004). Although originally conceptualised to elaborate the vulnerability paradigm, PAR also shows that disaster risk is a function of people's vulnerabilities and the hazards that affect them.

Wisner *et al.* (2012) further refined the PAR framework and highlighted the influence of human activities on the natural environment and on the occurrence and characteristics of some hazards. Human activities characterised by destructive use of natural resources could amplify the impacts or even prompt the occurrence of hazards. Negative impacts brought by society to the environment (e.g., anthropogenic climate change) could disrupt the productive base that sustains the basic needs of society such as food and water (see Hanjra and Qureshi 2010 for an analysis of the impacts of climate change on food and water supply). Coupled with unfair access to resources from the international (e.g., between developed and developing countries) to the national level (e.g., between the few elite and poor majority population) down to the community and household level (e.g., between men and women or adult and children), such a complex relationship between society and the environment shapes people's vulnerabilities.

Each paradigm demands focus on the specific factors that it deems important in risk analysis. To some extent, the methodologies include methods and tools influenced by the paradigmatic position and disciplinary focus of a researcher. Research methodologies in disaster studies are also applied in the physical and social science disciplines as well as in development research, and they are unique not in terms of the methods but for the context (e.g., different phases of a disaster) to which they are applied (Mileti 1987; Stallings 1997). In poverty research, Chambers (2007) proposed a typology of methodological approaches, which consists of (1) economic reductionism, (2) anthropological particularism, and (3) participatory pluralism. The following paragraphs regroup Chambers' (2007) typology of common methodologies, including methods and tools associated with the paradigms in disaster studies discussed above.

The hazard-centred paradigm dominated by physical scientists relies on rigid scientific methods characterised by experimentation and use of highly technological devices to understand the dynamics of natural hazards for the purpose of prediction, mitigation and prevention. The threat of climate change has further necessitated development of such kinds of methodologies in order to understand, for instance, the influence of global climate change on the characteristics of climate- or weather-related hazards such as typhoons (e.g., Yumul *et al.* 2011). Behavioural paradigms have relied on risk perception surveys and other quantitative methods in such a way that most of them are 'degenerated into standardized questionnaire surveys and "official" analyses (usually) applied in developing countries' (Dekens 2007, p. 9). Risk perception surveys and other related quantitative data collection methods associated with behavioural paradigms share the characteristics of economic reductionism in Chambers' (2007) typology of paradigmatic approaches wherein questionnaires are used as the most common method and outsiders and local people serve as data collectors and respondents respectively. The nature of such engagement is extractive because it usually benefits the researchers without contributing to the development of local knowledge. Such a methodology has also been highly criticised because of the imposition of knowledge and expertise by outside researchers.

Because of the necessity to understand the complex nature of vulnerability from a structuralist view, methods and tools (e.g., qualitative methods such as interview, participant observation and text analysis, amongst others) should produce a rich description of the historical and socioeconomic processes involved at different spatiotemporal scales such as those outlined in the PAR framework. Those methods are closer to anthropological particularism in Chambers' (2007) typology. In this view, the focus on quantitative methods is not to measure the risk perception of the people but to understand their access, or the lack of it, to the many resources (e.g., social, economic, political, natural, and cultural resources) that shape their vulnerabilities. These same resources are also the relevant indicators in development research (e.g., the Human–Development Index by the United Nations Development Programme, World Development Indicators by the World Bank and Community-based Management System by the Partnership for Economic Policy) (see Chambers 2007 for a review of approaches in measuring poverty and related development indicators). Yet such analysis of the social structures goes beyond any quantitative findings. It requires a more radical perspective to be able to question or address the unfair distribution of power in society and the dominance of certain ideologies and/or cultures.

The more balanced perspective on the conceptualisation of disaster risk offered by the complex and mutuality paradigm encourages a multidisciplinary nature of research (see Wisner *et al.* 2012). Hilhorst and Heijmans (2012) argue that paradigms co-exist and are influenced by each other. Such coexistence is necessary to enable the context-specific analysis of risk (i.e., more focus on either hazard or vulnerability, if necessary) without neglecting all other components. The complex and mutuality paradigm invites different but mutually beneficial perspectives and paradigms to foster the contribution of knowledge from different actors. Such flexibility, multidisciplinary emphasis, and knowledge sharing require the use of participatory methodologies (including methods and tools) that encourage dialogue among actors of DRR including CCA at different levels and scales (see the Integrated Disaster Risk Reduction Framework by Wisner *et al.* 2012, p. 2).

Central to the use of participatory methods and tools is the recognition of local knowledge (including traditional and indigenous) as an important component of people's capacities. Local forms of knowledge that are useful in DRR including CCA incorporate, but are not limited to, people's experiences of disasters, familiarity of their local territories, and many strategies embedded in local coping and adaptive mechanisms. This type of knowledge is largely derived from actual experiences and is thus specific to a particular context and can be considered endogenous to the local people. Local knowledge is often distinguished from scientific knowledge, which is derived from a strictly scientific methodology. These two bodies of knowledge are not necessarily conflicting, are sometimes the same, and must both contribute to reducing disaster risk.

Thus, efforts should be geared toward bridging the gaps between local and scientific knowledge by taking advantage of their specific strengths. The use of participatory methodologies, often utilised in Community-Based Disaster Risk Reduction (CBDRR) including the process for CCA, proved to be effective in combining both local and scientific knowledge. They allow local people, especially those from the marginalised sectors of society, to participate in different aspects and stages of DRR while encouraging the contribution of outside actors particularly scientists, higher government authorities, and NGOs (Cadag and Gaillard 2012) (see Box 18.1).

Box 18.1   Contribution of Non-Governmental Organisations to DRR including CCA in the Philippines

Mary Joy K. Gonzales[1]

[1] Tearfund, Manila, Philippines

Several international non-governmental organisations (INGOs) have been implementing programmes and projects in the Philippines for decades. A consortium led by Oxfam together with partners Handicap International, Christian Aid, and Plan International, through funding from the European Commission Humanitarian Aid programme (DIPECHO), recently concluded a ten-year project called Scaling Up Resilience in Governance (SURGE). This project aimed to increase resilience and reduce vulnerabilities in the local communities. While the project's primary focus is disaster preparedness, the adoption of climate change adaptation (CCA) in policies and practices has become integral to the implementation of the project, as both are closely linked to poverty and sustainable development.

The following are both current and past disaster risk reduction (DRR) including CCA interventions of INGOs in the local communities:

1   Promoting inclusive, community-based approaches in DRR including CCA especially among vulnerable or at-risk groups.
2   Creating an environment for DRR including CCA to be part of the development of local land use and development plans with clear mechanisms for implementation, and with particular attention to DRR including CCA financing.
3   Building champions not only in the local government units but also in the communities, emphasising local leadership.
4   Sharing the principles of ecosystem-based approaches in DRR including CCA with communities, especially in geographically isolated and disadvantaged areas.
5   Addressing policy gaps and influencing local, national, and international platforms.
6   Peer-to-peer learning or sharing of best practices and lessons learned from major disasters of recent and past years.
7   Improving knowledge management by utilising both local and scientific knowledge for a more informed and improved early warning system and contingency plan.

To ensure the project's sustainability and replicability, partnerships with both local and national government actors are imperative. While INGOs pave the way for these projects, it is widely recognised that civil society organisations (CSOs), including local non-government organisations (NGOs) and people's organisations should take the lead in developing the agenda for DRR including CCA in the years to come.

Connections, Disconnections, and Knowledge Production

For effective integration of DRR including CCA within development goals, it is necessary to further strengthen existing connections outlined in the earlier section. Frameworks should be geared toward the integration of CCA into DRR, under the overarching umbrella of development, in order to capture not only the climate-related hazard dimension of disaster risk but also non-climate-related-hazards, and people's vulnerabilities and capacities. DRR including CCA can then be integrated into larger development and environmental agendas to address root causes of people's vulnerabilities.

It is equally important to address disconnections discussed in the previous section as they complicate the goal of integration. The conceptual discrepancies and practical barriers discussed above imply that there is a lack of understanding of the real problems and issues that affect local communities. At the international level, stakeholders' priorities may appear different and can be addressed separately by well-defined objectives. Within local communities, however, the problems and opportunities associated with disaster risk, climate change, and several development and environmental issues are very much related and cannot be addressed separately. Further, difficulties experienced by local communities (e.g., environmental degradation) can be related to a complex set of issues instigated and sustained at different geographical and temporal scales. Current policies, scientific knowledge, and use of local knowledge remain inadequate on how to address such interrelatedness and multidimensionality of local issues and problems that affect local communities. Not surprisingly, when implemented at the local level, many programmes, especially those that cater to the needs of specific sectors (e.g., children, people with disabilities, and women), lack consideration of the needs of local communities and are implemented in isolation with other issues, thus leading to a lot of 'unsuccessful development interventions'. As a result, actions and solutions towards DRR including CCA especially those initiated at the international level, prove to be ineffective as they do not take into account the realities of local communities. For example, in Kiribati, aside from the ongoing discussions regarding the relocation of the population to neighbouring countries due to climate change, aid-driven policies for CCA have led to the formulation of measures that ignore the equally important but more frequent and urgent problems associated with water pollution, sanitation, and solid waste management in the local communities themselves (Gaillard 2012; Storey and Hunter 2010).

It is difficult to understand those disconnections without relating them to how knowledge to achieve the goals of DRR including CCA and development goals is produced, motivated, and structured. The review of different approaches and methodologies above suggests that there are different ways and means of knowledge production that shape actions and policies, for instance, in DRR. Supposedly, the ideal approach encourages co-production of knowledge through dialogue and collaboration among actors both horizontally (e.g., among scientists through multidisciplinary research) and vertically (e.g., between scientists, NGOs, and local communities through community-based approaches). In practice, it is difficult to see that there are genuine efforts to realise such co-production of knowledge as made evident by the lack of methodologies (e.g., including methods and tools) to facilitate dialogue and integration of local and scientific knowledge.

On the one hand, local communities often do not fully grasp or do not have access to the relevant scientific knowledge while NGOs sometimes consider such knowledge to be disconnected from reality. On the other hand, many scientists may remain confined to their rigid scientific methods and consider local knowledge 'too subjective and removed from scientific methodologies and rigorous protocols' (Gaillard and Maceda 2009, p. 109). And while the contribution of local knowledge to many aspects of development, encompassing DRR including CCA, has long been recognised, it remains excluded in shaping actions and policies. As Spiekermann *et al.* (2015, p. 96) put it, 'there is the vast knowledge related to the experience of communities, families and individuals that is not always capitalised on'.

The shortcomings to address local issues and problems, and thus actions and solutions, in the communities can be explained by how knowledge production is motivated and structured. Knowledge production for DRR including CCA and development goals are motivated by disciplinary or professional biases and western discourses, and some political and economic issues. Many global and national frameworks, in which biases and western discourses are embedded, are designed to maintain the structure that produced them. As a result, in the contexts of DRR including CCA and of development goals, many policies, actions, and solutions remain superficial, short-term, and supply-driven, all meant to foster dependency between less wealthy and more affluent countries, outside and local actors, and, in general, powerful and less powerful institutions, groups and individuals.

The dominance of hazard-centred approaches in dealing with so-called 'natural disasters' exemplifies the hegemony of western science, in much the same way that disease and poverty issues in the formerly colonised countries were dominated by western medicine and western investment and aid, respectively (Bankoff 2001, p. 28). With the emergence of climate change issues, much emphasis has been placed on the role of hazards in shaping disaster risk (both hazards and vulnerabilities), so the hazard-centred approach in DRR (including CCA) is again renewed. Gaillard (2010) argues that the hazard-centred interpretation of vulnerability (along with capacity and resilience) in the context of climate change allows those who are supposed to be accountable to use hazard or climate as a 'scapegoat' to evade their responsibilities (e.g., negligence by elected officials and government agencies).

Highlighting the role of hazards also allows for the evasion of some highly sensitive political and economic issues. DRR including CCA and development goals demand social reforms at different geographical scales which may not be of interest to those who hold political and economic power and resources. As a result, there is an apparent lack of acceptance of the real issues that shape people's vulnerabilities and thus disaster risk. Not surprisingly, knowledge production is rationalised according to the necessity of dominant social and economic structures and/or ideologies (e.g., capitalism). Ironically, those dominant structures are argued to be the main root causes of skewed development (or unfair development). And from that argument, the disconnections to the integration of goals (e.g., conceptual discrepancies and practical barriers) are simply the outcomes of a lack of sincerity in really addressing the problems by those who hold power and resources.

Towards Integration of DRR Including CCA in the Development Goals

This chapter has argued that enhancing connections and addressing disconnections to achieve a more inclusive DRR including CCA, and integrating it into development goals, requires co-production and sharing of knowledge. From the international level down to the local level, those goals should be considered as complementary objectives. The mainstreaming of goals must be supported by proper methodologies and translated into actions and policies. These methodologies should encourage dialogue among different actors involved. The importance of dialogue is justified by a widely accepted argument that effective DRR including CCA requires different knowledge from different actors. This same argument applies when integrating DRR including CCA into the development goals.

This chapter has also shown that the co-production and sharing of knowledge through dialogue can be realised using an appropriate research paradigm (e.g., complex and mutuality paradigm) and methodological approach (e.g., participatory pluralism). The use of participatory methods coupled with a sincere effort to engage all actors, particularly local communities and the most marginalised sectors of society, in the process could be a powerful mechanism to recognise the different issues surrounding DRR including CCA, and its integration to development goals. This could also be the key to identifying necessary solutions that are based on consensual decisions.

Further work needs to be done to encourage DRR actors to utilise participatory methodologies and to become engaged in a co-production and sharing of knowledge with other actors. Thus, any meaningful efforts that aim to integrate DRR including CCA into the development goals should aim to enhance 'connections' and to address 'disconnections' by deviating from a traditional and technocratic approach towards a more inclusive and people-centred approach.

References

Asian Disaster Preparedness Center. (2006) *Mainstreaming Disaster Risk Reduction into Development Policy, Planning and Implementation in Asia*, Bangkok: Asian Disaster Preparedness Center. Online http://www.preventionweb.net/files/2302_adpcdevelopmentnov06.pdf (accessed 12 December 2015).
Bankoff, G. (2001) 'Rendering the world unsafe: "Vulnerability" as western discourse', *Disasters* 25, 1: 19–35.
Begum, R.A., Sarkar, M.S.K., Jaafar, A.H. and Pereira, J.J. (2014) 'Toward conceptual frameworks for linking disaster risk reduction and climate change adaptation', *International Journal of Disaster Risk Reduction* 10: 362–373.

Blaikie, P., Cannon, T., Davis, I. and Wisner, B. (eds) (1994) *At Risk: Natural Hazards, People's Vulnerability, and Disasters, 2nd edition,* London: Routledge.

Cadag, J.R.D. and Gaillard, JC (2012) 'Integrating knowledge and actions in disaster risk reduction: The contribution of participatory mapping', *Area* 44, 1: 100–109.

Chambers, R. (2007) *Poverty Research: Methodologies, Mindsets and Multidimensionality,* Brighton: Institute of Development Studies. Online http://opendocs.ids.ac.uk/opendocs/bitstream/handle/123456789/399/Wp293%20web.pdf?sequence=1 (accessed 12 April 2012).

Dekens, J. (2007) *Local Knowledge for Disaster Preparedness: A Literature Review,* Kathmandu, Nepal: International Centre for Integrated Mountain Development (ICIMOD). Online http://www.preventionweb.net/files/2693_icimod8fc84ee621cad6e77e083486ba6f9cdb.pdf (accessed 11 November 2015).

Gaillard, JC (2010) 'Vulnerability, capacity and resilience: Perspectives for climate and development policy', *Journal of International Development* 22, 2: 218–232.

Gaillard, JC (2012) 'The climate gap', *Climate and Development* 4, 4: 261–264.

Gaillard, JC and Maceda, E.A. (2009) 'Participatory 3-dimensional mapping for disaster risk reduction', *Participatory Learning and Action* 60: 109–118.

Gaillard, JC and Mercer, J. (2013) 'From knowledge to action: Bridging gaps in disaster risk reduction', *Progress in Human Geography* 37, 1: 93–114.

Guha-Sapir, D. and Hoyois, P.H. (2012) *Measuring the Human and Economic Impact of Disasters, Foresight Project,* London: Government Office for Science.

Gutiérrez, J.M., San-Martín, D., Brands, S., Manzanas, R. and Herrera, S. (2013) 'Reassessing statistical downscaling techniques for their robust application under climate change conditions', *Journal of Climate* 26: 171–188.

Hanjra, M.A. and Qureshi, M.E. (2010) 'Global water crisis and future food security in an era of climate change', *Food Policy* 35, 5: 365–377.

Hilhorst, D. and Heijmans, A. (2012) 'University research's role in reducing disaster risk', in B. Wisner, JC Gaillard and I. Kelman (eds) *Handbook of Hazards and Disaster Risk Reduction,* London: Routledge, pp. 739–749.

Holden, W.N. and Jacobson, R.D. (2013) *Mining and Natural Hazard Vulnerability in the PHILIPPINES: Digging to Development or Digging to Disaster?,* London: Anthem Press.

IPCC (Intergovernmental Panel on Climate Change). (2014) *Climate Change 2014: Synthesis Report: Contribution of Working Groups I, II and III to the Fifth Assessment Report of the IPCC,* Geneva, Switzerland: IPCC.

Kelman, I. and Gaillard, JC (2010) 'Embedding climate change adaptation within disaster risk reduction', in R. Shaw, J.M. Pulhin and J.J. Pereira (eds) *Climate Change Adaptation and Disaster Risk Reduction: Issues and Challenges,* Bingley, UK: Emerald, pp. 23–46.

Kim, N. (2012) 'How much more exposed are the poor to natural disasters? Global and regional measurement', *Disasters* 36, 2: 195–211.

Lee, K.N. (1993) 'Greed, scale mismatch and learning', *Ecological Applications* 3: 560–564.

Lewis, J. (1999) *Development in Disaster-prone Places: Studies of Vulnerability,* London, UK: Intermediate Technology Publications.

Mercer, J. (2010) 'Disaster risk reduction or climate change adaptation: are we reinventing the wheel?', *Journal of International Development* 22, 2: 247–264.

Mileti, D. (1987) 'Sociological methods and disaster research', in R. Dynes, B. de Marchi and C. Pelanda (eds) *Sociology of disasters: Contributions of sociology to disaster research,* Milan, Italy: Franco Angeli, pp. 57–69.

O'Keefe, P., Westgate, K. and Wisner, B. (1976) 'Taking the naturalness out of natural disasters', *Nature* 260, 5552: 566–567.

Pielke Sr., R.A. and Wilby. R.L. (2012) 'Regional climate downscaling: What's the point?', *Eos Transactions of American Geophysical Union* 93, 5: 52–53.

Spiekermann, R., Kienberger, S., Norton, J., Briones, F. and Weichselgartner, J. (2015) 'The disaster-knowledge matrix – reframing and evaluating the knowledge challenges in disaster risk reduction', *International Journal of Disaster Risk Reduction* 13: 96–108.

Stallings, R.A. (1997) 'Methods of disaster research: Unique or not', *International Journal of Mass Emergencies and Disasters* 15, 1: 7–19.

Storey, D. and Hunter, S. (2010) 'Kiribati: An environmental "perfect storm"', *Australian Geographer* 41, 2: 167–181.

UNISDR (United Nations International Strategy for Disaster Reduction). (2015) *Reading the Sendai Framework for Disaster Risk Reduction,* Geneva: UNISDR. Online https://royalsociety.org/~/media/policy/Publications/2015/300715-meeting-note-sendai-framework.pdf (accessed 24 May 2016).

Wisner, B. (2001) *Vulnerability in Disaster Theory and Practice: From Soup to Taxonomy, then to Analysis and Finally Tool*, Holland: Disaster Studies of Wageningen, University and Research Centre Conference (29–30 June 2001).

Wisner, B., Blaikie, P., Cannon, T. and Davis, I. (eds) (2004) A*t Risk: Natural Hazards, People's Vulnerability and Disasters*, London: Routledge.

Wisner, B., Gaillard, JC and Kelman, I. (2012) 'Framing disaster: Theories and stories seeking to understand hazards, vulnerability and risk', in B. Wisner, JC Gaillard and I. Kelman (eds) *Handbook of Hazards and Disaster Risk Reduction*, London: Routledge, pp. 18–33.

Yumul, G.P., Cruz, N.A., Servando, N.T. and Dimalanta, C.B. (2011) 'Extreme weather events and related disasters in the Philippines, 2004–08: a sign of what climate change will mean?', *Disasters* 35, 2: 362–382.

PART III

Knowledges and Understandings

19

EDITORIAL INTRODUCTION TO PART III: KNOWLEDGES AND UNDERSTANDINGS

Towards Wisdom

Ilan Kelman, Jessica Mercer, and JC Gaillard

We question, therefore we are. Without critique, we cannot understand. Knowledge provides some power – although much power is exercised without knowledge.

Many different knowledge forms and many different levels of understanding exist. The chapters in Part III of the *Handbook* explore some aspects of multi-dimensional approaches to knowing and understanding. The focus tends to be on knowledges and understandings that are used and promoted less frequently in disaster risk reduction (DRR) including climate change adaptation (CCA).

Consequently, one chapter covers performing arts, but equivalent chapters are not provided for social sciences, physical sciences, and the humanities. Ideally, an exploration of interdisciplinarism, transdisciplinarism, and multidisciplinarism within different scientific fields would have advanced knowledges and understandings for DRR including CCA. Similarly, one chapter explores history as a foundation of knowledge and understanding, while there is no chapter on futures or foresight.

Education and training features prominently, begging the question regarding what we aim for in these actions. The answer is improvements in DRR including CCA. We 'educate', 'train', 'teach', and 'learn' in order to do better, to change the ways we think and behave, in the same way that fundamental human rights – such as permitting all adults to vote (although some countries prevent many prisoners from voting) and accepting that children have rights and responsibilities – required a change in the ways we think and behave in order to eliminate slavery and to permit due processes of justice and law. Yet we still have not achieved all this.

We do need to understand more about behaviour: influencing behaviour, changing behaviour, and retaining behaviour. A chapter would have been helpful on learning what understandings of wider risk-related behaviour could offer DRR including CCA, due to the significant overlaps and parallels, as well as recognising and explaining differences. Examples are smoking, drug use, alcohol consumption, safe sex, extreme sports, driving including seatbelt use, and cycling including helmet use.

Boundaries amongst different modes of behavioural influence require more work. How is education differentiated from brainwashing? Where is the line between training and indoctrination? How do laws and regulations lead to cultural and paradigmatic shifts – and vice versa?

How do perceptions influence knowledge and understandings – and the other way around? How do cultural constructs create perceptions while enhancing or blocking knowledge and understandings – and vice versa? What roles do values play and how are values constructed? Another handbook on DRR including CCA could be published exploring the interactions and non-interactions amongst attitudes, views, experiences, and behaviours for a gamut of risks and cultures.

A further handbook for DRR including CCA could be published on power. Power relations imbue all ideas, policies, and actions regarding DRR including CCA.

Often for DRR including CCA, attempts to separate out nature and to de-link human constructs from nature's forces and energies relate to human efforts to have power over nature. Sometimes we succeed, creating a significant level of control of our environment through air conditioning, base isolators for earthquakes, tying roofs to walls, and directing runoff or river flow. Other times, we try and fail, such as coastal and river engineering creating a false sense of security, so flood vulnerability reduction measures are neglected and warnings are sidelined. Senses of power of nature connect directly back to perceptions and behavioural changes.

Many individuals and societal sectors create and accrue power, making others vulnerable. The power might be political, social, financial, military, other forms, or a combination. Demographic characteristics such as age, gender, disability, sexuality, caste, religion, culture, and ethnicity are frequently used as the basis by which to acquire or deny power. Knowledge disciplinarism, as discussed in Part II, is one way of delineating and collecting power, where one knowledge field or paradigm excludes others in order, for instance, to highlight CCA over other DRR and development concerns.

One consequence of disciplinary-related power is fear of questioning and fear of critiquing. Some authors throughout this handbook openly challenge power, describing flaws in, explaining hypocrisies of, and providing ways forward beyond some dominating influences within DRR including CCA. Other authors of *Handbook* chapters were terrified about or hostile towards raising any forms of criticism against the powers-that-be, most notably the Intergovernmental Panel on Climate Change (IPCC), not just the 'Assessment Reports', but also Special Reports such as 'Managing the Risks of Extreme Events and Disasters to Advance Climate Change Adaptation (SREX)'.

No power structure should be eviscerated or venerated simply because it exists. All power structures should be questioned constructively, to advance knowledges, and to increase understandings. Explorations and challenges should be balanced, respectful, and open to explorations and challenges themselves. Failure to do so, and mindless protectionism of an elite or of an established process, is as much a disservice to ourselves as personal attacks, ostracising, and criticism for the sake of criticism.

As Junius long ago advised, 'The subject who is truly loyal to the Chief Magistrate will neither advise nor submit to arbitrary measures'. Paraphrasing yields 'Loyal subjects neither advise nor submit to arbitrary measures'. Our duty and responsibility is to be the most loyal subjects serving DRR including CCA.

Critiquing analysis and being open to the critiques from others defines healthy, respectful debate during which we can delve more deeply and more broadly into understanding and doing – through raising queries, answering them, accepting critiques, and responding thoughtfully to them. This is the way to improve, joining forces rather than creating barriers through our collective knowledges and understandings – in order to better integrate CCA into DRR.

And thus can this process be defined as 'wisdom'. Much more remains to be understood across DRR including CCA. Ignorance precludes neither action nor successful action. Wisdom is knowing how and when to apply the knowledges and understandings which exist while always seeking much, much more to succeed even more.

20

BUILDING ON THE PAST

Disaster Risk Reduction Including Climate Change Adaptation in the *Longue Durée*

Virginia García-Acosta

For me, History is the sum of all possible histories. If History is bound to pay privileged attention to the duration, the *longue durée* seems to be the more useful line for social sciences' observation and reflection (Braudel 1958) – and see Box 20.1.

Box 20.1   Climate in European History and Wine

Using evidence of crop yields and dates of grape harvests sourced mainly from French archives, Le Roy Ladurie (1972) reconstructed European climate for several centuries. His findings were compiled in his book entitled *A History of Climate since the Year 1000*, which was originally published in French in 1967 as *Histoire du climat depuis l'an mil*. It includes research sourced from German, Swiss and Luxembourgois archives, which he compares with documentary reports on wine quality.

All this phenological documentation was confronted with the long series of glacier records existing at the beginnings of the second half of the twentieth century. Le Roy Ladurie (1972) found that it converged with alpine records collected by those glaciologists.

The precision and accuracy of his research in the *longue durée*, looking not only at what he calls the 'terrible years', but also the processes before and after them, led him to identify that throughout the Western World since the fourteenth century, the cause of disasters was not the climate itself. He offers eloquent examples showing that the decline of viticulture in northern France during the sixteenth century was in fact due to the neglect and abandonment of the crop because they were not very profitable. Once again, it would seem that a close examination of unpublished texts suggests that the climate explanation alone is insufficient. Hence, Le Roy Ladurie (1972) insists that a good history of climate has to be interdisciplinary and comparative.

Introduction

The periodic floods of the largest river in Africa, the Nile, and its effects in Ancient Egypt dating back to the beginnings of that great civilisation, led to the development of specific knowledge surrounding climate adaptation and management. Through the construction of canals and levees and the invention of the *nileometer*, a measuring and recording of water levels was enabled, which permitted not only the control, but also the harnessing of benefit from major river floods, as silt deposits from its annual overflowing made the surrounding land highly fertile.

On the other side of the world, during the pre-Hispanic period of what is now Mexico, the city of Tenochtitlan, nowadays Mexico City, had been built on an island that was often flooded. Thanks to the knowledge of the topography of the lake and its environmental characteristics, Aztec authorities built an *albarrada* over 15 kilometres long and maintained a team of divers trained in construction, cleaning and maintenance of such works. Flooding was controlled. After the Spanish conquest (1521), these works and their maintenance were ceased, resulting in the city falling victim to some of the worst floods in its history. It quickly became apparent after the monumental 1555 event that flooding was not only the result of what primary sources refer to as 'torrential rains'.

All forms of human organisation represent a particular way of interacting with nature – society and nature are mutually involved in the shaping of one another. In this process, there are interpretive and material resources that stand between societies and the environment in which they settle: that is, the components that create culture(s). This complex, multifaceted relationship between nature and culture has, over the course of human existence, created diverse cultures and knowledges, and as such, multiple and diverse ways to cope with, prevent and adapt to adversity. What is today synthesised as the expressions 'Disaster Risk Reduction' (DRR) and 'Climate Change Adaptation' (CCA), by no doubt can be seen in countless demonstrations much before Western culture granted them a condition of contemporary, scientific or political interest.

This chapter reviews the construction and historical reconstruction of DRR, including the framework of CCA, by exploring what is referred to as 'coping or adapting historically' in the context of the emergence and development of historical disaster research.

Historical Disaster Research

The study of disasters, especially those associated with natural hazards, has occupied the attention of social scientists from various disciplines since the early twentieth century. Sociologists and geographers came first, later to be joined by anthropologists in the middle of that century, offering studies carried out as a result of the occurrence of disastrous events – typhoons, volcanic eruptions, earthquakes. Anthropological research, in particular, evolved mainly at the end of the 1970s.

Historical disaster research from a social perspective is a field of study developed over the last twenty years, but its identification as a specialised field is more recent. Its emergence has varied throughout the world and its evolution has been closely linked with the recognition that disasters are in themselves a process. The historical perspective in the study of disasters, that is, by using information from the past, has shown that hazards may act as triggers and, as well, play the role of revealers or disclosers of pre-existing critical conditions. Natural hazards and disasters may even shape societies. The study of disasters as a process constitutes a thread on which one can weave several histories (Bankoff 2003; García-Acosta 2002).

Historical research about disaster and climate (amongst other environmental characteristics) represents a focus on searching, selecting, reading and analysing documents coming from earlier times, and it is fundamental to re-dimension the interpretation of that historical information on the study of past disasters, which means fully understanding the reality of their context (Altez 2002). This required not only looking at events identified as disasters or catastrophes, but also seeking to understand the context in which they occurred, since 'disasters have to be understood as major forces shaping historical processes and therefore need to be studied not as isolated events but in their historical context' (Janku *et al.*'s chapter in Janku *et al.* 2012: 1–14).

It is clear that disaster analysis focusing on the past has not always included the central aspect of *The Routledge Handbook of Disaster Risk Reduction Including Climate Change Adaptation*. Nevertheless, we must accept that:

- disasters are historical processes;
- disasters involve a continuous and persistent social construction of risk;
- societies have not been passive agents in facing disasters; and
- communities have displayed culturally constructed adaptive strategies in the course of their interaction with the environment.

Furthermore, we must also accept, too, that societies have developed ways that shape new contexts that ultimately aid in helping with the reduction of the effects and impacts from recurring natural hazards. Studying disasters from a historical perspective therefore allows us to reconceptualise them and explore them with a holistic and global view.

Studies related to these problems are not unified epistemologically and neither do they represent a single theoretical scope. They include different analytical approaches, and ways to understand societies and their processes by using key variables like the disaster itself, risk, hazard and vulnerability(ies) that manifest throughout history.

Even the use of the term 'disaster' (often referred to before as 'calamity', 'accident' or 'plague') in discourse from the eighteenth and nineteenth centuries to the modern day has included principles of DRR, including CCA. Indeed, the ethos behind such terms did not simply begin in the twentieth century when the fields were formally born. Premises like global warming or climate change, while increasingly pressing in the modern day, are not new phenomena. The only variable that has changed, and thus offers historically different results, is the vulnerability in all its dimensions, and therefore the social and historical conditioning that affects people's lives rather than the natural phenomena itself (Altez 2011). That is why it is necessary to accept what Oliver-Smith (2013, p. 279) contends, that CCA and DRR are 'related undertakings' to introduce the historical perspective that avoid a 'concentration on climate change, instead directing attention to root causes and the fundamentals of vulnerability and resilience as long-term processes' (Kelman *et al.* 2015, p. 25).

One of the main interests of this chapter is to reinforce the idea that the 'past challenges our notions that contemporary ways are always better than techniques and practices developed by peoples and communities centuries ago to cope with the hazards that beset them' (Bankoff 2012, p. 40). Without doubt, understanding the historical roots of contemporary disasters, and then identifying and analysing the social construction of vulnerability(ies) and risk(s) within a society is essential to the reduction of the effects associated with natural hazards and to the prevention of future disasters.

Our starting point is therefore based on the following premise: societies have never been passive entities in facing hazards. They have, historically, formulated social and cultural ways to deal with potential risks and disasters, and have developed social strategies for prevention and adaptation in their interaction with nature. It is therefore necessary, then, to identify, recover, reinforce and update those cultural constructs, in particular, those ones that can be identified as best practices and/or lessons learned (García-Acosta's Introduction in García-Acosta *et al.* 2012, pp. 17–21).

Historical Interest in Climate Change Consequences

Human interest in climate, that is, understanding it in order to respond to it and its changes, goes as far as classic antiquity. The mention of changes in the climate, even in some cases identified and referred to as 'climate change,' can be found since the sixth century BC in pre-Socratic writings.

Later on, Aristotle himself reflects on the existence of large-scale climate changes in the transformations in the forms of relief in his work *Meteorologica*.

Inquiries into climate were not limited to identifying weather as changing elements however, but were also concerned with the nature, origin and consequences of extreme phenomena such as droughts. Connections between droughts and deforestation, for example, or the so -called *clareos* or clearing, appear as early as the first century AD in the work of scientist and naturalist Pliny the Elder (23–79 AD), who in his *Naturalis Historiae* talks about climate alterations due to deforestation for mining purposes. The Scottish philosopher David Hume (1711–1776) also acknowledged that changes in European climate were due to the action of man because the forests, which were once thick and shielded the land from the sun's rays, had been cleared. Count de Buffon, Georges Louis Leclerc himself (1707–1788), stated that if the France of his days was less cold than Gaul and Germania's two thousand years ago, it was because its forests had been cut, its swamps dried, its rivers controlled and its farmlands cleared of dead organic life remains. Buffon even surpassed the European borders by mentioning the example of changes in temperatures and humidity experienced in some areas of French Guiana because of the practised clearing.

Similarly, Alexander von Humboldt, at the beginning of the nineteenth century, in his *Essay on the Geography of Plants* (1805) indicated that forest logging carried out by farmers decreased humidity. Moreover, José de Echegaray (1832–1916), the Spanish civil engineer and mathematician in his *Memoria de Sequía (Memory of Drought*, 1850), referred to the relationship between lack of trees and lack of rain, and when analysing the causes of drought in the Southern Iberian Peninsula, said that deforestation was an important agent (Olcina-Cantos 2009).

Further understandings of the global climate system, particularly with the scientific basis of climatology consolidating from the end of the nineteenth century in Europe, help specify some of these ideas. For example, spaces cleared for agricultural purposes in flat terrain alter microclimatic conditions significantly, but not the general ones of the region where they are practised.

The majority of cases mentioned above come from the pre-instrumental climate period. Key moments in the past 1,300 years, within the Holocene (the geological period in which we are nowadays), like the 'Medieval Warm Period' (ca. AD 700 to 1300) named by the British climatologist Hubert H. Lamb in 1965, and the 'Little Ice Age' (ca. AD 1300 to 1850), including raw winters in Europe during the period known as the 'Maunder Minimum' (1645–1715) (Alberola 2014), strongly influenced the climate and cultural history of the world, particularly in Europe. More recently, Foster (2012) has addressed and evidenced their effects in North America. Extreme changes in the natural environment have led to different cultural responses that, in some cases, have also suggested similitudes in patterns of human response, as those documented by Foster himself in his remarkable seven-hundred-year time period book. Based on the information reviewed, 'the pattern of human response to climate change has been similar in the southern latitudes of the temperate zone of North America with each swing in the series of climatic oscillations over the past approximately four thousand years or more' (Foster 2012, p. 167).

This interest in understanding climate throughout history has even led to measuring it, using historical data from sources that had never been used nor even imagined functional before. Such information was mainly sourced through qualitative methods. Indeed, this can be seen in *rogativas* or prayers, devout processions, masses and sermons, as well as in the officially documented ecclesiastical archives and epistolary records, all of which have been the main source for an outstanding research led by Alberola (2003). The analysis of this information is complex as it often involves the request, collection, organisation and analysis of data on droughts (*rogativas pro pluvia*), heavy rainfall, floods, frosts, hails, earthquakes, and even invasive pests from ceremonies dating back to the first years of Christianity. It has been noted that even if the correlation between these religious events (i.e., their frequency, costs, concurrence and popularity) and the actual amount of rainfall is not accurate, general patterns are still evident and therefore essential for interpreting the scale or the extent of the disaster that occurred. The study of these documents therefore helps us to 'understand disasters from a cross-cultural perspective', as they are integral to a particular historical context (Bankoff 2003, p. 159). It is, however, useful to note that with the secularisation of societies in Europe and the Americas, these large-scale religious manifestations began to decline from the Enlightenment Age onwards.

Coping or Adapting Historically

An exploration into the 'significant endogenous capacities to cope with the threat and impact of natural phenomena' is an urgent task (Gaillard 2011, p. 95). This involves the understanding of the diversity of coping mechanisms, the different modes of disaster management, and the assorted systems of adaptation and prevention that aim to seek and achieve a better relationship between nature and people. Such practices, actions and/or strategies have been created and improved throughout time by societies exposed to climatic hazards, and can generally be generated and designed bottom-up by people themselves, individually or collectively (community-level). Alternatively, they are produced and generated top-down by authorities, either the State or the Church (at the institutional level). Very seldom is there an initiative that integrates these two approaches.

Given that many historical documents that researchers can access are from 'official' sources, it is common to find information of this type generated at the institutional level: political authorities building bridges or levees to prevent flooding, storing cereals in the city's warehouses (Rohland 2015), or legal regulations that were 'invented to manage disasters, prevent them, or protect against them' (Mauelshagen 2013, p. 65). Similarly, information from ecclesiastical authorities – organising monetary collections, religious processions or ceremonies – that is, the so-called *rogativas* that were mentioned before, arise from an association of 'calamities' with divine punishment of sinful humanity. In some cases, investing in infrastructure for disaster prevention has served as a strategy of political legitimation, as in the European Middle Ages (Janku *et al*.'s chapter in Janku *et al*. 2012, pp. 1–14).

In the Americas, these examples come from very ancient times. In the Basin of Mexico, even in a period prior to the Aztec Empire (ca. 1100 AD), the development of institutional measures to control and prevent frequent flooding, increase agricultural areas and facilitate transport, included important technological developments like river derivations, ridges, dams, dikes, bridges and docks, and hydraulic works (Carballal and Flores' chapter in García-Acosta 1997 v2, pp. 77–99).

Another institutional strategy, explicitly and directly related to climate change, is the one related to the transfer of cities, well documented by Alain Musset (2011) in what he calls the 'Old Spanish Empire'. The 'Empire', nowadays known as Latin America, was where 162 city transfers were conducted throughout three centuries of Spanish domination. The origin of these transfers was associated with the disastrous presence of natural hazards (mainly earthquakes but also hurricanes and floods), as well as with Indian uprisings and attacks from corsairs and pirates. Not all were successful. Musset (2011) relates it to the fact that in the period studied, the recognition of risk by the Spanish authorities was largely absent. Denis Coeur (2000), in his research on river floods in France during the eighteenth and nineteenth centuries, claims that the idea of risk is not a timeless concept but rather the fruit of a specific story where Earth sciences have played a determinant role. Here, he tries to prove this by looking at the correlation between the development of hydrology and hydraulics with the socio-economic problems arising from water management and flood protection. What emerges is an analysis of the origins of risk in modern-day France (Coeur 2000). The institutional and governmental response has sometimes been considered outdated and inadequate, such as when Janku (Janku's chapter in Janku *et al*. 2012,

p. 254) shows and states what happened during the Chinese hunger crises of 1928–30 – that the 'drought-related famine [was] simply overlooked'.

However, it is important to explore such practices and strategies at the community level. It requires a much more thorough search of historical documents, as it provides very revealing results about local or regional memory, or rather, memories of human adaptation to the environment and changing climate. The most obvious examples of this type can be found in the variety of manifestations of native housing, which shows a huge variety and diversity of options that human beings, coping with climate adversities, have imagined and developed. In seismically active areas, evidence dates back to Ancient Greece and Rome, 'where the classical temple façade of columns constituted a segmental (multi-block) rocking system for re-centring the structure's axial load during violent ground movements', or to Inca buildings in South America with 'the carefully bonded corners and alternate rows of headers and stretchers' as well as the use of the 'hatil (reinforcing beam)' in Byzantine and Ottoman houses of Turkey (Bankoff 2012, p. 41).

Similar examples relating specifically to climate can be found throughout the world, showing that the process of adaptation to climate variations is not the result of a sudden decision, but the product of a long process of trial and error, that communities have produced their habitat from traditions, cultures and even myths, and that vernacular housing at the community level, performed without architects, has proven its worth starting from a 'labour-intensive that requires planning, organisation, systematic knowledge of climate and technology [a task that] involves invention, innovation and adaptation as well as the oral transmission of knowledge', as evidenced by countless Mexican cases of housing adaptation to rainfall, hurricanes, floods, dry steppe or desert climate (Audefroy's chapter in García-Acosta et al. 2012, pp. 95–106). Many of them persist even nowadays, realising that local traditional knowledge has been useful and functional for DRR and CCA since ancient times.

At the community level, linking the concept of social capital with the idea of coping with natural hazards, including climate change, has led to interesting experiences. Actions, practices or adaptive strategies are culturally constructed over time, mounted upon the organisational structures of a community, based on an evaluation of the events by the group affected that has been developed via a system of networks that have ultimately resulted in the transmission of knowledge between generations. From these ideas stems the hypothesis that social capital developed by a group may represent a key element in the generation of adaptive strategies to recurring natural hazards, as has clearly been shown in Bankoff's work in the Philippines (2003, 2012). The further development of a group's social capital is subject to the reoccurrence of hydro-meteorological hazards or climate changes, and the ways in which the adaptations are intra- and intergenerational in scope. The trust and solidarity of a particular group and the integration of networks within that group allow for and assess the involvement of all its members, therefore developing a means of collectively determining the best practices within the cyclic presence of certain hazards and the effects and impacts caused by them. However, in no way is exploring the concept of social capital in these issues helpful in understanding how to manage groups' access to successful adaptive strategies in greater depth. Schenk's conclusions after comparing the management of natural hazards in Tuscany and the Upper Rhine Valley during the Renaissance show that

> both 'cultural' and 'natural' factors played formative roles in Tuscany and the Alsace. [. . .] The examples show that learning, sharing knowledge, communication, and cooperation [social capital] were invaluable when dealing with everyday problems, as well as with more or less frequent disasters. Learning from (varying) local experience paid off and a comprehensive coordination of infrastructural works was necessary. In both cases, natural and social factors characteristically interacted.
>
> (Schenk's chapter in Janku et al. 2012, p. 44)

In this sense, the case of Spain during the *Little Ice Age* (ca. 1300–1850) is well documented with what Alberola (2014) calls a serious problem of water excess, particularly in the Southeast peninsula. When floods turned from 'threatening' to 'dangerous', 'neighbourhood solidarity and the strong reaction from local authorities' (p. 257) emerged, while 'peasants applied rather modest secular technical solutions not exceeding the local level like strengthening *quijeros*, defences and piers, or the lifting of protection with walls or *tablachos*, and dams' (p. 257). These actions and practices were frequently accompanied by accusations of failing to diligently follow the recommendations of cleaning irrigation ditches, removing dams or varying the position of the hydraulic mills to make them less vulnerable.

The ways of interacting with the environment have produced a considerable wealth in knowledge that is reflected in languages, patterns of thinking, different religions, cults, mythologies and rituals, and tangible expressions in material culture, particularly those stemming from the pursuit of a harmonious relationship with the environment when climatic 'excesses' appear (Mentz 2012).

Examples of wars of conquest and invasion produced violent superimpositions of different cultures, each with separate historical origins. In the case of what is today considered Hispanic America, the Spanish conquest generated a profound rupture, given that it introduced a knowledge of purely European origin and devastated the pre-Hispanic organisation, including the spaces where scientific knowledge was transmitted (Broda's chapter in Mentz 2012, pp. 102–135). This destroyed many everyday practices, including those that inextricably tied man to his environment and the management of it. One of the best examples, mentioned at the beginning of this chapter, were the floods that occurred from the fifteenth to the nineteenth centuries in the Mexican Basin, beginning with the one in 1555, three decades after the Spanish conquest. This was an event identified by specialists as the first great colonial flood: water reached a height of two metres, houses were demolished and the population was forced to emigrate, in some cases never to return. The reason was that the Spanish authorities had ceased to maintain the pre-Hispanic hydraulic control works system. Many other historical examples highlight what we call 'constructed risks', or rather, environmental degradation – the destruction of basins, changes in agricultural practices, pollution of aquifers, urban concentration – all of which have been, with the passage of time, increasingly associated with the disasters witnessed throughout present day Latin America (García-Acosta 2007).

Another similar example, relating to the eighteenth century hurricanes in New Orleans during the French (1718–1762) and Spanish (1762–1802) colonial periods, shows how 'environmental inexperienced newcomers' were 'blind to local environment realities' which, in many cases, led to the reconstruction of risks associated with the loss of lessons learnt by previous generations (Rohland 2015, p. 158).

Many of these diverse practices, like those related to the ethnoclimatology of the area, are still present, active and practised by meteorological specialists using knowledge from pre-Hispanic times. Such people are known as *profetas da chuva* (prophets of rain) in Northeast Brazil or *graniceros* and *tiemperos* in Central Mexico, as well as experts in manipulating rain, wind, hail and their effects among the Aymara people in Peru, Bolivia, Ecuador and Chile and the Mam people in Guatemala. It is also notable in the case of the Arhuaco and Kogui people in Colombia, whose ancestral environmental authorities nowadays claim respect and recognition of their environmental management strategies, which are derived from an inherited knowledge of their collective territories, and increasingly threatened by processes of globalisation and the transmission of 'modern' control over nature. These have caused, and continue to cause, serious environmental disasters and cultural transformations affecting entire systems of beliefs, knowledge and practices related to climate in the area (Ulloa 2013). One cannot forget that knowledge and perception of climate is one of the key areas that explain the success of traditional cultures around the world. As Philippe Descola (2011) has stated, natural phenomena are apprehended through their translation into a kaleidoscope of practices and representations that underline, isolate, or hide specific characteristics, actions, analogies or contrasting relationships. Ignoring or denying the differences and diversities of those practices and representations will prevent the achievement of useful results in defining and applying public policies within DRR.

Climate Change Adaptation and Disaster Risk Reduction: The Link

It is not easy to detach CCA from DRR because, as has been explored throughout the chapter, CCA can be understood as one of many processes linked to DRR (Kelman *et al.* 2015). The two fields overlap, and this overlapping is with regard to higher frequencies and severities of disasters resulting from meteorological and climatological hazards that are expected to become more frequent with climate change. In addition to this, the vast majority of historical cases that are registered and documented usually relate to the event itself as the disaster.

Keeping in mind that disaster events are detonators and amplifiers of pre-existing critical situations and a product of continuous and persistent construction of risks (García-Acosta 2014), it is important to avoid the single-event approach; we are interested in identifying CCA throughout history as an inherent process, similar to what must be done in DRR. As Braudel asserts, 'conjunctural analysis, even when it is pursued on several levels, cannot provide the total undisputed truth [. . .] A conjunctural scaffolding helps to construct a better house of history', but there are problems that 'take us well outside the narrow confines of the conjuncture [nevertheless] it is a useful path by which to approach them' (Braudel 1972 v2, p. 900). That is why it is important to document disasters as events located in specific times and spaces, as this helps to weave the different threads of stories that aid us in understanding and apprehending what has happened throughout time. In our case, this helps relate DRR and CCA practices with different hazardous events in a *longue durée* perspective.

Researching recurrent and the expected extreme manifestations of climate throughout time has been particularly revealing. Indeed, let us return to Braudel (1972). By describing the events relating to the lack or excess of water in different parts of the Mediterranean, he refers to measures derived, in the vast majority of cases, from the reoccurrence of events. For example, the floods that were regular in that region, from Portugal to Lebanon including Mecca, which:

> disappear under torrential rains in some winters [. . .]. In winter, which is the rainy season, plains tend to flood. To avert disaster their inhabitants must take precautions, build dams and dig channels [. . .]. In the Balkans, the Turks built high hog-backed bridges without piles, to give as free a passage as possible to sudden rises in the water level.
>
> *(Braudel 1972 v1, pp. 62–63)*

Additionally, he addresses some practices – perhaps what we can call 'institutional type' strategies referring to the end of sixteenth century Italy:

> in 1590 widespread floods submerged the Maremma in Tuscany devastating the sown fields. At that time the Maremma was, with the Arno valley, the chief grain supplier of Tuscany. The Grand Duke was obliged to go to Danzig (for the first time) in search of grain, without which it would not have been possible to bridge the gap.
>
> *(Braudel 1972 v1, p. 63)*

In the Andean region and throughout pre-Columbian times, the frequent incidence of climatic disturbances associated with the *El Niño* phenomenon brought demographic rearrangements, changes in settlement patterns, variations in the food practices, architectural reconstructions, control flooding and agricultural intensification and, indeed, ideological alterations (Manzanilla's chapter in García-Acosta 1997 v2, pp. 33–58). This is also the case with tropical cyclones (termed hurricanes, typhoons, or cyclones depending on their location). Their annual occurrence, the study of their manifestations, effects and impacts as events in specific moments and over centuries, has shown that their periodic presence has allowed certain human groups to achieve cultural changes concerning their material life and organisation that have led to the application of certain survival and adaptation possibilities. Pre-Hispanic records left by the Peninsular Maya provide clues about this, or what Konrad (2003, p. 99) calls 'Mayan strategies of coping with environmental restraints and opportunities' which allow the minimisation of not only the damage caused by recurrent tropical cyclones, but for communities to even profit from them, showing that 'the environmental phenomena itself can be a hazard or a resource, depending on one's perspective' (Kelman *et al.* 2015, p. 23). Among those Mayan 'strategies', Konrad (2003) mentions the following: intensive agriculture in high fields, swidden agriculture, cultivation in terraces, hydraulic systems, backyard vegetable growing and technical forestry adapted to the tropical forest. Using 'Pre-Hispanic, colonial and contemporary texts as well as climatic data from the Caribbean region' suggests that 'effective adaptation to the ecological effects of tropical storms helped determine the success of pre-Hispanic Maya subsistence strategies' (Konrad 2003, p. 99).

The anticipation of what can happen in the presence of a natural hazard and the possibility of disaster occurrence is based on the assumption of recurrence, and this

> plays an important role in forming the respective culture, not only today but evidently also in the past. Natural disasters could be events, which reinforced this kind of mutual influence or even caused a shift in the nature-culture relationship of a given society [. . .] Human action reacts to natural processes, uses them, interferes and interacts with them.
> *(Schenk's chapter in Janku et al. 2012, p. 31)*

The element of vulnerability therefore plays a distinctive role in this relation.

Oliver-Smith (2016 forthcoming) has insisted on this mutuality, emphasising that the concepts of vulnerability and resilience encompass the duality of environment and society, therefore forming a mutually constitutive relationship. In one of his latest works (Oliver-Smith 2016 forthcoming), he expands on this by insisting that this mutual constitution becomes more and more evident in the context of disasters and climate change. People are vulnerable to hazards, but hazards are increasingly the result of human activity, of the increasingly social construction of risks and the concomitant intensification, and the multiplication and expansion of vulnerability(ies).

Lavell (1998) expresses that the history of a large part of the last 50,000 years of human existence on the planet is one of adaptation and adjustment to the natural environment; looking for their needs through the use of the elements of nature means harnessing natural resources for development and minimising as far as possible the dangers that this same nature presents. The experience and knowledge accumulated over the years has helped to find this balance during long periods of history. Disasters are, then, the result of the breakdown of this balance (Lavell 1998). Historical disaster research has demonstrated that if disasters have become more frequent over time, it is not because there are more natural hazards, but rather that throughout time our communities and societies have become more vulnerable (García-Acosta 2002).

Last Reflections

Identifying mid- to long-term processes related to DRR, including CCA, requires the review of a series of selected examples coming from different periods and regions throughout the world, which can enrich proposed typologies of disaster management, such as those proposed for pre-industrial Europe (Mauelshagen 2013). The aim has been to offer evidence that explores the past with a *longue durée* perspective. Indeed, one can find 'a vast wealth of knowledge in relation to "dealing with disaster", including "adapting to change", not currently [known and even] utilised to maximum effect' (Mercer 2010, p. 248). History has an extraordinary role in understanding the build-up of contemporary vulnerability, as well as in identifying traditional and ancient response mechanisms and the long-standing experiences involved in designing contemporary DRR 'policies'.

There are many good practice examples of adapting, responding, and reacting to natural hazards, including climate change, throughout history, but they have not been explored systematically through historical documents or other so-called *proxy records* as well as through ethnographies around the world. Indeed, 'coping with' means that DRR, including CCA, has existed since time immemorial, regardless of what has been called and framed in recent years as purely CCA.

References

Alberola, A. (2003) 'Procesiones, rogativas, conjuros y exorcismos: el campo valenciano ante la plaga de langosta de 1756', *Revista de Historia Moderna. Anales de la Universidad de Alicante,* 21: 382–410.

Alberola, A. (2014) *Los cambios climáticos: La Pequeña Edad del Hielo en España,* Madrid: Ediciones Cátedra.

Altez, R. (2002) 'De la calamidad a la catástrofe: aproximación a una historia conceptual del desastre', *III Jornadas Venezolanas de Sismología Histórica,* Caracas: FUNVISIS, 1: 169-172.

Altez, R. (2011) 'Prólogo', in R. Altez and A. De Lisio (eds) *Perspectivas Venezolanas sobre Riesgos: Reflexiones y Experiencias,* Caracas: Universidad Central de Venezuela, 2: 1–31.

Bankoff, G. (2003) *Cultures of Disaster: Society and Natural Hazard in the Philippines,* New York/London: Routledge.

Bankoff, G. (2012) 'Historical concepts of disaster and risk', in B. Wisner, JC Gaillard and I. Kelman (eds) *The Routledge Handbook of Hazards and Disaster Risk Reduction,* New York/London: Routledge, pp. 37–47.

Braudel, F. (1958) 'La longue durée', *Annales Economies Sociétés Civilisations* 4: 725–753.

Braudel, F. (1972) *The Mediterranean and the Mediterranean World in the Age of Philip II (2 volumes),* New York: Harper and Row.

Coeur, D. (2000) 'Aux origines du concept moderne de risque naturel en France. Le cas des inondations fluviales (XVIIIe–XIXe s.)', in R. Favier and A.M. Granet-Abisset (eds) *Histoire et mémoire des risques naturels,* Grenoble: Maison de Sciences de l'Homme-Alpes, pp. 117–137.

Descola, P. (2011) *L'écologie des autres. L'anthropologie et la question de la nature,* Versailles: Éditions Quae.

Foster, W.C. (2012) *Climate and Culture Change in North America AD 900–1600,* Austin: University of Texas Press.

Gaillard, JC (2011) *People's Response to Disasters: Vulnerability, Capacities and Resilience in Philippine Context,* Pampanga, Philippines: Holy Angels University.

García-Acosta, V. (ed.) (1997) *Historia y desastres en América Latina,* v2, Bogota: LA RED/CIESAS.

García-Acosta, V. (2002) 'Historical disaster research', in S.M. Hoffman and A. Oliver-Smith (eds) *Catastrophe & Culture: The Anthropology of Disaster,* Santa Fe/Oxford: School of American Research Press/James Currey Ltd, pp. 49–66.

García-Acosta, V. (2007) 'Risks and disasters in the history of the Mexico Basin: Are they climatic or social?', *The Medieval History Journal* 10, 1&2: 127–142.

García-Acosta, V., Audefroy, J.F. and Briones, F. (eds) (2012) *Social Strategies for Prevention and Adaptation,* Mexico: CIESAS/FONCICYT.

García-Acosta, V. (2014) 'De la construction sociale du risque a la construction sociale de la prévention: les deux faces de Janus', in C. Bréda, M. Chaplier, J. Hermesse and E. Piccoli (eds) *Terres (dés) humanisées: ressources et climat,* Investigations d'Anthropologie Prospective 10, Laboratoire d'anthropologie prospective, Louvain: Academia-L'Harmattan Editions, pp. 297–318.

Janku, A., Schenk, G.J. and Mauelshagen, F. (eds) (2012) *Historical Disasters in Context. Science, Religion, and Politics,* New York/London: Routledge.

Kelman, I., Gaillard, JC, and Mercer, J. (2015) 'Climate change's role in disaster risk reduction's future: Beyond vulnerability and resilience', *International Journal of Disaster Risk Science* 6: 21–27.

Konrad, H. (2003) 'Caribbean tropical storms. Ecological implications for Pre-Hispanic and contemporary Maya subsistence on the Yucatan peninsula', *Revista de la Universidad Autónoma de Yucatán* 224: 99–126.

Lavell, A. (1998) 'Un encuentro con la verdad: los desastres en América Latina durante 1998', *Anuario Social y Político de América Latina y El Caribe* (FLACSO) 2: 164–172.

Le Roy Ladurie, E. (1967) *Histoire du climat depuis l'an mil*, Paris: Flammarion.

Le Roy Ladurie, E. (1972) *Times of Feast, Times of Famine: A History of Climate Since the Year 1000*, London: Allen and Unwin.

Mauelshagen, F. (2013) 'Natural disasters and legal solutions in the history of State power', *Solutions*, January–February: 65–68.

Mercer, J. (2010) 'Disaster risk reduction or climate change adaptation: Are we reinventing the wheel?', *Journal of International Development* 22, 2: 247–264.

Mentz, B. (ed.) (2012) *La relación hombre-naturaleza. Reflexiones des de distintas perspectivas disciplinarias*, Mexico: Siglo XXI Editores/CIESAS.

Musset, A. (2011) *Ciudades nómadas del nuevo mundo*, Mexico: Fondo de Cultura Económica.

Olcina-Cantos, J. (2009) 'Percepciones de los cambios del clima a lo largo de la historia', in A. Alberola and J. Olcina (eds) *Desastre natural, vida cotidiana y religiosidad popular en la España moderna y contemporánea*, Alicante: Universidad de Alicante, pp. 433–470.

Oliver-Smith, A. (2013) 'Disaster risk reduction and climate change adaptation: The view from applied anthropology', *Human Organization* 72, 4: 275–282.

Oliver-Smith, A. (2016 forthcoming) 'Adaptation, vulnerability, and resilience: Contested concepts in the anthropology of climate change', in H. Kopina and E. Shoreman-Ouimet (eds) *Routledge Handbook of Environmental Anthropology*, Abingdon: Routledge, pp. 206–218.

Rohland, E. (2015) 'Hurricanes in New Orleans: Disaster migration and adaptation, 1718–1794', in B. Sommer (ed.) *Climate Change in North America*, Leiden: Brill, pp. 137–158.

Ulloa, A. (2013) 'Controlando la naturaleza: ambientalismo transnacional y negociaciones locales en torno al cambio climático en territorios indígenas en Colombia', *Iberoamericana* XIII, 49: 117–133.

21

PERFORMING ARTS FOR DISASTER RISK REDUCTION INCLUDING CLIMATE CHANGE ADAPTATION

Ellie Cosgrave and Ilan Kelman

Introduction

Much disaster risk reduction (DRR), including climate change adaptation (CCA), policy and practice is developed and progressed through linear depictions involving words. Examples are scientific papers, legislation, policy briefs, green papers, and white papers. Yet we know that non-linearities exist in expressing these topics and that all forms of communication, including diverse oral and written forms, have advantages and limitations. Notwithstanding its own limitations, artistic expression can seek to arrive at deeper explanation where language alone falls short. Fine art, poetry, sculpture, and creative literature forms, for instance, all reach for more than descriptive logic.

Other modes of expression, communication, and suggestion involve the performing arts, namely music (Box 21.1), dance, literature, theatre, and their combinations. On top of this ability of the arts to communicate different types of information, many neurological and psychoanalytic studies have pointed to the therapeutic effects of working through the body as a means of recovering from trauma as well as building resilience to potential future challenges (Hanna 2006; Harris 2009; Pennebaker 1997; Van Der Kolk 2014).

Box 21.1   Music as Risk Communication Tool for Disaster Risk Reduction Including Climate Change Adaptation in Madagascar

Bob Alexander[1]

[1] Independent, World Citizen

Music is being used to help communicate about disaster risk reduction (DRR) including climate change adaptation (CCA) in many countries. Songs, music videos, action dances, rhythmic poetry, theatre, and other expressions of music are being used by governments, NGOs, other agencies, and communities themselves to communicate both about current disaster risks and how these risks are changing due to creeping environmental problems such as climate change. Examples include the

---

'Sing for Preparedness' song contest in Grenada; LIPI's (Indonesian Institute of Sciences) disaster preparedness compilation CD; various village-based risk reduction song groups in Tanzania; and the ICPM (Initiative Commune de Plaidoyer pour la Réduction des Risques des Catastrophes à Madagascar) preparedness song contests along the eastern coast of Madagascar.

The political instability that began in 2009 in Madagascar has exacerbated disaster vulnerability through increased unemployment, crime, and food insecurity. Over a quarter of the Madagascan population has been exposed to some of the thirty-three experienced cyclones, droughts, floods, and landslides in the past ten years. Three or four cyclones make landfall during the annual rainy season with much of the resulting loss, damage, and disruption occurring along the eastern coast.

In Madagascar, music and history are highly intertwined. Often, people in villages themselves or professional troupes are asked to write and perform vakodrazana songs with clear, easily conveyed messages for sensitisation about social issues. On the cyclone-prone east coast, songs are sung in community meetings when people meet to discuss how to address problems. In this context, ICPM conducted an annual song contest about disaster preparedness and CCA activities. The process of song writing helped to reinforce what was learnt in training, increasing the effectiveness of two-way development approaches for local DRR including CCA.

In order to explore the potential of performative-based approaches to DRR including CCA, this chapter focuses on dance and theatre to apply the two lenses of 'embodied knowledge' and 'connection' inherent in performance practice, applying these lenses to both the individual and communal level. Without professing to accept strict delineations, and while recognising overlaps and combinations, this chapter does not deal with art forms which are more static (accepting that art is rarely entirely static) such as painting, sculpture, poetry, and prose.

The chapter argues that by leveraging the power of performing arts for accessing embodied knowledge and connection, action for DRR including CCA can be deepened. Embodied knowledge refers to knowledge of oneself (internal) and community (external). Connection also refers to connecting with oneself and one's body and mind (internal) and connecting with one's community (external). Different modes and mechanisms, of which performing arts represents one, are needed to internalise and externalise the need for action on DRR including CCA, particularly given the failure of many traditional attempts to incite effective action. Dance and theatre, in particular, support such action in two key ways, which will be continually alluded to in this chapter:

1   The performing arts are a powerful communication tool that can foster deep connection, in effect making everyone an 'expert' for themselves (internal) and community (external). This process lays the foundation for needed action.
2   The performing arts hold an ability to support survivors in coping with trauma as well as building individual and community resilience to potential future challenges.

Dance

Dance, universally adopted and deeply embedded in many cultural contexts, serves a variety of personal (individual) and social (community) purposes. From social dance that strengthens connection and cultural identity, to performative dance that can portray stories and challenge political paradigms, to more somatic approaches to movement that focus on processing internal experience, the potential application of dance-based approaches to DRR including CCA is both rich and wide ranging.

Dance is about energy, flow, feeling, and movement. It is also about connection, expression, and shared experience. For the dancers, mutual movement can be a development and expression of unity and common understanding, while for the audience, a choreographed piece can convey ideas, information, and facts. This art form can powerfully render messages about social issues that might not be expressible through traditional language-based communication (Shome and Marx 2009). Following Polanyi's (1966: 4) premise that 'we know more than we can tell', dance, it would seem, is an innate and embodied form of knowledge as much as it is a tool for communicating that knowledge (Parviainen 2002).

Dance as embodied knowledge

Dance is a way of 'being' in the body as opposed to the mind, it is a way of knowing that is not communicable through language, and is a way of accessing and processing physical experience. The idea that we 'know more than we can tell' is something that many of us can intuitively understand. We know the feeling in our stomach as we drive over a bump in the car, the sweaty palms and shortness of breath before a job interview, the ache in our feet after a long time standing. We know these feelings, yet we couldn't accurately describe them to someone who had never felt them. To understand them they must be experienced and embodied. Similarly, words are often not a sufficient mechanism for knowing or understanding the effect of a flood or the trauma of war, but those who have experienced it know it, and hold that knowledge in their bodies (Van Der Kolk 2014).

Through performance, dancers are able to translate this knowledge to an audience and press for social transformation. As Boal (2002: 16) asserts 'Theatre is a form of knowledge; it should and can also be a means of transforming society'. Indeed, Hayes (2006: 83) claims that embodied knowledge can be transferred through movement to create a ripple of change in the community explaining, '. . . when dancers are able to access their deepest emotions and spiritual connection through the body, they reach out to the corresponding levels of experience in the audience [. . .] the witnessing process ripples outwards'.

Dance has a unique ability to access depth of feeling within a dancer as it is itself embodied. It can access physical response, reignite it, and allows both the performer and the observer to access and process embodied knowledge (Rothschild 2000). Movement and touch also create a conduit for communication and connection between individuals and communities. They can foster trust, understanding, and a sense of community through a direct and shared physical experience.

Like many art forms, dance is an intuitive process as well as containing a 'teachable' technique. There are structures and frameworks that can be implemented, whilst at the same time allowing space and fluidity for the expression to be owned by the individual or community; directed by their individual and collective experience. Dance is culturally and politically influenced as well as being emergent from the dancer's unique inner experience and as such is an effective mechanism to unite personal as well as collective knowledge and experience. As such it can help create a common discourse and reinforce a communal identity essential to both a personal sense of belonging and a communal resilience.

On top of this, performative dance styles offer an opportunity to tell stories, experiences, lessons, and convey political messages. There is a plethora of professional dance performances from Christopher Bruce's 'Ghost Dances' which portrays the human rights violations of the Pinochet regime, to Hofesh Schecter's 'Political Mother' which explores more general ideas around nationalism, fear, control, and community.

Dance has also been used to communicate climate risks where the language and logic of science has fallen short. One of the key challenges in communicating risk is to be able to link the global challenge with the personal lived experience of individuals (Joffe 2003). As such, many artists have chosen to communicate climate change knowledge through movement.

Motivated by a frustration that the science around climate change was not spurring sufficient action, Karole Armitage created a dance work entitled 'On the Nature of Things', in collaboration with Stanford University biologist Paul Ehrlich. The show visualises the challenges to environmental sustainability, bringing the ideas to an audience through art. They explain 'we need a new method of presenting climate change as an issue we can't ignore, they attest. And that method should include art' (Brooks 2015). Professor Ehrlich, a collaborator on the project, calls for the social sciences and humanities to be reorganised and refocused to provide a better understanding of how to change patterns of human behaviour.

Dance has also been used globally as protest to spur action around climate change and other political and social goals. For example, the 2009 climate demonstrations in London were kicked off by a 'flash mob' (https://www.youtube.com/watch?v=BdSiTLPpXPY). More recently, dance has been used as protest by environmentalists at the Paris climate change negotiations in 2015 (http://www.trust.org/item/20151129182605-qzy1t).

Dance as connection

Drawing on an understanding of dance as a personal and social experience, dance and theatre can foster connection at these two levels and consequentially build community cohesion, resilience, personal strength, and connection to the environment (Figure 21.1) can:

a) Rebuild connection to 'self' after trauma and support survivors preparing for future disaster response, in effect supporting the individual in reducing their own physical and psychological vulnerability to disasters.

b) Foster connection between individuals by creating a shared physical experience, building trust and embodied empathy, a touted element for community-based DRR (including CCA) so that people help each other through participatory processes.

c) Communicate a story or experience to an audience, supporting performers in being witnessed and allowing learning to be passed on – very much along the lines of participatory theatre and role playing, as explored in the next section.

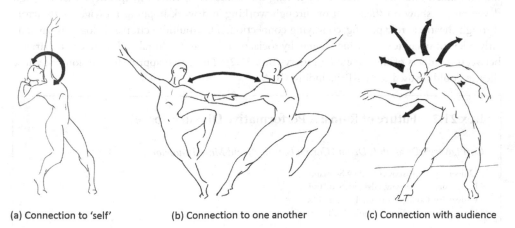

(a) Connection to 'self'   (b) Connection to one another   (c) Connection with audience

Figure 21.1   Three Levels of Connection Through Movement

(By Adrian Haak, Jr)

Dancing for Placing CCA into DRR

Dance and choreography can be used to integrate CCA into DRR by communicating the policies and actions needed at multiple governance scales. We can see examples of this type of integration and communication in projects across the world. 'Dance Your PhD', for example, communicates complex ideas and concepts through physical movement (Bohannon 2008). The 'Science Ceilidh' (http://www.scienceceilidh.com) uses Celtic dancing taught to the participants in order to explain basic science.

The experience of dancing together offers opportunities to foster communities of action and to build a collective purpose around DRR including CCA. Tikambilanie, an organisation that promotes communication methods as a key engagement and exploration tool in developing a global world from a local scale, uses dance practice to unite and engage communities. Their 'Let's Dance!' workshop series has supported:

- 'Earthdance – the global peace party' (http://www.earthdance.org);
- Dance for Freedom, an initiative that attempts to get the collective moving through working with people in small numbers; and
- Community participative drama for which local issues they feel strongly about are explored together and communicated to others.

Participatory dance emerges from each community's culture, ensuring that principles, knowledge, and wisdom about DRR are exchanged on the people's own dance terms and forges the connections within and outside of each individual and community which are needed for DRR including CCA.

Theatre

Theatre is another performing art that has long been used by humanity to convey social messages. Theatre can select from some or all performing arts, yielding different combinations of acting (role playing), dancing, music/sound, and painting and sculpture through set design, lighting, and props. Theatre has frequently been used in development settings in two ways.

First, communicating from actors to an audience by staging a show with messages, conveying embodied knowledge. Second, by having the 'audience' involved in theatre, either through developing a show on their own or through working and workshopping messages with actors through theatrical role playing, conveying connection. This multidirectional dialogue is particularly relevant in contexts in which hierarchy, social structure and cultural codes can create barriers between members of the same community (Box 21.2). These two approaches demonstrate how theatre could be used for DRR including CCA.

Box 21.2   Future of Ruins: A Performative Disaster Space

Carl Lavery[1], Deborah P. Dixon[2], Carina Fearnley[3], and Mark Pendleton[4]

[1] University of Glasgow, Glasgow, Scotland
[2] University of Glasgow, Glasgow, Scotland
[3] University College London, London, UK
[4] University of Sheffield, Sheffield, UK

The 'Future of Ruins: Reclaiming Abandonment and Toxicity on Hashima Island' pilot project (http://www.futureofruins.org.uk) explores how a multi-modal mapping of a specific disaster

---

site – Hashima island (端島), near Nagasaki, Japan – might offer alternative ecological futures. Mapping, thoughts, and representations are created through different performance forms, epitomising long-term disasters that could present analogies for climate change.

In the late nineteenth century, coal-mining started on Hashima, followed by the construction of a series of concrete dormitories and communal facilities for its workers, brought from all over Japan. Up to and during the Second World War, the island became a site of forced labour for Chinese and Korean prisoners. In the decades after, Hashima was increasingly domesticated as workers established their families on site and raised their children there.

With the shift in Japan from coal to petroleum, Hashima was abandoned in 1974, and its concrete tower blocks allowed to rot – eventually forming the backdrop for the hideout of the James Bond villain in *Skyfall* (2012). Current plans are to transform the island into a UNESCO World Heritage Site based on its industrial past and architectural remains.

The island compounds disasters: resource exploitation, pollution, an industrial cesspool, slavery, and torture. The island then represents a performative space in which to understand these disasters and moving beyond them. Using a variety of creative responses, from on-site performances to purgative postcards, the process asks: What happens when the transmogrification of matter becomes the condition for making a site hospitable?

This work culminated in the creation of a crossdisciplinary performance piece, a short film, topological maps and island transects, a website and blog, and academic publications. Through investigating how 'human time' and 'earth time' overlap and necessitate an expanded notion of history, the lessons can be applied for islands experiencing similar destruction, such as Nauru after phosphate mining, and those being affected by climate change – as well as moving beyond island locations.

Theatre as Embodied Knowledge

An example of theatre demonstrating embodied knowledge for development is from Fiji, through an annual school competition across the country called Tadra Kahani. Groups of school children are challenged to develop a music, theatre, and dance show conveying a development message. School groups have selected DRR messages, including one group communicating how a community should deal with a tsunami and another performing the consequences of sea-level rise due to climate change. By focusing on youth and children, the population learns the key material from an early age while educators themselves learn important messages. The children take the messages home reaching their family and neighbours, especially when they come to see the show. Tadra Kahani has also become a tourist attraction, reaching beyond Fiji, and appears on YouTube, reaching those who might not have school-age children in their family or social networks.

Involving youth and children in DRR including CCA demonstrates how much they can contribute (Cumiskey *et al.* 2015). Theatre gives freedom to express, freedom to experiment, and freedom to be. In Fiji and elsewhere, the process is as important as the product, with the children learning through doing, thereby realising their abilities to achieve and to fail in a make-believe world. With this experience, they can re-orientate themselves in the real world – with guidance offered by the theatre director, their teachers, their family, and their peers – and accept the importance of achieving rather than failing, thereby maturing in the real world.

Yet learning is lifelong. Adults also act and role-play, with theatre helping to embody and emote throughout one's life. Theatre for development has brought forward identity and has empowered audiences to recognise their own knowledge and interests, so that they can empower themselves for action (Banham *et al.* 1999). This ethos is applied to multiple theatre projects aimed at general audiences for DRR topics.

The Arctic Cycle is a sextet of plays by Chantal Bilodeau and produced by Clay Myers-Bowman which uses story telling to invoke passion and understanding on dealing with climate change in six Arctic countries: Canada, Greenland, Iceland, Norway, Russia, and the USA. The Cycle uses the power of storytelling to investigate and understand the many challenges posed by climate change. It melds science and society by combining the common themes of interconnectedness and urgency with the scientific jargon of adaptation and resilience. Feelings and knowledge of ice, identity, and 'The Right to Be Cold' (Watt-Cloutier 2015) need numerous media to be communicated, with theatre providing a combination of light, sound, roles, and settings that embraces and engages an audience without many other opportunities to be so involved and integrated into these topics.

Theatre as Connection

The initiatives in the previous section start the process from knowledge to connection. Theatre of the Oppressed (Boal 1979) is a suite of interactive techniques using theatrical forms and processes to empower people and communities. Examples are audience members leaping onto stage to take control of the performance and to change its outcome (Forum Theatre); acting out a show in a public setting in which those watching might not be aware that it is theatre, thinking instead that it is a real situation (Invisible Theatre); and participants forming their body along with the bodies of others to express an idea or emotion without words (Image Theatre). Image Theatre has clear parallels with some of the choreographing techniques explored in the previous section. Invisible Theatre has modern incarnations through flash mobs and pranks that have been used for pure entertainment as well as to raise awareness of social issues such as harassment and racism.

The overall direction is that those participating in Theatre of the Oppressed, as actors or as willing or inadvertent members of the public, learn how to take control of their own situations and change them through their own empowerment, with the stage (in a theatre or on the street) representing their communities where people would aim to enact change. These theatrical approaches parallel the dance approaches of taking control of one's own body so that emotions and interests flow out and are shaped by movement.

In the same way that dance is communal, not just individual, the connection made through theatre is individual as well as communal. Chagutah (2009) describes the theory and practice of using theatrical performance to portray role models for changing individuals' disaster-related behaviour, as applied to earthquakes in Armenia. In the Philippines, Tanner *et al.* (2009) detail how children become their communities' role models by leading their own theatre to indicate how to deal with floods and erosions as well as preventative actions including management of vegetation and waste.

Consequently, theatre plays a role in connecting within a community and connecting that community to disaster-related topics, including the potential hazard influencer of climate change. As with dance, theatre can build and rebuild connections to oneself and one's own community, including links severed or mangled due to a disaster. The shared physical experience and exchange of roles fosters trust and empathy, indicating how to assist each other with regard to disaster. Finally, the performance, the action of being an audience, and where the two intersect become spaces of communication and places of witness to actions and processes needed for DRR including CCA.

Theatre for Placing CCA into DRR

Embedding and embodying knowledge through theatre both internalises and externalises it for use while connecting the internal and external aspects. Internalising occurs by immersing oneself into the role being played. The role could range from being a righteous prophet of disaster and doom such as Stockmann in Ibsen's *An Enemy of the People* (1882) to a disaster survivor exemplified by the characters remaining at the end of Sondheim's *Into the Woods* (1986) banding together after their fairy tale collapsed in death and destruction. Externalising emerges through communicating how to deal with tragedy. Do audiences today respect or mock the upper class stiff upper lip in the face of adversity present in Noël Coward's *Cavalcade* (1931)? Is Rudetsky and Plotnick's *Disaster!* (2012) a night of comedy or a spur to DRR action – or both?

With the examples in the previous paragraph coming from only the Anglophone world, we also need to reflect on different languages and cultures portraying and applying theatrical endeavours to represent, communicate, and galvanise action on DRR including CCA. Combined with dance, theatre without words can bridge cultures and display emotions (Box 21.3). As a connector, theatre could potentially contribute to overcoming the separation between CCA and DRR, which exists in too many ways in too many venues. A performative space in which DRR-focused professionals playing roles of CCA officers are forced to work with DRR officers played by those trained in only climate change, might indicate the power of recognising CCA as a subset within DRR by applying the theory in practice.

Box 21.3 Dance as Communication of Political Messages: One Billion Rising

Ellie Cosgrave[1]

[1] University College London, London, UK

One Billion Rising is another example of dance as a conduit for communicating political messages. The campaign, launched on Valentine's Day 2012, calls communities across the world to come together to dance to raise awareness of violence against women and call for change. It is an international movement of solidarity for the 1/3 women (1 billion) globally who will be raped or beaten in her lifetime.

As a sign of strength, solidarity and unity in the fight to end violence against women, the movement calls people to dance together regardless of country, tribe, class or religion. A truly international movement, it unites cultures and communities around a single issue in a way that language never could. This international appeal and relevance has helped make One Billion Rising the biggest mass action to end violence against women in human history.

Since 2013 the movement has been focused on three themes: 'rising', 'justice' and 'revolution' – mirroring fundamental themes in disaster risk reduction including climate change adaptation. Further paralleling these topics, in 2016 it called participants to deepen the theme of revolution, arguing that revolution requires persistence, determination and courage, and encapsulates a spirit of positive action for lasting change. The campaign argues that most importantly the theme allows individuals to translate their personal anger, passion and desire for change into a collective one – to translate the personal to the political, the 'I' to the 'we' and call on people to rise for others, not just for themselves.

Dance is particularly powerful in building momentum around global social causes, including violence, vulnerability, resilience, and adaptation. The movement explains how dance connects people to the earth and allows women to re-enter their own bodies. One Billion Rising is a powerful example of how dance can be leveraged to raise awareness of a need for change; to foster empathy, solidarity and connection; and to demand political action. This is especially relevant to social causes that involve physical violation, but applies far more broadly.

Implications of the Performing Arts for DRR Including CCA

Both dance and theatre offer tools to access and communicate knowledge, as well as to foster connection. This has different manifestations and implications when taken at the individual and community level.

Personal Experience

Understanding Trauma – Somatic Experience

Many therapeutic processes and techniques used with trauma survivors involve accessing and processing emotion through working with the body. Formative psychotherapy, developed by Keleman (1981), is grounded on the premise that thoughts, actions, and emotions are held in the body. Therapists using this practice work with clients on an awareness of how they hold and use their body to alter hormone levels and influence thoughts, actions, and emotions in a helpful way. As part of this process, somatic exercises such as pushing, holding, and breathing, with awareness, are often adopted.

An example of this physical approach to working with psychological trauma can be seen in Gibney Dance's Community Action Programme, where dancers from the professional company run workshops with women living in domestic violence shelters. These workshops follow a four-part progression reflecting the Company's artistic process (Gibney Dance 2015): First, internal reflection; second, expressing through movement; third, collaborating through group dance; and fourth, self-care approaches. These four points reflect the wider ethos of performing arts suggested in the introduction with respect to DRR including CCA. Looking inside and reflecting is about reaching inwards, to accept that risk is something to be experienced. Expressing and speaking through movement accepts the importance of externalising risk as something to be dealt with. Collaboration entails connection while the fourth point covers caring.

Other non-performative dance practices are designed to support the dancer in connecting with their inner experience. For example 'contact improvisation', a technique developed by Paxton (1975), asks the dancers to pay attention to the feeling of their contact with another person and respond through movement. This type of movement can help the dancer to connect deeply with their bodies as well as to one another. By extension and in the context of CCA, this approach could also be applied to an individual's connection with their environment. Improvising movement with, for example, trees or a river could help to foster a better understanding of nature's forces. In particular, we often do not realise the power of moving water or the ground shaking nor do we necessarily realise the power of the slowly changing climate. Contact improvisation could potentially be used to introduce people physically and psychologically to these experiences, their environmental and social meanings, and ways of turning challenges into opportunities for DRR including CCA.

Connecting with the body is especially important for trauma survivors who have dissociated or disconnected from their body as a way of coping and staying safe. Bringing survivors back into their bodies can be an important part of recovery and also in preparing them to build psychological resilience. When applied to the trauma of disaster, reconnecting with embodied experiences can be a key to healing as well as an opportunity to prepare for potential upcoming disasters.

Preparing for and managing trauma

Drama has been used in post-disaster scenarios as a natural form of healing and in supporting survivors to develop personal and communal resilience, as well as to forestall post-traumatic stress. Landy (2010) describes how the drama therapy called 'Standing Tall' was applied to child survivors of the 11 September 2001 attacks on the World Trade Center in New York. The method, which broadly involves the exploration of different roles through imaginative storytelling, helped the children to make sense of the events they observed and to share their roles and stories with their community, leading to a mutual sense of support and hope.

Drama therapy also allows participants to model, experience and experiment with new ways of behaving and thinking. As Renee Emunah, Director, Drama Therapy Program, California Institute of Integral Studies, explains 'under the guise of play and pretend, we can – for once – act in new ways. The bit of distance from real life afforded by drama enables us to gain perspective on our real-life roles and patterns and actions, and to experiment actively with alternatives' (Emunah 2016). Here, we can see a rupture with pre-established roles, classes, social stratification, gender, age, or occupation conditioning. This process can support individuals preparing for disaster situations by pre-modelling ways in which they might choose to act in such scenarios. Role-play can also support participants in developing an understanding of other people's positions, responses and actions.

Community

DRR including CCA requires collective action based on shared experiences despite differences of opinion. Developing relationships alongside a sense of communal identity, purpose, and trust is an essential building block for such collective action to be fostered.

Connection and Social Dance

Dance and theatre offers a unique mechanism for community connection and cohesion. The physical connection with one another in space as well as a sense of communal achievement following a workshop or performance can engender a sense of communal identity, belonging and connection. This supports coming together to address the current disaster challenges and those projected under climate change, especially seeking ways to link them and to bring CCA within DRR without alienating groups.

Dancing outside of the therapeutic, workshop, or performative space can also be an effective way of building connection and resilient communities. Social dancing can help people to relax, unwind, and have a break from daily worries, as well as build trust, friendship, and communal identity. It can also bring in the 'fun factor' to DRR including CCA, rather than seeing it as a chore or added task, becoming an activity with which people naturally wish to engage. The Fun Theory has been used to promote environmentally friendly actions (de Valk *et al.* 2012) and could be applied to DRR including CCA; but see also the warning by Schaar (1970) – with shades of Huxley's *Brave New World* (1932) – about fun potentially being detrimental to happiness through being forced, superficial, and crowd-driven. By bringing it into their everyday lives, and indeed their bodies through dance, people can begin to feel connected to the reality of the risks they face and begin to take action.

Being Witnessed

A key part of drama and storytelling in therapy is to support the survivor in shared experience, connecting with others, feeling valued, expressing embodied knowledge, and validation. It can bring a level of acceptance, understanding, and self-confidence. This is particularly important in the development of a sense of worthiness. After a traumatic situation, previous 'belonging' groups (including family, friends, associations, and religious groups) might have disappeared. Creating or finding new groups with which to express experiences can be a useful mechanism by which to start the reconstruction work with oneself and others.

Conclusions

By combining the two lenses of 'embodied knowledge' and 'connection', dance and theatre may support communities recovering from trauma and particularly support wider processes of DRR including CCA.

When Joffe (2003) explored how individuals and communities perceive risk, which can be applied to DRR including CCA, she found they use three core mechanisms: distancing, stigmatisation, and blame. Risk is not necessarily something that all people wish to feel exposed to continually – notwithstanding extreme sports and the adrenaline junky culture, risk-taking behaviour from which positive environmental connections can be forged (Brymer *et al.* 2009) – so they tend to try to detach themselves from it and blame others rather than taking responsibility for action. Through this chapter we have explored how dance and theatre can help to reconnect participants with their inner experience, gain a physical understanding of scientific knowledge, develop empathy for others, and build a sense of community (Table 21.1). This is almost the directly opposite response to the constructed threat of disaster.

In fact, lack of control and choice regarding risk leaves us in a condition whereby we are made to be disconnected and lacking knowledge both as individuals and as communities. The lack of knowledge is not so much somatic obliviousness or systemic ignorance, as systematic ignore-ance (purposely choosing to ignore knowledge, e.g., Streets and Glantz 2000) in which societal constructs induce the preference to ignore what is known and should be acted upon. One consequence is the desire to distance oneself from these topics and to hide behind the

Table 21.1 The Power of Performing Arts

Feeling of Lack of Control and Choices Regarding Risk	Feeling of Dance/Theatre for Making Risk Choices
Distant	Embodied
Intangible	Physical
Fear	Safer
Isolating	Connected
Blame	Empathy
Hopeless	Hopeful

induced, ostensibly unchallengeable fear of risk, perhaps even fear of a dynamic environment. By bringing in the complementary practice of performing arts at individual and community levels, which brings with it connection and embodied knowledge, we lay the foundation for improved capability to act for DRR including CCA.

We can see that using the lens of dance and theatre in DRR including CCA is especially complementary. By morphing perceived risk into something that individuals and communities are able to connect with constructively, they are in a much better position to take positive action for positive change. And perhaps, in the midst of all the hard work needed to build DRR capacity, dance and theatre might help us to enjoy ourselves, de-mystify disaster (both hazard and vulnerability), and tackle the root causes that lead us to disaster-related problems, including climate change, in the first place.

References

Banham, M., Gibbs, J. and Osofisan, F. (1999) *African Theatre in Development*, Oxford: James Currey.

Boal, A. (1979) *Theatre of the Oppressed*, London: Pluto Press.

Boal, A. (2002) *Games for Actors and Non-actors (2nd ed.)*, New York: Routledge.

Bohannon, J. (2008) 'Can scientists dance?', *Science* 319, 5865: 905.

Brooks, K. (2015) *This Dance Project Is Out to Prove Climate Change Is an Issue We Can't Ignore*. Online http://www.huffingtonpost.com/2015/03/25/on-the-nature-of-things_n_6939826.html (accessed 25 January 2016).

Brymer, E., Downey, G. and Gray, T. (2009) 'Extreme sports as a precursor to environmental sustainability', *Journal of Sport and Tourism* 14, 2–3: 193–204.

Chagutah, T. (2009) 'Towards improved public awareness for climate related disaster risk reduction in South Africa: A participatory development communication perspective', *Jàmbá: Journal of Disaster Risk Studies* 2, 2: 113–126.

Cumiskey, L., Hoang, T., Suzuki, S., Pettigrew, C. and Herrgård, M.M. (2015) 'Youth participation at the Third UN World Conference on Disaster Risk Reduction', *International Journal of Disaster Risk Science* 6: 150–163.

De Valk, L., Rijnbout, P., Bekker, T., Eggen, B., de Graaf, M. and Schouten, B. (2012) 'Designing for playful experiences in open-ended intelligent play environments', in K. Blashki (ed.), *Proceedings of the IADIS International Conference Game and Entertainment Technologies 2012*, Lisbon, Portugal: International Association for Development of the Information Society 17–23 July 2012, pp. 3–10.

Emunah, R. (2016) *What is Drama Therapy?* New York: North American Drama Therapy Association. Online http://www.nadta.org/what-is-drama-therapy.html (accessed 25 January 2016).

Gibney Dance. (2015). *Training & Methodology*. New York: Gibney Dance. Online https://gibneydance.org/community-action/training-methodology/#methodology. (accessed 25 January 2016).

Hanna, J.L. (2006) *Dancing for Health: Conquering and Preventing Stress*, Lanham, Maryland: AltaMira Press.

Harris, D.A. (2009) 'The paradox of expressing speechless terror: Ritual liminality in the creative arts therapies' treatment of posttraumatic distress', *The Arts in Psychotherapy* 36, 2: 94–104.

Hayes, J. (2006) 'E-motion in motion', *Body, Movement and Dance in Psychotherapy* 1, 1: 81–84.

Joffe, H. (2003) 'Risk: From perception to social representation', *British Journal of Social Psychology* 42: 55–73.

Keleman, S. (1981) *Your Body Speaks its Mind*, Berkeley, California: Center Press.

Landy, R.J. (2010) 'Drama as a means of preventing post-traumatic stress following trauma within a community', *Journal of Applied Arts and Health* 1, 1: 7–18.

Parviainen, J. (2002) 'Bodily knowledge: Epistemological reflections on dance', *Dance Research Journal* 34, 1: 11–26.

Paxton, S. (1975) 'Contact improvisation', *The Drama Review* 19, 1: 40–42.

Pennebaker, J.W. (1997) *Opening Up: The Healing Power of Expressing Emotions*, New York: The Guilford Press, Rep Sub edition (8 August, 1997).

Polanyi, M. (1966) *The Tacit Dimension*, New York: Doubleday and Company.

Rothschild, B. (2000) *The Body Remembers: The Psychophysiology of Trauma and Trauma Treatment*, New York: W.W. Norton & Company.

Schaar, J.H. (1970) '. . . And the pursuit of happiness', *The Virginia Quarterly Review* 46, 1: 1–26.

Shome, D. and Marx, S. (2009) *The Psychology of Climate Change Communication*, New York: Center for Research on Environmental Decisions, Columbia University.

Streets, D.G. and Glantz, M.H. (2000) 'Exploring the concept of climate surprise', *Global Environmental Change* 10: 97–107.

Tanner, T., Garcia, M., Lazcano, J., Molina, F., Molina, G., Rodriguez, G., Tribunalo, B. and Seballos, F. (2009) 'Children's participation in community-based disaster risk reduction and adaptation to climate change', *Participatory Learning and Action Notes* 60: 54–64.

Van Der Kolk, B. (2014) *The Body Keeps the Score: Brain, Mind, and Body in the Healing of Trauma*, New York: Penguin Books.

Watt-Cloutier, S. (2015) *The Right to Be Cold*, Toronto: Penguin Random House.

22

LOCAL KNOWLEDGE FOR DISASTER RISK REDUCTION INCLUDING CLIMATE CHANGE ADAPTATION

Lisa Hiwasaki

Introduction

To date, much progress has been made in the field of disaster risk reduction (DRR) including climate change adaptation (CCA) in developing scientific knowledge and assessments, technical solutions and technological fixes as disaster response and preparedness mechanisms or adaptation measures. Efforts to mitigate impacts of natural hazards by national disaster management institutions usually focus on infrastructure development, such as constructing sea walls or breakwaters, or on high-tech solutions such as sophisticated early warning systems based on scientific data and modelling. The mainstream literature on hazards and disasters, and institutions responsible for DRR, had largely ignored local knowledge and practices until the mid-2000s (Dekens 2007a). In the aftermath of the 2004 Indian Ocean earthquake and tsunami, knowledge that helped indigenous communities survive the disaster was widely publicised. This played an important role in bolstering ongoing efforts that value the significant role local knowledge and practices can play in reducing natural hazard risk and climate change impacts. This is in contrast to the fields of development and natural resource management, where the important role local knowledge and practices can play for development and environmental conservation has received increased attention since the 1970s, with donors and development agencies promoting the protection, use, dissemination, and transmission of such knowledge since the early 1990s (Agrawal 1995).

Although social scientists working with local communities have been documenting local knowledge, observations, and practices related to DRR including CCA since the 1970s, such work 'remained as marginal as many of the peoples whose knowledge it was' (Dekens 2007a: 3). This was because local knowledge was often placed in contrast to Western scientific knowledge and regarded by scientists and development experts as 'non-knowledge', 'primitive', and maladaptive to the modernised world, and an obstacle to development (Agrawal 1995). Local knowledge related to disaster preparedness – most notably early warning signals based on observations of the environment – was often seen by scientists and DRR including CCA practitioners with scepticism, considered as mere 'mysticism' or 'superstitions'.

The growing evidence that local knowledge and practices can improve DRR including CCA has been recorded more in the developing world than in developed countries (Dekens 2007a).

Many examples of such local knowledge have been documented in Asia (e.g., Bangladesh, China, India, Indonesia, Japan, Malaysia, Nepal, Pakistan, the Philippines, Taiwan, Thailand, Timor-Leste) and the Pacific (e.g., Fiji, Papua New Guinea, Solomon Islands, Tonga, and Vanuatu), but also in other parts of the world (e.g., Burkina Faso, Mexico, and Zimbabwe). Similarly, in climate change research, social scientists have studied local knowledge and its relevance for our understanding of climate change and developing adaptation strategies since the 1970s, but recent years have witnessed an explosion of research on the topic, much of it in the Arctic and the Pacific.

Local Knowledge for DRR Including CCA: Definition and Context

Local knowledge refers to the understandings, skills and philosophies developed and maintained by peoples with long histories of interaction with their natural surroundings. Such knowledge, know-how, practices and representations provide the basis for decision-making about fundamental aspects of day-to-day life, such as agriculture, hunting, fishing, and gathering; coping with crises such as disease, injury, and disasters; and changing environments (Nakashima and Roué 2002). This broad definition encompasses the various terms that are frequently used in different contexts, such as: local and indigenous knowledge, indigenous knowledge, traditional ecological knowledge, traditional knowledge, indigenous science, and folk knowledge. Although these terms are often used interchangeably, 'indigenous knowledge' typically refers to knowledge held by indigenous people that is unique to their culture or society, while 'indigenous science' tends to be used in contexts to advocate the use of indigenous knowledge. 'Traditional ecological knowledge' is the term used more frequently in natural resource or environmental management, to signify indigenous knowledge handed down through generations by cultural transmission. 'Local knowledge' or 'local and indigenous knowledge' are terms used more broadly to refer to knowledge of local people regardless of whether they are officially recognised as indigenous people or not (see Raymond *et al.* 2010 for a discussion on definitions of different knowledges).

The whole array of local knowledge and practices related to DRR including CCA can be categorised as follows (Hiwasaki *et al.* 2015):

1 Observations of environmental and climatic changes, and other natural phenomena (e.g., prediction of storms or volcanic eruptions based on observations of changes in the sky, sea, wind, atmosphere and smoke, etc.), which help communities predict hazards and adapt to climate change;

2 Local food, materials and structures (e.g., construction of disaster-resilient houses using local materials and methods; food processing and preservation methods), which enable communities to adapt to, mitigate, and prepare for hazards;

3 Customary laws that govern behaviour (e.g., customary law that bans logging in certain areas and maintains vegetation on river banks; prohibition against building houses in certain areas), which engender and reinforce respect for the environment, strengthen social cohesion, and prevent and mitigate disasters; and

4 Folklore, rituals and ceremonies (e.g., ceremonies and festivals that commemorate divine beings, demonstrate appreciation for nature, and ask for protection from hazards), which increase people's awareness of hazards, reinforce respect for nature, and strengthen community resilience.

In a study conducted in coastal and small island communities in three countries in Southeast Asia, it was found that local knowledge belonging to the third and fourth categories above was particularly prevalent in all sites. For example in Timor-Leste, killing sacred snakes (*Rainain Samea*

Lulik), cutting down sacred trees (e.g., teak, bamboo and gmelina) or removing sacred stones (*fatuk lulik*) is forbidden, especially from river banks. If landslides occur, it is believed that such hazards were brought on by people breaking these rules, and rituals to 'apologise' to nature for the human behaviour that caused the hazard – such as *Monu ain ba lulik* ('apologise to nature') – are held, in addition to reforestation of the affected area. Local scientists explained that the ground becomes fragile and prone to landslides after heavy rain when there are many empty burrows after snakes are killed, absence of tree cover and roots due to deforestation, and lack of stones stabilising and covering the ground. Although people are in awe of natural hazards, they also see them as something that they can attempt to prevent or mitigate through customary laws (*Tara Bandu*) that make certain things 'sacred' and prohibit behaviour thought to cause hazards, thus ensuring respect for nature. The significance of such prohibitions based on folklore and rituals to the local communities that practise them was clearly demonstrated during discussions with communities. They considered such knowledge just as effective in mitigating disasters as other knowledge – such as observations of various changes in the environment (such as the sun, moon, stars, waves, clouds, and winds) and/or observations of animal, insect, and plant behaviour – is in predicting climate-related hazards (Hiwasaki *et al.* 2015). Box 22.1 below provides examples of other categories of local knowledge and practices.

Box 22.1 Local Knowledge that has Helped Communities Mitigate, Prevent, or Survive Disasters, and to Adapt to Climate Change

Lisa Hiwasaki[1]

[1] World Agroforestry Centre (ICRAF), Hanoi, Vietnam

Predicting storms: combination of observations of the environment – dark towering clouds at the horizon and their upward movement from winds, position of beehives in trees, calm sea-weather during transition period (according to the traditional calendar), and rancid smells from the sea – have helped communities predict and prepare for *Angeen Badee* (strong winds and high waves). From local knowledge recorded in Aceh, Indonesia (Hiwasaki *et al.* 2014).

Surviving earthquakes and tsunamis: construction of beachfront houses from local, light material such as bamboo and thatch, so that when they collapse after an earthquake, occupants will not be killed. Retreating water along shores after an earthquake is a sign of tsunami striking, so drop everything and run for higher ground. From oral history passed on by the Mokens in southern Thailand and Myanmar (Rungmanee and Cruz 2005).

Anticipating and surviving flash floods: observations of signals such as appearance or unusual behaviour of wildlife, colour of the clouds, heavy rainfall, unusual sounds, and changes in water flow, have helped communities predict flash floods. Building houses away from river beds and debris flows is another survival strategy observed in Chitral, Pakistan (Dekens 2007b).

Adapting to drought: communal digging and maintenance of canals, strict enforcement of water use, planting of crops in selected land based on its moisture-holding capacity, and intercropping edible ground cover crops are some of the local knowledge adapted by Somali farmers who moved to dry areas of Eastern Kenya (Wisner *et al.* 1977).

From the wide range of local knowledge described above and in Box 22.1, it is clear that at the local level, little distinction can be made between knowledge used to help communities cope with natural hazards and that used to deal with hazard drivers such as climate change. In recent years, the important role local knowledge can play in increasing community resilience to natural hazards and climate change has been increasingly acknowledged. Local knowledge can play a particularly important role for marginalised and vulnerable communities whose resources – natural, geographical, human, economic, political, cultural and social – are often compromised and thus for whom impacts of hazards can be accentuated (Wisner *et al.* 2012). Vulnerability is the consequence of a long-term process that results from the social, political, and geographical marginalisation of some communities (or in the case of internal marginalisation, members within communities) (Wisner 1993). Local knowledge, skills, technologies, and traditional systems of governance and social networks are key components that can strengthen the capacity of a community, along with other resources and assets that people use to mitigate, cope with, and recover from disasters (Dekens 2007a). It is easier to enhance the capacity of a community than to reduce their vulnerability, hence the use of local knowledge is especially important (Wisner *et al.* 2012).

For DRR including CCA efforts to be successful, all the inter-related factors affecting a community need to be included: not just its geophysical circumstances, meteorological conditions, and availability of hazard prevention and mitigation measures, but also a community's access to human, social, political, and economic resources, which are closely linked to the wider issue of sustainable development. Strengthening and valorising local knowledge and practices can empower the holders and users of such knowledge, and play a positive role in their sustainable development. There is now a general consensus that DRR including CCA should be integrated into wider development planning (Mercer 2010).

It is important not to over-romanticise local knowledge, and to recognise that local knowledge does not always reduce communities' vulnerability to natural hazards and climate change impacts. The Convention on Biological Diversity (CBD) cautions against romanticisation of local knowledge, especially by non-governmental organisations (NGOs) and indigenous rights organisations, since it can reduce its reliability (CBD 2003). First, local knowledge may not adapt fast enough to changing environments, climatic and other changes, globalisation, and population growth. An oft-cited case is that of the ill-fated villagers who did not evacuate at the time of the 2010 Mt Merapi eruption in Indonesia (Donovan 2010). Another example, also from Indonesia, is the traditional practice in Sayung in Central Java to use sand from the beach to elevate houses as a mitigation measure against the impacts of climate change such as sea-level rise and coastal abrasion, and other climate-related hazards. The recently introduced stilt houses are a more effective way to mitigate impacts of coastal flooding, especially in light of increasing population density (Hiwasaki *et al.* 2015). As explained by Wisner *et al.* (1977; see also Box 22.2), external, disruptive pressures placed on local knowledge systems are precisely what can lead to 'irrational' or 'non-adaptive' behaviour of local communities. In the climate change literature, it has been noted that traditional forecasting techniques are becoming less reliable (Weatherhead *et al.* 2010).

Box 22.2   Tapping into 'People's Science' for Disaster Risk Reduction

Lisa Hiwasaki[1]

[1] World Agroforestry Centre (ICRAF), Hanoi, Vietnam

Evidence shows that there has been a statistically significant increase in the number of large-scale disasters with devastating impacts, with the increase in death tolls taking place mostly in developing countries. Wisner *et al.* (1977) explain that this is due to the increasing vulnerability of poor people. Considering that the probability of the occurrence of hazards induced by physical events – such as droughts, floods, hurricanes, earthquakes, tsunamis, and volcanic eruptions – occurring has been stable for centuries, Wisner *et al.* (1977) argue for the need to stop referring to disasters as 'natural'; rather, it is necessary to focus on the increased vulnerability that the poor are experiencing. As the poor become trapped in a 'vicious circle', they become more and more vulnerable to hazards of lesser severity, are pushed aside to live in places that are more exposed to hazards, and become unable to resist diseases that often follow hazards. Furthermore, such conditions negatively impact their local knowledge systems, making it difficult for them to tap into their own knowledge and practices to help them survive disasters. This, in addition to the lack of respect shown by outsiders, is precisely what leads to 'irrational' or 'non-adaptive' behaviour – such as refusing to evacuate after a disaster warning – and thus, to their increased vulnerability. The only way to mobilise 'people's science' is through participatory research which can result in emergence of 'hybrid technology' which draws on both local and outside scientific knowledge, which broadens the choices that people have. It is only after we tap into 'people's science', and ensure that all information relevant to disaster preparedness is mobilised, that we can reduce local people's vulnerability.

Second, local knowledge is often unevenly distributed within a community, reflecting the existing power dynamics, based on factors such as gender, age, and clan affiliations. This results in unequal access to local knowledge within communities, which can increase the vulnerability of certain groups, most notably women and the poor, who are often the most marginalised within a community (Wisner 1993). Efforts need to be made to ensure that inclusion of local knowledge in DRR including CCA strategies does not further reinforce existing power dynamics that contribute to increased internal marginalisation of some members of a community.

Current Trends: Increased Recognition of the Importance of Local Knowledge for DRR Including CCA

In response to the shift in recent years that recognises the important role of local knowledge to deal with uncertainties and risk in DRR including CCA, we have witnessed some positive steps that have been taken to harness local knowledge by governments and educational institutions at different levels: from local through to national and international.

Internationally, in the field of climate change, the Intergovernmental Panel on Climate Change (IPCC) recognised, in its Fourth Assessment Report, indigenous knowledge as 'an invaluable basis for developing adaptation and natural resource management strategies in response to environmental and other forms of change' (Anisimov *et al.* 2007: 673–674). The Fifth Assessment Report went further by stating that there is 'robust evidence' that '... indigenous, local, and traditional knowledge systems and practices, including indigenous people's holistic view of community and environment, are a major resource for adapting to climate change' (IPCC 2014: 87). Despite such progress, local knowledge is still not sufficiently recognised in climate change policy and science, and scientific processes such as the IPCC reports have yet to include observations and assessments by local and indigenous people.

Furthermore, tremendous efforts have been made by indigenous peoples and others to promote collaboration between indigenous peoples, scientists and policy-makers. International Indigenous People's Forum on Climate Change organised prior to the United Nations Framework Convention on Climate Change (UNFCCC)'s 15th Conference of the Parties (COP) in 2009 was an important event that resulted in *The Anchorage Declaration*, which called for, among other things, 'Parties to the UNFCCC to recognise the importance of our Traditional Knowledge and practices shared by Indigenous Peoples in developing strategies to address climate change' (IPGSCC 2009: 2), and similar events have been organised since then, including most recently, an International Conference titled 'Resilience in a time of uncertainty: Indigenous peoples and climate change' organised in 2015 as a contribution to COP21 (http://indigenous2015.org/). Global initiatives that focus on or that have strong emphasis on local knowledge have emerged, such as Indigenous Peoples' Biocultural Climate Change Assessment Initiative (IPCCA: http://ipcca.info/), Climate Frontlines (http://www.climatefrontlines.org/), and the Many Strong Voices Programme (http://www.manystrongvoices.org/). At the national level, various indigenous groups have been involved with co-developing CCA plans or in other similar discussions with national governments. For example, Indian and Northern Affairs Canada (INAC) has worked with various First Nations to develop renewable energy plans and climate change risk assessment and adaptation plans (INAC 2010).

In the international DRR discussions, the Hyogo Framework for Action (HFA) (2005) highlighted the importance of local knowledge. This ten-year action plan acknowledged 'traditional and indigenous knowledge and cultural heritage' as one source of 'knowledge, innovation and education to build a culture of safety and resilience at all levels' (UNISDR 2007: 9). This culminated in the designation of the International Day for Disaster Reduction (IDDR) in 2015 as 'Knowledge for Life', whose focus was upon local knowledge and its contribution to modern science and resilience building. The successor to the HFA, the Sendai Framework for Disaster Risk Reduction (SFDRR) (2015–2030) (UNISDR 2015), contains many references that indicate the need for a strong focus on local knowledge, most notably:

> 24. (i) Ensure the use of traditional, indigenous and local knowledge and practices, as appropriate, to complement scientific knowledge in disaster risk assessment and the development and implementation of policies, strategies, plans and programmes of specific sectors, with a cross-sectoral approach, which should be tailored to localities and to the context;
> 36. (v) Indigenous peoples, through their experience and traditional knowledge, provide an important contribution to the development and implementation of plans and mechanisms, including for early warning;

At the national level, official recognition by governments of the importance of local knowledge in their national DRR plans is particularly significant, and is integral to support the numerous efforts being undertaken by NGOs and international development agencies. Among the numerous stories submitted on the occasion of the IDDR 2015, several highlighted the importance of local knowledge, for example, Rwanda's Minister of Disaster Management and Refugee Affairs called Rwandans to 'apply local knowledge in disaster prevention and preparedness' (MIDIMAR 2015), and the Vietnamese Vice Minister of Agriculture and Rural Development underlined 'the importance of correctly utilising traditional knowledge and experiences to complement scientific knowledge in assessing natural disaster risks and in developing and implementing policies and plans to help mitigate such risks' (UNDP 2015).

At the local level, communities have been using their own knowledge to deal with natural hazards, including climatic extremes, for centuries. Climate change is often perceived by local communities as one of many factors that contribute to their increased vulnerability to natural hazards. In this regard, it is important to embed CCA within DRR, and identify local knowledge that can be used to deal with all hazards, including climatic events. Local communities can then use their knowledge in DRR strategies which include CCA strategies to increase their resilience (Mercer 2010).

Agenda for the Future: Integration of Local Knowledge and Science for DRR Including CCA

It has been shown above that the importance of local knowledge, observations, and practices related to DRR including CCA has been officially recognised at different levels. Unfortunately, local knowledge has yet to be fully harnessed by scientists, practitioners, and policy-makers into mainstream scientific research, policy-making, and planning. Much more needs to be done, especially in policy and practice – most notably education and awareness-raising efforts – to actually use local knowledge and practices to increase communities' resilience against the impacts of natural hazards and hazard drivers such as climate change. In order to facilitate the use of local knowledge, it is first necessary for the whole array of local knowledge and practices related to DRR including CCA to be recognised. Depictions of local knowledge in the popular press, and examples of local knowledge mentioned in international and national discussions, tend to highlight local knowledge and responses to natural hazards and natural hazard influencers including climate change, which are embedded in traditional myths or oral traditions. These popular depictions are restrictive and do not do justice to the wide range of practices, skills, and knowledges related to DRR including CCA, as shown above in the four categorisations. Only after the concept of local knowledge is widely understood, can the next step to integrate local knowledge with science and technology take place.

This next step of knowledge integration is important because local knowledge, in combination with other knowledges, has been known to help communities manage various crises and conflicts, be it natural hazards, economic problems, or political conflicts (Ellen 2007). Moreover, it is now generally recognised in the DRR and CCA literature that integrating local knowledge and scientific knowledge can lead to successful DRR including CCA strategies (e.g., Cronin *et al.* 2007; Williams and Hardison 2013). In combination with the latest technology and scientific assessment, local knowledge can provide communities and decision-makers with a very good knowledge base to enable them to make decisions about the environmental issues they face, and develop strategies to cope with natural hazards and drivers including climate change, now and in the future.

The term 'hybrid knowledge' is often used to refer to knowledge co-produced by bringing together different knowledge systems – for example, local knowledge, knowledge from different scientific disciplines, and knowledge of policy-makers and practitioners. The use of this term acknowledges that such knowledges are complex and heterogeneous, and their production is a result of contested processes of social negotiation involving multiple actors. Hybrid knowledge is thus adaptive, enabling communities to deal with complex systems, diverse changes, and uncertainties (German 2010; Raymond *et al.* 2010).

The process of producing hybrid knowledge to help local communities overcome vulnerabilities to impacts of natural hazards and natural hazard influencers including climate change is not simply a matter of randomly putting local knowledge into preparedness plans that have been drafted by technical and scientific experts. The knowledge integration process must be fully mindful of the criticisms made by some scholars on such processes: the existing power relations between local knowledge holders and scientists, lack of clarity on who owns the local knowledge and the integrated knowledge, lack of understanding of the adaptability and dynamism of local knowledge, and the compartmentalisation and narrow interpretation of local knowledge (Nadasdy 1999; Williams and Hardison 2013).

The process of integrating local and scientific knowledge must engage multiple stakeholders – such as key community members, scientists from across disciplines, educators, and decision-makers – to draw on different knowledges, and to use them. A range of participatory planning methodologies has been used to integrate local and external knowledge for DRR including CCA. An example is the community organising processes adopted in Community-Based Disaster Risk Reduction (CBDRR), which includes mobilising community leaders, implementing awareness-raising activities, strengthening local organisations, and establishing linkages with local government officials. The process can be undertaken through participatory action research where local community researchers are trained and mentored to do the research and go through the process on their own.

Regardless of the method used, it should be acknowledged that integrating different knowledges is a long-term process that requires the building of trust and relationships, and the commitment of all stakeholders involved. It also requires community engagement and open communication, and close linkages between local knowledge-holders and external scientists and researchers, who agree to work together on an equal footing or on communities' terms (e.g., Gratani *et al.* 2011). Scientists need to recognise the significant contributions local knowledge can make in DRR including CCA, as well as the important role scientists can play in complementing local knowledge and science. Such a process thus makes it possible to address some of the criticisms mentioned above. Embedding CCA within DRR, which necessitates breaking down barriers between different scientific disciplines and other knowledge holders, can go further by enabling the design and development of context-specific and appropriate DRR efforts that incorporate climate risks.

Box 22.3 gives examples of some collaborative research projects that have successfully integrated local knowledge and science and technology to increase the resilience of local communities to natural hazards and natural hazard drivers including climate change. Knowledge integration cannot be realised if local and national government entities do not support these endeavours and enact policies to promote local knowledge, and research on such knowledge, as priorities in their DRR including CCA strategies (Hiwasaki *et al.* 2014). Thus, official recognition by national governments of the importance of local knowledge such as that mentioned above is a step in the right direction, but this needs to go further to put into practice policies that support the use of local knowledge and hybrid knowledge through appropriate processes. Successful integration of local and scientific knowledge enables practitioners and policy-makers to use it in policies, education, and actions related to DRR including CCA.

Box 22.3 Examples of Successful Integration of Local Knowledge and Science that have Helped Communities Mitigate, Prevent, or Survive Disasters, and to Adapt to Climate Change

Lisa Hiwasaki[1]

[1] World Agroforestry Centre (ICRAF), Hanoi, Vietnam

Blending volcanology with traditional knowledge for disaster preparedness planning: participatory workshops with communities in Ambae Island, Vanuatu, helped communities better understand scientific warnings and volcanic monitoring tools, and at the same time, preserve traditional knowledge related to volcanic eruption warning signs, and cultural practices related to volcanoes. Emergency response plans, including a hazard map, that were developed as a result of these workshops took into consideration both knowledges. This resulted in efficient, locally driven responses to a volcanic eruption on the island in November 2005 (Cronin *et al.* 2007).

Combining indigenous and scientific knowledge to address adaptation to climate change: a collaborative study involving scientists and indigenous communities' traditional knowledge filled knowledge gaps in historical weather monitoring data in Nova Scotia, Canada, and provided insights into potential climate change impacts and adaptive strategies, based on experiences of the Wagmatcook First Nations community. This resulted in incorporating CCA planning into its community development strategy (INAC 2010).

Implementing a flood warning system that combines indigenous and scientific knowledge: the use of a bamboo musical instrument that was traditionally used to call community members to assemble at the village hall for meetings, alert people to incidents, or call children home was revived to become part of a flood early warning system in Dagupan City, the Philippines. In combination with placement of flood markers in strategic locations and monitoring of flood levels, the instrument is used based on agreed warning signals, and relayed from point to point. The system is part of the City's Emergency Response and Disaster Risk Management Plan (Victoria 2008).

Teaching children local and scientific knowledge on traditional calendar and storms prediction: a comic book was developed by scientists in collaboration with local communities, which provides information on the traditional seasonal calendar used in Aceh and other knowledge that helps communities predict storms. Using an easy-to-understand story line, and images that are set in a culturally appropriate context, the book informs children and youth about local knowledge that helps predict storms, with the science behind local knowledge outlined to demonstrate the importance and legitimacy of local knowledge. The book thus demonstrates why local knowledge is a valid tool for DRR including CCA (Hiwasaki *et al.* 2014).

Conclusion

The recognition of the importance of local knowledge and practices to improve DRR including CCA represents a welcome shift away from the situation when such knowledge was largely ignored, or worse, discredited as superstition. In response to this shift, many efforts to research and document local knowledge have been undertaken worldwide. We have also witnessed some positive steps taken to officially acknowledge the importance of local knowledge by international bodies and governments at different levels. Unfortunately, however, local knowledge has yet to be fully harnessed by scientists, practitioners, and policy-makers into mainstream scientific research, policy-making, and planning. Much more needs to be done, especially in policy and practice, to use local knowledge and practices to increase communities' resilience against the impacts of natural hazards and natural hazard drivers such as climate change. Mobilising their own knowledge and practices is particularly important to enhance the capacity of marginalised and vulnerable communities to mitigate and recover from disasters.

The agenda for the future is to ensure integration of local knowledge and science through appropriate collaborative research processes involving multiple stakeholders, to enable local knowledge to play a key role in increasing the resilience of local communities. This is important for three reasons: first, local communities will be empowered to use their knowledge combined with outside scientific knowledge, to continue to make informed decisions about managing their DRR including CCA strategies. This in turn enables practitioners and scientists to further implement activities and research to support local knowledge for DRR including CCA. Second, DRR including CCA policies and educational programmes that value local knowledge will not only increase local communities' capacity and reduce their vulnerability to hazards, but also, are more readily accepted by them, due to their familiarity.

Third, integrating local and scientific knowledge can contribute to transmission of local knowledge to the younger generations. The important role of participatory knowledge integration in transmission of local knowledge has already been noted (e.g., Gratani *et al.* 2011), and is a key component that positively contributes to a community's capacity. Undertaking knowledge integration and ensuring that such knowledge continues to be handed down is crucial, especially considering the challenges that local knowledge faces in transmission to the younger generation. As was noted by local communities in Timor-Leste, it is up to 'the good will of new generations' to ensure that rituals and ceremonies will continue to be effective in helping them deal with hazards (Hiwasaki *et al.* 2015). It has been noted elsewhere that local knowledge is not static, nor embedded only in traditional rituals or customs. Local knowledge is dynamic and complex, produced and reproduced during day-to-day social encounters; thus, it should be considered as a process, rather than simply as content (Sillitoe 1998). Local knowledge and practices can only continue to play an important role in strengthening community resilience if such knowledge remains dynamic and relevant by being valued, maintained, and used for DRR including CCA.

References

Agrawal, A. (1995) 'Dismantling the divide between indigenous and scientific knowledge', *Development and Change* 26: 413–439.

Anisimov, O.A., Vaughan, D.G., Callaghan, T.V., Furgal, C., Marchant, H., Prowse, T.D., Vilhjalmsson, H. and Walsh, J. (2007) 'Polar regions (Arctic and Antarctic)', in M.L. Parry, O.F. Canziani, J.P. Palutikof, P.J. van der Linden and C.E. Hanson (eds), *Climate Change 2007: Impacts, Adaptation and Vulnerability*, Contribution of working group II to the fourth assessment report of the Intergovernmental Panel on Climate Change (IPCC). Cambridge: Cambridge University Press, pp. 653–685.

CBD (Convention on Biological Diversity). (2003) *Composite Report on the Status and Trends Regarding the Knowledge, Innovations and Practices of Indigenous and Local Communities Relevant to the Conservation and Sustainable Use of Biodiversity*, UNEP/CBD/WG8J/3/4, Montreal: United Nations Environment Programme (UNEP).

Cronin, S.J., Nemeth, K., Charley, D. and Thulstrup, H.D. (2007) 'The day Mount Manaro stirred', *A World of Science* 5, 4: 16–20.

Dekens, J. (2007a) *Local Knowledge for Disaster Preparedness: A Literature Review*, Kathmandu: International Centre for Integrated Mountain Development (ICIMOD).

Dekens, J. (2007b) *Herders of Chitral: The Lost Messengers? Local Knowledge on Disaster Preparedness in Chitral District, Pakistan*, Kathmandu: International Centre for Integrated Mountain Development (ICIMOD).

Donovan, K. (2010) 'Doing social volcanology: Exploring volcanic culture in Indonesia', *Area* 42: 117–126.

Ellen, R. (ed.) (2007) *Modern Crises and Traditional Strategies: Local Ecological Knowledge in Island Southeast Asia*, Oxford: Berghahn Books.

German, L.A. (2010) 'Local knowledge and scientific perceptions: Questions of validity in environmental knowledge', in L.A. German, J.J. Ramisch and R. Verma (eds), *Beyond the Biophysical: Knowledge, Culture, and Power in Agriculture and Natural Resource Management*, Dordrecht: Springer, pp. 99–125.

Gratani, M., Butler, J.R.A., Royee, F., Valentine, P., Burrows, D., Canendo, W.I. and Anderson, A.S. (2011) 'Is validation of indigenous ecological knowledge a disrespectful process? A case study of traditional fishing poisons and invasive fish management from the wet tropics, Australia', *Ecology and Society* 16: 25.

Hiwasaki, L., Luna, E., Syamsidik and Shaw, R. (2014) *Local & Indigenous Knowledge for Community Resilience: Hydrometeorological Disaster Risk Reduction and Climate Change Adaptation in Coastal and Small Island Communities*, Jakarta: United Nations Educational, Scientific and Cultural Organisation (UNESCO).

Hiwasaki, L., Luna, E., Syamsidik and Marçal, J.A. (2015) 'Local and indigenous knowledge on climate-related hazards of coastal and small island communities in Southeast Asia', *Climatic Change* 128: 35–56.

INAC (Indian and Northern Affairs Canada). (2010) *Sharing Knowledge for a Better Future: Adaptation and Clean Energy Experiences in a Changing Climate*, Ottawa, Canada: Minister of Public Works and Government Services.

IPCC (Intergovernmental Panel on Climate Change). (2014) *Climate Change 2014: Impacts, Adaptation, and Vulnerability. Part A: Global and Sectoral Aspects. Contribution of Working Group II to the Fifth Assessment Report of the Intergovernmental Panel on Climate Change*, Geneva, Switzerland: IPCC.

IPGSCC (Indigenous Peoples' Global Summit on Climate Change). (2009) *Anchorage Declaration: Declaration Agreed by Consensus of the Participants in the Indigenous Peoples' Global Summit on Climate Change*, Anchorage, Alaska: IPGSCC.

Mercer, J. (2010) 'Disaster risk reduction or climate change adaptation: Are we reinventing the wheel?', *Journal of International Development* 22, 2: 247–264.

MIDIMAR (Ministry of Disaster Management and Refugee Affairs Rwanda). (2015) *Rwandans Urged to Apply Local Knowledge in Disaster Prevention.* Online http://midimar.gov.rw/index.php?id=45&tx_ttnews[tt_news]=94&cHash=c429a91c7b4a2ed31278688c36deca93 (accessed 16 April 2016).

Nadasdy, P. (1999) 'The Politics of TEK: Power and the "integration" of knowledge', *Arctic Anthropology* 36, 1/2: 1–18.

Nakashima, D. and Roué, M. (2002) 'Indigenous knowledge, peoples and sustainable practice', in P. Timmerman (ed.), *Encyclopedia of Global Environmental Change: Volume 5: Social and Economic Dimensions of Global Environmental Change*, New Jersey, USA: Wiley, pp. 314–324.

Raymond, C.M., Fazey, I., Reed, M.S., Stringer, L.C., Robinson, G.M. and Evely, A.C. (2010) 'Integrating local and scientific knowledge for environmental management', *Journal of Environmental Management* 91: 1766–1777.

Rungmanee, S. and Cruz, I. (2005) 'The knowledge that saved the sea gypsies', *A World of Science* 3, 2: 20–23.

Sillitoe, P. (1998) 'The development of indigenous knowledge: A new applied anthropology', *Current Anthropology* 39, 2: 223–252.

UNDP (United Nations Development Programme). (2015) *Traditional Knowledge and Experience in Natural Disaster Prevention.* Online http://www.vn.undp.org/content/vietnam/en/home/presscenter/pressreleases/2015/10/13/traditional-knowledge-and-experience-in-natural-disaster-prevention.html (accessed 16 April 2016).

UNISDR (United Nations International Strategy for Disaster Reduction). (2007) *Hyogo Framework for Action 2005–2015: Building the Resilience of Nations and Communities to Disasters*, Geneva, Switzerland: UNISDR.

UNISDR (2015) *Sendai Framework for Disaster Risk Reduction 2015–2030*, Geneva: UNISDR (United Nations International Strategy for Disaster Reduction).

Victoria, L.P. (2008) 'Combining indigenous and scientific knowledge in the Dagupan city flood warning system', in R. Shaw, N. Uy and J. Baumwoll (eds), *Indigenous Knowledge for Disaster Risk Reduction: Good Practices and Lessons Learned from Experiences in the Asia-Pacific Region*, Bangkok: United Nations International Strategy for Disaster Reduction (UNISDR) Asia and Pacific, Bangkok, pp. 52–54.

Weatherhead, E., Gearheard, S. and Barry, R.G. (2010) 'Changes in weather persistence: Insight from Inuit knowledge', *Global Environmental Change* 20: 523–528.

Williams, T. and Hardison, P. (2013) 'Culture, law, risk and governance: Contexts of traditional knowledge in climate change adaptation', *Climatic Change* 120: 531–544.

Wisner, B. (1993) 'Disaster vulnerability: Scale, power and daily life', *GeoJournal* 30, 2: 127–140.

Wisner, B., Gaillard, JC and Kelman, I. (2012) 'Framing disaster: Theories and stories seeking to understand hazards, vulnerability and risk', in B. Wisner, JC Gaillard and I. Kelman (eds), *Handbook of Hazards and Disaster Risk Reduction*, Abingdon: Routledge, pp. 18–33.

Wisner, B., O'Keefe, P. and Westgate, K. (1977) 'Global systems and local disasters: The untapped power of people's science', *Disasters* 1, 1: 47–57.

23

EDUCATION AND TRAINING FOR DISASTER RISK REDUCTION INCLUDING CLIMATE CHANGE ADAPTATION

Emmanuel M. Luna

The Need for an Integrated Framework on Education for Disaster Risk Reduction and Climate Change

Building awareness and competence are fundamental tasks in disaster risk reduction (DRR) including climate change adaptation (CCA). While there are several education-related agendas for DRR, the education arena for climate change 'lacks a coherent framework that articulates how education can combat climate change' (Anderson 2010, p. 6). Anderson (2010) identified five education agenda and communities of practice (Table 23.1).

The imperative for education and training in DRR including CCA is recognised and moulded by key stakeholders from the local communities and schools, the academe, governments, NGOs, agencies and the international community (e.g., Action Aid 2011; Selby and Kagawa 2012; Wisner 2016). Three international agreements forged in 2015 affirm education's importance: The 'Sendai Framework for Disaster Risk Reduction 2015–2030' (UNISDR 2015), the Paris Agreement on climate change adopted by the United Nations Framework Convention on Climate Change (UNFCCC) in 2015, and the Sustainable Development Goals (SDG). Formulated during the Third World Conference on Disaster Risk Reduction in March 2015, the 'Sendai Framework for Disaster Risk Reduction 2015–2030' (UNISDR 2015) that followed the 'Hyogo Framework for Action 2005–2015' stated as its Priority 1:'Understanding disaster risk' (p. 14). Education and training are important for it (UNISDR 2015, p. 15):

(g) To build the knowledge of government officials at all levels, civil society, communities and volunteers, as well as the private sector, through experiences, lessons learned, good practices and training and education on disaster risk reduction, including the use of existing training and education mechanisms and peer learning . . .

(l) To promote the incorporation of disaster risk knowledge, including disaster prevention, mitigation, preparedness, response, recovery and rehabilitation, informal and non-formal education, as well as in civic education at all levels, as well as in professional education and training:

(m) To promote national strategies to strengthen public education and awareness in disaster risk reduction, including disaster risk information and knowledge, through campaigns, social media and community mobilization, taking into account specific audiences and their needs.

Table 23.1 Education Agenda and Communities of Practice (Anderson 2010, pp. 7–9)

Education Agenda and Communities of Practice	Proponents	Features	Remarks
Education for All (EFA)	United Nations Educational, Scientific and Cultural Organisation (UNESCO)	A global education movement that aims to meet the rights and learning needs of all children, youth, and adults by 2015 as part of the Millennium Development Goals (MDGs).	The EFA has been primarily concerned with access to organised, formal education.
Education for Sustainable Development (ESD)	UNESCO	Promotes relevant and interdisciplinary education that fosters critical thinking and problem posing; encompasses formal and non-formal approaches, from early childhood development to tertiary and adult education. Aims to develop individuals to make informed and responsible decisions and to take actions in the context of sustainable development.	Reaffirmed that climate change is a key action theme; constituents are educators who come from arenas of development education, environmental education, peace education, and global citizenship.
The quality of learning agenda	Network of actors from the United Nations (UN), Non-governmental organisations (NGOs), academia, and governments.	Focuses on acquiring knowledge and development skills through learners' engagement in critical analysis of problem solving. It promotes teaching and learning methodologies that are participatory, experiential, critical, and inclusive.	Its added value is that it complements ESD by focusing on essential life skills and learning, plus testing the knowledge and development skills acquired.
	The United Nations International Strategy for Disaster Reduction (UNISDR) Thematic Platform for Knowledge and Education (TPKE), as well as its members and many others, such as United Nations Children's Fund (UNICEF), UNESCO, Plan, Action Aid, the Inter-Agency Network for	Aims to plan for educational continuity in times of disasters and strengthen education systems and learning for disaster reduction and prevention. These are also CCA components.	By investing in DRR through education, human perceptions and patterns of behaviour can change, resulting in the reduction of disaster risks and costs. Having disaster-resilient school buildings and facilities and having evacuation plans can decrease disaster risk.

(*Continued*)

Table 23.1 Continued

Education Agenda and Communities of Practice	Proponents	Features	Remarks
DRR	Education in Emergencies (INEE), the Coalition for Global School Safety and Disaster Prevention Education, the InterAgency Standing Committee (IASC) Education Cluster, local NGOs, and regional networks.		Education in DRR can reduce vulnerability.
Environmental and climate change education	Education ministries, schools, non-formal education programmes, NGOs, and UN agencies such as United Nations Environment Programme (UNEP), UNESCO, and UNICEF.	Seeks to teach about the effective integration of environmental stewardship and climate change education into educational programmes and school curricula, including understanding the causes and consequences of climate change.	It promotes knowledge about the environment and climate change as well as attitude and motivation to make informed decisions and to take responsible action.

The Paris Agreement had lengthy provisions on action points and recommendations in most of its Articles. Article 12 stands distinctly regarding education:

> Parties shall cooperate in taking measures, as appropriate, to enhance climate change education, training, public awareness, public participation and public access to information, recognizing the importance of these steps with respect to enhancing actions under this Agreement.
>
> *(UNFCCC 2015, Article 12)*

Having this single statement underscores the significance of education and training in pursuing actions towards DRR including CCA (and climate change mitigation). Ensuring inclusive and equitable quality education opportunities for all is one of the seventeen Sustainable Development Goals (SDGs). SDG4 'Quality Education', SDG13 'Climate Action' and other goals relate to attaining safe environments, enacting vulnerability and risk reduction, and supporting capacity building. The challenge is how to link these goals with education for DRR including CCA.

Linking the SDG goals with DRR (including CCA) education and training starts with the review of current programme policies and practices. For example, Save the Children first identifies the SDG and the specific provisions related to DRR including CCA. Then, they assess which of their approaches and programmes are linked to the SDG. The Comprehensive School Safety (CSS) Framework provides an anchor for a strong national and sector-based policy that provides guidance on the institutionalisation of education for DRR including CCA. The CSS approach is organised along three overlapping lines, namely: 1. Safe learning Facilities; 2. School Disaster

Management; and 3. Risk Reduction and Resilience Education. The CSS goals are (Save the Children 2013):

- protect learners and education workers from death, injury, and harm in school;
- plan for educational continuity in the face of expected hazards;
- safeguard education sector investments; and
- strengthen climate-smart disaster resistance through education.

Challenges and Opportunities for Education for DRR Including CCA

Education for DRR including CCA is for all people: all ages, all races, all genders, all ethnicities, and all socio-economic, cultural, environmental, and political contexts in which people find themselves. Climate change is a global concern but it will have differential impacts on disaster risk due to varying levels of vulnerabilities and capacities. The approaches and strategies for DRR including CCA are vast and are at different levels – from the community, to local, national, regional, and international levels – with different functions. This complexity gives rise to a plurality of educational and training strategies for DRR including CCA. The innovations and creativity can be seen in the educational and related programs and services explored in this section, which can be adapted and modified to fit one's needs and resources.

Capacity Development for the Education Sector

Mainstreaming DRR including CCA education into the formal setting starts with capacity development of all key players in the formal education sector. This includes teachers, students, policy makers, school managers, other school staff, and other decision makers and programme implementers. One cannot share what one does not have. Policy makers must know what policies are appropriate and how to develop such policies. School managers need to know how to develop and manage DRR including CCA programmes in schools. Students and the community must be equipped to reduce their vulnerabilities and risks, and to enhance their capacities. There are formal and non-formal ways of moving forward through professional, academic, and community education and training.

In all cases, resource materials are needed. Several resource materials and tools have been developed. One of these is the resource manual produced by UNICEF (2012, p. 6) which is:

> primarily a capacity development tool to support governments and their development partners in guaranteeing the right to quality education for all children . . . Focusing on equity and rights, this resource manual aims to enhance the climate change adaptation, mitigation, resilience and risk reduction capacities of children and their communities in response to changing physical environments.

Disciplinary Integration for DRR Including CCA Education

One barrier to integrated DRR including CCA education is having separate frameworks, approaches, and programmes for DRR and CCA. Specialists in DRR and CCA can have the tendency to think separately from each other, if not competing as to which one is more relevant, more comprehensive, and more encompassing. An integrated education approach requires appreciation of both concerns, and necessitates a common framework in which CCA sits within DRR rather than creating separation. Social and natural scientists must cooperate in formulating integrated frameworks, modules, and educational methodologies. Rather than highlighting the differences, education would be more effective if there were enhanced communication and collaboration amongst different sectors and disciplines. This 'will allow for the identification and exploitation of the co-benefits between CCA, Disaster Risk Management (DRM), development and environmental protection and other areas' (UNISDR and UNDP 2012, p. 21).

The benefits of collaboration amongst experts from different fields are exemplified by the project Capacity Development for Hazard Risk Reduction and Adaptation (CATALYST), which brought together 130 experts who are practitioners, policy advisers, and academics from diverse sectors and countries around East and West Africa, Central America, the Caribbean, the European Mediterranean, and South and Southeast Asia. Through these experts, the CATALYST project was able to compile, synthesise, and disseminate existing knowledge on DRR including CCA. The professional experts provided their views on various DRR including CCA themes, identified practices and measures vital for adoption, and identified gaps in scientific knowledge and networks. They also gave recommendations for fostering capacity development in the various regions. Their output was a 'Best Practice Notebook for Disaster Risk Reduction and Climate Change Adaptation: Guidance and Insights for Policy and Practice' (Hare *et al.* 2013).

Schools and Universities

Integrating with Existing Subjects

DRR including CCA is a more recent formalised subject matter compared with conventional subjects in schools such as natural sciences, social sciences, technology, arts, and management. Most standard curricula are developed without much on CCA or DRR. Problems could emerge in creating a new subject of DRR including CCA, because the required number of units within a curriculum are already at their maximum and laws or regulations might need to be changed.

In such situations, a conventional yet effective strategy would be incorporating DRR including CCA into existing subject matters and activities. There are countless and creative ways of integrating DRR including CCA into these subjects (Sharpe and Kelman 2011) such as lectures on the specific subject matter, research work, school projects, or supplementary activities (Luna 2012). The relationship of climate change and other hazard drivers to food security, ecology, energy, water, solid waste management, infrastructure, and health can be extensively discussed without creating a separate topic of climate change. The science of climate change can enhance the discussion in chemistry, biology, physics, and meteorology. This approach seeks to integrate and connect DRR including CCA into existing topics and curricula, rather than creating something entirely new and separate.

An example of how to integrate DRR including CCA into a curriculum without adding new subjects is the experience of Save the Children in linking DRR including CCA into the school curriculum in the Province of Koh Kong in southwest Cambodia. Prior to the programme, the residents expressed concerns about situations in their community such as environmental degradation, climate change, and their lack of knowledge on environmental protection. Save the Children partnered with the Ministry of Education, Youth and Sports and developed a teaching manual which the teachers can use in teaching DRR including CCA concepts to the children through practical approaches such as mangrove conservation and gardening. The project reached 6,783 children and 250 adults, with 1,600 trees being planted around the school grounds. This experience allowed the children and the communities to better understand DRR including CCA concepts. They also became advocates by urging the local authorities to protect their environment and the natural resources (Save the Children 2013).

Instituting a New Subject

Instituting DRR including CCA as a new subject is another way of placing them into formal education. The review and change of school curricula provides opportunities for doing so. In the Philippines, the addition of two more years to secondary education starting in 2016 necessitated the development of new subjects to be taught during those two years. One of the subjects instituted was on disaster preparedness and risk reduction.

In Vanuatu, the development of the curriculum for kindergarten to grade 6 with climate change and DRR education was done through a workshop involving curriculum writers. The Curriculum Development Unit of the Department of Education in Vanuatu conducted a workshop where the curriculum writers defined the specific messages that should be taught from kindergarten to grade 6 (Curriculum Development Unit, Department of Education Vanuatu 2011). They agreed on integrating DRR including CCA into the curriculum and on the sequencing of the messages.

In universities and colleges, instituting subjects on DRR including CCA could be more viable, since curricula can often be revised in a more flexible way, depending on the autonomy the institution has in modifying its tertiary courses (Leal Filho 2010; Shaw *et al.* 2011). Holloway (2009) describes this process for the University of Cape Town, South Africa. The University of the Philippines has a Master of Community Development Programme in which a course on 'Community-Based Disaster Management' has been taught as a seminar course since the 1990s and was formalised as part of the curriculum in 2006. In 2016, it was revised to include climate change as a consideration in the teaching of community-based DRR and disaster management.

Professional Degrees

Another development in the academe is the establishment of professional courses in DRR and climate change related topics. Thus far typically offered at the graduate level (with scattered examples of exceptions of undergraduate programmes), these courses prepare adults for professional careers in this field, as practitioners, policy and programme officers, researchers, and educators. Universities in several countries have developed and offered DRR and/or climate change related education in formal settings, with different areas of emphasis.

Examples are the science aspects of climate change for the biological, meteorological, and agricultural sciences; technological and engineering aspects of DRR including CCA; and through different social science lenses such as the 'Master of Arts Program in Climate and Society' at Columbia University, New York, USA (http://climatesociety.ei.columbia.edu). Not all are focused on one dimension. University College London's Institute for Risk and Disaster Reduction runs three Masters of Science and three Postgraduate Diplomas which integrate natural and social sciences (https://www.ucl.ac.uk/rdr/teaching/graduate). Other examples are:

- Master of Disaster Management, University of Copenhagen, Denmark http://www.mdma.ku.dk/masters_programme
- Master of Climate Change, Australian National University, Australia, http://programsand-courses.anu.edu.au/program/MCLCH
- Master of Disaster Risk Management and Climate Change Adaptation, Lund University, Sweden, http://www.lunduniversity.lu.se/lubas/i-uoh-lu-TAKAK
- Masters in Disaster Risk Management and Development Studies, Centre for Disaster Risk Management and Development Studies, Federal University of Technology, Minna, Nigeria, http://www.preventionweb.net/academic/view/32580

- Master's Elective Course on Disasters, Environment and Risk Reduction by Partnership for Environment and Disaster Risk Reduction (PEDRR) in collaboration with the Cologne University of Applied Sciences' Center for Natural Resources and Development (CNRD), http://pedrr.org/activities/graduate-course
- Graduate Course on Disasters, Environment and Risk Reduction, United Nations Environment Programme (UNEP) and the Center for Natural Resources and Development (CNRD), http://www.unep.org/newscentre/Default.aspx?DocumentID=2718&ArticleID=9539
- United Nations University, Institute for Environment and Human Security, http://ehs.unu.edu/education/non-degree

As shown in the list, the degree titles vary, from a single focus on DRR or climate change through to an integrated degree such as the Master's Programme in Disaster Risk Management and Climate Change Adaptation in Lund University, Sweden. There are also degree programmes that link DRR to other fields interdisciplinarily such as the Master's degree in Disaster Risk Management and Development Studies at the Federal University of Technology in Minna, Nigeria. It is interesting to note that there are academic institutions which, instead of instituting a full degree programme in DRR and/or climate change, offer single courses that can serve as electives for professional degree courses. These elective courses range from 50 hours to ten days. The formal master's degree programmes have durations lasting from 12 to 36 months. Much, including the number of academic units for credit, depends on the academic requirements of the university and country.

Other Institutional Education and Training for DRR Including CCA

Stakeholders and key players in DRR including CCA have various needs and many are no longer in the position to do formal education. They might never have been in the position to do formal education, because the most important stakeholders and key players in DRR including CCA are people in communities, namely those most affected by vulnerability. Everyone can always learn more, so other options for education and training are useful for enhancing capacities to better integrate DRR including CCA into their policies, programmes, services, behaviours, and actions. Thus, beyond the formal academic setting of schools and universities, there are multiple and creative platforms and approaches by various other service providers. These include the offering of DRR including CCA training courses by international organisations, the UN, NGOs, and the private sector as well as conducting country- or community-specific education and training programmes and incorporating DRR including CCA material and training within development programmes and projects.

UN Agencies

The United Nations Institute for Training and Research (UNITAR) has a Climate Change Programme that offers a range of services, including executive training, capacity development for education and training institutions, support for national learning strategies, learning methodology development, and knowledge sharing. Activities are carried out through partnerships with other UN organisations, bilateral development partners, learning institutions, and think tanks. Box 23.1 shows the training resources of the United Nations Institute for Training and Research (UNITAR).

Box 23.1   United Nations Institute for Training and Research (UNITAR)

Emmanuel M. Luna[1]

[1] University of the Philippines, Diliman, the Philippines

UNITAR's portfolio of climate change related education and training includes:

- UN CC:Learn: Servicing a One UN approach to climate change learning and skills development including collaborative work in eight partner countries (http://www.uncclearn.org)
- Climate Change Capacity Development Network (C3D+): Applied research and knowledge sharing through a global network of Centers of Excellence (http://www.c3d-unitar.org)
- National Adaptation Planning for Least Developed Countries (LDCs): Capacity development support for LDCs for identifying, financing, and implementing adaptation actions (http://www.undp-alm.org/projects/naps-ldcs)
- National Budgeting for Climate Change: Targeted skills development to promote the effective allocation and use of public finance in support of national climate change objectives (http://www.unitar.org/ccp/portfolio-projects/national-budgeting-climate-change)
- National Engagement with the UNFCCC: Development and execution of tailor-made courses that address priority learning needs in partner countries (http://www.unitar.org/ccp/portfolio-projects/national-engagement-unfccc).

Source: http://www.unitar.org/ccp/what-we-do

International Training and Humanitarian Organisations

One approach in training for DRR including CCA is through a public offering of courses. One well-respected and long-running training programme is from the Asian Disaster Preparedness Center (ADPC) with headquarters in Bangkok, Thailand. Established in 1986, ADPC is an independent NGO. It has well-established networks with government ministries and agencies as well as strong partnerships with regional organisations and development agencies, providing a strong foundation for ADPC's work. The training programmes of ADPC are designed for specific target audiences and range from five days to three weeks. Several other NGOs have also developed education and training programmes on DRR including CCA, including many which are incorporated into their development projects (Table 23.2).

Integrating DRR and CCA Education and Training in Development Programmes and Projects

An approach in DRR including CCA education and training outside of schools and universities is by making DRR including CCA part of a joint agenda in development programmes and projects, including action research. This is done by ensuring that DRR including CCA education and training is incorporated as components for the staff planners, implementers, and beneficiaries on the ground. Graduates of various forms of education and training in DRR including CCA apply their learning by mainstreaming DRR including CCA into their day-to-day work and by putting educational and training services into their programmes and projects. Box 23.2 shows

Table 23.2 Samples of DRR Including CCA Education and Training Providers

Examples of Training Providers	Examples of Training Modules and Programmes with a DRR Including CCA Component
UN Country Specific Training Disaster relief and reconstruction. UNISDR Joint Training Programme – Mainstreaming Climate Change, South Korea http://www.koica.go.kr/english/board/whats_new/1321727_3545.html	UNISDR Joint Training Programme – Mainstreaming Climate Change Adaptation and Disaster Risk Reduction (21 days) Implemented by the National Civil Defence and Disaster Management Training Institute, South Korea Module 1: Disaster Management System Module 2: Natural and Social Disaster Management and System Module 3: Mainstreaming CCA and DRR into Development
International Training Provider Asian Disaster Preparedness Center http://www.adpc.net	Regional Training Course on Disaster Management (46th in 2016) This training course aims to provide necessary and useful fundamental knowledge and skills in disaster risk management in order to enhance the capabilities of disaster managers who wish to reduce the impact of disastrous events in communities (3 weeks). Regional Training Course on Community-Based Disaster Risk Reduction in a Changing Climate (fifth in 2016). Provides an opportunity for practitioners to learn, upgrade and share essential skills and knowledge to systematically address DRR challenges at the community level and to facilitate the processes to reduce disaster risk of vulnerable communities (ten days). International Training Course On Climate Risk Management In A Changing Environment (seventh in 2015). Offers a unique opportunity for these professionals to enhance their knowledge, expertise, and skills on the topic. The aim of the course is to harmonise climate risk management, DRR, and development planning into a holistic approach to sustainable development (ten days).
NGO: Plan Integrating DRR and CCA http://careclimatechange.org/tool-kits/drr_cc_emodules/	Five e-learning modules about integrating CCA into DRR programmes and plans. Developed by CARE's Strengthening Community-Based Disaster Risk Management in Asia (SCDRM+) Project. The modules form part of a comprehensive regional learning curriculum for key government, civil society, and community representatives from the SCDRM+ project countries. The five e-learning modules are self-paced, interactive PowerPoint presentations. The E-LEARNING DOWNLOAD zip file includes: Module 1: Concepts in DRR and CCA Module 2: Understanding vulnerability Module 3: Introduction to DRR Module 4: Introduction to CCA Module 5: Guiding principles for integrating adaptation into DRR

a case of how an agricultural development programme was enhanced with the inclusion of a training component to help the farmers reduce potential risks brought by climate-related hazards. Box 23.3 illustrates how an action research project on local and indigenous knowledge on DRR and CCA (keeping them separate) in coastal and small island communities pursued education- and awareness-related activities in communities for building resilience against meteorological and hydrological hazards. This work not only helped in educating community leaders, development workers, and policy makers, but also generated lessons on DRR including CCA that could be integrated into formal education at primary, secondary, and tertiary levels. Note that the examples in both boxes could have been even further improved by taking a vulnerability-orientated perspective which would explicitly place CCA as a subset of DRR to cover all hazards and all hazard drivers.

Box 23.2   DRR and CCA Training in Development Programme: The Climate Resiliency Field School (CrFS)

Emmanuel M. Luna[1]

[1] University of the Philippines, Diliman, the Philippines
Based on Rice Watch and Action Network (2016).

To illustrate a training programme for DRR including CCA that is built into a development project, the agriculture sector is taken as an example highlighting where the impact of climate change appears to have become more pronounced. In the Philippines, the increasing impacts of climate-related phenomena on agriculture have become more evident, some of which might be related to climate change.

The Rice Watch and Action Network (R1), a network of NGOs and individuals, has been doing research and advocacy for sustainable agriculture production and fair trade of rice. As the R1 people worked with rice farmers and their communities, they saw the rainfall patterns have become less predictable while tropical storms have increased in frequency and intensity. Droughts have become longer and more frequent. These changes have resulted in lower rice yields, crop losses, and pest infestations, threatening the farmers' livelihoods. This situation is aggravated by little awareness on climate change amongst the farmers and those in the local governments.

In searching for a practical response, R1 launched the project 'Integrating Climate Risk Management into Local Agriculture Planning' in partnership with the local government, the national climate agency, and other NGO partners. They initially implemented the project in two municipalities, forming a multi-disciplinary team comprising agriculturists and agricultural technicians, a climate scientist, and development workers from the partner NGOs and R1. The team drew up baseline information based on existing climate data, scenarios, and observations from the farmers and other community members generated during community visits. Workshops were organised, generating climate-sensitive agriculture plans. Each plan was discussed with the mayors and legislative councils of the municipalities for adoption. One component of the plan is the establishment of an early warning programme for agriculture, later named the Climate Resiliency Field School (CrFS). As of 2016, the programme has expanded and been implemented in more than thirty municipalities.

The programme aimed at enhancing the ability of farmers and farming communities to adapt to climate change, thus building their resilience to natural hazards and hazard influencers, including climate change. The programme aimed to:

- enhance farmers' sensitivity to and knowledge of weather and climate;
- improve farmers' ability to anticipate extreme climate events and assist them in interpreting climate and weather information for their decision-making regarding farm adjustments and crop contingency plans;
- promote a higher level of organisation for DRR; and
- promote sustainable agriculture as a CCA programme.

A key component of the programme is the development and institutionalisation of early warning systems for agriculture through the localisation of climate services by the Local Governments Units. It established the Municipal Climate Information and Monitoring Center that performs climate services, such as local weather data recording, climate risks and impact monitoring, localisation of farm-weather advisories, dissemination of these advisories, and regular liaising with the Philippine Atmospheric Geophysical and Astronomical Services Administration (PAGASA). The localised and precise climate information is also used in formulating a more climate-informed municipal CCA plan.

The central platform of the programme is the Climate Resiliency Field School (CrFS) that has been instrumental in the education and training of the farmers on climate change and DRR, although unfortunately separating the topics to some degree. The CrFS is a season-long field-based training which utilises a learner-centred, participatory, and experiential learning approach. The CrFS was modelled on the Farmer Field School which was designed for Integrated Pest Management.

Box 23.3 Incorporating DRR Including CCA into an Action Research Programme

Emmanuel M. Luna[1]

[1] University of the Philippines, Diliman, the Philippines
Based on Hiwasaki *et al.* (2014).

The convergence of educational and training activities for DRR including CCA is illustrated by a UNESCO action research on 'Local and Indigenous Knowledge (LINK) for community resilience: hydro-meteorological DRR and CCA in coastal and small island communities'. This was a three-year programme (2012–2014) implemented in Indonesia, Timor-Leste, and the Philippines. The first year was devoted to the inventory and assessment of local knowledge. The second year focused on the integration of science and technology with local and indigenous knowledge, and the production of information and educational materials such as booklets, audio-visual production, posters, teaching aids, maps, and calendars. The third year was action-orientated involving advocacy and information dissemination, education, and training activities on DRR including CCA amongst community leaders, teachers, local government officials, and staff from NGOs and government agencies. Policy papers for national and local government entities, academics, and practitioners were developed that can help promote the use of LINK in increasing coastal community resilience against disaster impacts, including where the hazard is influenced by climate change.

One of the policy briefs was on LINK and education. It presented the steps for developing information, education, and communication materials. The policy actions suggested include the mainstreaming of the use of LINK into DRR including CCA; dissemination of information, education, and communication materials; and the incorporation of these materials into formal, non-formal, and informal education. The school curriculum can be enhanced by integrating LINK for DRR including CCA into the subject matter.

The project developed action points for incorporating LINK into DRR including CCA education and training, such as:

- Developing policies that promote the integration of LINK related to DRR including CCA into primary and secondary education curricula.
- Supporting efforts by teachers to include LINK related to DRR including CCA in specific subjects.
- Encouraging extra-curricular activities in schools that promote the identification, documentation, and validation of LINK by communities.
- Promoting research by universities and research institutes that focuss on documenting, validating, and integrating LINK with science.

Current Concerns and Ways Forward

DRR including CCA is a comparatively new area of study and emerging discipline, but there seems be a wide range of scope of learning contents available. The multi-disciplinary nature of DRR including CCA and the upsurge of studies and programmes related to the topic have contributed towards the advancement of education and training activities. The strategy of integrating DRR including CCA into existing academic disciplines and courses has enhanced the flourishing of the contents as shown in the various degree programmes that focus on DRR, CCA, or both. There are policy briefs on education for DRR including CCA that feature relevant content knowledge, critical thinking skills, safe and adaptive schools, and green schools, thereby also demonstrating how to connect with climate change mitigation and wider environmental concerns (see also Petal 2015; Zint 2001). Such education and training is further enhanced with the development of teaching aids such as audio-visual productions, cases showing good practices, handbooks, and manuals for teaching (Sharpe and Kelman 2011; Wisner 2006).

The active separation of climate change from other topics remains of concern. Work on climate change education can nonetheless inform wider scopes. In particular, cross-cutting issues continue regarding the active participation of the community as agents of change and the enhancement of linkages amongst climate change researchers (Anderson 2010) which also apply to concerns beyond climate change. Recommendations to overcome these challenges from Anderson (2010, pp. 13–14) are:

- '1: Expand the climate change agenda to include education as a tool in adaptation and mitigation strategies'.
- '2: Promote an Education for Sustainable Development agenda that incorporates DRR, quality learning, and environmental and climate change education'.
- '3: Finance education to combat climate change'.
- '4: Strengthen the knowledge base, information sharing and networking'.

There is the need to review, develop, and strengthen education curricula and practice to integrate education for DRR including CCA and to build the capacities of teachers and education personnel in these topics. The enhancement of climate change communication and public awareness must be continually pursued (UNESCO 2015), as must be the case for wider topics, notably for integrating DRR including CCA into education.

Climate change can and should also learn from DRR. Analyses have been completed of DRR education material including evaluation approaches (Petal 2007, 2015; Wisner 2006) which provide a baseline for adding CCA-focused education material into the DRR compilations. Many examples exist of analysing education and training for DRR, CCA, and DRR including CCA (Leal Filho 2010; Shaw *et al.* 2011). Beyond formal means, Glantz (2007) proposed a 'spare-time university' which would use any means and mechanisms available to support people in learning information about DRR including CCA which has since been developed (Kelman *et al.* 2015).

As well, understandings of what happens after education and training are important. It has long been known that information, awareness, knowledge, and understanding do not necessarily lead to appropriate behaviour and action for DRR including CCA (e.g., Sims and Baumann 1983). Education and training for DRR including CCA is an essential component of any strategy, but it is not the final step in implementing successful action.

References

Action Aid. (2011) *Disaster Risk Reduction through Schools Final Report*. Online http://www.actionaid.org/sites/files/actionaid/drrs_final_report_to_dfid.pdf (accessed 1 October 2016).

Anderson, A. (2010) *Combatting Climate Change through Quality Education*, Washington DC: The Brookings Institution.

Curriculum Development Unit, Department of Education, Vanuatu (2011). *Workshop Results*, Port Vila, Vanuatu: Curriculum Development Unit, Department of Education Vanuatu.

Glantz, M.H. (2007) 'How about a spare-time university?', *WMO Bulletin* 56, 2: 1–6.

Hare, M., van Bers, C. and Janoslav, M. (eds) (2013) *A Best Practices Notebook for Disaster Risk Reduction and Climate Change Adaptation: Guidance and Insights for Policy and Practice From the CATALYST Project*. Online http://www.catalyst-project.eu/doc/CATALYST_D65_Best_Practices_Policy_Notebook.pdf (accessed 01 October 2016).

Hiwasaki, L., Luna, E., Syamsidik and Shaw, R. (2014) *Local and Indigenous Knowledge for Community Resilience: Hydro-Meteorological Disaster Risk Reduction and Climate Change Adaptation in Coastal and Small Island Communities*, Jakarta: United Nations Educational, Scientific and Cultural Organisation (UNESCO).

Holloway, A. (2009) 'Crafting disaster risk science: Environmental and geographical science sans frontiers', *Gateways: International Journal of Community Research and Engagement* 2: 98–118.

Kelman, I., Petal, M. and Glantz, M.H. (2015) 'Using a spare-time university for disaster risk reduction education', in H. Egner, M. Schorch and M. Voss (eds) *Learning and Calamities: Practices, Interpretations, Patterns*, London: Routledge, pp. 125–142.

Leal Filho, W. (ed.) (2010) *Universities and Climate Change: Introducing Climate Change to University Programmes*, Berlin and Heidelberg, Germany: Springer-Verlag.

Luna, E.M. (2012) 'Education and disasters', in B. Wisner, JC Gaillard, and I. Kelman (eds) *The Routledge Handbook of Hazards and Disaster Risk Reduction*, Abingdon, UK and New York, USA: Routledge, pp. 750–760.

Petal, M. (2007) 'Disaster risk reduction education material: Development, organization and evaluation', *Regional Development Dialogue* 28, 2: 1–20.

Petal, M. (2015) 'Critical reflection on disaster prevention education', in H. Egner, M. Schorch and M. Voss (eds) *Learning and Calamities: Practices, Interpretations, Patterns*, London: Routledge, pp. 159–179.

Rice Watch and Action Network. (2016) *The Climate Resiliency Field School and Localization of Climate Services Program: Program Brief*, Philippines: Rice Watch and Action Network.

Save the Children. (2013) *Practitioners' Guidelines on the Integration of Disaster Risk Reduction and Climate Change Adaptation into Sector-Based Programs, First Edition*, Melbourne, Australia: Save the Children. Online http://resourcecentre.savethechildren.se/sites/default/files/documents/sc_practitioner_guidelines_on_the_integration_of_drr_and_cca.pdf (accessed 1 October 2016).

Selby, D. and Kagawa, F. (2012) *Disaster Risk Reduction in School Curricula: Case Studies from Thirty Countries,* Paris: UNICEF and UNESCO. Online http://www.unicef.org/education/files/DRRinCurricula-Mapping30countriesFINAL.pdf (accessed 1 October 2016).

Sharpe, J. and Kelman, I. (2011) 'Improving the disaster-related component of secondary school geography education in England', *International Research in Geographical and Environmental Education* 20, 4: 327–343.

Shaw, R., Shiwaku, K. and Takeuchi, Y. (eds) (2011) *Disaster Education,* Bingley, UK: Emerald Group.

Sims, J.H. and Baumann, D.D. (1983) 'Educational programs and human response to natural hazards', *Environment and Behavior* 15, 2: 165–189.

UNESCO (United Nations Educational, Scientific and Cultural Organization). (2015) *Not Just Hot Air: Putting Climate Change Education into Practice,* Paris: UNESCO.

UNFCCC (United Nations Framework Convention on Climate Change). (2015) *Adoption of the Paris Agreement,* Bonn: UNFCCC. Online http://unfccc.int/meetings/paris_nov_15/meeting/8926.php (accessed 1 October 2016).

UNICEF (United Nations Children's Fund). (2012) *Climate Change Adaptation and Disaster Risk Reduction in the Education Sector Resource Manual,* New York: UNICEF. Online http://www.unicef.org/cfs/files/UNICEF-ClimateChange-ResourceManual-lores-c.pdf (accessed 1 October 2016).

UNISDR (United Nations International Strategy for Disaster Reduction). (2015) *Sendai Framework for Disaster Risk Reduction 2015–2030,* Geneva: UNISDR

UNISDR (United Nations International Strategy for Disaster Reduction) and UNDP (United Nations Development Programme). (2012) *Disaster Risk Reduction and Climate Change in the Pacific: An Institutional and Policy Analysis,* Suva, Fiji: UNDP. Online http://www.unisdr.org/files/26725_26725drrandccainthe-pacificaninstitu.pdf (accessed 1 October 2016).

Wisner, B. (2016) *Let Children Teach Us: A Review of the Role of Education and Knowledge in Disaster Risk Reduction,* Bangalore, India: Books for Change.

Zint, M.T. (2001) 'Advancing environmental risk education', *Risk Analysis* 12, 3: 417–426.

24

BUILDING INCLUSIVE DISASTER RISK REDUCTION INCLUDING CLIMATE CHANGE ADAPTATION

Virginie Le Masson and Emma Lovell

Introduction

Disasters impact countries and people in different ways, yet the diversity of affected populations is not systematically recognised in disaster risk reduction (DRR), including climate change adaptation (CCA). Many people are socially, culturally, economically and/or politically marginalised, often discriminated against on the basis of their gender, age, disability, ethnicity, culture, caste and other social identities. Those groups tend to be the most at risk from hazards, suffer disproportionately from disasters and are excluded from decision-making and planning that affects their lives. This recognition calls for DRR, including CCA, to be more inclusive.

The working definition provided by the Expert Group meeting of the United Nations Department of Economic and Social Affairs (UNDESA) describes social inclusion as:

> the process by which efforts are made to ensure equal opportunities – that everyone, regardless of their background, can achieve their full potential in life. Such efforts include policies and actions that promote equal access to (public) services as well as activities to enable citizens' participation in the decision-making processes that affect their lives.
>
> *(2007: 20)*

Social inclusion in DRR, including CCA, requires addressing the root causes of disasters, taking into account the different needs, vulnerabilities, strengths and capacities of those at risk, as well as the equal participation of all individuals and groups in DRR planning and activities. A recent report published by the Inclusive Community Resilience for Sustainable Disaster Risk Management (INCRISD) project in South Asia highlights that although communities at risk of hazards may have benefitted from existing DRR efforts, many people within those communities have been side-lined from opportunities to expand their knowledge and capacities to cope better with disaster risks (Ferretti and Khamis 2014); they may also have been excluded from DRR planning and management, or from accessing resources and aid equally.

The new international agreement for DRR: the Sendai Framework for Disaster Risk Reduction 2015–2030 (SFDRR) explicitly highlights the inclusion of different perspectives and the participation of at risk groups in DRR (UNISDR 2015: Section III, 19d). Although there have been great improvements from the Hyogo Framework for Action 2005–2015: Building the Resilience of Nations and Communities to Disasters (HFA) in this regard, the SFDRR does not specify how the inclusion and participation of vulnerable and at risk groups in DRR plans, policy and programming will be promoted and accounted for, and by whom. The final document still misses the requirements that would enforce stronger accountability for action on social inclusion and adequate attention to social vulnerability. This paper attempts to address some of these challenges, looking at existing best-practice approaches to create inclusive DRR practices and programming.

Several studies have analysed the inadequacies of 'top-down' approaches to DRR, including CCA, that ignore social dynamics and deep-seated power relations, while others have considered the illusion of terms such as 'inclusion', 'participation', 'stakeholder engagement', and 'bottom-up' processes (Cooke and Kothari 2001). Acknowledging the diversity of conceptualisations of inclusion, this chapter unpacks the meaning of inclusion in relation to DRR, including CCA, addressing questions such as the inclusion of whom and to what?

A useful starting point is therefore to look at the reasons why some groups, and some individuals within those groups, are excluded. The first section of this chapter will examine the opposite of inclusion – i.e., social exclusion – and how it relates to concepts of vulnerability and capacity when facing disaster risks and the adverse impacts of climate change. Addressing the question of why exclusion exacerbates disaster risks will help examine the purpose of making DRR inclusive. The section will answer questions such as exclusion from what, and the exclusion of whom? The second section will explore how inclusion and DRR are closely linked when fostering the goal of sustainable development. It will consider lessons from development studies pertaining to inclusion to inform DRR practices. This exercise has the advantage of focusing the analysis on the reasons behind people's exclusion, which often relates to wider development issues rather than solely studying the exposure of excluded people to hazards. Having assessed why inclusion matters for DRR, including CCA, the third section will focus on existing approaches to create inclusive DRR, including CCA, practices effectively and equitably.

Why does exclusion exacerbate disaster risks?

Exclusion from What?

Social exclusion describes a condition or a process whereby excluded individuals or groups are unable to actively participate in their society, on the basis of their social identity (Beall and Piron 2005). This exclusion might occur on the basis of people's gender, age, ethnicity, caste, religion or other manifestation of cultural identity. It may also be based on their location, for example people who live in remote areas, or regions that are stigmatised or suffering from war and/or conflict. Drawing on Beall and Piron's typology (2005: 10), social exclusion prevents individuals or groups from asserting their rights and their full participation in (i) the economy (for instance exclusion from labour markets, employment and enterprise opportunities, assets or livelihood strategies), (ii) social life (for instance access to infrastructure and basic services, social services, security and protection, public safety and social cohesion), and (iii) political affairs (for instance restricted access to organisations, exclusion from political representation, decision-making and the rights and responsibilities of citizenship).

Although the manifestation of exclusion varies across geographical, social, economic, cultural and political contexts, people who face a denial of their rights because of their social identities are marginalised in their daily lives. DRR literature examines the causes of disasters based on people's vulnerabilities and has documented why and how people who are the most marginalised are also the most vulnerable to disaster risks (Lewis 1999; Wisner *et al.* 2012). Whether they suffer from social, economic, geographical, political or cultural marginalisation, or a combination of these forms of exclusions, these people are less likely to access strong and secure livelihoods. For instance, discrimination on the basis of ethnicity means that many indigenous peoples and ethnic minorities are marginalised within the country in which they live, resulting in limited rights, social protection, support or political recognition (Cadag 2013).

The lack of access to basic services and employment may result in those who are excluded being trapped in areas prone to hazards without access to warning systems or other preventative measures. In addition, people who are marginalised may be forced to live in areas that are based in high-risk locations on cheaper land (Wisner *et al.* 2012), or may choose to live in areas that are at risk of hazards, due to the livelihood opportunities available in the area. For instance, in St Vincent and the Grenadines, many illegal settlers and farmers who are illegally cultivating marijuana have chosen to live on the flanks of La Soufrière volcano due to its fertile soil. As a result, they are not included in the census data collection, despite being the most at risk of volcanic activity (Lowe 2010).

Exclusionary processes create and/or sustain socio-economic inequalities which in turn exacerbate the vulnerabilities of marginalised people to hazards and hazard drivers such as climate change. In other words, exclusion restricts people's access to adequate living conditions and infrastructure, services (such as sanitation, education and healthcare), income diversification, organisations (including employment and welfare schemes) and decision-making (including political representation) that would otherwise support them to prepare for, cope with and recover from disasters.

Approaches differ in terms of components or indicators of social exclusion. Factors such as housing tenure, household composition, religious affiliation and critical life events (for example death in the family, divorce, separation or pregnancy) are sometimes identified as risks that may trigger exclusion. Exclusion also occurs at different scales: at the individual level: for instance, one family member may be excluded by other family members due to his/her sexual orientation; at the national level: for instance, Roma communities face discrimination in most countries where they migrate; and at the global level: for instance, women are still marginalised at the highest levels of decision-making processes. Lessons from the INCRISD project highlight that to foster inclusive disaster risk management (DRM), a better understanding of the root causes of exclusion in disaster contexts is needed, as well as the identification and meaningful involvement of excluded groups in reducing disaster risks.

Exclusion of Whom?

Identifying who is excluded, in terms of individuals or groups, is part of the process of assessing who is vulnerable to disaster risks and why. The literature emphasises that a large number of people considered excluded are invisible from those who hold power and who influence decision-making processes, meaning their needs and views are not recognised or taken into account. Hence, DRR scholars and practitioners concerned with documenting the perspectives and needs of those most marginalised, unseen or unheard, have produced reports that tend to categorise people into groups based on the social characteristics that contribute to their exclusion. The most 'disadvantaged', 'at risk' or 'vulnerable groups' are typically considered to be women, children, older people, people living with disabilities, ethnic minorities, indigenous people, Lesbian, Gay, Bisexual, Transgender (LGBT) and minority religious communities, although this list is not exhaustive. The common denominator is often discriminatory norms that prevent these groups from being included in decisions or from receiving assistance from those they rely on for help.

Children, particularly those who are separated from their parents after a disaster, are more prone to illness, malnutrition, illegal adoption, trafficking, labour and abuse, and may suffer life-threatening consequences as a result of deprivation (Peek and Stough 2010). Disasters can also prevent children from going to school. This may be as a result of lack of access to schools (for instance blocked roads) or the schools having been destroyed, damaged or used as shelters during a crisis. It may also be a consequence of children being expected to help their families post-disaster, in terms of relief and recovery. In Kenya for instance, droughts and famine have affected school attendance, with children dropping out of school to find work or to get married in the case of many girls (North 2010).

Such vulnerabilities are exacerbated by the lack of effective social pension and wider welfare systems in many low and middle income countries. These factors also undermine the capacities of people with disabilities who may be more dependent on other household or community members to fulfil their daily basic needs; thereby making them more at risk if the levels of supportive infrastructure and social relationships are limited. Priestley and Hemingway (2006) stress that people with disabilities are consistently amongst the poorest members of communities and face greater risk of death, injury, discriminatory attitudes and destitution, or loss of autonomy following a disaster. In the 2013 United Nations (UN) global survey of persons living with disabilities on how they cope with disasters (UNISDR 2013), only 20 per cent of respondents stated that they could evacuate immediately without difficulty in the event of a sudden disaster event, while 6 per cent said that they would not be able to evacuate at all.

On the one hand, the labelling of 'excluded' and 'vulnerable' serves the efforts of non-governmental organisations (NGOs) and civil society organisations (CSOs) who target categories of people they work with in order to make them more visible and advocate for their rights and needs. On the other hand, labelling someone as vulnerable can be disempowering, as the term does not recognise the different knowledge, skills and capacities that individuals or groups have. Those who are most at risk of disasters are vulnerable not because of their biological characteristics, but because they live in a society that does not promote social and cultural diversity, respect their rights, or value their participation, meaning that they are not supported to build their resilience to disasters.

In addition, these groups are not homogeneous entities, and gender, age or ethnic-based categorisations might not recognise the differences which exist within these groups (Box 24.1). For instance, the term 'children' could be used to describe new-born babies through to adolescents. Similarly, 'women' could be used to refer to girls and older women, women from low-income backgrounds or women within rich families; all of whom are likely to face different challenges and mobilise distinct capacities, according to the place they live and the social norms which exist. Moreover, people may experience multiple exclusions. For instance, older women living with disabilities are likely to be even more vulnerable to the effects of disasters because, in addition to facing the gender-based inequalities, responsibilities and social restrictions mentioned above, they are less likely to have engaged in paid work and will therefore be less likely to have their own assets or receive a pension, meaning that their self- and social-protection will be further limited.

Box 24.1 Gender in Disaster Risk Reduction beyond the Man–Woman Binary

JC Gaillard[1]

[1] The University of Auckland, Auckland, New Zealand

The study and integration of gender in disaster risk reduction (DRR), including climate change adaptation (CCA), has almost exclusively revolved around the vulnerabilities and capacities of women, and more recently but to a much lesser extent those of men and Lesbian, Gay, Bisexual, Transgender (LGBT) individuals. This approach that draws upon the Western man–woman binary and its extension to LGBTs hardly reflect the realities of a wide range of many gender minorities in the non-Western world.

Indeed, how gender minorities deal with disaster cannot be explained by their biologically determined sex (male or female) or their sexual orientation. Rather, gender minorities' response to disaster reflects their day-to-day position and role in their household and society. For example, the *bakla* of the Philippines and *fa'afafine* of Samoa are biologically male individuals who usually perform tasks, notably house chores, most often associated with women in their respective societies. Nonetheless, they are also able to carry out male-orientated tasks in a disaster, when they shift from one role to another, for example to fetch water and firewood or to harvest root crops. Furthermore, *bakla*'s and *fa'afafine*'s community leadership and ability to organise both relief operations and evacuation centres stem from their ability to organise community events in everyday life.

Despite such capacities, gender minorities often suffer from discrimination and disrespect as well as lacking access to resources available to men and women. In Indonesia, *waria* cannot get access to evacuation centres. In the Philippines, *bakla* teenagers are sometimes deprived of sufficient food, unlike their brothers and sisters, while in Samoa, *fa'afafine* feel uncomfortable using toilet facilities where they feel rejected by men and women. Ultimately, the underlying drivers of gender minorities' vulnerability mirror changing societies, especially under the increasing influence of global ideologies emphasising Western heteronormative values.

Gender minorities' vulnerability and capacities are most often overlooked in DRR, including CCA, policies and practices designed after homogenising Western standards and values based upon the man–woman binary. At the international level, the latest version (2011) of the Sphere's *Humanitarian Charter and Minimum Standards in Humanitarian Response* requires that resources (e.g., toilet facilities) be provided according to the number of affected men and women only. In Indonesia, the 2010–2012 National Action Plan for DRR claims to 'mainstream gender' whilst, in practice, *waria* are still ignored, if not discriminated against.

Recognising gender minorities is nonetheless critical and underlines the need for a culturally grounded approach to DRR, including CCA. Gender minorities indeed claim identities that reflect unique cultures that need to be recognised as much as Western values. Non-Western gender identities are, however, neither homogeneous nor static nor unchallenged, especially in a world that is increasingly globalised. They should thus not be romanticised.

Gender minorities' identities, including their vulnerability and capacities, rather mirror diverse and complex historical, cultural, social as well as political constructs, which are often very difficult to comprehend from an outsider's perspective. Gender-sensitive DRR, including CCA, should therefore be considered beyond men and women, and ultimately emerge from local perspectives that encourage the genuine participation of gender minorities in all their diversity.

Overall, excluded individuals and groups are not only more vulnerable to disaster risks, they are also less likely to participate in DRR interventions, thus making them even more vulnerable to the impacts of disasters.

Inclusive DRR, including CCA, through the lens of social development

The fact that many people are excluded from DRR is a reflection of their wider marginalisation from the economy, social life and institutional sites of power. Exclusionary processes not only limit DRR interventions to adequately strengthen people's capacities to face disaster risks, but they also impede those marginalised attaining human development and equal citizenship, thus undermining the full enjoyment of their human rights (Beall and Piron 2005). The concepts of rights and equality are therefore central to the idea of inclusion.

Social inclusion is concerned with the promotion of the full participation of individuals and groups that are currently, or at risk of being, disadvantaged, in all aspects of community life, including in DRR. Hence, while exclusion exacerbates disaster risks, the goal of inclusion to rectify imbalances and support people's access to equal services, resources and opportunities adheres to the purpose of DRR to reduce the vulnerabilities of people most at risk. The INCRISD report states that inclusion promotes equity and rights in DRM and is a condition for community resilience (Ferretti and Khamis 2014: 1).

Building equal opportunities as part of DRR cannot be separated from promoting the full participation of excluded people in social, economic and political life, if one views disaster risk with a sustainable development lens. For instance, DRR efforts might target women in order to support their access to income-generating opportunities, as a way to increase their financial resources and enhance their capacities in times of shocks and stresses. Nevertheless, in patriarchal societies where men primarily control the financial assets of the family and the community, such interventions might not lead to the desired outcomes if there is no longer-term engagement with community members, particularly men, to address the social norms that restrict women's control of the money they earn. Hence, it is useful for inclusive DRR to draw on lessons and experiences of development practices aimed to promote social inclusion. The next sections highlight development efforts that address (i) excluded people's lack of access to, and control of, resources and rights; (ii) the need for supportive governance and social services; (iii) excluded people's high vulnerability to human rights abuse at all levels; and (iv) cultural differences between places and contexts that influence how risks are perceived by different people.

First, the general lack of access to, and control of, resources and rights of disadvantaged groups creates or maintains conditions of poverty and exclusion. Roldán (2004) documents pervasive land-based exclusion in Latin American countries including Ecuador, Colombia and Brazil, and argues that not addressing the causes surrounding land-grabbing and protecting land rights can hinder people's integration or even fuel segregation and conflict in the area. Agarwal *et al.* (2012) suggest territorial recognition of land titles to excluded groups as a way of assuring inclusion by access to cultural rights and productive resources. This would ultimately help secure their autonomous subsistence of social and economic livelihoods and therefore increase their capacities to face shocks and stresses. Issues pertaining to land access affect DRR, including CCA, because many interventions target excluded groups who rely on farming and on ecosystem services to sustain their daily needs, which are sensitive to natural hazards and climate change.

Second, excluded people tend to suffer from limited resources, which affects their adequate access to basic services thereby exacerbating their vulnerabilities to disaster risks. Nevertheless, if supportive local and national governance are in place, they may support both service provision and social inclusion. Acosta and Clark (2010) present examples of operational improvements in public services as an approach to achieve inclusive development. Their research highlights examples in China and India, where public and private actors made efforts to empower migrants and include them in service delivery processes. The benefits of these measures led to excluded communities becoming entitled to services, as well as being actively integrated into their delivery.

Education is also seen as crucial for promoting social inclusion. In the case of Jordan, for instance, a large-scale government education programme managed to visibly increase girls' enrolment in educational institutions and improve female literacy rates in the area (World Bank 2013). Grimes and Bagree (2012) note, however, the longer-term effects of discriminatory social norms in perpetuating girls' limited access to key services, which in turn contributes to their longer-term disempowerment. In addition, the literature examining social inclusion and the links with gendered service access tends to simplify deprivations or barriers that maintain social exclusion to one dimension. Mohanty (2012), for instance, documents how women face multiple deprivations causing low maternal care outcomes in India and suggests that different experiences of social exclusion (as opposed to one form of exclusion) must inform integrated policy solutions. This example supports the idea that DRR, including CCA, must cross-cut different sectors (for instance access to water and sanitation, education, food, security, shelter, adequate governance and so on) in order to complement development practices that address the root causes of people's vulnerabilities to disasters.

Third, many excluded groups have a long track-record as victims of human rights abuses at all levels: from young women sentenced to rape as a punishment by unelected all-male village councils in rural India (Amnesty International 2015), to stateless Rohingya Muslims facing discriminatory religious laws passed by the Burmese parliament (Human Rights Watch 2015). Women and girls suffer disproportionately from gender-based violence (GBV): 30 per cent of women globally reported having experienced physical and/or sexual intimate partner violence (IPV) at some point in their lives, and 45 per cent of women in Africa reported that they had experienced physical and/or sexual violence (WHO 2013). This violation of women's rights undermines a myriad of capacities (such as physical and mental health, access to financial resources, trust in local institutions and so on) which must be recognised in DRR if interventions are to meaningfully reduce the vulnerabilities of GBV survivors.

Legal and social protection therefore becomes a necessity if excluded groups are to be protected from all sorts of physical violence and discriminatory practices. This includes both measures to grant social and cultural autonomy, whilst at the same time improving the legal protective status of the community in order to rebuild their confidence in state institutions and overcome distrust on both sides. The issue of justice and associated reform is particularly critical in post-conflict contexts. For instance, there have been visible improvements of women standing in the informal justice system in Bangladesh (World Bank 2013), after a number of NGOs collaborated to improve the representation of women in the shalishs – Bangladesh's informal court system. This aimed to improve jurisprudence for this disadvantaged group through their increased involvement in tribunals. The strategy was fruitful, as a significant change in jurisprudence was observed, especially in mediation cases dealing with dowries or domestic violence (World Bank 2013). Golub (2003) warns, however, that once NGOs' work ends due to budget constraints or other external developments, fragile improvements may not be sustainable and may even be reversed.

Fourth, cultural factors define what risk means for people (IFRC 2014). Communities and organisations that are involved in DRR may disagree about 'how they should act in relation to risk', and therefore what preventative or responsive measures should be taken (IFRC 2014: 12). For instance, community priorities are often given to day-to-day challenges including livelihoods rather than hazards (Lewis 1999) and a different logic is often followed in managing various levels of risk. Consequently, based on their culture, people's beliefs and values shape their perceptions and response to risk and often help them to 'live with risks and make sense of their lives in dangerous places' (IFRC 2014: 8).

Different members of communities have different perceptions of events, timings, priorities and of actions to take. For example, older people will have longer memories of past disasters and development trends than younger members of the community; women will often have different roles from men and therefore different perspectives on how to meet their needs and those of their family; and indigenous people might value different priorities from the rest of society including, for instance, their cultural heritage. This diversity must be acknowledged by DRR, including CCA, interventions, as it can help draw on development practices that aim to promote inter-cultural dialogue and social inclusion. For instance, to address situations of cultural isolation amongst disadvantaged communities, the official recognition of minority languages was highlighted as a best practice by Agarwal *et al.* (2012). Ecuador is used as an example within this study, where after decades of oppression, the recognition of indigenous languages led to the significant cultural empowerment of the community, as well as raising national awareness of their place within society (Agarwal *et al.* 2012). In addition, people who know how to speak, read and write in the language used by national institutions and outside organisations are more likely to engage in DRR, including CCA, practices and benefit from increased support to prepare for, cope with, manage and respond to disasters.

Implementing Inclusive DRR Including CCA in Practice

One key purpose of making DRR, including CCA, inclusive is for different individuals and members of communities to be involved in planning and decision-making processes, to feel valued and respected, and to have equal access to rights and opportunities (Sharma *et al.* 2014). Nevertheless, there does not seem to be one unique recipe to promote and achieve inclusive DRR, and guidance tools often focus on one category of disadvantaged people, such as disability inclusive DRR.

At the broad level, the INCRISD project, mentioned above, highlights that DRR must not only (i) identify the most excluded and vulnerable groups, but also (ii) recognise the causes of exclusion to promote inclusive allocation of resources and (iii) involve all stakeholders to ensure resilience through accountable risk governance (Sharma *et al.* 2014). These are all measures that have been described in previous sections, and are needed to help foster sustainable development.

Nevertheless, such objectives may become more challenging at the practitioner level. Promoting social inclusion often leads to activities that challenge the status quo of unequal power relations. Many NGOs are, therefore, often reluctant to implement activities that will challenge cultural norms and transform local practices even though they might target the most excluded groups to enhance their capacities. A typical consensual approach is to separate these groups from dominant ones in order to create a space for the former to express themselves more freely and engage in development activities. Yet, if the process of separating different groups is a useful practice to make excluded groups more visible, it does not necessarily lead to increased awareness and acceptance amongst dominant groups of the reallocation of resources and power.

Community-based DRR and Participatory Approaches

Recognising that the most marginalised groups are often the most vulnerable to disaster risks and the least able to influence community decision-making, requires practitioners to reach and engage with these groups. Communities comprise individuals who have different needs, interests, views, experiences and capacities that need to be acknowledged and utilised in order to help reduce vulnerability to disasters. There is also growing awareness of the value of integrating indigenous and traditional skills, knowledge and practices within DRR, alongside scientific and technological contributions.

Community-based DRR (CBDRR) and the use of participatory approaches is a way to involve communities in the assessment and the remedies of the problems they face (Delica-Willison and Gaillard 2012) and a means to tailor interventions to address the priorities of specific groups. The integration of local knowledge in DRR can be achieved through 'participatory capacity assessment and horizontal planning', which encourages people to use their knowledge, skills and practices to inform and develop DRR strategies that are likely to be more sustainable in the long term (Shaw *et al.* 2008).

Participatory approaches rely on specific tools such as participatory rural appraisal (PRA) methods which aim to support practitioners in conducting risk assessments and vulnerability/ capacity analyses through collecting information from a multitude of perspectives and different community members (Chambers 2007). Such approaches also promote the involvement of different groups in decision-making related to all aspects of planning, action and monitoring of DRM interventions, which therefore support the goal of inclusivity.

However, CBDRR requires social mobilisation and community organisation in order to ensure sustainability is achieved (Delica-Willison and Gaillard 2012). This becomes challenging when exclusionary processes occurring at the household and community level restrict certain people from engaging in decision-making processes in the immediate and long-term, including after NGOs have left. It thus becomes crucial for development and DRR, including CCA, practitioners to understand the political context of areas where they operate, as well as the local social norms and the strategic interests of dominant groups. In practice, however, this often clashes with NGOs financial resources and time frame. Such challenges echo critics stating that participatory approaches do not necessarily serve the interests of local communities, and of excluded people within these communities (for instance, see Cooke and Kothari 2001).

Addressing and transforming power relations and societal inequalities requires time and the commitment of various stakeholders to engage with both excluded and dominant groups. For instance, NGO workers might have to arrange separate meetings at different times in order to involve various groups. They will also need to engage local people to work as facilitators and interpreters, and may need to recruit staff with diverse gender and ethnic backgrounds in order to gain more trust and encourage more honest and accurate responses from different groups. Moreover, practitioners might have to request permission from local leaders to interact with certain community members.

Practitioners are required to encourage listening, dialogue, debate and consultation with all groups, and may report back results from assessments to build momentum and ownership. Providing an enabling environment for this dialogue is important to help negotiate different people's perspectives and cultural values effectively, and to enable the most excluded groups to articulate their concerns, and build their capacity to deal with these. Nevertheless, such activities necessitate long-term engagement with communities, CSOs and with authorities. This engagement is not yet widespread in DRR, particularly in emergency contexts, where many practitioners stress that there are more urgent priorities than dealing with the inclusion of marginalised groups in assessing risks and conducting responses (Le Masson *et al.* 2016).

Practical strategies of inclusive DRR, including CCA, can be highlighted in a three-year programme funded by the Department for International Development (DFID), aimed at Building Resilience and Adaptation to Climate Extremes and Disasters (BRACED). The BRACED programme supports fifteen NGO consortia to build the resilience of people to climate and weather extremes in thirteen countries in the Sahel, East Africa, South and South-east Asia. Box 24.2 summarises the guidelines from the consortium operating in Myanmar to their field staff on conducting inclusive resilience assessments.

Box 24.2   Ensuring Inclusion in Resilience Assessment

From BRACED Myanmar Alliance (2015)

- Create spaces for plurality of voices and narratives of community including physical access
  - Where are meetings/activities being held?
  - Can all members of the community physically access the venue?
  - When (day/time) are meetings held and are all members available during the time?
  - Can all members afford to travel to the venue?
  - Are facilities provided for elderly people/pregnant women, e.g. toilets, seating?
- Divide the community into groups to better manage input and discussions
- Do not predetermine groups; however, judge who is participating
  - Who may have different perceptions and opinions?
  - Why and how do you split up the groups e.g. women, people with disabilities, children, elderly?
- Opportunity to participate
  - Are people aware that activities are taking place?
  - How will you inform all members of the community?
- Meaningful participation
  - Is there space for people to speak up?
  - Are activities facilitated in such a way, e.g. in smaller groups/gender or age homogeneous groups, as to bring out the opinions of the most vulnerable?
  - Are people's suggestions listened to?
  - How will language and use of words differ for children/elderly/women/indigenous people?
- Confidence to participate
  - Do people have the confidence to attend meetings and participate, including to speak up? How can we increase their confidence?

Conclusion

Using a social development lens as an entry point to inclusive DRR, including CCA, has the advantage of highlighting that those who are most vulnerable to disaster risks tend to face multiple and intertwined exclusions from wider social, economic, political and cultural contexts. Marginalised groups are more likely to suffer disproportionately from disasters, yet disasters can also exacerbate their existing vulnerabilities and inequalities. Being excluded from protective social safety nets and decision-making processes that affect their lives means that individuals' and groups' different needs, vulnerabilities, capacities and strengths are not taken into account; thereby undermining the effectiveness and sustainability of DRR, including CCA, efforts. In contrast, recognising patterns of exclusion as part of addressing the root causes of disasters will help to tackle people's underlying vulnerabilities and increase their capacities to cope with the effects of disasters.

Systematically promoting social, economic, cultural and political inclusion in DRR, including CCA, and development decision-making will help tackle social inequalities and unjust power relations. This will also enhance international efforts to eradicate poverty by 2030 as part of the Sustainable Development Goals (SDGs). Furthermore, monitoring and evaluation processes need to be disaggregated by sex, age, disability, ethnicity and socio-economic status in order to adequately integrate different gender perspectives, and promote social, economic and cultural diversity within decision-making processes and projects implementation.

References

Acosta, A.M. and Clark, J. (2010) 'Inclusive local governance for poverty reduction: A review of policies and practices', *Learning Project on Socially Inclusive Local Governance*, Bern, Switzerland and Brighton, UK: The Decentralisation and Local Governance Network (DLGN) of the Swiss Agency for Development and Cooperation, and the Institute for Development Studies.

Agarwal, N., André, R., Berger, R., Escarfuller, W. and Sabatini, C. (2012) *Political Representation and Social Inclusion: A Comparative Study of Bolivia, Colombia, Ecuador and Guatemala*, New York: The Americas Society and The Council of Americas.

Amnesty International. (2015) *Dalit Sisters Ordered to be Raped in India, 24 August 2015*. Online https://www.amnesty.org/en/documents/asa20/2316/2015/en/ (accessed 8 September 2016).

Beall, J. and Piron, L.H. (2005) *Department for International Development (DFID) Social Exclusion Review*, London: London School of Economics and Overseas Development Institute.

BRACED (Building Resilience and Adaptation to Climate Extremes and Disasters) Myanmar Alliance. (2015) *Community Resilience Assessment and Action Handbook*, Myanmar: BRACED Alliance.

Cadag, J.R.D. (2013) *A l'ombre du géant aigre-doux. Vulnérabilités, capacités et réduction des risques en contexte multiethnique: le cas de la région du Mont Kanlaon (Philippines)*, Montpellier: History Department, Université Paul Valéry.

Chambers, R. (2007) 'Poverty research: Methodologies, mindsets and multidimensionality', *Institute of Development Studies (IDS) Working paper 293*, Brighton: Institute of Development Studies.

Cooke, B. and Kothari, U. (2001) *Participation. The New Tyranny?*, London: Zed Books.

Delica-Willison, Z. and Gaillard, JC (2012) 'Community action and disaster', in B. Wisner, JC Gaillard and I. Kelman (eds), *The Routledge Handbook of Hazards and Disaster Risk Reduction*, New York: Routledge, pp. 711–722.

Ferretti, S. and Khamis, M. (2014) 'Inclusive disaster risk management: a framework and toolkit for DRM practitioners', *Inclusive Community Resilience for Sustainable Disaster Risk Management (INCRISD) South Asia*. Online http://incrisd.org/toolkit/frameworkpdf/Final%20version.pdf (accessed 8 September 2016).

Golub, S. (2003) *Non-state Justice Systems in Bangladesh and the Philippines*, London: Department for International Development (DFID).

Grimes, P. and Bagree, S. (2012) *Equity and Inclusion for All in Education*, London: Global Campaign for Education (GCE).

Human Rights Watch. (2015) *Burma: Discriminatory Laws Could Stoke Communal Tensions, August 23 2015*. Online https://www.hrw.org/news/2015/08/23/burma-discriminatory-laws-could-stoke-communal-tensions (accessed 8 September 2016).

IFRC (International Federation of the Red Cross and Red Crescent Societies). (2014) *World Disasters Report: Focus on Culture and Risk*, Geneva: IFRC.

Le Masson, V., Mosello, B., Le Masson, G., Diato, E. and Barbelet, V. (2016) 'Integrating gender equality in WASH emergency response: The case of the Central African Republic', in C. Frohlich, G. Gioli, F. Greco and R. Cremades (eds), *Water Security Across the Gender Divide*, London: Springer book series 'Water Security in a New World', forthcoming.

Lewis, J. (1999) *Development in Disaster-Prone Places: Studies of Vulnerability*, London: Intermediate Technology Publications.

Lowe, C.J. (2010) 'Analyzing vulnerability to volcanic hazards: application to St. Vincent', PhD dissertation, London: University College London.

Mohanty, S.K. (2012) 'Multiple deprivations and maternal care in India', *International Perspectives on Sexual and Reproductive Health* 38, 1: 6–14.

North, A. (2010) 'Drought, drop out and early marriage: Feeling the effects of climate change in East Africa', *The Equals Newsletter, Issue 24 (February 2010)*, New York: United Nations Girls' Education Initiative (UNGEI).

Peek, L. and Stough, L.M. (2010) 'Children with disabilities in the context of disaster: A social vulnerability perspective', *Child Development* 81, 4: 1260–1270.

Priestley, M. and Hemingway, L. (2006) 'Disabled people and disaster recovery: A tale of two cities?', *Journal of Social Work in Disability and Rehabilitation* 5, 3–4: 23–42.

Roldán, R.O. (2004) 'Models for recognizing indigenous land rights in Latin America', *Biodiversity Series, Paper No. 99*, Washington DC: The World Bank Environment Department.

Sharma, A., Singh, H. and Biswas, C. (2014) *Making Disaster Risk Management Inclusive*, London and Oxford: ActionAid International, Handicap International and Oxfam.

Shaw, R., Uy, N. and Baumwoll, J. (2008) *Indigenous Knowledge for Disaster Risk Reduction: Good Practices and Lessons Learned from Experiences in the Asia-Pacific Region*, Bangkok: United Nations International Strategy for Disaster Reduction (UNISDR).

UNDESA (United Nations Department of Economic and Social Affairs). (2007) *Final Report of the Expert Group Meeting on Creating an Inclusive Society: Practical Strategies to Promote Social Integration*, Paris: UNDESA.

UNISDR (United Nations International Strategy for Disaster Reduction). (2013) *UN Global Survey Explains Why So Many People Living with Disabilities Die in Disasters*, Geneva: United Nations Office for Disaster Risk Reduction.

UNISDR (United Nations International Strategy for Disaster Reduction). (2015) *Sendai Framework for Disaster Risk Reduction 2015–2030*, Geneva: United Nations Office for Disaster Risk Reduction.

WHO (World Health Organization). (2013) *Global and Regional Estimates of Violence against Women: Prevalence and Health Effects of Intimate Partner Violence and Non-Partner Sexual Violence*, Geneva: World Health Organization.

Wisner, B., Gaillard, JC and Kelman, I. (eds) (2012) *The Routledge Handbook of Hazards and Disaster Risk Reduction*, New York: Routledge.

World Bank. (2013) 'Inclusion matters: The foundation for shared property', in *The World Bank: New frontiers of social policy (Advance Edition)*, Washington DC: World Bank.

25

INTEGRATING DISASTER RISK REDUCTION INCLUDING CLIMATE CHANGE ADAPTATION INTO THE DELIVERY AND MANAGEMENT OF THE BUILT ENVIRONMENT

Ksenia Chmutina, Rohit Jigyasu, and Lee Bosher

Introduction

Recent disasters across the world have highlighted the fragility of the built environment to a range of natural hazards, including those that may be influenced by climate change. Moreover, the rapid pace of urbanisation has increased concerns about the resilience of cities in light of disaster risks including those influenced by climate change; with contemporary discussions considering how physical/protective interventions can be integrated into the built environment or, indeed, what types of interventions are most effective.

Too often Disaster Risk Reduction (DRR) and Climate Change Adaptation (CCA) have been treated as separate issues. Despite a shift to more pro-active and pre-emptive approaches to managing disaster risk, DRR appears to have been overly influenced by more reactive emergency management practices (UNISDR 2015). At the same time, CCA activities have typically fallen within the realm of environmental sciences. As a result there appear to be critical disconnects between policies for CCA and DRR; often centred in different departments with little or no coordination. Moreover, there is a lack of integration of these policies within building regulations, the scope of which is largely limited to rigid restrictions in height and volume and specifications of materials and technology. Most often these building regulations are focused on the mitigation of a single hazard such as earthquakes, floods or cyclones.

It is becoming clear that DRR and CCA must go hand in hand – particularly when it comes to the planning, design, construction and operation of the built environment, with references to both areas increasingly appearing in international guidance and reports. This is hardly surprising, considering that the built environment generates significant amounts of greenhouse gas emissions and can play a major role in the vulnerability of communities. For instance, the impacts of climate change on the built environment are likely to be experienced through climate variability and extreme weather events, with both linking climate change to DRR. Ideally, urban planners, architects, builders and other decision makers who influence the delivery and management of the built environment should be increasingly asked to respond simultaneously to the challenges posed by DRR, which by definition includes CCA. Thus, it is critical to enhance resilience and sustainability of the built environment through addressing multiple hazards across spatial and temporal scales.

Considering the built environment at a regional scale can help ensure that locally available natural resources are appropriately utilised so that an ecological balance can be found in light of disaster risks including those influenced by climate change. In fact, many of these lessons for DRR, including CCA, can be learnt from local knowledge in planning and building design and construction, and therefore deserve sympathetic consideration when drawing up new policies and regulations.

This chapter will discuss the above mentioned issues by highlighting the lack of integration between DRR and CCA in built environment related policies and regulations. It will highlight how policy and regulations can be used to make DRR including CCA inputs from key built environment stakeholders more proactive and thus more effective.

CCA, DRR, and the Built Environment

As pointed out by Wisner *et al.* (2012: 31), the 'natural environment is neither a hazard nor resource until human action makes it one or the other (or both)'. Vulnerability is thus created not by the environment but by poor decision-making, practices (including construction practices) and planning. Natural hazards only become disastrous if a settlement (or any kind of a built environment) is located in a hazard-prone area, poorly constructed and/or does not have a warning system in place.

The notion of a built environment is quite recent; it is conveyed through a system's perspective and emphasises relationships between various built elements (Moffat and Kohler 2008). The built environment is made up of dwellings, neighbourhoods, public spaces, waterways, and infrastructure and transport systems. Their specific characteristics, such as location, structure and density, can make their residents and assets more or less vulnerable to hazards and threats (Boyd and Juhola 2014). The built environment is one of the largest contributors to greenhouse gas emissions worldwide (Anderson *et al.* 2015) and at the same time it can be extremely vulnerable to the effects of climate change. This emphasises the increasing importance of the role of the built environment in reducing its negative contributions to climate change by making the building stock more energy efficient, and in adapting to the negative impacts of climate change by increasing resilience through investment in DRR measures (Lizarralde *et al.* 2015). However, while the concepts of climate change and DRR are widely discussed, it is not always clear to what extent these notions are interrelated. There appear to be fundamental conflicts between perspectives dominated by eco-efficiency (minimising the use of resources) and long-term resilience (robustness of built assets) to the impacts of climate change. This, however, does not mean that both these perspectives cannot be addressed simultaneously as will be discussed later in this chapter.

The impacts of climate change can affect the built environment directly and indirectly. Direct impacts occur through increased frequency and intensity of extreme weather events that could lead to destruction of physical assets and property, and widespread displacement of people. Indirect impacts are the consequences of the direct impacts correlated with demographic, economic and political stressors that increase vulnerability (e.g., poverty and political instability).

Climate change and DRR are closely linked to urbanisation. Combined together, the drivers of urbanisation and the risk factors induced by climate and other hazards and threats create a diverse range of vulnerabilities unique to urban environments. Vulnerability is often discussed in the context of urbanisation, referring to the exposure of a city (and its inhabitants and systems) to disturbances, such as natural hazards, economic crises or political unrest, exacerbated by population dynamics, informal settlements and inappropriate governance and planning. According to Davies *et al.* (2013), there are three main factors that multiply the risks generated by urbanisation:

- Geographical location with respect to extreme weather events and human-induced threats;
- Dependence on the complex systems that are vulnerable to various threats and hazards;
- The level of resilience and the governance of resilience.

In the ever-expanding cities, the governance capacities and state are often unable to regulate urban development or to provide the necessary infrastructure to adequately support the increase in populations, which leads to increase in vulnerabilities. Whilst there are a large number of advantages for the inhabitants of large cities (e.g., improved economic development, easier access to basic services, a comparatively rich cultural life), with increasing social polarisation, segmentation and fragmentation, the number of people that are excluded from these benefits is growing (Butch *et al.* 2009).

An increasing number of international and national policy documents acknowledge climate change as a 'risk multiplier', although it can also diminish risks, and as a result a large number of climate change mitigation strategies aimed at reducing greenhouse gas emissions (mainly by reducing fossil fuel consumption and introducing new renewable energy technologies) have been introduced in recent decades. Being a global challenge – and one which can only be addressed globally – climate change has become a distraction from other equally important concerns or 'creeping environmental problems' (Glantz 1994), such as resource overexploitation or inequality. As stated by Kelman and Gaillard (2010: 32), 'Climate change has been changing the characteristics of weather and climate phenomena, but did not cause the vulnerability to them'. Therefore, whilst it is not appropriate to ignore climate change, it is important to bear in mind other hazards. CCA efforts should be seen as a part of the DRR agenda, with climate change being treated as one of the hazards (Kelman 2015), although it is equally important not to overlook climate change mitigation.

The impacts of climate change on disaster risks are not only relevant to the increase in frequency and severity of a hazard, but also to encompassing vulnerabilities, as climate change rapidly affects local environments changing them in such a way that local knowledge becomes less applicable (Kelman 2015). Taking into consideration the possible effects of hazards and threats related to climate change and disasters that may affect the built environment presents a great challenge to both policy-makers and built environment professionals. They have to make a choice of either taking as a basis the upper limits of uncertainties provided by the projection scenarios, or continue with current practices, therefore potentially reducing the lifetime of a structure. Whilst the former is a more effective adaptation strategy, it may be less cost-effective.

A large number of cities have introduced and applied numerous mitigation measures aimed at greenhouse gas emissions and energy consumption reduction; however, only a few cities have been creative and productive in the realm of adaptation (Jabareen 2015) (for an example of an effective resilience strategy see Box 25.1). This suggests that built environment professionals and policy-makers do not act enough to mitigate uncertainties from climate change and other natural hazards and human-induced threats. Instead of developing strategies for coping with risks, the vulnerabilities are often increased by decisions that do not take local context into account or are not appropriately enforced (Bosher 2014).

Box 25.1 Addressing Flooding in Surat

Ksenia Chmutina[1], Rohit Jigyasu[2], and Lee Bosher[1]

[1] Loughborough University, UK
[2] Ritsumeikan University, Kyoto, Japan
Based on Bhat et al. (2013).

Surat is one of India's most economically successful cities. However, with much of the city and its surrounds being less than 10 m above mean sea level, it is also prone to fluvial and pluvial flooding. Flooding is a recurring event, bringing major disruption to the city's economy. A vulnerability assessment has highlighted the vulnerability of households, particularly of low-income dwellers residing in riverine areas. It was therefore decided that it is important to address and decrease Surat's vulnerabilities. The modelling, sector studies and vulnerability assessment together informed the development of the city's resilience strategy, and highlighted the short-term, mid-term and long-term approaches for addressing the various issues. The short-term strategies included developing an end-to-end early warning system and improved information and data management. Mid-term strategies identified mapping of flood-risk areas and the regulation of construction in floodplains, while long-term strategies included the diversion of floodwaters from the Tapi River and the construction of a balloon barrage system. These flood-specific measures were identified alongside other sector-specific strategies that would also build resilience to flood risks, such as improving wastewater and sanitation systems to reduce health risks from flooding, as well as improving the health surveillance system. This city resilience strategy now forms part of the city's plans for preparing for climate change impacts.

Regulations and policies that address how the built environment is designed, planned and operated are critical for DRR including CCA, as the ways in which land is used and buildings and infrastructure are designed and operated influence exposure to hazards and threats. Once the investment in built assets in a risk-prone location has been made, it will remain there for a long period of time; in addition, once in place it is more expensive and less effective to correct and add new DRR measures than it would have been to avoid the creation of the risk in the first place (UNISDR 2011). It is therefore clear that building regulations and planning policies can be a primary prevention, mitigation and adaptation mechanism.

During the past twenty-five years, building regulations and codes have been developed for virtually every type of construction; there are also an increasing number of informal guidance documents for the construction sector. They are constantly revised and improved, and the evidence shows that in those countries where building codes have been effectively applied, there is a dramatic improvement in performance of new construction (Krimgold 2011). The majority of the current building codes and regulations and land-use planning policies take into account various hazards and threats (e.g., floods, storms and earthquakes). However, whilst these policies and regulations have shifted towards addressing the root causes of vulnerabilities to disasters such as structural integrity of a building, they do not often do so explicitly and tend to focus only on a single hazard or one part of the problem – this will be demonstrated later in this chapter. In addition, mandatory built environment policies are based on the historical trends and previous events thus neglecting future projections that are critical for effectively embedding CCA within DRR.

Challenges and Opportunities for Including CCA in DRR

DRR and climate change are addressed in separate policy arenas at international and national levels. However, starting with the Hyogo Framework for Action 2005–2015 and 2007 Bali Action plan, a number of efforts have been made to point out the importance of addressing DRR and CCA together (UNISDR 2008). This has also been reemphasised in the Sendai Framework for DRR introduced in 2015, and further strengthened during the COP21 meeting in Paris in 2015. For instance, the building code reviews, which usually reflect the most recent impact of a disaster event (be that natural hazard (e.g., an earthquake) or a human-induced threat (e.g., terrorism)), will now likely be made based also on future projections of change in wind speeds or height of storm surges, as well as other climate impacts.

However, despite recent debates for integrating CCA into DRR, there is hardly any evidence about technical and institutional challenges in practice (Davies *et al.* 2013). Around the world, solid frameworks for CCA and DRR exist; however, these frameworks are not easily included in the built environment-related regulations and policies. There is a disconnection in the way that DRR and CCA are treated: for instance, both CCA and DRR are often preparedness and response oriented, thus paying less attention to prevention considerations in a country's development and planning practices, and consequently not sufficiently mainstreaming DRR and CCA into policy-making.

Whilst the issues addressed under CCA and DRR policies relate to the built environment, the interventions are often planned and implemented by different ministries. Neither DRR nor CCA is a sector, as they require informed action across a number of sectors (from education to health to utilities). DRR is often handled by civil defence and emergency management departments, which do not have links with environmental or economic ministries that overlook national planning and climate-change related policies. In addition, DRR and CCA are not the sole responsibilities of these departments and therefore tend not to be at the top of their priority lists. This creates further challenges for the built environment when building regulations, codes, and planning policies are introduced, as often the contribution of both DRR and CCA into these policies is negligible. Moreover, professional training of the built environment professionals does not mainstream DRR and CCA as these competencies are not required in order to follow the existing regulations.

Building regulations and planning policies present an excellent opportunity for incorporating CCA into DRR. However, there are some challenges that can diminish the role of building regulations and codes in DRR. For instance, land use planning may be ineffective if it is implemented at a local level but a given risk crosses legislative boundaries of that locality. In addition, planning processes are often long-winded and inconsistent with the rapid development of a city (this is particularly an issue in the middle- and low-income countries). Similarly, building codes and regulations often do not take local specifics into account, and their implementation is often hindered by a lack of required expertise and manpower within the local government to monitor and enforce the regulations (UNISDR 2011).

Governments are often reactive and slow in responding to the issues related to DRR including CCA, and although new improved regulations are introduced, there is often a lack of incorporation of older buildings' and infrastructure upgrade. The lack of government initiative also drives market barriers, as often risk-averse construction professionals are reluctant to invest in new technologies and practices that could be more appropriate in terms of CCA and DRR (van Heijden 2014). Another issue is lack of implementation of these regulations and policies. In countries like India, heavy bureaucracy also hinders effective implementation. Moreover, these regulations and policies are not designed to address specific design and construction technologies as prevalent in various regions; their contextualisation thus indeed being a challenge.

Another important challenge is a lack of stakeholder engagement, particularly in the private sector. DRR is often seen as a responsibility of emergency managers; however, multi-stakeholder participation can increase the capacity and capability of those who take part in DRR. Involvement of various public and private stakeholders can also lead to and facilitate knowledge and experience sharing. It is essential to identify those stakeholders who can have a positive influence over DRR in the built environment at various stages of the design, construction and operation processes, including commissioning, operation and maintenance, as effective decision making requires an integrated understanding of how to avoid and mitigate the effects of disasters (Chmutina *et al.* 2014).

Tensions Created by CCA and DRR Policies

Whilst complementary, CCA and DRR policies also create some tensions when addressing the challenges faced by the built environment. This can be due to differing interpretations of terminology, institutional responsibilities and contextual differences, such as:

1   Specific vs. broad scope: CCA policies largely focus on what can be achieved in terms of adapting to climate change-induced threats, in particular storms and floods. DRR policies put emphasis on the capacities that are (or should be) available in order to cope with a wider range of risks and threats, both natural and human-induced often regardless of their connection to the impacts of climate change. For instance, in India, the National Building Code mainly contains administrative regulations, development control rules and general building requirements, such as fire safety requirements, stipulations regarding materials, structural design and construction (including safety), and building and plumbing services. However, it has been reviewed a number of times in recent decades in order to include lessons learnt in the aftermath of a number of disasters associated with devastating earthquakes and super cyclones witnessed by the country. The salient features of the revised National Building Codes include the changes with regard to further enhancing response to meet the challenges posed by disasters and reflecting the state-of-the-art and contemporary applicable international practices. These include criteria for earthquake resistant design of structures (including guidelines and a code of practice specifically for reinforced cement concrete and earthen structures), cyclone resistant structures and fire protection. However, when it comes to CCA, the code only relates to mitigation of climate related hazards such as floods and cyclones but does not deal with adaptation challenges posed by climate change. Flood risk is emphasised in IS 13739:1993 'Guidelines for estimation of flood damages'. This standard lays down a detailed scientific procedure for collection of flood damages (other than loss of human life) data under various categories and also methods of translating them to monetary terms (GoI 2008).

2   Efficiency vs. redundancy: The overarching climate change agenda that informs CCA policies often endorses a lean approach to development and streamlining processes that goes hand in hand with climate change mitigation; i.e., to reduce consumption and minimise environmental impacts. DRR policies are more open to the potential benefits of over-designing (i.e., using more material resources to increase robustness) in order to avoid damages and prevent disasters. Hurricane Sandy hit New York hardest right where it was most recently redeveloped: Lower Manhattan, which from a political and economic perspective should have been the least vulnerable part of the island. But it was rebuilt to be 'sustainable', not resilient: the buildings were designed to generate lower environmental impacts, but not to respond to the

impacts of the environment; e.g., by having redundant power systems. After the terrorist attacks of 11 September 2001 in the US, Lower Manhattan contained the largest collection of LEED-certified, green buildings in the world. In the aftermath of Hurricane Sandy the Building Resiliency Task Force Report was issued; it provided 33 actionable proposals for making New York buildings and residents better prepared for the next extreme weather event taking into account that the weather events in the future will be more extreme due to the impacts of climate change (Urban Green Council 2013). The Task Force provided key recommendations on how to make various types of buildings more resilient.

3   Emphasis on standards vs. emphasis on potential: CCA policies have been informed by, and focused on, globally accepted standards often neglecting local context. DRR policies are often driven at the local level and encourage the identification and reinforcement of local potentials and capacities of the system. For instance, whilst Barbados does not have a dedicated national policy that covers CCA, the draft of the Building Code in Barbados emphasises the importance of addressing the effects of climate change on the country, and 'takes into account the climate and geological conditions of Barbados, especially the nature of the Caribbean environment and the region's susceptibility to hurricanes and earthquakes' (BNSI 2013: 23). Whilst based on international standards (Scottish Executive Technical Standard and the UK Building Regulations), it nevertheless focuses on local forms of construction such as the 'chattel house' (traditionally a small moveable wooden house). The Code covers (to an extent) some natural hazards: flooding receives a dedicated section (Part 2, section A-2); and hurricanes and earthquakes are mentioned in the section on Structure in the context of the installation of windows and hurricane shutters, and roof coverage (which should resist the uplift from a minimum wind speed of 208 km/h), and the calculation of the earthquake loads (Chmutina and Bosher 2015).

4   Reactive vs. proactive: CCA policies acknowledge that climate change will have a negative impact on the built environment and therefore suggest the ways of adapting to these impacts. DRR policies (at least on a theoretical level) acknowledge the importance of a more pro-active approach to dealing with risks. This is clearly demonstrated in UK policy. CCA policies fall under the sustainability agenda, and DRR policies fall under the resilience agenda. The National Adaptation Plan emphasises the importance of CCA for the built environment, with its main focus being on adaptation measures to flooding and heatwaves. The resilience agenda in the UK underpins the development of all DRR-related work, including plans for the protection of critical infrastructure and for the prevention of violent extremism. National policy recognises that a risk cannot be fully eliminated (and in some cases it may also provide new opportunities). Resilience on the other hand is thus seen as a way of building capacity to respond to extreme events and return to normality, notably to guarantee business continuity (Lizarralde *et al.* 2015).

Main Areas in which Synergies Could and Should be Created

These tensions are important to consider; however, a number of areas in which synergy can (but does not necessarily do so yet) complement both CCA and DRR is in relation to the challenges faced by the built environment.

1   Similar goals: CCA and DRR policies implemented at the local level essentially address the same issues. For instance, the UK National Adaptation Plan focuses on natural hazards that may have an impact on the functionality of the built environment; this is in line with the UK National Risk Assessment, although these two documents seldom acknowledge each other.

2   Synergising CCA and DRR can provide a basis for the much needed multi-stakeholder engagement: currently CCA is mainly addressed by environment-related departments, whereas DRR is a responsibility of emergency managers, with the private sector and communities in many cases not being involved in decision-making at any stage. Multi-stakeholder engagement can bridge disconnected policy and practice by putting those at risk (e.g., businesses and vulnerable sections of society) to the forefront.

3   Knowledge sharing: Multi-stakeholder engagement will allow for the integration of scientific knowledge of environmental professionals (and others), local knowledge of communities that is prevalent in the DRR, and practical context-specific knowledge of the built environment professionals. In addition, CCA can draw on some of the tools developed within DRR (e.g., risk monitoring).

4   Overarching DRR plans can employ a holistic approach by emphasising natural resource protection, land-use planning and building codes that also address reduced energy consumption.

5   Time scales: synergies between CCA and DRR would allow for the expansion of DRR's efforts time horizon by utilising future projections developed as part of CCA. By doing this it could be easier to justify investment in pre-emptive risk reduction considerations for future developments.

6   Budget allocation will be more effective if it is aimed at both DRR and CCA, thus helping to reduce doubling efforts and increasing institutional effectiveness.

However, in order to create these synergies, some basic challenges need to be overcome. These include existing institutional gaps and lack of coordination between various departments/ministries linked to DRR and CCA. Also, there is the challenge of using commonly understood vocabulary for DRR and CCA. Another common issue is the nature of financial allocations that are made under separate budget heads for DRR, CCA and other related areas thereby making it difficult to pool the resources for integrated planning and implementation. Last but not least is the challenge of integrating CCA into DRR policies and programmes at national, district and local levels.

The following framework for DRR including CCA can be suggested for the reduction of the vulnerabilities in the built environment (Figure 25.1). The framework suggests that DRR including CCA should be incorporated at the earliest possible stage and be based on multi-stakeholder engagement throughout the process of building and zoning regulations development. Such a framework incorporates cooperation at all levels and takes into account national priorities as well as local context-specific needs and vulnerabilities. Whilst it partially decentralises efforts for DRR including CCA, it on the other hand helps share the financial burden of its implementation as local priorities become national and vice versa.

This framework includes regulatory interventions based on collaboration with government, businesses and civil society. It establishes direct regulatory mechanisms that help in creating enforcement approaches for new and existing buildings addressing the needs and requirements of the climate change agenda and by incorporating CCA into DRR. However, such a regulatory regime allows for learning from the built environment and other professionals, therefore ensuring that regulatory bodies keep up with the pace of changing construction practices.

DRR including CCA Contributions to Climate Change Mitigation

Deeper understanding of synergies of DRR and CCA also presents an opportunity for incorporating climate change mitigation into the development of the built environment. A number of initiatives worldwide, such as offshore wind farms located in Flevoland flooded areas in the

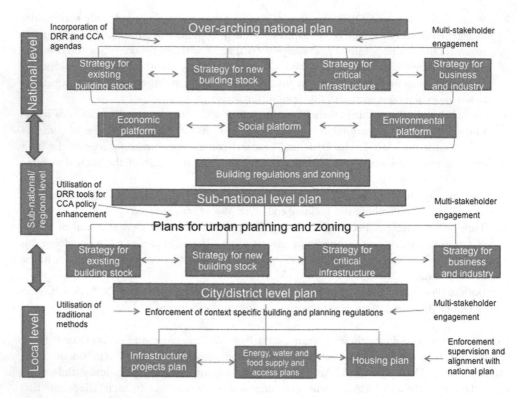

Figure 25.1 Framework for Incorporating DRR Including CCA into Built Environment Policies
(By Authors)

Netherlands, to protect coasts from flooding or security features that simultaneously act as sustainable urban mechanisms are becoming more and more common. Other solutions have also been summarised by Coaffee and Bosher (2008), including:

- Pavements that can act as sustainable urban drainage systems (SUDS) and harvest rainwater that can then be used for functions that do not require treated water from the mains (e.g., flushing toilets and irrigation), which may contribute to water efficiency.
- Cool roofs that reflect the sunlight and therefore reduce overheating: they help keep spaces cool during heatwaves especially if there is a blackout, and simultaneously reduce indoor temperatures thus reducing air-conditioning and therefore energy consumption.
- Energy co-generation system: incorporation of off-site renewable energy technology can provide electricity during blackouts as well as contribute to an overall decrease of energy from fossil fuels.
- Use of window shutters: protects glazing from smashing during storms and hurricanes as well as protects spaces from overheating thus reducing energy consumption for air-conditioning.

The main challenge, however, is to recognise the importance of identifying the barriers that restrict the opportunities to integrate DRR, CCA and mitigation measures into the built environment as have been outlined within this chapter.

Conclusions

As demonstrated in this chapter, the contribution of the built environment to climate change and CCA is well accepted in current building policies and regulations; however, the risk reduction rationale in these regulations originates mainly from the past. This sets a challenge of expanding the current existing focus of building regulations: there is a need to incorporate a wider holistic ecological approach that looks at regional impacts and vulnerabilities and is not just limited to the performance of the built environment.

CCA and DRR initiatives currently work in silos, neglecting and underestimating their commonalities and goals, or being unable to overcome political constraints. Such a lack of synergy should not be ignored as it increases the risk of unsuccessfully reducing vulnerabilities of the built environment in the long run. Whilst there is enough understanding about how to place CCA within DRR, there is a lack of appropriate governance approaches and tools. This leads to multiple negative consequences, including duplicating efforts that lead to organisational inefficiencies and ineffective use of resources as well as counter-productive efforts, in particular by reinventing older approaches (Mercer 2010).

In order to achieve a truly sustainable and resilient built environment it is critical to achieve an effective scale of hierarchically interdependent built elements. If such hierarchy is weak, the vulnerability of a built environment increases and therefore an impact of one hazard may exacerbate the impact of another hazard, thus creating a complex/compound hazard. Vulnerability continually increases in many places because the size and complexity of the built environment is increasing, with systems and networks planned, designed, constructed and operated without appropriate attention to the potential risks. Climate change presents an additional challenge and opportunity; therefore, what were previously considered reasonable margins of safety in the traditional engineering approaches may no longer be relevant or effective.

Climate change has become a part of the built environment's political agenda nationally and internationally in many countries, and it therefore could act as a mechanism to attract attention of policy makers to DRR. This, however, has to be done carefully in order not to shift the agenda to climate-induced hazards only, but instead it is critical to make DRR part of the sustainability agenda. Whilst it is important to build a structure that is energy efficient and constructed using materials that have minimal impacts on the environment, it is equally important to make sure that it is not in a risk-prone area and is not going to be destroyed by the next earthquake or flood. DRR including CCA should play a bigger role in building regulations and planning policies.

Structural measures can predominate in DRR, but this is also appropriate for CCA. Incorporation of CCA into DRR in the context of the built environment can be imposed through effectively implementing, monitoring, and enforcing building regulations and codes and land use planning and zoning requirements, ensuring that responsibility for preventive, protective and mitigation actions lies with engineering and planning professionals. It can also contribute towards climate change mitigation. Planning policies also present a unique opportunity to integrate policies of mitigation, adaptation, land use and other sustainability-related measures in one legally binding document. However, it is important to incorporate ecological perspectives through adaptable design, which increases flexibility and durability of the built environment. Better integration of CCA into DRR can promote more structured and coordinated planning, construction and operation mechanisms and simultaneously provide support for overall sustainable development.

References

Anderson, J.E., Woldhurst, G. and Lang, W. (2015) 'Energy analysis of the built environment – a review and outlook', *Renewable and Sustainable Energy Reviews* 44: 149–158.

Bhat, G.K., Karath, A., Dashora, L. and Rajasekar, U. (2013) 'Addressing flooding in the city of Surat beyond its boundaries', *Environment and Urbanisation* 25, 2: 1–13.

BNSI (Barbados National Standard Institution). (2013) *Barbados National Building Code* 2013 BNSSP1, St Michaels, Barbados: BNSI.

Bosher, L.S. (2014) '"Built-in resilience" through disaster risk reduction: Operational issues', *Building Research and Information* 42, 2: 240–254.

Boyd, E. and Juhola, S. (2014) 'Adaptive climate change governance for urban resilience', *Urban Studies* 25, 7: 1234–1264.

Butch, C., Etzold, B. and Sakdapolrak, P. (2009) *The Megacity Resilience Framework*, Tokyo, Japan: UNU-EHS. Online https://collections.unu.edu/view/UNU:1830 (accessed 4 April 2016).

Chmutina, K. and Bosher, L. (2015) 'Risk reduction or risk production: The social influences upon the implementation of DRR into construction project in Barbados', *International Journal of Disaster Risk Reduction* 13: 10–19.

Chmutina, K., Ganor, T. and Bosher, L. (2014) 'The role of urban design and planning in risk reduction: Who should do what and when', *Proceedings of ICE – Urban Design and Planning* 167, 3: 125–135.

Coaffee J. and Bosher L.S. (2008) 'Integrating counter-terrorism resilience into sustainable urbanism', *The Proceedings of the Institution of Civil Engineers: Urban Design and Planning* 161, 2: 75–83.

Davies, M., Bene, C., Arnall, A., Tanner, T., Newsham, A. and Coirolo, C. (2013) 'Promoting resilient livelihoods through adaptive social protection: lessons from 124 programmes in South Asia', *Development Policy Review* 31, 1: 27–58.

Glantz, M.H. (1994). 'Creeping environmental problems', *The World and I*, June: 218–225.

GoI (Government of India). (2008) *National Disaster Management Guidelines: Management of Floods*, National Disaster Management Authority, GoI. Online http://www.ndma.gov.in/images/guidelines/flood.pdf (accessed 4 April 2016).

Jabareen, Y. (2015) 'City planning deficiencies and climate change – The situation in developed and developing cities', *Geoforum* 63: 40–43.

Kelman, I. (2015) 'Climate change and the Sendai framework for disaster risk reduction', *International Journal of Disaster Risk Science* 6, 2: 117–127.

Kelman, I. and Gaillard, JC (2010) 'Embedding climate change adaptation within disaster risk reduction', *Environment and Disaster Risk Management* 4: 23–46.

Krimgold, F. (2011) 'Disaster risk reduction and the evolution of physical development regulation', *Environmental Hazards* 10, 1: 53–58.

Lizarralde, G., Chmutina, K., Dainty, A. and Bosher, L. (2015) 'Tensions and complexities in creating a sustainable and resilient built environment: Achieving a turquoise agenda in the UK', *Sustainable Cities and Society* 15: 96–104.

Mercer, J. (2010) 'Disaster risk reduction or climate change adaptation: Are we reinventing the wheel?', *Journal of International Development* 22: 247–264.

Moffat, S. and Kohler, N. (2008) 'Conceptualising the built environment as a social-ecological system', *Building Research and Information* 36, 3: 248–268.

UNISDR (United Nations International Strategy for Disaster Reduction). (2008) *Links between Disaster Risk Reduction, Development and Climate Change*, Geneva: UNISDR.

UNISDR (United Nations International Strategy for Disaster Reduction). (2011) *Global Assessment Report on Disaster Risk Reduction*, Geneva: UNISDR. Online http://www.preventionweb.net/english/hyogo/gar/2011/en/home/index.html (accessed 4 April 2016).

UNISDR (United Nations International Strategy for Disaster Reduction). (2015) *Global Assessment Report on Disaster Risk Reduction*, Geneva: UNISDR. Online http://www.preventionweb.net/english/hyogo/gar/2015/en/home/ (accessed 4 April 2016).

Urban Green Council. (2013) *Building Resilience Task Force*, New York: Urban Green Council. Online http://urbangreencouncil.org/sites/default/files/2013_brtf_summaryreport_0.pdf (accessed 4 April 2016).

Van Heijden, J. (2014) *Governance for Urban Sustainability and Resilience*, Cheltenham: Edward Edgar Publishing.

Wisner, B., Gaillard, JC and Kelman, I. (2012) 'Framing disasters: Theories and stories seeking to understand hazards, vulnerability and risk', in B. Wisner, JC Gaillard and I. Kelman (eds), *The Routledge Handbook of Hazards and Disaster Risk Reduction*, Abingdon: Routledge, pp. 18–34.

26

CONNECTING KNOWLEDGE AND POLICY FOR DISASTER RISK REDUCTION INCLUDING CLIMATE CHANGE ADAPTATION

Loïc Le Dé

The Importance of Knowledge in DRR Including CCA Policies

Large scale, as well as smaller and/or more creeping, disasters continue to affect humanity worldwide. Furthermore, in recent years additional global challenges have arisen with climate change expected to increasingly impact lives and livelihoods. As a result, policy-makers, researchers, and practitioners are urging society to find ways to reduce disasters' impacts including through adapting to climate change. It is only recently that the need for combined efforts, so not assuming a separation of climate change and disasters, was emphasised. Yet, while disaster and climate change issues are intricately linked, the policies of these two fields have evolved separately. This is partly because they draw on two different research approaches or bodies of knowledge to inform their respective policies.

Disaster Risk Reduction (DRR) policies draw on over a century of research on disaster-related issues. Disaster research has long been considered from physical science perspectives, including volcanology, hydrology, seismology and meteorology. These approaches mostly focused on natural hazards' intensity, scale, and frequency as the cause of harm and losses (Heijmans 2009). This dominant viewpoint, also called the hazard paradigm, perceives disasters as a result of extreme and rare natural hazards. Nonetheless, the importance of the human dimension in disaster had long been identified, such as Prince (1920) on the 1917 Halifax shipping explosion. In the late 1970s, the assumption that disasters are natural and the sole results of extreme events became widely criticised. What was then termed the vulnerability paradigm emphasised that disasters are rather political, historical and socio-economic in their origin and that unequal access to resources among society members may create the conditions for a hazard to become a disaster (Hewitt 1983; Watts and Bohle 1993).

Until the 1990s, the focus on the hazard paradigm greatly influenced international DRR policies, which were largely based on the transfer of technical knowledge from industrialised countries towards non-industrialised countries – exactly how development work was also framed. For example, four of the five goals from the International Decade for Natural Disaster Reduction adopted by the General Assembly of the United Nations for the 1990s were focused on reducing disasters through the transfer of scientific and engineering knowledge. The significance of humanity in disaster research and the concept of vulnerability in the understanding of disaster risk started to be integrated at policy level in the mid-1990s at the World Conference on Natural Disaster Reduction (WCNDR 1994).

Nowadays, academics, practitioners and policy-makers agree that disaster risk is part of everyday life and thus, reducing the risk of disaster relates strongly with providing the conditions for improved personal and social resources and choices within a wider context of sustainable development (UNISDR 2004). Strategies aimed at reducing the risk of disaster require identifying the capacities and vulnerabilities of local communities, which means involving those directly concerned in the development of policies aimed at lifting their well-being. This approach was stressed in the review of the Yokohama Strategy and used in preparation of the Hyogo Framework for Action (HFA) for the 2005–2015 decade. The review identified the need for 'involving people in all aspects of disaster risk reduction in their own local communities' (UNISDR 2005: 2). This implies building on people's knowledge as recently emphasised in the Sendai Framework for Disaster Risk Reduction (SFDRR) (UNISDR 2015: 15) through the following goal: 'To ensure the use of traditional, indigenous and local knowledge and practices, as appropriate, to complement scientific knowledge in disaster risk assessment and the development and implementation of policies'.

Climate change science and policies are more recent, thus reflecting the emergence of challenges to be faced at global and local levels. The IPCC (2014) highlights that fast land use changes and industrialisation have been generating the burning of more fossil fuels and reducing their uptake leading to severe effects on the climate. The consequences include altered weather and increased coastal inundations because of sea-level rise. While DRR research tends to build on past or present experiences from local communities while simultaneously considering and planning for a wide range of possibilities in the future, climate change research focuses on scenarios and models designed by scientists to address possible future risks. Nonetheless, these different scenarios and models encompass uncertainties including impacts that may vary from one location to the other. For example, rainfall may be more intense in certain areas resulting in more floods and fewer droughts, but more scarce in others leading to fewer floods and more droughts.

Although being more recent than disaster research, climate change and its related effects have received prominent attention from policy- and decision-makers at both international and national levels. Climate change policies mainly focus on the need for mitigation and adaptation to climate change, both identified as complementary approaches for reducing risks of climate change impacts (IPCC 2014). The fifth assessment report from the IPCC (2014: 125) defines mitigation to climate change as 'A human intervention to reduce the sources or enhance the sinks of greenhouse gases' and Climate Change Adaptation (CCA) as:

> The process of adjustment to actual or expected climate and its effects. In human systems, adaptation seeks to moderate or avoid harm or exploit beneficial opportunities. In some natural systems, human intervention may facilitate adjustment to expected climate and its effects.
>
> *(IPCC 2014: 118)*

For local people, this may mean switching crops or diversifying agricultural activity to adapt to environmental changes, education and awareness programmes, protection of natural resources, and risk and impact assessments (UNFCCC 2007).

While the two research approaches often do not overlap, increasingly the need for joint policy efforts to tackle disaster and climate change issues is emphasised. For some, climate change science is perceived as being a re-invention of what is already known in disaster studies, being a shift back to the hazard paradigm (Mercer 2010). CCA's definition by exclusively focusing on 'actual and expected climate and its effects' reinforces the hazard paradigm. It is suggested that greater ability to forecast the frequency, magnitude and severity of climate change effects will generate successful adaptation at local scale. The consequences in terms of policies are a great focus on technical and structural measures (often with a transfer of knowledge from rich to poor countries) to the detriment of non-structural measures such as poverty reduction, equitable access to resources such as land and water, and better delivery of social services from governments. Yet, climate change is just another hazard driver, and CCA policies should be embedded in DRR policies (Kelman and Gaillard 2010).

Different academics, policy-makers and practitioners have identified and debated the similarities and differences between DRR and CCA policies. There is a consensus that a diversity of knowledge is needed to strengthen policies targeting CCA and DRR. However, despite evolution in disaster research, policy-makers still give prime consideration to the hazard paradigm approach and its related recommendations (Gaillard and Mercer 2013). Most national DRR policies are largely based on 'top down' and 'command and control' frameworks that advocate scientific knowledge and government interventions to the detriment of local efforts to address disaster risk. In addition, CCA follows a political agenda framed at global level, with little attention given to community-level adaptation and integration of such knowledge at the policy level (Shaw *et al.* 2010).

For centuries, people have demonstrated abilities to respond to recurrent natural hazards including adapting to environmental changes, with mechanisms such as food security techniques, traditional housing construction, crop resilience, reciprocity and cooperation (not an exhaustive list) (Campbell 2006). These mechanisms are based on their understanding of their place, social organisation, traditional values and historical processes and constitute what can be termed local, vernacular, or traditional knowledge. A common challenge for DRR including CCA policies lies in integrating the knowledge of local communities into the policies aimed at tackling disaster and climate change issues (Mercer 2010; Shaw *et al.* 2010). This may imply recognising the validity of local knowledge with issues linked to DRR including CCA, and developing tools and frameworks to enable interaction between different forms of knowledge. It may also require implementing policies that address in a holistic way disaster risk and climate change issues particularly by integrating CCA into DRR (Gaillard and Mercer 2013).

DRR Including CCA Policies: Why is it Important to Combine Local and Scientific Knowledge?

DRR including CCA policies rely largely on scientific knowledge with the notion of 'expertise' having a central role. Volcanologists, geologists, climatologists, and economists are generally given the status of 'expert', providing them with authority and power over other forms of knowledge that exist within local communities, who in turn may perceive scientific knowledge as superior (Agrawal 1995). Local and indigenous knowledge has proved effective in dealing with disasters, including the effects of climate change. Besides, local knowledge constantly evolves and adapts to environmental changes at large, constraints and opportunities (Mercer *et al.* 2010). As a result, many scholars and practitioners have argued that effective DRR, including CCA policies, can only occur if local knowledge is combined with scientific knowledge (Gaillard and Mercer 2013) (see Box 26.1). This view reflects an approach that tends to move away from the dominant expert-focused and top-down measures towards the integration of place-based methods of adaptation and mechanisms.

Box 26.1   Combining Local and Expert Knowledge about Remittances to Strengthen Disaster Risk Reduction Including Climate Change Adaptation Policies

Loïc Le Dé[1]

[1] Auckland University of Technology, New Zealand

Remittances are the money and goods sent by migrants back home. In the last two decades they have become a main source of revenue for those living in low-income countries. Remittances are also increasingly important for local people to deal with disasters including those involving hazards influenced by climate change. Remittances are twice as important in Small Island Developing States (SIDS) as in other so-called 'developing' countries. Both UNISDR (2005; 2015) and the IPCC (2014) have expressed the need for particular attention towards SIDS and their vulnerabilities, but it is also important not to neglect SIDS' own abilities and resources to implement DRR including CCA.

Although remittances are a central mechanism for local people, they are seldom included in CCA or DRR policies. This is largely because policy-makers and practitioners have little information about remittances at household and community level. Most of the existing research on remittances has focused on the knowledge of experts, generally economists, that do not or limitedly integrate the knowledge of local people receiving and sending remittances. They commonly use national data from central banks of states and apply econometric methods that are often mixed with questionnaire-based surveys. While providing information relevant for informing policies (e.g., large remittance flows, migration fluxes following disasters, and reaction of remittance flows in function of international aid), this dominant approach proves somewhat limited for understanding certain, though central, aspects of remittances in disaster.

A three-year project was developed in Samoa, one of the Pacific SIDS, to bridge this knowledge gap. Participatory tools were developed with disaster-impacted people to produce qualitative and quantitative information on remittances. The visual strength of activities permitted the involvement of people with little understanding of scientific concepts or who were innumerate. People themselves quantified remittances in relation to their incomes, measured remittance increase following disasters, indicated the channels used and difficulties faced to receive them, and how they were utilised. This information helped to produce percentages and trends on remittances, which were presented in the form of tables and graphs to be easily communicable and tangible to economists, practitioners, and policy-makers (Le Dé *et al.* 2015). Although the knowledge produced was contextual (e.g., about a specific area, focusing on particular events), combined with the knowledge from external experts (e.g., national level of remittances before and after disasters, along with migration patterns), it helps to improve the mainstreaming of remittances within DRR, including CCA, policies. This is because the data were more detailed, accounted for unrecorded/informal remittances, and provided qualitative input (according to local people's perceptions) that is often lacking to policy-makers dealing with remittances.

While no universal definition exists, local or traditional knowledge encompasses customs, beliefs, cultural values, and world views of local people. Knowledge from local people has been named in different ways, including indigenous knowledge, traditional knowledge, traditional ecological knowledge, local knowledge, and folk knowledge. Each term carries different connotations (Chambers 1983). The word 'indigenous' refers to the knowledge of those who are generally first to settle a place (although this definition is not strict). Yet, it may overlook the knowledge from other people now living in the same place but who are not identified as indigenous. The term 'traditional' focuses particularly on the notion of knowledge transmission from one generation to another alongside cultural continuity, but may overlook the capacity of traditional societies to adapt to rapid changes (Mazzocchi 2006). The word 'local' emphasises the knowledge of a local environment without necessarily being specific about the nature of this knowledge. In this chapter, local knowledge will be used for its simplicity.

Local knowledge frequently constitutes the local basis for decision-making in the mechanisms used to deal with hazards, including the effects of climate change, management of natural resources, agriculture practices, food preparation, housing construction, health care, and other activities that are central to people's livelihoods. Local knowledge has a basis in concrete experiences and intuitions from everyday realities of living, but can also be rooted in folklore, myths, custom, and religion. It is therefore deeply rooted in its context. Chambers *et al.* (1989: 162) argue that:

> Modern scientific knowledge is centralised and associated with the machinery of the state; and those who are its bearers believe in its superiority. Indigenous technical knowledge, in contrast, is scattered and associated with low prestige rural life; even those who are its bearers may believe it to be inferior.

Knowledge and power are intrinsically linked and, as such, knowledge is a key factor in shaping policies. Those who produce knowledge hold power since they set the public agenda and can include or exclude some voices (or other forms of knowledge) in action upon it (Freire 1970). Thus, knowledge and its production are not only about the expertise of the scientists and local people, but also about shaping awareness of the policies in the first place and capacities for actively influencing these policies. For Gaventa and Cornwall (2001: 72), 'Knowledge determines definitions of what is conceived as important, as possible, for and by whom. Through access to knowledge, and participation in its production, use and dissemination, actors can affect the boundaries and indeed the conceptualization of the possible'. Therefore, whose knowledge, how it is produced and used are central aspects in shaping policies and actions, such as for DRR including CCA (Wisner 1995).

Increasingly it is recognised that integrating local knowledge into existing policies constitutes a prerequisite to DRR, including CCA (Gaillard and Mercer 2013). People who are directly affected when disasters happen are generally the first responders before any external assistance arrives. Nobody is therefore more interested than those affected in DRR – and to ensure that CCA is part of DRR. It is thus indispensable to involve them in the design of policies aimed at lifting their well-being. Besides, locals usually know more than anyone external to the community about their place and conditions of life. This might include knowledge about major hazards that occurred in the past, trends in environmental patterns, and the development of strategies to deal with such events and changes. Local knowledge is a valuable and cost-effective resource that can strengthen DRR, including CCA policies. Moreover, integrating local knowledge within existing policies can help to generate solutions that are culturally and socially compatible with local people's values and norms – keeping in mind that any community is composed of different sets of local people and hence with differing sets of values and norms. This approach provides conditions that assist with the acceptance of policies by local communities in the long-term. Additionally, it contributes to preserving relevant knowledge and practices, generates local pride, and may lead to empowerment of local communities (Cornwall 2011).

This is, of course, not to deny or reduce the relevance of disaster science, including that related to climate change. Indubitably, scientific knowledge has contributed to a better understanding of issues linked to natural hazard, vulnerability, and risk patterns (including climate, climate change, and climate change adaptation) helping to inform policy-makers. This includes the development of inclusive datasets and the use of the Geographic Information System (GIS) for better identification of areas prone to tsunami or floods, utilisation of instruments to better monitor volcanic eruptions, development of warning systems, progress in construction techniques of buildings that resist earthquakes, identification of gender or ethnic inequities to support those who are more vulnerable, or communication system improvements that enable, for example, the transfer of remittances during and after large-scale disasters.

Nonetheless, it is necessary to recognise that any single knowledge form cannot provide all the information required to deal with disaster risk effects, including those created by climate change. For example, scientific projections about climate change are limited and can be inaccurate and imprecise. Likewise, databases such as the Centre for Research on the Epidemiology of Disasters' Emergency Disaster database (CRED EM-DAT), used to inform DRR policy-makers, comprise very important volumes of information on disasters. To keep it manageable, only a few indicators are incorporated. The database is therefore limited or even inaccurate, such as in assessing the indirect cost of disaster impacts (Méheux *et al.* 2007). The knowledge input from local people, including quantitative information, could improve the quality and accuracy of the CRED EM-DAT database.

Alternatively, the DesInventar database uses a larger set of indicators and different information sources, including official data, newspapers, academic information and institutional reports. The information gathered highlights small-scale and creeping disasters (not only large-scale events) that affect the livelihoods of the most vulnerable (Marulanda *et al.* 2010). This approach enables dialogue amongst local institutions, national governments and other relevant stakeholders on ways to reduce the risk of future disasters, including CCA. These examples underline that scientific knowledge cannot provide a one-size-fits-all solution but must account for contextual specificities that require insights from local communities.

On the other hand, local knowledge should not be over-romanticised, since it has not always been successful in dealing with all hazards and hazard influencers including climate change (Mercer 2010). While DRR including CCA policies based on solid scientific evidence may fail for not being adapted to local perceptions and values, DRR including CCA strategies based on knowledge developed by local people may fail to adjust to the fast pace of change currently experienced. Different academics and practitioners point out the need to bridge the gap between local and scientific knowledge in order to support a more balanced decision-making process needed for DRR, including CCA, policies. Local knowledge, when combined with scientific knowledge, would generate more sustainable solutions, including on issues involving technical responses (Cornwall 2011). This implies recognising the value of all knowledge forms without labelling any single knowledge form as being superior.

Current Difficulties in Integrating Local and Scientific Knowledge in Policies

Integrating different forms of knowledge in DRR, including CCA, policies is a difficult task. Local knowledge is largely context-specific and does not necessarily match scientific formats and standards, thus making it difficult for researchers and decision-makers to know how to deal with it. Meanwhile, the knowledge and strategies that are relevant to DRR, including CCA, are not always identified as such by people because they are embedded within their everyday livelihoods and community life. A major challenge is therefore to make local knowledge both tangible and communicable to scientists and policy-makers. This is indispensable to facilitate knowledge sharing and to foster dialogue on DRR, including CCA issues. Another significant obstacle to integrating local knowledge lies in the hazard-focused and technocratic character of existing institutional frameworks, particularly CCA policies.

Current hazard assessments and climate change projections mainly focus on identifying probabilities of environmental events occurring and thus rely greatly on modern technology such as Global Positioning Systems, remote sensing, probabilistic models, and radar to name a few. Assessments of risk perception and vulnerability carried out by social scientists generally use questionnaire-based surveys, interviews, and focus groups together with secondary data sources such as demographic, economic, and governance data to build indices. These assessment methods are intended to match the universal standards of science and produce information that is easily communicable to government agencies, international organisations, and donors.

Surely, these methods produce a wealth of data indispensable for informing policies. However, the tools utilised are often based on western technology and/or have been designed by researchers who are outsiders to the local communities. Most of the time the data is analysed away from those at risk, is intended to be quantitative to fit policy-makers' presumed needs, and thus accounts for the local contexts in a limited manner. Moreover, scientific knowledge, due to its complexity, is not necessarily accessible to locals, which ultimately might generate an unequal power relationship between scientific experts and local communities. Consequently, a valid critique is that such assessment methods should be more balanced with a bottom-up process that better integrates local knowledge.

CCA policies have mostly been treated as a top-down process with responses largely framed through the UNFCCC. National Adaptation Programmes of Action (NAPAs) have been developed to provide less wealthy countries with a UNFCCC-endorsed process to identify priority activities that respond to their urgent and immediate needs to adapt to climate change. Yet, NAPAs do not provide space for integrating the perspectives of local and indigenous people such as current experiences of climate change or lessons learnt from the past. They are framed in a top-down fashion and informed largely by methods of assessments based on scientific knowledge. There is therefore a strong focus on developing technical devices able to forecast weather hazards and develop early warning systems. Indeed, the framework relies largely on the transfer of scientific knowledge from more wealthy countries towards less affluent ones. NAPAs could benefit from having assessments done by community members themselves, producing data informed by local knowledge and geographical specificity.

In turn, DRR policies formulated at international level have increasingly emphasised the importance of local knowledge to inform policies (UNISDR 2005; 2015), although national-level processes do not necessarily convert into local-level action. Many national policies are still framed around command-and-control and top-down approaches that are often not fitted to incorporate local knowledge into the policy-making process or to address the root causes of vulnerability that are very contextual (Gaillard and Mercer 2013). At the same time, many researchers and practitioners, coming from DRR research and the broader field of development, have used bottom-up approaches and have developed different tools to counterbalance the distribution of power in order to make local knowledge more tangible to policy-makers.

In Ambae Island, Vanuatu, a bottom-up approach allowed improved volcanic hazard management guidelines together with an alert system and maps that fit the local communities better than the previous 'top-down' plans imposed by outside governmental and scientific institutions (Cronin *et al.* 2004). The bottom-up approach was catalysed and facilitated by external scientists, using a modicum of top-down control within the local context. Some successful initiatives have also contributed to the improvement of policies at national level. For example, Shah (2013) described how in Rwanda participation of locals in the production of knowledge helped identify those who are the poorest and most marginalised. This bottom-up approach strengthened the health insurance system and agricultural policies at country level. These examples reflect DRR efforts to combine knowledge from the grassroots with initiatives from the top down, so DRR policies and actions can be strengthened and can address broader goals of sustainable development. Certainly, CCA could draw on DRR to inform policies and be better integrated within it.

Towards Integrating CCA within DRR?

There is clearly the need to combine scientific and local knowledge to strengthen DRR, including CCA, policies and actions. While scientific knowledge provides indispensable technical information, local knowledge, too, is indispensable to ensure that DRR, including CCA, policies address local specificities and are compatible with local communities' values. Furthermore, CCA needs to be embedded within DRR in a common framework so vulnerability reduction, alongside sustainable development, can be addressed. To date, the two fields of research operate separately to a large degree, with few coordinated efforts among scientists to work with local communities, and without a common framework to implement policies.

Climate change is increasingly used as a 'catch all' expression to explain the occurrence of many disasters (Kelman and Gaillard 2010). Where local communities are experiencing climate change effects, adaptation policies and actions are required in a broader development context. CCA policies cannot be framed as stand-alone issues disconnected from other factors shaping vulnerability. Rather, they should be integrated into a DRR framework, which not only focuses on physical aspects of vulnerability and risk, but also considers social, economic, cultural, and political components. Earlier disaster research focusing greatly on the hazard paradigm has led to policies and solutions based on the transfer of technical knowledge from industrialised nations to low-income countries (Hewitt 1983). Such policies reflected western construction of knowledge with limited to no consideration for local perspectives and specificities (Bankoff 2001). These policies often led to unsuccessful or even counterproductive outcomes, failing to reduce vulnerability since they did not address the root causes of disasters. Disasters, whether they are climate-related or not, obstruct development and addressing their root causes requires an understanding of other vulnerability aspects that are part of the everyday life of people (Mercer 2010). The lessons learnt from disaster research and DRR policies should be used to inform CCA.

Embedding CCA within DRR is therefore necessary. This would avoid reverting to the hazard paradigm which tends to emphasise local people as 'victims' impacted by events that go beyond their control and without much focus on their capacities and knowledge. A key outcome for embedding CCA within DRR would include reduced vulnerability of local communities through the consideration of DRR and CCA simultaneously without tension between them. They both demand identifying and acting upon urgent and long-term challenges. Having a framework that encompasses disaster risk and climate change issues together would be beneficial to local communities since the CCA mechanisms they develop are, in effect, the same as their DRR mechanisms (Campbell 2006; Mercer 2010).

Such a framework would incorporate a bottom-up perspective that enables people to actively participate in the production of knowledge. This first requires recognising that local communities are not helpless and passive in the face of disasters – which includes disasters involving hazards influenced by climate change – but hold knowledge that is relevant to policy-makers and complementary (not inferior) to scientific knowledge. This also entails recognising that local people are capable of producing rich data, including quantitative information. It finally requires tools making local knowledge tangible and communicable to external stakeholders, especially so that CCA becomes integrated into DRR. While CCA has evolved through top-down methods of assessment, DRR has commonly been based upon both top-down and bottom-up approaches, some involving local communities' knowledge for which a wide range of tools have been developed. Drawing on Participatory Learning Action (PLA) and Participatory Rural Appraisal (PRA) approaches, amongst others, these tools include ranking, scoring, Venn diagrams, timelines, problem trees, and participatory mapping. Toolkits like the Vulnerability and Capacity Analysis (VCA) have also been developed and are now widely used at community level. They resulted effectively in fostering dialogue and combining local and scientific knowledge in ways that strengthened DRR policies. Therefore, it would be more effective to embed CCA within existing DRR tools (Mercer 2010).

The utilisation of mutual tools might, for example, enable the comparison of hazard assessments between local communities and external experts. This would help in fostering dialogue amongst stakeholders that rarely work together. Moreover, scaling up such information at national level would help increase the quality and consistency of data reported to relevant agencies in charge of DRR, and, usually separate, CCA, thus strengthening the quality of current and future policies (Méheux *et al.* 2007). Additionally, the development of mutual tools under a framework that embeds CCA in DRR would help establish cooperation between scientists from climate change and disaster-related fields.

The use of common tools would therefore avoid duplication of efforts and would enable knowledge sharing on intricately related fields of research such as environmental studies, social science, volcanology, hydrology, gender studies, development studies, and economics. This may improve scientific knowledge and thus contribute to better informing policy-makers on ways to reduce disaster risks and generate sustainable development (Innocenti and Albrito 2011). In addition, mutual tools may not only enable scientists from disaster and climate change research to join together, sharing experiences and lessons learnt, but would also contribute to identifying gaps in knowledge. For national government agencies that generally operate separately, this may imply a better utilisation of budgets through connecting informed policies.

Concluding Statement

This chapter has reviewed how research and knowledge have contributed to shaping CCA and DRR policies. It emphasised that diversity of knowledge helps strengthen policies and that a greater consideration of local knowledge is required for DRR including CCA. This first requires recognising that local knowledge is valuable and not inferior to scientific knowledge – and vice versa. This also demands developing tools to make local knowledge tangible and communicable to scientists and policy-makers – and vice versa – so dialogue can occur, allowing local communities to participate in the framing of policies. Moreover, the chapter argues that CCA needs to be embedded within DRR in a common framework and use common tools, so diversity of knowledge can be mainstreamed in existing policies, and policies can be more effective in addressing vulnerability reduction and sustainable development goals.

References

Agrawal, A. (1995) 'Dismantling the divide between indigenous and scientific knowledge', *Development and Change* 26: 413–439.

Bankoff, G. (2001) 'Rendering the world unsafe: "Vulnerability" as western discourse', *Disasters* 25, 1: 19–35.

Campbell, J. (2006) *Traditional Disaster Reduction in Pacific Island Communities*, Wellington, New Zealand: GNS Science.

Chambers, R. (1983) *Rural Development: Putting the Last First*, London: Longman.

Chambers, R., Pacey, R. and Thrupp, L. (1989) *Farmer First: Farmer Innovation and Agricultural Research*, London: Intermediate Technology Publications.

Cornwall, A. (2011) *The Participation Reader,* London: Zed Books.

Cronin, S. J., Gaylord, D.R., Charley, D., Alloway, B.V., Wallez, S. and Esau, J.W. (2004) 'Participatory methods of incorporating scientific with traditional knowledge for volcanic hazard management on Ambae Island, Vanuatu', *Bulletin of Volcanology* 66: 652–68.

Freire, P. (1970) *Pedagogy of the Oppressed,* New York: Continuum.

Gaillard, JC and Mercer, J. (2013) 'From knowledge to action: Bridging gaps in disaster risk reduction', *Progress in Human Geography* 37, 1: 93–114.

Gaventa, P. and Cornwall, A. (2001) 'Power and knowledge', in P. Reason and H. Bradbury (eds), *Handbook of Action Research: Participative Inquiry and Practice,* Thousand Oaks, CA: Sage, pp. 70–80.

Heijmans, A. (2009) *The Social Life Of Community-Based Disaster Risk Reduction: Origins, Politics and Framing. Disaster Studies Working Paper 20,* London: Aon Benfield University College London Hazard Research Centre.

Hewitt, K. (1983) 'The idea of calamity in a technocratic age', in K. Hewitt (ed.), *Interpretations of Calamity from the Viewpoint of Human Ecology,* Boston: Hewitt, Allen and Unwin, pp. 3–32.

Innocenti, D. and Albrito, P. (2011) 'Reducing the risks posed by natural hazards and climate change: The need for a participatory dialogue between the scientific community and policy makers', *Environmental Science and Policy* 14: 730–733.

IPCC (Intergovernmental Panel on Climate Change). (2014) *Climate Change 2014: Synthesis Report,* Geneva: IPCC.

Kelman, I. and Gaillard, JC (2010) 'Embedding climate change adaptation within disaster risk reduction', in R. Shaw, J.M. Pulhin and J.J. Pereira (eds), *Climate Change Adaptation and Disaster Risk Reduction: Issues and Challenges,* Bingley: Emerald, pp. 23–46.

Le Dé, L., Gaillard, JC, Friesen, W. and Smith, F.M. (2015) 'Remittances in the face of disasters: A case study of rural Samoa', *Environment, Development and Sustainability* 17, 3: 653–72.

Marulanda, M.C., Cardona, O.D. and Barbat, A.H (2010) 'Revealing the socioeconomic impact of small disaster in Columbia using the DesInventar database', *Disasters* 34, 2: 552–70.

Mazzocchi, F. (2006) 'Western science and traditional knowledge', *European Molecular Biology Association* 7, 5: 463–468.

Méheux, K., Dominey-Howes, D. and Lloyd, K. (2007) 'Natural hazard impacts in small island developing states: A review of current knowledge and future research needs', *Natural Hazards* 40: 429–446.

Mercer, J. (2010) 'Disaster risk reduction or climate change adaptation: Are we reinventing the wheel?', *Journal of International Development* 22, 2: 247–264.

Mercer, J., Kelman, I., Taranis, L. and Suchet-Pearson, S. (2010) 'Framework for integrating indigenous and scientific knowledge for disaster risk reduction', *Disasters* 34, 1: 214–239.

Prince, S. (1920) *Catastrophe and Social Change, Based upon a Sociological Study of the Halifax Disaster,* PhD Thesis, New York: Columbia University Department of Political Science.

Shah, A. (2013) 'Participatory statistics, local decision-making, and national policy design: Ubudehe community planning in Rwanda', in J. Holland (ed.), *Who Counts? The Power of Participatory Statistics,* Rugby, UK: Practical Action Publishing, pp. 137–146.

Shaw, R., Pulhin, J. and Pereira, J.J. (eds) (2010) *Climate Change Adaptation and Disaster Risk Reduction: Issues and Challenges,* Bingley, UK: Emerald Group Publishing Limited.

UNFCCC (United Nations Framework Convention on Climate Change). (2007) *Climate Change: Impacts, Vulnerabilities and Adaptation in Developing Countries,* Bonn: UNFCCC.

UNISDR (United Nations International Strategy for Disaster Reduction). (2004) *Living with Risk: A Global Review of Disaster Reduction Initiative,* Geneva: UNISDR.

UNISDR (United Nations International Strategy for Disaster Reduction). (2005) *Hyogo Framework for Action 2005–2015: Building the Resilience of Nations and Communities to Disasters,* Geneva: UNISDR.

UNISDR (United Nations International Strategy for Disaster Reduction). (2015) *Sendai Framework for Disaster Risk Reduction 2015–2030,* Geneva: UNISDR.

Watts, M.J. and Bohle, H.G. (1993) 'The space of vulnerability: The causal structure of hunger and famine', *Progress in Human Geography* 17, 1: 43–67.

WCNDR (World Conference on Natural Disaster Reduction). (1994) *Yokohama Strategy and Plan of Action for a safer World. Guidelines for Natural Disaster Prevention, Preparedness and Mitigation,* Yokohama, Japan: United Nations Office for Disaster Reduction.

Wisner, B. (1995) 'Bridging "expert" and "local" knowledge for counter-disaster planning in Urban South Africa', *GeoJournal* 37, 3: 335–338.

PART IV

Governance

27

EDITORIAL INTRODUCTION TO PART IV: GOVERNANCE

Disaster Risk Reduction Including Climate Change Adaptation for All?

Ilan Kelman, Jessica Mercer, and JC Gaillard

Governance is systems, processes, and actions for developing, enacting, monitoring, enforcing, and evolving the rules and regulations used by people and social structures to function within society. Disaster risk reduction (DRR) including climate change adaptation (CCA) is a governance system that affects and is affected by other governance systems at multiple space and time scales.

The governance system of DRR including CCA is enacted by multiple players, with varying levels of interaction and coherence. Government, referring to a collection of institutions that is charged with formalising governance, usually over a spatially delineated area, is only one player amongst many. Other governance players are non-governmental organisations, which can range from charities helping people to terrorist groups; the private sector, which can be for-profit or not-for-profit since many businesses have not-for-profit divisions; community leaders and non-government governance structures such as elders' councils; households; and individuals. Even government, elected or not, has different branches, often described as being executive (writing, executing, and enforcing laws), legislative (writing and passing laws), and judicial (interpreting, upholding, and striking down laws). In some systems, these branches might overlap or coincide, rather than being entirely independent. Members of any branch might be elected, unelected, or a combination.

Governance scales, whether for government or not, can overlap and are not necessarily consistent within a jurisdiction. In Port of Spain in Trinidad and Tobago, elected councillors then elect a mayor, compared with Toronto, Canada where the mayor is directly elected by the people as are councillors. Savo in the Solomon Islands has a system of 'Bigmen' (chiefs) and elders who govern alongside decision-making from democratically elected representatives sitting in the provincial parliament. Scotland and Wales each currently have their own elected assemblies in addition to being governed by the parliament elected for the entire UK – as does Northern Ireland, but the Republic of Ireland also has some say in governing Northern Ireland. These jurisdictions have also been governed by the supranational entity, the European Union, which is likely to continue as a governance regime for the Republic of Ireland. Many countries have sub-national jurisdictions called provinces, states, territories, or commonwealths (amongst other terms) with diverse degrees of responsibility and power.

In places, governance scales concatenate. Many small countries have only one government, at the national level, which is smaller than many municipal governments in larger countries.

Governance nonetheless can happen locally, especially in some island states where a single government at the national level is responsible for many dispersed islands or communities.

Ultimately, the principle for DRR including CCA is frequently articulated as being subsidiarity: governance as locally as possible. In practice, local governance for DRR including CCA needs to be balanced with multi-scalar approaches. Not all jurisdictions have the resources, knowledge, or ability to deal with all principles, policies, and practices. Many baselines for DRR including CCA emerge from universal principles, such as human rights and duties. Any governance level can abuse power without oversight or checks and balances on the power.

Additionally, if a local institution implements governance with the full support of the community but without due regard to non-local considerations, then problems can result. This situation represents the classic upstream/downstream problem in environmental management and development: a town addresses its waste problem by dumping it downstream in the river, yet further downstream sits another town which receives the waste from the upstream town. For disaster risk, flood and drought risk management measures upstream, such as dams and channel engineering, impact downstream towns' options for governing their own flood and drought regimes.

On the topic of human rights, local attitudes might engender cultural approaches that do not support long-term DRR including CCA, such as discrimination. A local majority might decide that inequality based on gender, sexuality, caste, religion, ethnicity, disability, or other characteristics is desirable, even though it violates human rights, tends to perpetuate disaster risk, and frequently creates conflict. Participatory approaches and macro-governance principles entail treating all people equitably, with everyone contributing to and gaining from DRR governance including CCA. Where local preferences interfere with such principles, non-local approaches might be essential, not necessarily imposed forcefully, but wilfully directed with incentives and disincentives offered to adopt and enforce the principles.

Yet some argue that human rights are a Western construct and a form of neo-colonialism. Many DRR practitioners have been confronted with the classic dilemma of fostering 'universal' human rights to the detriment of local culture – and vice versa – despite respect for local culture and traditions now being a priority put forward by many development and DRR agencies. For example, in patriarchal societies, should food aid in emergencies be channelled through male elders or male heads of households in order to respect local culture, knowing that it would be detrimental to women and children who may then receive smaller amounts than they need and are due? Or should food aid bypass traditional leadership and values in order to target those who might be the most vulnerable from an outsider's point of view? Should DRR-related interventions include challenging/stopping or respecting 'cultural' practices such as female genital mutilation, sexual assault, and child abuse? In these circumstances, whose values matter most?

The chapters in this part do not presume that a single governance approach or scale for DRR including CCA would be universally successful. Instead, a balance is sought – always ensuring that checks and balances exist. Both governmental and non-governmental governance intersect, as would different scales of governance. At times, aspects of government may need to be bypassed to achieve successful governance for DRR including CCA. At other times, success would not be possible without full governmental support.

Involving everyone in DRR including CCA, and ensuring that the process serves everyone, means taking on board all the governance institutions covered by the chapters. International organisations including UN bodies plus regional intergovernmental and supranational institutions, national and sub-national level authorities, communities and neighbourhoods, non-governmental organisations, and the private sector are all part of the work. They are all involved in the executive, legislative, and judicial processes of writing, promulgating, monitoring, and enforcing laws, rules, and regulations, both formal and informal. As are groups with governance responsibilities that are not covered by individual chapters, such as the media, religious institutions, and sports, arts, and science councils.

Other chapters could have covered individual and household responsibilities for DRR including CCA alongside specific techniques for reducing power abuse and effecting power balances. No matter how careful the governance structures and processes, groups are frequently excluded; for instance prisoners, homeless, itinerants, and armed forces conscripts. Moral arguments are put forward regarding different degrees of rights to DRR including CCA. Prisoners have already lost some rights to freedom and livelihoods by violating laws – which presupposes that the laws are just and that the judicial system is fair. Should prisoners for petty theft out of hunger and for political activities be treated the same as rapists and murderers? It is also an ethical judgement regarding whether or not different citizens, irrespective of their status or past actions, should be accorded different levels of rights for DRR including CCA.

Several groups are labelled as being 'vulnerable', although options and opportunities vary within them. The elderly are often lumped together and said to be a 'vulnerable group', but the majority of governance leaders – political, financial, business, religious, and military – are far older than the average population, giving them power coupled with a lifetime of experience, even if they choose not to apply their experience towards appropriate use of power.

The overall governance lesson is that no institution, no authority, no group, and no community is homogeneous. Power relations and disagreements are ubiquitous. DRR including CCA for all, building from the local outwards, might work well in theory. The *Handbook* chapters in this part provide some details on policy and practice.

28

INTERNATIONAL ORGANISATIONS DOING DISASTER RISK REDUCTION INCLUDING CLIMATE CHANGE ADAPTATION

Juan Pablo Sarmiento

Intergovernmental organisations (IGOs) are entities created by states to facilitate cooperation, provide a forum for discussion, resolve disputes, and exchange information among nations (Brahm 2005). IGOs are formed when several lawful state representatives provide legal personality to the entity, or group, through a process of ratification. Brahm (2005) differentiates IGOs from other multilateral mechanisms, such as groups, coalitions (e.g., G8, G20, and treaties) or agreements (e.g., North American Free Trade Agreement). Others (YIO 2015) consider an IGO to be established by a treaty or agreement that acts as a charter creating the group and generates obligations between governments, whether the agreement is eventually published or not. The Yearbook of International Organizations 2015–2016 reports that there are 2,412 organisations under multilateral treaties and intergovernmental agreement categories (YIO 2015).

The United Nations (UN) is a prominent global intergovernmental organisation. The UN acts through six principal bodies: General Assembly, Security Council, Economic and Social Council, Secretariat, International Court of Justice, and the UN Trusteeship Council. Other UN specialised agencies, funds, programs, and entities cover multiple development and environmental themes.

The UN intergovernmental mechanisms that lead the discussion on disaster risk and climate change include the UN Office for Disaster Risk Reduction (UNISDR), the Intergovernmental Panel on Climate Change (IPCC), and the UN Framework Convention on Climate Change (UNFCCC). Other UN intergovernmental mechanisms that play a minor role but are still important in the disaster risk and climate change discussion include the World Meteorological Organization (WMO), UN Environment Programme (UNEP), and the UN Development Programme (UNDP).

By analysing IGOs within a broader context than those solely associated with disaster risk conditions, including climate change, we can visualise the various IGO dimensions: 1) the ideal and deliberative realm where the mechanism facilitates active participation among member states and other stakeholders (e.g., UNFCCC); 2) the technical or technocratic dimension generated within the mechanism itself, which does not necessarily involve member states (e.g., 'The UN

System Delivering as One'); and 3) other bilateral or multilateral mechanisms, not necessarily subordinated to or associated with the previous two (e.g., the US–Chinese agreement on emissions reductions announced in November 2014).

This chapter has four sections. The first section begins with a description of the context in which IGOs operate by addressing the topics of international disaster risk reduction (DRR) and climate regimes, the power balance within the international mechanisms, the common but differentiated responsibility, and the quest for flexible regimes. The second section on the UN system's role addresses the DRR and climate change cases. The third section deals with an inside analysis of the UN, the mandate, and the challenge of moving the system in one direction. Finally, a reflection on the coming future, looking for an integrated and comprehensive approach.

The IGOs and the International DRR and Climate Regimes

Some of the intergovernmental mechanisms associated with the UN deliberately took actions aimed at confronting disasters long before the UN General Assembly decided to move forward as a system with specific strategies to address them, as will be seen later in this chapter. Four agencies deserve mention: (1) The UN Children's Fund (UNICEF, formerly UN International Children's Emergency Fund). The end of World War II led to a humanitarian crisis where European children suffered from illness and starvation. UNICEF was established in 1946 to address this problem, becoming a permanent part of the UN in 1953. Since that time UNICEF has dealt with humanitarian needs, of children in particular, generated by conflict and natural hazards, and risk drivers exacerbated by the combined effect of climate change and urbanisation; (2) The World Food Programme (WFP). WFP was conceived in 1961 as a three-year experimental project, but began operations sooner than expected due to the occurrence of three major events: the 1962 Iran earthquake, a hurricane that hit Thailand the same year, and the resettling of 5 million refugees in the newly independent Algeria. WFP provided the urgently needed food at these disaster sites; (3) The World Meteorological Organization (WMO). WMO launched two pioneer projects related to disaster management, the World Weather Watch in 1963, and the Tropical Cyclone Project in 1971, which later became the Tropical Cyclone Programme; lastly, (4) the World Health Organization (WHO), which has responded to emergency situations within its portfolio since the late 1960s.

Two decades later, climate change was identified as a problem that compromises the entire planet, in which humanity is involved in both contributing to the problem and experiencing the consequences. Official UN mechanisms implemented to lead the action and face the problem began in 1988. These mechanisms focused on the study of the phenomenon itself through the IPCC. The efforts were then expanded, through the creation of the UNFCCC in 1992 at the Rio Summit, to deal with the social, economic, and political factors associated with climate change. Despite abundant scientific evidence on climate change, proposed efforts to tackle the problem have been inconsistent with the scale and urgency of the problem.

The institutionalisation process experimented with climate change differs from that of disaster risk. In the latter, the traditional approach, known as disaster management, involved acting in response to disasters. It evolved into a process focused on anticipating the consequences, identifying and characterising all types of hazards, determining the factors associated with the conditions of vulnerability, and creating probable risk scenarios, becoming the so-called disaster risk reduction. But the progress did not end there.

Risk reduction from the UN initially centred on a 'corrective' or 'compensatory' approach, where action was concentrated on intervention in existing vulnerabilities and the existing built environment. Today, the UN accepts what scientists had been publishing since the 1970s of the need to go beyond this compensatory focus and evolve into 'prospective' risk reduction, structurally modifying the patterns of development. This could be done enforcing disaster risk reduction (DRR) to ensure safety and sustainability of new settlements, expansion of existing settlements, and generally speaking, mainstreaming DRR in all public investment.

IGOs, and particularly the UN, have embraced the DRR strategy with the endorsement of the bilateral and multilateral cooperation. There is a strong consensus at all levels: global, regional, national and local. In the last five years, there has been strong cooperation from the private sector that has joined the global DRR effort through the ARISE mechanism, the Private Sector Alliance for Disaster Resilient Societies, building on and integrating two prior efforts: the Private Sector Partnership (PSP) and the Disaster Risk-Sensitive Investments (RISE) Initiative.

Power Balance within the International Mechanisms

The intergovernmental mechanisms are far from being perfect power-sharing schemes; they undergo a process of asymmetric participation that should be analysed, and in doing so, there is a need to address the North–South debate and the aspects of environmental justice, equity, and sustainability. The so-called North–South debate is characterised by positions and arguments at each end that reflect diametrically opposed philosophies. Two different dimensions and trajectories of this debate can be identified, the first related to disasters, and the second to climate change.

In the first case, disasters, the North – representing the donor community, with ostensibly robust emergency response and logistical capabilities – leads the response mechanisms offered through the UN, particularly the International Search and Rescue Advisory Group (INSARAG), the UN Disaster Assessment and Coordination (UNDAC), and the On-Site Operations Coordination Centre (OSOCC). While the South – representing the recipient community, overwhelmed by the disaster impacts and emerging needs – accepts this foreign collaboration with concern, which under the pretext of reaching a prompt and efficient response often duplicates and weakens the affected country's structures and competencies.

There are efforts seeking to change the imbalanced situation through capacity building in disaster response and coordination in countries exposed to different types of hazards, e.g., earthquakes, flash floods, tornadoes, and tsunamis. 'Recipient' nations, which have benefited from this capacity building, have recently participated in the above-mentioned UN response mechanisms in emergencies that have occurred in neighbouring countries. This trend could result in a change in the equation from North–South to South–South, North–North, and South–North where a response can be provided in a quick, efficient, effective and more sustainable way. Nevertheless, despite these advances, an 'assistentialism' approach prevails in the UN mechanisms mentioned, dominated by countries in the North.

In the case of climate change, the situation is more complex. The North – representing the wealthy world, ostensibly with accrued benefits, convening capacity, and geopolitical power – relies on the provision in which an old rule continues to apply to some existing situations, while a new rule will apply to all, known as the grandfathering rule. The South – represented by the less affluent world, ostensibly exposed to environmental challenges and climate change impacts, fragile, with limited resources and capacities to cope and adapt – takes a demanding position, entitled to compensation.

These two philosophies are vividly present in the IGO mechanisms, and strongly rooted in the issues of environmental justice, equity and sustainability. According to Ikeme, environmental justice is a 'term coined and extensively used in the scholarship that has focused on different exposure of minorities to environmental stresses and risks' (2003: 197). Inevitably, the mention of environmental justice is associated with less wealthy countries, which are often labelled as more exposed and susceptible to the effects of climate change than more affluent countries (Burton 1996; Smit and Pilifosova 2001; Thomas and Twyman 2005). This condition confirms the existence of an inequitable and uneven distribution pattern of adverse climate change impacts (Miller 1992) compounded by the existence of limited resources and a poor capacity to cope with and adapt to those changes. These are aspects considered procedural in less wealthy countries (Anand 2001). For Thomas and Twyman (2005), the major issue in environmental justice within the climate change debate is not limited to the distribution of impacts and costs, but includes the distribution of benefits, and with it, the distribution of responsibility.

Equity in the context of climate change has been strongly associated with the distribution of its effects and the need to ensure that the most vulnerable people are relieved of the burden of the impacts. However, the traditional IGO approach has fallen short of surpassing such 'assistentialism', in search of empowerment that transforms 'victims' into actors who make decisions and implement actions to properly adapt, contributing to their well-being and sustainable development.

It is clear that climate change occurs within a broader context in which societies in the less affluent world operate and that there are interactions between the prevailing socioeconomic conditions and climate change. Existing inequalities could be reinforced or, on the contrary, they could present an opportunity to achieve some level of progress, which is translated to equitable development, lessening the susceptibility to climate change (Beg *et al.* 2002; IPCC 2001; IPCC 2013). This last statement is supported by the fact that communities in less wealthy countries can be highly resourceful in facing external factors and disruptions.

IGOs and the Common but Differentiated Responsibility

Continuing with the Thomas and Twyman (2005) argument, the paradox is compounded when the North block takes the 'global common good' as its cause, imposing a new regime of conditions and restrictions on others, but after having achieved higher levels of wealth, economic growth, and quality and quantity of goods and services available. This position is usually exercised through the intergovernmental mechanisms. The South sees the new requirements limiting their opportunities for development, denying them the opportunity to grow without restrictions (which is only one development mode, but is often assumed to be the only one), unlike the North, which positioned itself globally as strong economic entities at the expense of a high environmental cost for the entire planet. In another contradiction, the North claims that past actions were made by actors without access to the scientific knowledge we have today. The South, in turn, claims that the rationale for the current discussion is based on the theme of sustainability, which implies the responsibility of one generation to the next. As such, the North must bear the consequences of the actions made by past generations through the compensation of the ecological debt by the depreciation of the common atmospheric capital base.

The intergovernmental mechanisms have difficulty maintaining an impartial position in the debate, and as already mentioned, they often reflect the imbalances given by their members. The differences between the constructions of the environmental justice of the North and South contribute to the polarisation and perpetuate the antagonistic positions within the IGO mechanisms, blocking the way to address the issue of a climate change consensus. However, as Göğüş (2014) indicates, the danger of oversimplifying the stalemate in climate change negotiations, attributing it to the rhetoric of the North–South conflict, blurs the understanding of the multiple causes and possible factors and barriers that influence the negotiations. Therefore, she suggests that the North–South lens is used as a means of informing and improving our understanding of the issue, but not as an end in itself.

The Quest for a Flexible Regime

The climate regime is the mode or system used to rule the global climate effort. So far, we have experienced a climate regime with a hybrid approach that seeks to balance national flexibility and an international discipline through bottom-up and top-down approaches. This hybrid characteristic aims to foster broader participation through flexibility, by giving countries enough space and autonomy in defining their policies, including their contributions and commitments. Simultaneously, the setup of an international framework, which provides a transparency and accountability mechanism, helps us to acknowledge and encourage advances beyond the commitments. The 2015 Paris Agreement met these criteria, and it became a unique milestone for the UN system and the Member States.

That being said, it is necessary to maintain flexibility in the climate regime, so it can be adjusted according to new scientific knowledge. It must also be based on mutual trust, with the conviction that everyone is contributing their fair share and that there is a clear political will to carry it out in a sustainable manner.

From another perspective, the DRR framework also requires a very flexible approach, which allows the consideration of DRR policies, guidelines and tools that facilitate their adaptation and implementation within local contexts subject to permanently evolving circumstances. This is well understood by most of the stakeholders involved, from the international community and IGOs to the regional and national authorities. Although there is a wide dissemination of success stories, best practices, and lessons to be learned, there is a mutual agreement on the need to better understand those DRR experiences and their contexts before one moves to adjust, adapt and then, adopt them.

While there are significant differences between DRR and climate change, one should also recognise the similarities shared by the two approaches, such as the need for innovation, adoption of new practices, and experimentation, through a contextualised and permanent learning process. Additionally, DRR's contribution to the strengthening of adaptive capacity to climate change, and the special attention required to foster interventions targeting the most vulnerable groups, has been vital. DRR and climate change each have valuable means that the other can benefit from. The climate change regime can learn the importance of equitable representation, accountability, and learning and action related to new opportunities. DRR can learn from the process of systematic and long-term adjustments, acting not only on risk conditions, but in a context where change is already happening (SEI 2014).

The UN System's Role

The DRR Case

The paths followed by the IGOs on issues of disasters and climate change were very different. The DRR path began at the local level, driven by events, involving practitioners and institutions, and relying on social economic, political, academic and scientific sectors. However, the climate change path commenced from and stayed at the academic–scientific level, reaching out to other sectors from there.

The UN initiatives geared to meet the challenges associated with disaster impacts have been typically reactive (Table 28.1). As already mentioned, UN agencies (UNICEF, WFP, WMO, and WHO) have undertaken concrete measures to face disasters since the mid-1940s. However, it was not until the mid-1960s that the General Assembly officially took direct responsibility, as indicated in the 'Milestones in the History of Disaster Risk Reduction' (UNISDR n.d.). The first action was after the 1962 Buyin-Zara (Iran) and the 1963 Skopje (Yugoslavia) earthquakes, and the devastating 1963 hurricane season in the Caribbean region. At that time the General Assembly (GA) requested member states to inform the Secretary-General about the relief and technical assistance that they could provide.

Table 28.1 UN Disaster Risk Reduction Milestones

Year	Disaster	UN Resolution	Topic of Action
1946	World War II		Post–World War II, UNICEF was established to help European children affected by famine and disease.
1962	Iran earthquake, Thailand Tropical Storm, and the Refugee crisis in the newly independent Algeria.		The World Food Programme (WFP) provided the urgently needed food at these disaster sites.
1963			The World Meteorological Organization (WMO) launched the World Weather Watch.
1962	Iran, Buyin-Zara earthquake. About 12,000 deaths.		
1963	Yugoslavia, Skopje earthquake. About 1,200 deaths.		
	Territories of Cuba, Dominican Republic, Haiti, Jamaica and Trinidad and Tobago struck by a hurricane.		
1965		Res. 2034	UN requests member states to inform the type of emergency assistance they are in a position to offer.
1968	Iran earthquake. About 10,000 deaths.	Res. 2378	UN specialised agencies support the reconstruction of devastated areas in Member States.
1970	Peru, Huascarán avalanche, estimated 20,000 deaths.		
1971		Res. 2816	The United Nations Disaster Relief Office (UNDRO) is established.
1974	Bangladesh, famine caused by massive flooding along the Brahmaputra river, about 27,000 deaths.		
1976	China, Tangshan earthquake, about 650,000 deaths.		
1985	Chile and Mexico, severe earthquakes.		
	Colombia, eruption of El Ruiz volcano. 23,000 deaths.		
	Ethiopia, famine (1984–1985). About 400,000 deaths. The high mortality is associated with the civil war, human rights abuses, and widespread drought conditions.		
1986	Ukraine, Chernobyl nuclear accident.		
1987		Res. 42/169	UN designates the 1990s as The International Decade for Natural Disaster Reduction.

(Continued)

Table 28.1 Continued

Year	Disaster	UN Resolution	Topic of Action
1991	Bangladesh cyclone, estimated death toll of 139,000.		
		Res. 46/182	International humanitarian coordination system.
1994		Res. 49/22	First World Conference on Disaster Reduction was held in Yokohama, Japan.
1998	1997–1998 El Niño Southern Oscillation Phenomenon. North Korea famine (1994–1998).		
2000		Res. 52–55/2000	The United Nations system designs an El Niño strategy.
	Severe famine impacts Ethiopia (1998–2000). Situation worsened by Eritrean–Ethiopian War.		
2004	Indian Ocean earthquake and tsunami, estimated death toll of 280,000. Congo (1998–2004), 3.8 million people died from starvation and disease during the Second Congo War.		
2005	Pakistan earthquake, about 75,000 deaths.		
		Res. 60/195	The Hyogo Declaration and the Hyogo Framework for Action 2005–2015: building the resilience of nations and communities to disasters.
2007		Res. 62/192	First session of the Global Platform on Disaster Reduction
2008	Sichuan earthquake, estimated death toll of 87,500. Myanmar, cyclone Nargis. About 84,500 deaths.		
2010	Haiti earthquake, estimated death toll of 160,000.		
2011	Tōhoku earthquake and tsunami, which generated the 2nd largest nuclear disaster after Chernobyl (1986).		
2012	Famine in Somalia, associated with the 2011 East Africa drought. Famine in West Africa, caused by the 2012 Sahel drought.		
2015		Res. 69/283	The Sendai Disaster Risk Reduction Framework 2015–2030 is adopted.
		Res. 69/284	UN establishes an intergovernmental expert working group on indicators and terminology relating to DRR.

Adapted from 'Milestones in the History of Disaster Risk Reduction', UNISDR https://www.unisdr.org/who-we-are/history

After an intense earthquake hit Iran in 1968, the GA requested the Secretary-General and heads of specialised agencies to support the reconstruction of devastated areas in Member States.

In 1971, the General Assembly established the UN Disaster Relief Office (UNDRO) as the specialised agency for disaster related affairs. For the first time, there were clear policies to promote the study of disasters, and to provide guidance on pre-event planning at national and international levels.

After severe disasters such as Mexico's earthquake and Colombia's eruption of Nevado del Ruiz, both in 1985, the decade of 1990 was identified by the GA as a time period in which the international community, led by the UN, would promote international cooperation in the field of the so-called 'natural disasters'. The Yokohama Strategy and its Plan of Action were adopted in 1994 at the first World Conference on Disaster Reduction which was held in Yokohama, Japan.

In 2000, as a reaction to the severe 1997–1998 El Niño Southern Oscillation phenomenon, the GA convened the UN system (i.e., Intergovernmental Oceanographic Commission of UNESCO, WMO, WHO, FAO, UNEP, UNDP, the World Climate Research Programme, and the International Council of Scientific Unions), to work on better understanding of El Niño, and to strengthen cooperation among regions and countries exposed to ENSO. Eventually, at the 2005 World Conference on Disaster Reduction held in Kobe, Japan, the Hyogo Declaration and the 'Hyogo Framework for Action (HFA) 2005–2015: Building the Resilience of Nations and Communities to Disasters' were adopted. The HFA was then endorsed by the International Strategy for Disaster Reduction. The ten-year strategy sought to mainstream DRR within development programmes at national and local levels; enhancing disaster risk studies and early warning initiatives; strengthening disaster risk knowledge and a resilience culture; reducing disaster risk drivers; and bolstering disaster preparation and response. The first session of the Global Platform for Disaster Risk Reduction took place in 2007. This permanent mechanism provides guidance for the HFA implementation, and a space for exchange of knowledge, capabilities, and experiences. The Global Platform has held biannual meetings since then, replacing the Inter-Agency Task Force for Disaster Reduction. The global platform is preceded by regional meetings, which in turn convene national platforms, focusing their attention on the HFA implementation.

Within the UN system the ISDR has been an articulating political and technical mechanism, which defines the UN role and scale of operations, embracing the coherence and coordinated action, including accountability and reporting (see Box 28.1), through the UN system called 'Delivering as One'.

Box 28.1   Disaster Risk Reduction in the United Nations

Juan Pablo Sarmiento[1]

[1] Florida International University, Miami, Florida, USA

The United Nations (UN) has committed to address disaster and climate risk throughout all of its development and humanitarian work.

The UN system has committed to:

- Making disaster risk reduction a priority for UN system organisations.
- Ensuring timely, coordinated and high quality assistance to all countries where disaster losses pose a threat to the health and development of the people that reside in these locations.
- Ensuring that disaster risk reduction for resilience is central to the post-2015 development agenda.

The contributions that drive the UN system's support to countries and communities to implement the Sendai Framework for Disaster Risk Reduction 2015–2030 are found in the UN Plan of Action on Disaster Risk Reduction for Resilience.

UN Role and Scale of Operations

The UN is one of the world's largest multilateral development partners, channelling 17 per cent of the total official development assistance. The UN is leading efforts to integrate disaster risk reduction into key sectors, such as agriculture, health, tourism and water.

The UN contributes to:

- Building national, regional and local/city capacities for disaster risk reduction in support of development and disaster recovery efforts.
- Undertaking research, producing earth observations and monitoring hazards, exposure and vulnerability.
- Generating weather and seasonal predictions to support preparedness and early warning systems.
- Setting norms, managing awareness campaigns, addressing underlying risk factors, and making risk informed investments.

Twenty-nine specialised organisations in the UN system contribute with their respective expertise, networks and resources to the reduction of disaster risk and collaborate to deliver as one at the global, regional and country level.

Thirteen UN organisations – FAO, UNDP, UNEP, UNFPA, UNHABITAT, UNICEF, UNOPS, WFP, WMO, WHO, UNESCO, UNV and the World Bank – have prioritised disaster risk reduction in their 2014–2017 strategic work plans and included disaster risk reduction in their results-based monitoring frameworks. This represents a 70 per cent increase, in comparison with the previous work planning cycle.

Coherence and Coordinated Action

In 2014, over 70 per cent of countries with UN programmes implemented all or some pillars of the Delivering as One approach. This approach gives the UN a more holistic capacity to address complex issues, such as disaster risk reduction at the country level. UN Resident Coordinators ensure that disaster risk reduction is effectively incorporated in the country-level programming and regularly report on the progress. In countries where a disaster risk reduction advisor was deployed to support the UN Resident Coordinator's Office, the UN provided more coherent and effective support to the national and local authorities. The pooling of capacities within UN Country Teams and the strengthening of regional inter-agency cooperation on disaster risk reduction are proving to be effective, particularly when these efforts are framed by regional intergovernmental policies and institutions.

The work of the UN Regional Commissions and Regional UN Development Groups, supported by the Regional Offices of UNISDR, is contributing to more effective and aligned assistance to countries. UNISDR is strongly committed to UN coherence and its primary objective: to achieve sustainable results in an effective and efficient way. UNISDR is the focal point in the UN system for the coordination of disaster reduction; it ensures synergy among the disaster reduction activities of the UN system and regional organisations (UN General Assembly resolution 56/195).

The UN High-level Committee on Programmes promotes system-wide cooperation, coordination and knowledge sharing across UN system organisations. The UN High Level Committee on

Programmes Senior Management Group on Disaster Risk Reduction supports the roll-out of the UN Plan of Action on Disaster Risk Reduction for Resilience, reviews progress against the UN Plan of Action and reviews the effectiveness of the UN Plan of Action implementation and delivery.

Accountability and Reporting

UN organisations have increased their accountability by adopting a single set of indicators to measure progress as they accelerate and mainstream disaster risk reduction into their operations. The indicators include minimum requirements for agencies to report on the extent to which disaster risk reduction is integrated in work plans, strategic frameworks, governing body agendas and agencies' own results based management systems.

The UN reports on its progress and impact of its work in countries in implementing disaster risk reduction in its annual Secretary-General Report to the UN General Assembly on the topic and against the UN's Quadrennial Comprehensive Policy Review (QCPR), which monitors the UN systems' operational development work.

Adapted from 'Disaster Risk Reduction in the United Nations Roles, Mandates and Results of Key UN Entities', by UNISDR © 2013. United Nations. Reprinted with the permission of the United Nations.

The Climate Change Case

As mentioned earlier in the chapter, the UN system has been working on the climate change adaptation (CCA) agenda through different mechanisms: (1) using direct channels with key specialised organisations (i.e., IPCC, UNFCCC, and ISDR); and (2) organising an internal, cohesive, and comprehensive strategy, which in the case of CCA is called 'Acting on Climate Change: The UN System Delivering as One', where the UN system has put together the broad spectrum of expertise and knowledge available within its own components to centre its action on priority areas and particular outcomes.

The 'Delivering as One' approach is part of a broader effort, introduced in the late 1990s, to revisit the UN as an institution and examine the operational fragmentation which had come about over decades. UN engagement was initially conceptualised as three mutually reinforcing pillars: (1) peace and security (hence the importance of the war and conflict background), (2) development in the sense of economic and social development, and (3) human rights (Gallagher 2015).

The Acting on Climate Change initiative encompasses forty UN specialised agencies, funds, programmes, and other entities involved in climate change. The mechanisms they use vary, from internal partnerships within the UN system to involving other international organisations, civil societies, private sectors, academic institutions and scientific institutions. The strategy provides broad depth and scope on the core development themes, in addition to widespread coverage at the national, regional, and global levels.

Most of the UN activities associated with climate change fall under the Working Group on Climate Change, led by the WMO. Established in 2007, this working group was part of the High Level Committee on Programmes, one of three pillars mentioned under the UN system Chief Executives Board for Coordination, directed by the UN Secretary-General. There is a close coordination with UNFCCC and other UN system bodies (Box 28.2). The Working Group on Climate Change also cooperates with regional and national structures.

Box 28.2   United Nations System Mandates on Climate Change

Juan Pablo Sarmiento[1]

[1] Florida International University, Miami, Florida, USA

- Caring for Climate advances the role of business in addressing climate change.
- The Economic and Social Commission for Asia and the Pacific promotes low-carbon green growth as a key strategy for addressing climate change.
- The Food and Agriculture Organization helps to achieve food security for all by making agriculture, forestry and fisheries more productive, sustainable and resilient in the face of climate change.
- The International Atomic Energy Agency publishes reports on the potential role of nuclear energy in climate change mitigation and the use of nuclear science in assessing climate change impacts.
- The International Civil Aviation Organization climate mandate and environmental goal is to limit and reduce GHG emissions from international civil aviation.
- The International Fund for Agricultural Development is committed to building up the climate resilience of smallholder farmers in developing countries through managing competing land-use systems, while at the same time reducing poverty, enhancing biodiversity, increasing yields and lowering emissions.
- The International Labour Organization works with its tripartite constituency (workers, employers and governments) to link the eradication of poverty, sustainable development, climate change and green jobs.
- The International Maritime Organization contributes to international efforts to reduce atmospheric pollution and address GHG emissions from international shipping, including through the effective implementation of mandatory energy-efficiency measures for ships.
- The International Organization for Migration focuses on human mobility, climate change, disaster-risk reduction and adaptation.
- The International Telecommunications Union promotes the use of information and communications technologies (ICTs) to address climate change through technical standards, ICT-enabled applications and radio communications for climate monitoring.
- The Intergovernmental Oceanographic Commission, through international cooperation, aspires to help its Member States to increase resilience to climate change and variability and enhance the safety, efficiency and effectiveness of all ocean-based activities through scientifically founded services, adaptation and mitigation strategies.
- The United Nations Convention to Combat Desertification builds the resilience of ecosystems and communities by fostering adaptation at the landscape level based on sustainable land-management practices.
- The United Nations Development Programme provides programming and policy support that addresses the impacts of climate change, putting countries on the path towards low emissions and climate-resilient development.
- The United Nations Educational, Scientific and Cultural Organization enhances and applies the climate change knowledge base for building green societies through climate change education, science, culture and communication.
- The United Nations Environment Programme supports climate resilience, low-emission pathways, ecosystem-based adaptation, clean and renewable energy and technologies, and the greater awareness of climate change science for policymaking and action.

- The United Nations High Commissioner for Refugees leverages evidence and enhances knowledge on, and understanding of, human mobility prompted by climate change, including the protection of the most vulnerable populations.
- The United Nations Institute for Training and Research empowers people, organisations and countries to respond to the challenges of global climatic change through the design and delivery of individual learning, backed by strategic advice and capacity development for national education and training institutions.
- The United Nations Office for the Coordination of Humanitarian Affairs brings together humanitarian actors to ensure a coherent response to climate-related emergencies and disasters.
- The United Nations Population Fund supports countries in integrating population dynamics and data into climate action and helps build the resilience of individuals and communities, including through achieving universal access to sexual and reproductive health and gender equality and female empowerment.
- UN Women works with partners to ensure gender-responsive climate action through norm-setting, policies and programmes. It helps strengthen the capacity of females to cope with climate change impacts.
- The World Bank Group works to leverage both public and private sources of climate finance to support climate-smart policies and investments and help countries and businesses adapt to a changing climate.
- The World Food Programme has a climate-change focus on building the resilience of the most food-insecure people and countries against increasing climate risks.
- The World Health Organization provides evidence and technical guidance and directs approaches to strengthen health resilience to climate risks, and gain health benefits from climate mitigation.
- The World Meteorological Organization is the UN system's authoritative voice on the state and behaviour of the Earth's atmosphere, its interaction with the oceans, the climate it produces and the resulting distribution of water resources.

Adapted from 'How the United Nations System Supports Ambitious Action on Climate Change', by UN-CEB © 2014. United Nations. Reprinted with the permission of the United Nations.

This coordination focuses on areas such as adaptation, technology transfer (an old paradigm), emissions reduction, financing mitigation and adaptation measures, and capacity building. So far, the agenda has primarily been defined by UN agencies with a technocratic approach, reconciling the missions of each of the different UN agencies and programmes.

A Mandate Mosaic

Core Themes Addressed

In the intersection between disasters and climate there are several criteria to define the themes to be addressed by the IGOs, including: (1) the processes involved; (2) topics of negotiation; (3) priority areas of intervention; or (4) associated issues. According to the processes involved, the topics include: observed changes and their causes, future climate change, risk and impacts, disaster risk reduction and adaption, and sustainable development.

Topics of negotiation can be seen in several fora, such as the 'Durban Platform', which emerged from the UN climate conference held in Durban in December 2011. The core elements of the Durban Platform negotiations included: mitigation, adaptation, finance, technology, capacity building, and the transparency of action and support. Priority intervention areas comprise: the sustainability of water resources, global energy economy, conventional and alternative energy sources, technology, policy, rapid climate change including changes in the mean and variation in temperature and precipitation, effects on managed and natural ecosystems, biodiversity, agroeco-systems, and disaster risk reduction.

Associated labelled issues encompass: understanding of the climate system, anticipating impacts of climate, adapting – reducing human vulnerability to the impacts of climate change–implementing disaster risk reduction and risk transfer strategies, mitigating climate change or lessening its sever-ity, strategies to reduce greenhouse gas emissions, global water cycle, and sea level rise.

The Durban Platform is an example of the intergovernmental mechanisms' limitations, allow-ing interesting technical discussions on core topics but with serious political difficulties to reach binding agreements. Although the documents generated constitute 'legal instruments', the way they were written allows signatory countries to have their own interpretations, heavily influenced by domestic political developments and agendas.

The Challenge of Moving the System in One Direction

According to Gallagher (2015), the international cooperation is much more complex now than before the 2009 Copenhagen Summit, when the 15th annual UN Climate Change Conference took place. The complexity of the issue lies not only in the uncertainty of scientific forecasts, but also in the unpredictability and instability of countries' positions and policies. The private sector, including companies and investors, have lost confidence in their governments to implement and maintain policies and to fulfil the commitments and proposed objectives. This instability and loss of governance sends mixed signals to the various stakeholders, creating a 'political risk' with higher cost of capital, resulting in a lower overall investment, accompanied by a change in the rules that govern the trade of high- and low-carbon emission assets (Gallagher 2015). Nevertheless, the 2015 Paris Agreement showed the convenience of the hybrid approach to rule the global climate effort, where it is possible to balance national flexibility and an international discipline through bottom-up and top-down approaches – although most provisions in the 2015 Paris Agreement are not legally binding. We will now see if the required endorsement process at national levels happens, in order to comply with their contributions and commitments. The IGOs became the forum for discussion, and they will be in charge of the monitoring and evaluation process, within the transparency and accountability required.

In contrast with the traditional top-down approaches, where the policy discourse associated with climate change aims to generate a behavioural change, extreme events, interpreted as facts of climate change materialisation, have been elements of persuasion for society, allowing the association of the scientific discourse with the lay person and everyday realities, generat-ing powerful bottom-up approaches. A good example is what happened in 2012 on the east coast of the United States, when Hurricane Sandy struck the northeast coastline, resulting in a strong popular support for the climate change policy promoted by the Obama Administration, irrespective of the reality of how climate change did or did not influence Hurricane Sandy. As noted by Gallagher (2015: 9), 'This realization of the impacts of climate change can empower atypical actors to become more politically engaged'. Parallel to the mentioned support for environmental policies after Hurricane Sandy, the US government decided to push through its DRR policy with a change in the hurricane warning system, based so far on the hurricane wind component, to now incorporate the more dangerous effect of tropical storms: the storm surge. This task, carried out by the National Oceanic and Atmospheric Administration (NOAA) agency, has generated a profound transformation in several aspects of emergency management, and will have serious implications for DRR strategies associated with local development, such as land use, zoning, location and design of critical facilities and lifelines, and building codes, among others.

As has been shown in the 2015 Paris Conference, shifts in national discussions of strategic countries have an impact on the international negotiations, strengthening regional and global governance mechanisms (UNFCCC 2015).

But it is not appropriate to focus the discussion exclusively on climate hazards and climate change. Other hazards reinforce the importance of processes that initiate and materialise at local levels. Just analyse the accelerated urban risk construction, which through unregulated growth, occupies river beds, wetlands, mangroves, unstable slopes, and nature reserves, resulting in a high exposure of the population to hazards such as landslides, floods, and geological faults. Frequently, the environmental impact exceeds the carrying capacity of the ecosystem, a fact that is accompanied by other risk factors such as segmentation, marginalisation, livelihoods' fragility, overcrowding, poor access to services, etc. In cases like those described, IGOs advocate and support policies and mechanisms aimed at strengthening decentralisation, deconcentration, devolution, and empowerment at local level. The Sendai framework 2015–2030 (UNISDR 2015) is an excellent example of the IGOs' role in this type of approach.

Looking for an Integrated and Comprehensive Approach

The UN system is experiencing permanent and continued adjustments by trying to keep pace with the world's changes. The so-called post-2015 agenda has serious implications involving the UN system and governments, as well as the private sector, civil society and individuals. This can be seen with the High-Level Panel on the Post-2015 Development Agenda named the 'global partnership'.

Several reform processes of the UN Development System took place simultaneously in 2015:

- The Sendai Disaster Risk Reduction Framework, adopted in Sendai in March 2015;
- The Addis Ababa Action Agenda on Financing for Development, adopted by the General Assembly in July 2015;
- The UN summit for the adoption of the post-2015 development agenda, held in New York in September 2015; and
- The 21st session of the Conference of the Parties to the UN Framework Convention on Climate Change (COP21) organised in Paris from 30 November to 11 December 2015.

Although there has been a serious attempt to connect these efforts, segregation and disarticulation persists, where these processes are not directly linked to each other. Consequently, the challenges, the complexity of the problems, and the shortcomings of international governance mechanisms will be a constant in the coming years.

The outcomes of the 2015 negotiations should now be followed by policy- and decision-makers, including urban and infrastructure planners, financiers, investors, businesses, industrial entrepreneurs, regulators, and local administrators, among others. All of them are ultimately responsible for viable policies through their implementation, which implies a real transformation of development processes, a low carbon world limiting global warming and unquestionable progress in building resilience, particularly at the urban level. This multiplicity of actors and initiatives addressing disaster risk and climate change at national and international levels demonstrates a significant change in the dynamics of public policies.

The international governance system, particularly the UN system, has been strongly challenged in recent decades in an atmosphere where interdependence and globalisation break into vital areas of the world economy generating risk issues (e.g., financial stability, resource trades, and food prices). Then comes the dilemma for governments, whether these risks should be managed through multilateral mechanisms, or through unilateral approaches (Lee *et al.* 2012).

More recently, a minilateralist approach is proposed (Falkner 2015), involving reducing the number of actors involved in multilateralism and dealing with defined subject matters, as an alternative solution for the international power asymmetries. While the minilateralist approach does not eliminate the structural barriers to global and ambitious agreements, it offers a space to intensify the political dialogue within the multilateral negotiations, allowing advancement in agreements and coalitions, and reducing the dangers of free-riding. To a certain extent, it is a legitimation of the global issues as climate and DRR regimes, characterised by a complex context of profound and continuous changes of power. The minilateralist approach could inject a political momentum in stalled international processes, providing collective leadership options and innovative approaches that contribute to global governance.

Finally, it is critical to understand the heterogeneous nature and the power imbalances that govern the intergovernmental institutions, and the current global social, political and economic uncertainties that they need to face. These power imbalances and the uncertainties that IGOs are often not willing to face might be the biggest challenges for the UN and other IGOs fully implementing DRR including CCA.

References

Anand, P. (2001) 'Procedural Fairness in Economic and Social Choice: Evidence from a Survey of Voters', *Journal of Economic Psychology* 22: 247–70.

Beg, N., Morlot, J.C., Davidson, O., Afrane-Okesse, Y., Tyani, L., Denton, F., Sokona, Y., Thomas, J.P., La Rovere, E.L., Parikh, J.K. and Rahman, A.A. (2002) 'Linkages between Climate Change and Sustainable Development', *Climate Policy* 2: 129–44.

Brahm, E. (2005) 'Intergovernmental Organizations (IGOs). Beyond Intractability', in G. Burgess and H. Burgess (eds), *Conflict Information Consortium*, University of Colorado, Boulder: Conflict Information Consortium. Online http://www.beyondintractability.org/essay/role-igo (accessed 6 November 2015).

Burton, I. (1996) 'The Growth of Adaptive Capacity: Practice and Policy', in J. Smith, N. Bhatti, G. Menzhulin, R. Benioff, M.I. Budyko, M. Campos, B. Jallow and F. Rijsberman (eds), *Adapting to Climate Change: An International Perspective*, New York: Springer, pp. 55–67.

Falkner, R. (2015) *A Minilateral Solution for Global Climate Change? On Bargaining Efficiency, Club Benefits and International Legitimacy*, London and Leeds: Centre for Climate Change Economics and Policy (Working Paper No. 222) and Grantham Research Institute on Climate Change and the Environment (Working Paper No. 197).

Gallagher, L. (2015) *Political Economy of the Paris Climate Agreement*, Third Generation Environmentalism's (E3G's) Climate Diplomacy programme. Online http://act2015.org/ACT_2015_FINAL_Political_Framing.pdf (accessed 9 October 2015).

Göğüş, S. (2014) *Understanding Impasse in Climate Change Negotiations: The North-South Conflict and beyond*, University of East Anglia, UK: Climate Exchange. Online https://climate-exchange.org/2014/02/06/understanding-impasse-in-climate-change-negotiations-the-north-south-conflict-and-beyond/ (accessed November 9, 2015).

Ikeme, J. (2003) 'Equity, Environmental Justice and Sustainability: Incomplete Approaches in Climate Change Politics', *Global Environmental Change* 13, 3: 195–206.

IPCC (Intergovernmental Panel on Climate Change). (2001) *Climate Change 2001*. Cambridge: Cambridge University Press.

IPCC (Intergovernmental Panel on Climate Change). (2013) *Climate Change 2013 – The Physical Science Basis: Working Group I Contribution to the Fifth Assessment Report of the Intergovernmental Panel on Climate Change.* Cambridge: Cambridge University Press. Online http://ebooks.cambridge.org/ref/id/CBO9781107415324 (accessed 28 December 2015).

Lee, B., Preston, F., Kooroshy, J., Bailey, R. and Lahn, G. (2012) *Resources Futures*, London: Chatham House. Online https://www.chathamhouse.org/sites/files/chathamhouse/public/Research/Energy,%20 Environment%20and%20Development/1212r_resourcesfutures.pdf (accessed 8 September 2016).

Miller, D. (1992) 'Distributive Justice: What the People Think', *Ethics* 102, 3: 555–93.

SEI (Stockholm Environment Institute). (2014) *Climate Change and Disaster Risk Reduction. Background Paper prepared by SEI for the 2015 Global Assessment Report on Disaster Risk Reduction*, Geneva, Switzerland: UNISDR. Online http://www.preventionweb.net/english/hyogo/gar/2015/en/bgdocs/SEI,%202014. pdf (accessed 3 January 2016).

Smit, B. and Pilifosova, O. (2001) 'Adaptation to Climate Change in the Context of Sustainable Development and Equity', in IPCC (ed.), *Climate Change 2001. Impacts, Adaptations and Vulnerability*, Cambridge: Cambridge University Press, pp. 879–967.

Thomas, D.S.G. and Twyman, C. (2005) 'Equity and Justice in Climate Change Adaptation amongst Natural-Resource-Dependent Societies', *Global Environmental Change* 15, 2: 115–24.

UN-CEB (United Nations System Chief Executives Board for Coordination). (2014) *How the United Nations System Supports Ambitious Action on Climate Change, The United Nations System Delivering as One on Climate Change and Sustainable Development*, New York: UN-CEB. Online http://www.unsceb.org/ content/how-un-system-supports-ambitious-action-climate-change (accessed 20 December 2015).

UNFCCC (United Nations Framework Convention on Climate Change). (2015) *Conference of the Parties: Adoption of The Paris Agreement, FCCC/CP/2015/L.9*, Paris: UNFCCC. Online http://unfccc.int/ resource/docs/2015/cop21/eng/l09.pdf (accessed 13 December 2015).

UNISDR (United Nations International Strategy for Disaster Reduction). (2013) *Disaster Risk Reduction in the United Nations: Roles, Mandates and Results of Key UN Entities*, Geneva: UNISDR. Online http://www. unisdr.org/files/32918_drrintheun2013.pdf (accessed 8 September 2016).

UNISDR (United Nations International Strategy for Disaster Reduction). (2015) *Sendai Framework for Disaster Risk Reduction 2015–2030*, Geneva: UNISDR.

UNISDR (United Nations International Strategy for Disaster Reduction). (n.d.) *Milestones in the History of Disaster Risk Reduction*, Geneva: UNISDR. Online https://www.unisdr.org/who-we-are/history (accessed 3 January 2016).

YIO (Yearbook of International Organizations). (2015) *Yearbook of International Organizations 2015–2016, Vol. 1 (A and B): Organization Descriptions and Cross-References*, Brussels, Belgium: Brill.

29

UN INSTITUTIONS DOING DISASTER RISK REDUCTION AND CLIMATE CHANGE ADAPTATION

UNISDR, UNFCCC, and IPCC

Christophe Buffet and Sandrine Revet

Introduction

Disaster risk reduction (DRR) became a subject of strong, intense, and contentious international political negotiations for perhaps the first time in its history during the Third World Conference on Disaster Risk Reduction organised by the United Nations in Sendai, Japan, in March 2015. Some of the key notions of the Sustainable Development Goals and the climate change agenda appeared substantively in the DRR agenda. An example is the 'Common but differentiated responsibilities' (CBDR) principle, enshrined in the 1992 United Nations Conference on Environment and Development's Rio Declaration and mentioned in the UNFCCC principles (Article 3.1). Though the DRR meetings had been rather consensual in the past (Revet 2012), the Sendai negotiations witnessed tensions between less wealthy and more affluent countries.

These tensions surprised many observers. They may be a logical consequence of the growing interactions between climate change and DRR issues that have been a subject of many discussions and controversies among scientists, practitioners and policy-makers across both fields. Conceptually, some consider climate change adaptation (CCA) as a subset of DRR, both being placed under sustainable development (Kelman 2015; Kelman and Gaillard 2010).

A transformation of this conceptual perspective into institutional reforms would be sensitive. First, it may imply that climate change be placed under weaker legal regimes than the UNFCCC, with the risk that commitments to reduce emissions become emptier than ever. Second, the coalitions of less wealthy countries – such as G77+China, Least Developed Countries (LDCs), and the Alliance of Small Island States (AOSIS) – would accept losing their bargaining power gained in the UNFCCC negotiations. Third, the tensions that arose in Sendai represent a current danger that climate-related political dissent would spread and undermine the entire sustainable development umbrella regime. A deeper integration of CCA and DRR in a unified institutional arrangement would raise complex questions related to the legal status of a new international regime and its capacity to offer a deliberative and effective space that takes into account asymmetries of power and a diversity of interests – without being jammed by geopolitical dissents.

To explore these issues, this chapter examines the roles played by three DRR and CCA international institutions: UNISDR, UNFCCC, and IPCC. It examines advantages and limitations of their separation and differing mandates.

UNISDR: From 'Natural' Disaster Prevention to
Disaster Risk Reduction

Historically, 'natural' disaster management has been a national matter, led by emergency experts specialising in saving lives and managing crises (Knowles 2011). When disaster prevention started to emerge, it was framed for a long time mainly through efforts to predict natural hazards and to protect from them, especially through constructing infrastructure (e.g., dams and dikes). The institutionalisation at an international level of what nowadays is called DRR is comparatively recent, starting in the 1970s, at the crossroads of different historical moments.

First, a moment when the international humanitarian sphere was developing, with its actions becoming more complex as the number of actors on the scene increased. Coordination of disaster management became the keyword, and an Office of the United Nations Disaster Relief Co-ordinator (UNDRO) was created in 1971 in order to address this challenge. During the same period, the media were developing. New modes of information were appearing, images were becoming more present and were producing their effects on the world. Big crises such as the 1970 Bhola Cyclone in Bangladesh and the 1970 earthquake and rockslide in Peru caused international reactions from both governments and the public.

With its quite weak mandate to 'coordinate' relief actions of other UN agencies and its low budget, UNDRO had no capacity to work on disaster prevention. Yet, social scientists, most of whom were working in less wealthy countries, started significant work on disasters in the 1970s, with new perspectives shedding light on the vulnerability of societies, showing that root causes of disasters were as important as, if not more important than, the natural phenomena and should therefore be confronted in order to reduce disaster impacts (O'Keefe *et al.* 1976).

Yet, it was not until the end of the 1980s that the International Decade for Natural Disaster Reduction (IDNDR running from 1990 to 1999) began, voted by the United Nations General Assembly in 1989 (Resolution 44/236, December 1989) with the objective to 'reduce through concerted international action, especially in developing countries, the loss of life, property damage and social and economic disruption caused by natural disasters' (Annex, Article A.1). IDNDR started with the strong influence of earth scientists' and engineers' framing. In 1994, the first World Conference organised by IDNDR was held in Yokohama, Japan, with its title including the phrase 'Natural Disaster Reduction'. It proposed and adopted the 'Yokohama Strategy and Plan of Action for a Safer World: guidelines for natural disaster prevention, preparedness and mitigation' (UNIDNDR 1994) in which the influence of earth sciences, engineers and technology was still strong and the focus remained on hazard reduction. NGOs and social scientists involved in disaster prevention and vulnerability reduction programmes strongly criticised this conference for its technical perspective (Davis and Myers 1994). These critiques and others opened the door for a first turning point at the international level.

The idea of DRR progressively replaced the idea of Natural Disaster Reduction in these institutional spheres. Thanks to the mobilisation of some social scientists and activists, changes started to appear in the involvement of NGOs and other local actors in the international processes. The UN International Strategy for Disaster Reduction (UNISDR) was launched in Geneva in 1999, at the end of IDNDR, by the UN General Assembly, as an 'inter-agency platform, inter-agency task force and inter-agency-secretariat for disaster reduction' (Resolutions 54/219, article 4 and 56/195, article 3). Placed under the direct authority of the Under-Secretary-General for Humanitarian Affairs, UNISDR had a low budget and the mandate to 'serve as a main forum within the United Nations system for devising strategies and policies for disaster reduction and ensure complementarity of action by agencies involved in disaster reduction, mitigation and preparedness' (Resolution 56/195, p. 2). In 2003 and 2004, UNISDR's Secretariat carried out a review of the Yokohama Strategy.

The next important turning point occurred in 2005. The review of the Yokohama Strategy, together with the 26 December 2004 tsunamis that preceded the second World Conference on Disaster Reduction held in Kobe in 2005, led to a strong critique and an evaluation of the first years of the strategy. The Hyogo Framework for Action 2005–2015 (HFA; UNISDR 2007) voted in during the Kobe conference was the first UN framework related to DRR. Its objective was to 'build the resilience of nations and communities to disasters' (UNISDR 2007). Global Platforms have been organised every two years in Geneva since 2007, in order to create a 'forum for Member States and other stakeholders to assess progress made in the implementation of the Hyogo Framework for Action, enhance awareness of disaster risk reduction, share experiences and learn from good practice, identify remaining gaps and identify actions to accelerate national and local implementation' (Resolution 62/192, article 15). From 2009 onwards, a global report called the Global Assessment Report (GAR) on DRR has been produced for each Global Platform, presenting both new UNISDR campaigns and thematic ideas, often based on country reports and scientific studies. The weight of scientific expertise within the GAR has been increasing since 2009, and the 2015 GAR has been elaborated on scientific bases with peer-reviewed papers and scientific teams in charge of reviewing each part.

UNFCCC: From Scientific Alerts to Institutional Proliferation

The institutionalisation of climate change occurred later but faster than that of DRR. Climate change began to emerge as a major issue during the 1970s, such as through Charney's report to the US National Academy of Sciences (Oreskes *et al.* 2008). The topic became institutionalised at the international level with the first World Conference on Climate in 1980 in Geneva, organised by the World Meteorological Office (WMO) and United Nations Environment Programme (UNEP). After scientists raised concerns about 'global warming', a series of international conferences transformed the topic from a scientific issue into a policy issue. Intergovernmental negotiations led to the creation of both the Intergovernmental Panel for Climate Change (IPCC) in 1988 and the United Nations Framework for Climate Change Convention (UNFCCC) at the Rio Earth Summit in 1992.

From the beginning, continuous dialogue has characterised climate governance through the UNFCCC. In addition to the annual Conferences of Parties (COPs), at least four intersessional sessions are organised during the year. In comparison, major international conferences for DRR are held every 10 or 15 years, with biennial Global Platforms taking place in Geneva since 2007. Contrary to the DRR frameworks, which are voluntary, COP sessions aim at reaching legally binding agreements such as treaties and protocols.

The consensus rule prevails, which tends to lead to agreements corresponding to the lowest common denominator. Each country is granted one vote, but each is organised through different groups, the biggest being G77+China (132 countries). Contrary to UNISDR conferences prior to the Sendai conference, COPs are the scenes of high geopolitical tensions and an underlying atmosphere of defiance, with delegations often suspecting others of playing a 'free rider' game by trying to benefit from the other's efforts.

The Progressive Emergence of CCA

While the ultimate objective of the UNFCCC, specified in Article 2, remains a 'stabilization of greenhouse gas concentrations in the atmosphere at a level that would prevent dangerous anthropogenic interference with the climate system', the Convention also stipulates that wealthy countries grouped in Annex I are committed to supporting countries considered to be 'particularly vulnerable'. Article 4.8 draws a vague list of countries (coastal territories, small island states, disaster-prone countries, among others) that have 'specific needs and concerns (. . .) arising from the adverse effects of climate change and/or the impact of the implementation of response measure's. CCA progressively gains importance from the beginning of the 2000s through an arduous path that combines scientific and political arguments.

The successive reports of IPCC demonstrate a change of framing that influences the COP agendas. From 'an unacceptable, even politically incorrect, idea' (Burton 1994: 14) that might divert attention from the core issue of mitigation, adaptation becomes the subject of a 'realist' perception considering that it is intrinsically linked with mitigation efforts (Schipper 2006). From the IPCC's Fourth Assessment Report (IPCC 2007), mitigation and adaptation appear as the two faces of the same coin: it becomes necessary to avoid the unmanageable and manage the unavoidable.

In the negotiations, pressures from G77+China, AOSIS and NGOs led Annex I Countries to overcome their fear that CCA costs could become a bottomless pit, and to discuss institutional arrangements and finance for adaptation. Although G77+China often presents a unified front, it faces complex games for non-Annex I countries to improve their bargaining power. LDCs benefit from the weight of G77+China, but 'emancipation' impulses appear, from states considering that their own survival becomes progressively uncorrelated with the interests of the so-called 'emerging economies', all the more so since Brazilian, South African and Chinese per capita emissions now surpass those of some Annex I countries such as France. The 'Most Vulnerable Countries' Forum or the 'V20' group – a deliberate comparison to the G20 – thus try to highlight their particular interests, without being official Groups within the UNFCCC.

'Climate Justice' or 'Solidarity'?

Though they admit their 'historical responsibilities', affluent countries continuously strive to frame CCA as a matter of 'solidarity', which echoes with an imaginary of Western countries generously sharing a part of their wealth with the poorest ones. The IPCC nevertheless demonstrates that the countries least responsible for causing climate change will actually be the most affected by it, initiating a 'climate justice' framing. On their side, G77+China refer to a 'polluter pays' principle that would correlate finance for CCA with impact costs, even though these are challenging to assess robustly.

This approach progressively led to what can be summed up as 'chequebook diplomacy': the transaction consists of balancing insufficient reduction measures from Annex I countries with funding pledges for CCA. Whereas adaptation finance appears much larger than for DRR, CCA nevertheless remains the 'poor relative' of the negotiations compared to mitigation. OECD/CPI (2015) estimates that current overall climate financing was as high as USD 62 billion in 2014, but only 16 per cent is devoted to CCA. This figure is then far from the commitment of Annex I countries to provide less wealthy countries with USD 100 billion per year by 2020, which is meant to be balanced between climate change mitigation and CCA. Moreover, climate funds are not supposed to substitute for current development aid, but their real additionality is questioned, especially since assessment methodologies vary widely. It also appears that a majority of these funds consists of loans that will be reimbursed by beneficiary countries, instead of grants.

In 2015, methodologies to assess current climate funding caused fierce debates during COP21 in Paris. Consequently, UNFCCC's Subsidiary Body for Scientific and Technological

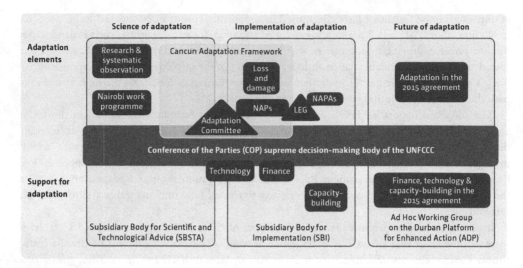

Figure 29.1 Institutional Structure on Adaptation

(By UNFCCC 2013)

Advice (SBSTA) has been mandated to propose a common methodology by 2018. If this work leads to a stricter accounting, then it may create a divergence between actual adaptation finance flows and the Paris Agreement's pledge of 'a new collective quantified goal from a floor of USD 100 billion per year' (FCCC/CP/2015/10/Add.1/53) after 2025, exacerbating political differences.

An institutional architecture of CCA was also progressively set up, with a complexity directly linked to continuous and politically tense negotiations. Twenty years of COPs have thus led to a proliferation of entangled structures under the UNFCCC, dealing with research, policy advice (SBSTA, LDC Expert Group), work programmes, capacity building and funding (Figure 29.1).

This institutional proliferation under the aegis of UNFCCC contributes to a broader adaptation network in which coordination, transparency and effectiveness capacity come into question (Smith *et al.* 2011). It also raises questions about institutional inertia. Built on the mitigation–adaptation dichotomy and on successive geopolitical compromises, this complexity makes it all the more difficult to bridge the gap between UNFCCC and UNISDR.

The Role of Expertise: IPCC and Global Assessment Reports

UNISDR and UNFCCC built up largely in parallel, each following their own respective paradigms. A major and often underlined difference between the two institutions is in that UNFCCC deals with only climate-related phenomena, whereas UNISDR considers a much broader range of hazards (including earthquakes, tsunamis, volcano eruptions, and those which are weather- and climate-related amongst many others). Conversely, both institutions share characteristics that are similar to all international processes, such as harmonisation, standardisation, global data and knowledge production, elaboration of some shared vocabulary, and mobilisation of the international community. UNFCCC and UNISDR bear other similarities and have other differences that can be underscored, in particular regarding the type of science and expertise that they mobilise.

UNFCCC and IPCC: A Complex Science–Policy Nexus

As regards climate change, the institution labelled as having scientific expertise (IPCC) and its first report in 1990 precede UNFCCC, whereas the first Global Assessment Report (GAR) for DRR was published in 2009, almost twenty years after the launch of IDNDR and almost ten years after UNISDR began. IPCC responded to some degree to the necessity of producing sound scientific assessments so as to face 'merchants of doubt' (Oreskes and Conway 2010) who were (and still are) trying to deny any human responsibility for witnessed climate change. It also provided political impetus to act. According to Article 7 of UNFCCC, obligations of the parties and institutional arrangements should periodically be examined in COP sessions 'in the light of the objective of the Convention, the experience gained in its implementation and the evolution of scientific and technological knowledge'. Decisions (or lack of decisions) in COPs are thus systematically put in perspective with the IPCC reports. Those reports do not propose new scientific work, but a synthesis of the 'state of the art' of climate change knowledge, essentially from peer-reviewed publications, associated with degrees of confidence and the identification of research gaps to be filled.

IPCC has established itself as the major reference for all matters related to climate change science. Working Group I deals with climate and biosphere science, Working Group II deals with impacts on the biosphere and socio-economic system (vulnerability and adaptation), and Working Group III deals with climate change mitigation. Its position of reference was gained thanks to a long and detailed process of reviewing leading to regular procedural changes in response to mistakes and critics. North–South balance among Lead Authors was improved after criticisms about a Science of the North dealing with a problem of the North; climatologists were included in Working Group II; and stricter rules about including grey literature were implemented for the Fifth Assessment Report published in 2013 and 2014 in response to several mistakes that had tarnished the Fourth Assessment Report.

The successive IPCC reports have punctuated the evolution of different framings of climate change challenges and accompanied major decisions in COPs. For example, the first chapter about CCA in the Third Assessment Report (IPCC 2001) supported the Marrakesh Agreement that launched the Adaptation Fund. The Fourth Assessment Report (IPCC 2007) contributed to reinforcing the CCA theme, which was included as one of the four pillars of the Bali Roadmap launched the same year at COP13. The UNFCC/IPCC tandem thus contributes to a science–policy nexus that is characteristic of the climate change regime.

Consequently, can it be considered that the climate regime is driven by science? The answer remains ambivalent. Following its motto to be 'policy-relevant but not policy-prescriptive', the IPCC officially claims to remain distant from political issues. This role of a 'boundary organisation' between science and policy is specifically devoted to the Subsidiary Body for Scientific and Technological Advice (SBSTA) under UNFCCC. However, the IPCC cannot pretend to embody 'pure science' fully detached of any political issue. Science and Technology Studies (STS) have shown that the interplay between science and policy represents a dynamic that is much more complex than a linear model of science speaking truth to power which is often put forward by both scientists and policy-makers (Dahan 2008). Whether the management of the science–policy boundary has been effectively secured by SBSTA remains debatable, as well as its effectiveness on policies (Hulme and Mahony 2010).

Experts are appointed by their respective governments. Although IPCC assessments themselves are produced outside of direct influence from member governments, each chapter is subject to a review by scientists and governments. The IPCC Summaries for Policy Makers are approved line-by-line by government representatives. The objective to limit the rise of global mean temperature to below 2 °C above pre-industrial levels is then less a result of IPCC work than a political decision, in particular from the EU (Tol 2007). This gap between science and policy in the climate regime is widening even after the Paris Agreement affirming that Parties will pursue 'efforts to limit the temperature increase to 1.5 °C above pre-industrial levels' (FCCC/CP/2015/10/Add.1/Art.2). Successfully supported by countries that relied on IPCC to claim that they would suffer in a temperature rise of above 1.5 °C, this objective is perceived as unrealistic by many scientists who doubt that a 2 °C threshold itself is still within reach (Jordan *et al.* 2013).

In this sense, the IPCC can be considered as a half-success. It has contributed to demonstrate scientifically the reality of anthropogenic climate change and how to avert major consequent problems. On the other hand, political commitments move away from sustainable pathways sketched in IPCC reports, demonstrating that both the climate change mitigation gap and the CCA gap (UNEP 2014) are widening. In spite of these limits, scientific works remain an overarching source for advocacy from all driving forces of the negotiations ('vulnerable' countries, civil society, countries willing to engage, etc.) in order to spur on greater ambitions in the climate regime.

UNISDR and Global Assessment Reports: Integrating Science and Policy

UNISDR edits a 'Global Assessment Report' (GAR) every two years with the objective of 'monitoring risk patterns and trends and progress in disaster risk reduction' (from https://www.unisdr.org/we/inform/gar). Experts and scientists have participated in the GAR from the beginning, yet their increasing weight and role is notable. The 2015 edition (UNISDR 2015) was written with the help of many scientists, and has been inspired by the IPCC process of peer-reviewed papers and scientific committees.

Scientific work has changed the general framing of UNISDR. During the Cold War era and the first steps of internationalising DRR prevention in the late 1980s, a strong influence of seismology and American hazard-oriented disaster studies was seen on international understandings of disasters. This can be explained by the role played by some seismologists – such as Frank Press – and hazard-centred social scientists – such as the members of the Disaster Research Centre at the University of Delaware and the Natural Hazards Centre at the University of Colorado, Boulder – in the first international scientific committees put in place by the US National Science Foundation in order to consider an IDNDR (Cabane and Revet 2015).

Social sciences embedded in development studies (Maskrey 1993; Wisner *et al.* 1977) nevertheless made a significant contribution in theorising the concept of vulnerability, which showed that disasters are not the result of natural hazards, but depend on the level of human development and the processes leading to and maintaining these levels. This research progressively contributed, throughout the 1980s and mainly in the 1990s, to transforming the definition of disasters and disaster-related policy instruments, leading to a stronger focus on poverty and development, adding to the natural and technical sciences (Revet 2012). The integration of some of the key researchers of this social science trend into international networks culminated in the 2000s with their participation in international institutions such as UNISDR, the World Bank, UNDP, UNU (United Nations University), and ICSU (International Council for Science).

This social and environmental paradigm has been inserted into national and international DRR policies from the 1990s, whereas in the climate change field, bottom-up CCA began to emerge only at the beginning of the 2000s, questioning the top-down, impact-based and technocratic paradigm that had prevailed so far. Bottom-up CCA, presented as the second generation of CCA approaches (Burton *et al.* 2002), is anchored in local contexts, knowledge and socio-political aspects and emphasises contextual vulnerability and adaptation. Through this approach, CCA progressively became closer to the 'social paradigm' that DRR researchers developed in the 1970s. Conceptual debates have contributed to enriching both fields, by wondering whether CCA is reinventing the DRR wheel (Mercer 2010), or through comparisons of core definitions like vulnerability in UNISDR's and IPCC's glossaries (Kelman and Gaillard 2010). These debates stay vivid and were particularly fierce for the 2012 IPCC report on 'Managing the Risks of Extreme Events to Advance Climate Change Adaptation' that Norway and UNISDR requested IPCC to write.

Those debates fundamentally reflect crucial differences between the types of knowledge that are mobilised, more particularly the status of models in the different epistemic communities.

Models and Human Dimensions for DRR and CCA

General Circulation Models (GCMs) emerged as the privileged tool for climate studies as early as the 1970s and have played a major role for climate change alerts and framing. The community of climate modellers succeeded in imposing its methodology as being the most 'credible' or even the most 'rational' and met the politically globalised scale of COPs (Demeritt 2001). This globalising approach was fiercely criticised for its homogenisation of the world by climate science. GCMs convey an 'impersonal, apolitical and universal imaginary of climate change projected by science [that] comes into conflict with the subjective, situated and normative imaginations of human actors engaging with nature' (Jasanoff 2010: 233). Therefore, GCMs' globalising impetus reconstructs a past climate and projects it into the future while bypassing history, culture, and local knowledge forms.

For a long time, UNISDR work was fundamentally based on data about past disasters at a national scale. The CRED EM-DAT (http://www.emdat.be) database was the main provider of this data, although Latin American social scientists, gathering data on smaller and more common disasters, had developed Desinventar (http://www.desinventar.org) based on such data during the 1990s (Revet 2015). Recently, a new model has been developed by UNISDR, and has appeared in the last two GARs (GAR 2013 and 2015) which, rather than gathering data on past disasters, aims to give a 'global vision' of forthcoming risks and their costs.

In this probabilistic vision, potential future disasters are modelled in each country in order to build a global database of disaster risks based on standardised data (UNISDR 2015). Data are then translated into three main metrics: Loss of Human Life Years, describing the time required to produce economic development and social progress; Probable Maximum Loss, 'which represents the maximum loss that could be expected within a given period of time' (UNISDR 2015: 56); and Average Annual Loss, which is calculated based on 'all the events that could occur over a long time frame' (UNISDR 2015: 56).

This focus on modelling to provide a global vision of forthcoming risks in GAR seeks to match the apparent effectiveness of climate models in reaching policy-makers. Nevertheless, it also imports into DRR major drawbacks that STS has highlighted since the 1990s. Models can be perceived as predictive and apolitical truth machines, whereas they should remain a knowledge base to be placed in a large array of motivations and meaning, and in social, ethical and political preoccupations (Wynne and Shackley 1994). Should they become as prominent as they previously were in climate framing, models in GAR could become instrumental and depoliticise DRR. It would represent a step backwards compared with the social and environmental paradigm, while IPCC is making inroads in following the opposite direction, partly under DRR's influence. Although modelling keeps an overbearing status in climate knowledge, the fifth IPCC report (2014) particularly insists on social and political factors of vulnerability, such as inequalities and marginalisation often produced by uneven development processes.

Conclusion: Bridging the Gap?

Growing interactions between CCA and DRR at a conceptual level underline the artificial institutional divide at the international scale. UNISDR, UNFCCC and the Development Goals process (the Millennium Development Goals from 2000–2015 and the Sustainable Development Goals from 2015–2030) largely work as and in silos despite increasing communication among them. Some steps forward have been made. The Hyogo Framework for Action (2005–2015) acknowledged that climate change can be an underlying threat in relation to disasters, and the Bali Action Plan adopted in 2007 during COP 13 refers to enhancing action on adaptation, including disaster reduction strategies.

The year 2015 demonstrated significant cross-over, being marked by the Sendai Framework for DRR (SFDRR) in March, the Sustainable Development Goals in New York at the UN General Assembly in September, and COP21 in Paris in December. It could have provided a greater integration of sustainable development, DRR, CCA and climate change mitigation. However, though SFDRR merely recommends more 'coordination' and 'coherence' among the various processes, it fails to propose concrete methodologies to achieve them. By reaffirming UNFCCC's pre-eminence on climate change issues (as do the Sustainable Development Goals), SFDRR even entrenches artificial differences and divisions in the process. Similarly, the Paris Agreement barely quotes the SFDRR and only refers to 'comprehensive risk management strategies' in the Loss and Damage section.

So as to overcome the phenomenon of institutional silos, a pathway could be to consider CCA as a subset of DRR, with both being placed under the umbrella of sustainable development (Kelman 2015; Kelman and Gaillard 2010). Operationalising this conceptual view into an international framework would be an arduous task in light of the respective characteristics of CCA and DRR regimes, especially that they are both underpinned with different legal statuses, procedures, objectives, architectures, finance flows and bodies of expertise.

In addition to the complexity in bridging the institutional gap, the thorniest issue towards a more unified regime could come from international relations. It is highly unlikely that any state would agree to place all sustainability goals, from poverty alleviation to justice, as well as their DRR policies, into an international legally binding agreement similar to what UNFCCC seeks. These topics are primarily presumed to be a matter of national prerogatives as illustrated by the challenges that UNFCCC has faced in achieving an international, legally binding treaty.

In the hypothetical case in which an extended development goals regime would follow UNFCCC procedures, the rules on consensus and one-country-one-vote would give the possibility for any country to scrutinise and criticise other countries' commitments – and, most likely, eventually block an agreement. The Sendai conference in 2015 exemplified how climate change-related disagreement can spread. It is easy to imagine endless debates about respective responsibilities in such an international regime aimed at dealing with all aspects of sustainability in a legally binding manner. The capacity of such a regime to reach any agreement would be at stake.

Yet, it is unclear how mitigation, adaptation, and their overlaps would move forward if they were considered within a voluntary Sustainable Development Goals regime, with meetings such as world conferences held every 10 or 15 years and deprived of real negotiations between countries.

For all these reasons, designing a unified development goal regime incorporating DRR, CCA and climate change mitigation would not only mean reconciling deep institutional differences between UNISDR and UNFCCC. It would also imply a global political will for launching a long-term process in which each country is ready to assess and re-negotiate its interests for reaching a strong, fair, effective and wide-ranging regime.

References

Burton, I. (1994) 'Deconstructing adaptation . . . and reconstructing', *Delta* 5, 1: 14–15.

Burton, I., Huq, S., Lim, B., Pilifosova, O. and Schipper, E.L. (2002) 'From impacts assessment to adaptation priorities: The shaping of adaptation policy', *Climate Policy* 2, 2–3: 145–159.

Cabane, L. and Revet, S. (2015) 'La cause des catastrophes. Concurrences scientifiques et actions politiques dans un monde transnational', *Politix* 28, 111: 47–67.

Dahan, A. (2008) 'Climate expertise: Between scientific credibility and geopolitical imperatives', *Interdisciplinary Science Reviews* 33, 1: 71–81.

Davis, I. and Myers, M. (1994) 'Observations on the Yokohama World Conference on Natural Disaster Reduction, 1994', *Disasters* 18, 4: 368–372.

Demeritt, D. (2001) 'The construction of global warming and the politics of science', *Annals of the Association of American Geographers* 91, 2: 307–337.

Hulme, M. and Mahony, M. (2010) 'Climate change: What do we know about the IPCC?', *Progress in Physical Geography* 34, 5: 705–718 .

IPCC (Intergovernmental Panel on Climate Change). (2001) *IPCC 3rd Assessment Report: Climate Change 2001*, Geneva, Switzerland: IPCC.

IPCC (Intergovernmental Panel on Climate Change). (2007) *IPCC 4th Assessment Report: Climate Change 2007*, Geneva, Switzerland: IPCC.

IPCC (Intergovernmental Panel on Climate Change). (2014) *IPCC 5th Assessment Report: Climate Change 2014*, Geneva, Switzerland: IPCC.

Jasanoff, S. (2010) 'A new climate for society', *Theory, Culture & Society* 27, 2/3: 233–253.

Jordan, A., Rayner, T., Schroeder, H., Adger, N., Anderson, K., Bows, A., Quéré, C.L., Joshi, M., Mander, S., Vaughan, N. and Whitmarsh, L. (2013) 'Going beyond two degrees? The risks and opportunities of alternative options', *Climate Policy* 13, 6: 751–769.

Kelman, I. (2015) 'Climate change and the Sendai Framework for Disaster Risk Reduction', *International Journal of Disaster Risk Science* 6, 2: 117–127.

Kelman, I. and Gaillard, JC (2010) 'Embedding climate change adaptation within disaster risk reduction', in R. Shaw, J.M. Pulhin and J.J. Pereira (eds), *Climate Change Adaptation and Disaster Risk Reduction: Issues and Challenges*, Bingley: Emerald, pp. 23–46.

Knowles, S.G. (2011. *The Disaster Experts. Mastering Risk in Modern America*, Philadelphia, PA, USA: University of Pennsylvania Press.

Maskrey, A. (ed.) (1993) *Los Desastres No Son Naturales*. Panama: La Red de Estudios Sociales en Prevención de Desastres en América Latina (La Red).

Mercer, J. (2010) 'Disaster risk reduction or climate change adaptation: Are we reinventing the wheel?', *Journal of International Development* 22, 2: 247–264.

O'Keefe, P., Westgate, K. and Wisner, B. (1976) 'Taking the naturalness out of natural disasters', *Nature* 260, 5552: 566–567.

OECD/CPI (The Organisation for Economic Co-operation and Development/Climate Policy Initiative). (2015) *Climate Finance in 2013–14 and the USD 100 billion goal*, Paris, France: OECD/CPI.

Oreskes, N. and Conway, E.M. (2010) *Merchants of Doubts: How a Handful of Scientists Obscured the Truth on Issues from Tobacco Smoke to Global Warming*, New York, NYC, USA: Bloomsbury Press.

Oreskes, N., Conway, E.M. and Shindell, M. (2008) 'From Chicken Little to Dr. Pangloss: William Nierenberg, global warming, and the social deconstruction of scientific knowledge', *Historical Studies in the Natural Sciences* 38, 1: 109–152.

Revet, S. (2012) 'Conceptualizing and confronting disasters: A panorama of social science research and international policies', in F. Attina (ed.), *The Politics and Policies of Relief, Aid and Reconstruction. Contrasting approaches to disasters and emergencies*, London: Palgrave Macmillan, pp. 42–55.

Revet, S. (2015) 'Compter et raconter les catastrophes', *Communications* 96: 81–92.

Schipper, L. (2006) 'A conceptual history of adaptation in the UNFCCC process', *Review of European Community and International Environmental Law (RECIEL)* 15, 1: 82–92.

Smith, J., Dickinson, T., Donahue, J., Burton, I., Haites, E., Klein, R. and Patwardhan, A. (2011) 'Development and climate change adaptation funding: Coordination and integration', *Climate Policy* 11, 3: 987–1000.

Tol, R.S. (2007) 'Europe's long-term climate target: A critical evaluation', *Energy Policy* 35: 424–432.

UNEP (United Nations Environment Programme). (2014) *The Adaptation Gap Report 2014*. Nairobi, Kenya: UNEP.

UNFCCC (United Nations Framework Convention on Climate Change). (2013) *The State of Adaptation under the United Nations Framework Convention on Climate Change,* Bonn, Germany: United Nations Climate Change Secretariat.

UNIDNDR (United Nations International Decade for Natural Disaster Reduction). (1994) *Yokohama Strategy and Plan of Action for a Safer World: Guidelines for Natural Disaster Prevention, Preparedness and Mitigation,* Geneva, Switzerland: UNIDNDR.

UNISDR (United Nations International Strategy for Disaster Reduction). (2007) *Hyogo Framework for Action 2005–2015: Building the Resilience of Nations And Communities to Disasters, Final Report of the World Conference on Disaster Reduction (A/CONF.206/6),* Geneva: UNISDR.

UNISDR (United Nations International Strategy for Disaster Reduction). (2015) *Global Assessment Report on Disaster Risk Reduction 2015,* Geneva, Switzerland: UNISDR.

Wisner, B., O'Keefe, P. and Westgate, K. (1977) 'Global systems and local disaster: The untapped power of people's science', *Disasters* 1, 1: 47–57.

Wynne, B. and Shackley, S. (1994) 'Environmental models: Truth machines or social heuristics?', *Globe* 21: 6–8.

30

REGIONAL ORGANISATIONS DOING DISASTER RISK REDUCTION INCLUDING CLIMATE CHANGE ADAPTATION

Ian O'Donnell

Introduction

Although there is not a long-standing set of literature on the role of regional organisations in supporting disaster risk reduction (DRR), including addressing climate risks through climate change adaptation (CCA), a number of comparative studies have been published in the last several years, looking at the roles that inter-governmental regional organisations have played in support-ing DRR / CCA capacities at national level.(Ferris and Petz 2013; Hollis 2015). In addition, the United Nations International Strategy for Disaster Reduction (UNISDR) has published a series of reports outlining regional achievements toward the Hyogo Framework for Action (HFA) that also highlight action and contributions of inter-governmental regional organisations on DRR (UNISDR online collection of regional progress reports).

While the term 'regional organisation' is often used to refer to inter-governmental organ-isations within specified geographies, this chapter takes a wider view of the term, looking at the DRR, including CCA, role of regional organisations that may be governmental, scientific, technical, business and trade, or civic in nature as well as regional organisations that exist outside the context of traditional geographic regions. Such organisations also range along the continuum between individual institutions and networks.

This chapter explores the following aspects of regional organisations:

- Rationales or affinities for regional collaboration
- Organising structures of regional organisations

Rationales or Affinities for Regional Collaboration

Catchment Areas

Much of the focus on regional organisations to date has looked at regions that are contiguous geographic areas and share commonalities or relations in language, culture, trade, and politics. This aspect of regionality has provided a strong relevance to DRR, including CCA, by addressing

Table 30.1 Examples of Regional Coordination of DRR / CCA

Common efforts	Examples
• risk analysis and assessment	Central America Probabilistic Risk Assessment (CAPRA)
• training and preparedness	Training curriculum offered to member states by the Caribbean Disaster Emergency Management Agency (CDEMA)
• development of joint mitigation strategies	Mekong River Commission
• mutual support in disaster response	Coordinating role played by the Association of Southeast Asian Nations (ASEAN) in the Cyclone Nargis response in Myanmar, which was later formalised in the establishment of the ASEAN Coordinating Centre for Humanitarian Assistance on disaster management (AHA Centre)

the transboundary impacts and consequences of hazards that cross national boundaries. Natural hazards are often related to features in the environment (e.g., earthquake faults along mountain ranges, floods along river basins, etc.) that are natural catchment areas for local ecosystems that may cross national boundaries but are linked in common environmental and social systems. Trade and transport links have extended the size of such ecosystems, providing increased opportunities for growth and development. However, they also create interdependencies that increase exposure to some hazards. The SARS outbreak in the early 2000s and H1N1 Avian Flu outbreaks in 2009–10 demonstrated how modern trade and transport links have extended exposure pathways for pandemic risks (Ear and Campbell 2012).

The transboundary aspects of such hazards provide a direct rationale for collaboration for the regional coordination of DRR / CCA among countries linked by shared hazards and shared vulnerabilities due to regional inter-dependencies (Table 30.1). For other stakeholders as well, regional planning and coordination on DRR including CCA concerns may align with protecting the operation of regional ecosystems, whether they are environmental, social, or trade-based. Corporations in particular are likely to rely on strategies and business models that align to regional trade patterns, supply chains, and markets. Similarly non-governmental organisations targeting environmental or social issues are interested to build collective capacity to address hazards that span across borders and vulnerable groups that may be affected by regional threats or may themselves cross borders to escape threats. Regional organisations or networks bring an important context as a mediating layer for organising between national/local and global levels. This is especially important with globalisation and the rise of transnational corporations and organisations, including international non-governmental organisations and religious organisations.

Regional Public Goods

In essence, delivering DRR / CCA services in these types of transnational catchment areas provides a regional public good. As globalisation increases and emphasises transnational links that are not state-centred, the relevance of national sovereignty becomes less important, increasing interest in global or regional public goods. Regional public goods provide a mechanism for transnational problem-solving for issues that have distinct regional characteristics as well as for challenges that have been neglected at global level (Hettne and Söderbaum 2006). The climate change agenda is an example where regional progress has sometimes outpaced global commitments over the last twenty years.

Regional public goods operate either to maximise the distribution or 'spillover' of positive benefits to participating countries or to limit or reduce the distribution of negative impacts from a threat or hazard. Often, achieving public goods is not an economic problem but a political problem, which sometimes may be easier to solve at regional level (Hettne and Söderbaum 2006).

Regional public goods may accrue through the actions of regional organisations to either support national members to pursue national goals (with aggregated benefits, economies of scale, etc.) or to pursue distinct regional goals through regional cooperation. The value of regional cooperation can be amplified in organisations or networks that draw in multiple stakeholders and provide for complementary vertical cooperation between global, regional, national, and local levels and not just horizontal cooperation between countries (Hettne and Söderbaum 2006).

The reason that this aspect of regional cooperation is so important is that it responds to the systems-based nature of many of the issues of risk to which DRR, including CCA, are addressed. To the extent that regional organisations or networks are able to bring together actors across levels and from diverse institutional backgrounds, they can provide the mix of stakeholders that are necessary to use systems approaches to both analyse risks and identify and design solutions.

In this context, effective DRR, including CCA, will be an essential part of the toolkit for regional organisations to address:

a   direct service gaps in humanitarian emergencies – e.g., as in Cyclone Nargis;
b   market failures and opportunities to optimise risk management resources – which may be addressed through e.g., shared training or risk pooling;
c   gaps in political will and conflicting political agendas that limit opportunities for joint advocacy and action to reduce risks – which may be addressed e.g., through development of dialogue mechanisms like the Asian Ministerial Conference on DRR (AMCDRR) – to reduce the siloisation that has existed between DRR, CCA, and other humanitarian and development sectors;
d   interdependencies and vulnerabilities within new 'networked geographies' like city clusters that are not just poles within their own countries but increasingly nodes with in regional and global networks, connected by trade and supply chains, transport links, migration patterns, and cultural affinities.

Collective action through regional organisations can help to strengthen funding channels; address market gaps (e.g., in insurance and training provision); enable alignment in planning and coordination, addressing externalities that exist beyond national borders (such as natural hazards that affect wider catchment areas and cross-border population displacements); and address regional catchment areas, e.g., sensor networks for tsunamis, earthquakes, and flooding and storm forecasting and tracking networks.

Sovereignty and Subsidiarity

The principles of sovereignty and subsidiarity are just as important in understanding the way that risk decision-making is governed as in other areas of collective governance. Sovereignty is recognised at the national level, meaning that national governments are ultimately accountable for protecting their citizens from significant risks. The principle of subsidiarity on the other hand states that governance and collective decision-making should take place as close as possible to the local level.

The regional level is important because, as discussed earlier, regions are often catchment areas for risks either due to natural conditions (e.g., shared borders within a river basin) or socio-economic factors (e.g., shared pandemic risks due to trade and transit routes). While there are cases where subsidiarity is appropriate at the regional level and requires supra-sovereign solutions, the value of bringing a regional approach to subsidiarity is not necessarily in finding the single most appropriate level but promoting increased cooperation between levels to address regional issues and concerns (Hettne and Söderbaum 2006).

Such solutions are usually developed through cooperation among states rather than formal transfer of decision-making by treaty to regional organisations. Examples include the research and planning projects that have been undertaken by the Mekong River Commission and Zambezi Watercourse Commission (Box 30.1). In this sense regional organisations can be important catalysts to solutions whether they have executive decision-making authority ('actorness' as described by Hettne and Söderbaum 2006) or rely on consensus decision-making among their members.

Box 30.1   Mekong River Commission and Zambezi Watercourse Commission

Ian O'Donnell[1]

[1] Red Cross Red Crescent Global Disaster Preparedness Center, Washington, DC, USA

These two river basin organisations were established to bring together neighbouring states to monitor and implement integrated water resource management (IWRM) strategies for their respective river basins. The Mekong River Commission (MRC) was established officially in 1995, although precursors existed as far back as 1957 as the Mekong Committee and Interim Mekong Committee. The Zambezi Watercourse Commission (ZAMCOM) came into being in 2011.

Like river basin organisations (RBOs) in other parts of the world, these organisations were founded around the idea that river basins are a single ecological unit irrespective of the national borders that may cut through them (Hettne and Söderbaum 2006). They balance a variety of programme implementation and coordination roles to improve flood prevention and protection, improve water quality, and protect groundwater resources in their respective basins (Schmeier 2010).

The MRC itself has largely functioned as a technical institution to implement regional research and planning projects. Much of the technical expertise has been housed within the MRC secretariat. The majority of the funding resources for MRC's activities since its inception have come from donor governments and multi-lateral organisations outside the region. As its member states have developed richer resource and income bases, there is growing interest in evolving the MRC and to potentially involve member states to a greater extent in policy as well as technical cooperation (Schmeier 2010).

ZAMCOM is charged with the coordination of river basin policy among its member states and the development of a basin-wide master plan. The organisation was only recently ratified and made operational, so its track record on promoting regional cooperation is still quite young.

While such organisations have helped to identify regional environmental and ecological interdependencies and even highlight areas of unrealised potential for regional cooperation in IWRM, there are still significant incompatibilities among states' interests in energy production – which can also affects flood risk and CCA, e.g., when drawing down reservoir levels to reduce flood risk conflicts with power generation needs from hydropower plants connected to the same reservoir. Thus, the mandate for regional RBOs is often relatively narrowly focused on sustainable management of water resource (Linn and Pidufala 2008) rather than broader 'decision-making, participation and enforcement mechanisms to facilitate the common good' (Hettne and Söderbaum 2006: 221).

Virtualised Regions

Just as the argument is increasingly made that local communities are not just geographically defined but may also include other types of communities of identity, so 'regions' may not always be geographic as well. While regions may be organised to address common interests around natural ecosystems, they may also be organised around trade, linguistic, and cultural ecosystems that are dispersed instead of localised. Regional identity is also challenged by the growing existence and influence of diaspora communities.

The Commonwealth countries and G8 and G20 groups represent such virtualised regional networks and even function as quasi-regional organisations, as is the case with the proposed New Development Bank by the so-called BRICS states (Brazil, Russia, India, China, and South Africa).

Regional networks can be interpreted usefully as a layer of organising hierarchy between global and national levels that is not strictly geographic in nature. There are an increasing number of networks in the world that have regional aspects that are important to consider, and connectivity of people through multiple networks and communication channels is a growing trend (World Economic Forum 2015). A specific example in the context of DRR, including addressing climate risks, is the Alliance of Small Island States (AOSIS), a coalition of small island and low-lying coastal countries that share a common set of risk exposure and vulnerability factors and work together to advocate within the United Nations system.

Organising Structures of Regional Organisations

Inter-governmental Regional Organisations

The inter-governmental regional organisations created in many geographic regions as membership organisations among nation States to address a wide range of political and economic cooperation issues and opportunities have recognised the importance of DRR / CCA. However, they have tended to address these issues through voluntary mechanisms, focusing typically on non-binding commitments and the provision of policy and technical guidance and capacity strengthening to member States. Table 30.2 outlines the scope of relevant policy documents adopted by some of the larger and long-established governmental regional organisations.

Policy commitments adopted by these inter-governmental regional organisations often do have links to the legally binding treaty mechanisms of the organisations, but typically the DRR / CCA commitments themselves are non-binding and lack compliance mechanisms. One exception is ASEAN, whose members ratified the ASEAN Agreement on Disaster Management and Emergency Response (AADMER) in 2009 (Ferris and Petz 2013).

Operating through non-binding coordination, these organisations are still able to exercise an important convening power and to bring DRR, including CCA, to regional political and cooperation agendas. To help anchor the provision of services, several inter-governmental regional organisations have developed their own regional centres on disaster risk management. These include:

- the Pacific Islands Applied Geoscience Commission (SOPAC), established in 1972 as a United Nations Development Programme (UNDP) Regional Project and then, in 1990, as an independent inter-governmental agency affiliated with the Council of Regional Organisations of the Pacific (CROP) and now organised as part of the Secretariat of the Pacific Community (SPC);
- el Centro de Coordinación para la Prevención de los Desastres Naturales en América Central (CEPREDENAC), a specialised institution of the Central American Integration System for disaster prevention, mitigation, and response, established in 1988;
- the Caribbean Disaster Emergency Management Agency (CDEMA), an offshoot of the Caribbean Community (CARICOM) established in 1991;

Table 30.2 DRR / CCA Policy Commitments of Inter-governmental Regional Organisations

Regional Organisation	Disaster Risk Reduction Framework	Disaster Management Framework	Climate Change Adaptation Framework
AFRICA			
AU – African Union	2004	In development	2010 (agriculture), 2014
ECOWAS – Economic Community Of West African States	2007	2012 (humanitarian policy)	2010
SADC – Southern African Development Community	2005-6	2001	2011 (water sector), 2011[1]
AMERICAS			
OAS – Organisation of American States	2003		2009[2]
SICA – Central American Integration System	1999		2010
CARICOM – Caribbean Community	2001		2009
CAN – Community of Andean Nations	2004		2006[3]
MERCOSUR – Southern Common Market	2009		2009 (health)[4]
ASIA			
LAS – League of Arab States	2010	1990	2007[5]
SAARC – South Asian Association for Regional Cooperation	2007		2008
ASEAN – Association of Southeast Asian Nations	2005		2012[6]
APEC – Asia-Pacific Economic Cooperation	2009		2007[7]
EUROPE			
EU – European Union	2009	2001	2000
Council of Europe	1987		2010[8]
PACIFIC			
Regional Pacific Framework	2005		2005

1   Tripartite agreement among the Common Market for Eastern and Southern Africa (COMESA), the East African Community (EAC), and the Southern African Development Community (SADC).
2   Recognised in mandates of the Summits of the Americas although not yet as a dedicated policy framework specifically on CCA.
3   Andean Environmental agenda 2006-2010.
4   Estrategia de Acción MERCOSUR para proteger la Salud Humana de los efectos del Cambio Climático.
5   The Arab Ministerial Declaration on Climate Change, adopted by the Council of Arab Ministers Responsible for the Environment (CAMRE) in its 19th session, 5–6 December 2007.
6   ASEAN Action Plan on Joint Response to Climate Change.
7   APEC leaders' 'Declaration on Climate Change, Energy Security and Clean Development'.
8   Recommendation 2010–1 of the Committee of Permanent Correspondents on reducing vulnerability in the face of climate change, adopted at the 12th Ministerial Session of the European and Mediterranean Major Hazards Agreement (EUR-OPA).

Source: DRR and disaster management framework data from Ferris and Petz (2013). CCA framework compiled for this chapter. A further list with additional organisations is also provided in Hollis (2015).

- the SAARC Centre for Disaster Management and Preparedness established under the South Asian Association for Regional Cooperation (SAARC) in 2005;
- ASEAN Coordinating Centre for Humanitarian Assistance on disaster management (AHA centre) established in 2011.

These centres have played an important role in providing services to member States, particularly training and capacity development opportunities for smaller countries in the Pacific, Caribbean, and Central America, and in providing a forum for regional planning for hazards that affect multiple countries. These centres have typically demonstrated better success in the provision of specific services than in pursuing wide ranging policy mandates.

The impact of the centres has also been limited for several reasons, including that:

- the centres' activities have a tendency to focus on disaster risk management counterparts rather than being able to draw in other departments within member State governments. So this has been a missed opportunity to mainstream DRR, including CCA, within the wider development activities of member State governments.
- member States have usually had opportunities to obtain the same types of services through bilateral relations and other types of donor support rather than through the centres or their host governmental regional organisations. This has, in effect, limited demand and reduced the anticipated economies of scale accruing through regional organisations for such activities (SAARC 2013; White 2015).
- member States' attention is often more 'focused on setting up their own national systems than on investing in regional arrangements' (White 2015: 27).
- overlapping memberships between regional organisations have also been a challenge at times dividing regional momentum across several forums (Ferris and Petz 2013), for example during the Asian Financial Crisis and SARS crisis (Hettne and Söderbaum 2006), although the links in membership between regional organisations may also hold advantages by increasing opportunities for sharing and exposure.

Other Types of Regional Organisations or Networks

A range of other types of regional organisations are providing similar or complementary services, sometimes to the same national governments and sometimes to other national stakeholders who will also ultimately need to be part of DRR, including CCA. This is an example of the new multi-dimensional regionalism described by Hettne and Söderbaum. Rather than being state-centred, this new regionalism 'involves state, market and civil society actors in many institutional forms' (Hettne and Söderbaum 2006: 183).

Examples of these types of regional organisations or networks include regional local government associations and the regional components of global networks like the International Federation of Red Cross and Red Crescent Societies (IFRC). CityNet is a regional network of cities in Asia and the Pacific that has committed to UN-Habitat's Safer Cities programme to support the development of a regional strategy for urban safety in the Asia-Pacific Region (UN-Habitat, CityNet, and Seoul Metropolitan Government, 2015). The Inter-American Framework for the IFRC similarly outlines a regional strategy for strengthening humanitarian efforts at national and local level and prioritises advocacy with states as a key mechanism for advancing the DRR / CCA agenda in the Americas.

The DRR / CCA activities of these other regional organisations increase the overall supply and availability of services to national governments and their counterparts and also offer opportunities for leveraging global and local links, for example by channelling global and international resources for investment in DRR, including CCA, into the regions and providing additional channels to reach national, provincial and local stakeholders.

In general these wider sets of regional organisations have already recognised the relevance of risk management to their members' concerns and agendas and have committed their organisations to invest increased attention in DRR / CCA in their regional policies and services. However, like the centres discussed above, there often exists significant unrealised potential for further integration of risk-informed approaches within the breadth of their policy, partnering, and services.

These opportunities for expanded roles may be advantageous for maximising the adaptive role of regional organisations in addressing risk issues like changing risk patterns, new threats (e.g., pandemic), etc. which take place in complex ecosystems of actors with opportunities for links at many levels. With no fixed structural or programmatic pattern from region to region, the structure and agenda of regional organisations are often determined by opportunities and gaps in relation to what other national and international actors are doing (or not doing) and are thus potentially able to pivot to areas of concern and current risk.

Regional Multi-lateral Development Banks

Like the World Bank, regional multi-lateral development banks typically have units dedicated to providing technical assistance on DRR or CCA to programme teams working with member States to design and implement socio-economic development projects that will be financed with loans from the development banks (Table 30.3). In these institutions, increasingly DRR / CCA concerns are being addressed alongside existing safeguards to ensure that issues of risk related to climate change and other hazards are being factored either directly into the environmental impact assessments and cost–benefit analyses used to screen and assess programme impact and feasibility (as in the case of the Caribbean

Table 30.3 DRR / CCA Policy Commitments of Multi-lateral Development Banks

	DRR policy	CCA policy	Dedicated structures
African Development Bank (AfDB)	Policy Guidelines and Procedures for Emergency Relief Assistance, 2010	Climate Risk Management and Adaptation Strategy (2009)	ClimDev-Africa Special Fund (still being capitalised)
Asian Development Bank (ADB)	Disaster and Emergency Assistance Policy, 2004 Further reinforced in ADB's Operational Plan for Integrated Disaster Risk Management 2014–2020, linking disaster risk management, CCA, and disaster risk financing		Integrated Disaster Risk Management Fund (covering Southeast Asia)
Caribbean Development Bank (CDB)	Disaster Management Strategy and Operational Guidelines, 2009	Climate Resilience Strategy, 2012	Community Disaster Risk Reduction Trust Fund
Inter-American Development Bank (IADB)	Disaster Risk Management Policy, 2008	Integrated Strategy for Climate Change Adaptation and Mitigation and Sustainable and Renewable Energy, 2010	Disaster Prevention Fund, Multi-Donor Disaster Prevention Trust Fund, and Sustainable Energy and Climate Change Initiative Fund, Contingent Credit Facility
World Bank	OP/BP 8.00, Rapid Response to Crises and Emergencies, 2007		Global Facility for Disaster Reduction and Recovery

Source: World Bank (2012) and each organisation's website.

Development Bank) or through dedicated project screening tools (as in the case of the Asian Development Bank and the African Development Bank) (World Bank 2012). These efforts to mainstream DRR, including CCA, are helping the organisations and their members to expand beyond stand-alone DRR / CCA projects to embed DRR / CCA in a wider range of development projects (ADB 2014). To support and sustain these mainstreaming efforts, the regional multi-lateral development banks and the World Bank also offer grant-funded training and capacity strengthening opportunities to member States as well as support to the development of regional services (e.g., risk pooling or assessment) as public goods and services to address market gaps that may exist at national level.

Regional Platforms

Regional platforms within the United Nations or its agencies have also played key roles in linking DRR / CCA to wider socioeconomic development agendas among member States in their regions, especially to ensure development programmes are risk-informed and to protect development gains. By virtue of the States' membership in the larger UN body, the regional platforms function much like inter-governmental regional organisations. The Economic Commission for Latin America and the Caribbean (ECLAC), the Economic and Social Commission for Asia and the Pacific (ESCAP), and the Economic Commission for Africa (ECA) have been among the most active of such regional platforms on DRR and CCA. In some cases this has been done using research on regional risks as an entry point to develop common tools and approaches. The ECLAC methodology for assessment of disaster damage is an excellent example of this. In other cases these platforms have helped to fill gaps in regional coordination among states; e.g., the formation of the ESCAP Trust Fund for Tsunami, Disaster and Climate Preparedness, which was originally created to support tsunami early warning through a multi-hazard approach in the Indian Ocean basin. UNISDR has also used regional platforms for DRR to good effect to inform the strengthening of national policies for DRR / CCA. In Asia the ISDR Asia Partnership has helped to anchor the AMCDRR, to promote the role of parliamentarians in developing DRR / CCA legislation, and to broadly champion and advance national progress toward the HFA (Box 30.2).

Box 30.2   The Potential for State-led Regional Platforms: The Asian Ministerial Conference on Disaster Risk Reduction (AMCDRR)

Ian O'Donnell[1]

[1] Red Cross Red Crescent Global Disaster Preparedness Center, Washington, DC, USA

The AMCDRR was first convened as the Asian Conference on Disaster Reduction (ACDR) in Beijing, China in 2005 and brought together delegations from forty-two Asian and Pacific countries as well as thirteen UN agencies and international organisations. With thirty-three countries represented at the ministerial level, the conference provided a significant platform for leaders from Asian countries to exchange best practices in DRR, elaborate priorities for action to be considered by countries for implementation under the HFA, and promote enhanced regional cooperation to implement the HFA (UNISDR 2005).

Subsequent AMCDRR meetings have taken place every two years. In Delhi in 2007, the AMCDRR was officially designated as the Regional Platform for DRR for Asia under the HFA, with the following objectives to:

i.   Review the action taken by the national governments and other stakeholders for the implementation of the HFA.

ii. Take stock of initiatives taken in various sub-regions of Asia for promoting and enhancing cooperation among the nations within and outside the governments for disaster risk reduction.

iii. Share and exchange best practices and lessons learned from disaster risk reduction in various fields including application of science and technology, community based disaster preparedness, public–private partnership etc.

iv. Further enhance regional cooperation for disaster risk reduction.

With the adoption of the Sendai Framework for Disaster Risk Reduction (SFDRR) in 2015, the efforts of AMCDRR will now also be oriented toward the global targets and priorities for action outlined in the SFDRR as part of the post-2015 development agenda.

The AMCDRR has achieved high levels of both vertical and horizontal cooperation. With UNISDR as its secretariat and functioning as a regional platform under the HFA, the AMCDRR has been well anchored in the global DRR agenda. Yet, at the same time the network has been driven by its national governments and their continued ministerial presence at AMCDRR events, as evidenced in the way in which hosting countries have shaped the agenda for each of its bi-annual conferences (although this has also challenged long-term regional policy cohesion). Representatives of intergovernmental regional organisations have also played a key role in AMCDRR, including the ASEAN Committee on Disaster Management, the SAARC Disaster Management Centre, and SOPAC. Finally, civil society organisations, scientific and research institutions, and the private sector have participated enthusiastically in the regionally owned process.

Supra-national Networks

The Global Resilience Partnership, sponsored by the Rockefeller Foundation, United States Agency for International Development (USAID), and Swedish International Development Agency (SIDA), includes a set of regional hubs for South and Southeast Asia, the Horn of Africa, and the Sahel. The hubs are intended to provide regional fora for linking innovation in technology, measurement, financing, policy, and learning.

Local government associations and city networks, such as United Cities and Local Governments (UCLG) and Local Governments for Sustainability (ICLEI), are also often organised in regional networks to promote and facilitate peer sharing and to organise capacity development services in the same catchment areas within which governmental regional organisations typically form and operate. These associations and networks have recognised the centrality of risk management and resilience to their membership in terms of protecting and sustaining local development gains. Often, this progress has been stronger on the CCA side and offers further potential for more integrated approaches to risk management that include DRR, CCA, and improved contingency planning and management of residual risks and consequences (ICLEI 2011).

Private Sector

Similarly, the APEC Business Advisory Council (ABAC), a private sector body established to advise APEC leadership, has recognised the economic impacts of DRR / CCA and the need for coordination between national governments and the business community. In 2009 ABAC developed a Strategic Framework for Food Security in APEC to achieve food security in Asia and the Pacific which recommends that APEC refocus on a comprehensive approach to food security that addresses 'in a holistic way, access to food, availability of food, supply reliability, trade liberalisation, food safety, dietary health, environmental security, climate change, and sustainability' (Ear and Campbell 2012: 74). While just one case, this example demonstrates the potential for business associations or chambers of commerce to champion regional attention to risk and resilience.

Non-governmental, Scientific or Academic Organisations

Starting as far back as 1993, the Network of Social Studies in the Prevention of Disasters in Latin America (La Red) brought together a range of non-governmental organisation (NGO) and academic partners in Latin America and created a programme of comparative research, documentation, and information sharing on key themes relating to risk management in the region. Over the years La Red has offered a wide variety of training courses on local risk management and even developed a Master's degree programme in Risk Management and Disaster Prevention for working professionals and researchers in collaboration with FLACSO (the Latin American Social Science Faculty). La Red has also been instrumental in fielding coordinated assessments after disasters in the region and supported the development of the DesInventar database, use of which has since been picked up in other regions of the world with support from UN-ISDR.

Other regions also have strong examples of regional networks among NGOs. For example, the Asian Disaster Reduction & Response Network (ADRRN) consists of 34 national NGOs from 16 countries across the Asia-Pacific region and was formed in 2002 with the support of the Asian Disaster Reduction Center (ADRC) and the UN Office for the Coordination of Humanitarian Affairs (OCHA) (Ferris and Petz 2013). The ADRRN specifically aims to foster 'multilevel networking and collaboration' among its members and has provided a structure for enhanced cooperation among NGOs both in disaster response activities as well as in joint preparedness and DRR training and capacity development. Similarly, the Duryog Nivaran network was established in 1995 as a research, training, and advocacy network among NGOs and media organisations to fill a void in cross border dialogue and experience sharing in South Asia.

Regional scientific or academic organisations offer another strong potential venue for advancing cooperation on DRR, including addressing climate risks. A number of examples exist. Launched in 2010 the Southern Africa Society for Disaster Reduction (SASDiR) was developed as a community of practice for DRR practitioners and researchers within the regional context of the SADC. SASDiR has been active in advocating in both regional and global fora for national and sub-national policy and legislative reforms to advance DRR / CCA progress benefitting local communities (SASDiR 2014). Peri-Peri U is a partnership of African universities that was established in 2006 to build local disaster risk management capacity. Peri-Peri U has been instrumental in strengthening academic programmes for 'integrated risk scholarship' in Africa (Peri-Peri U 2015).

In addition the International Council for Science (ICSU) in collaboration with the International Social Science Council (ISSC) and UNISDR has supported through its Regional Committees and International Centres of Excellence (ICoE) coordination with regional platforms to implement integrated approaches to research and action in DRR and resilience building.

The United States Agency for International Development (USAID 2010) mentions the idea for a Mekong Panel on Climate Change as a regional spin-off to the Intergovernmental Panel on Climate Change (IPCC), which could promote research on climate change science, impacts, and adaptation in the region. At the global level the IPCC is an intergovernmental body that has drawn extensively on the contributions of scientists and other experts to increase the understanding of the risk of climate change to ground policy and programme development. Regional versions of such institutions could expand existing research services to provide more specific regional data and analysis, allow governments and other institutions in the region to rely less on outside technical support, and strengthen regional voice in international forums.

USAID has also worked with a wide range of university partners in Africa to establish sub-regional resilience innovation labs as part of the Resilient Africa Network (RAN). Existing institutions like the African Academy of Science or Association of African Universities could also play a role.

Military

There is a long history of regional military cooperation that has been applied to DRR / CCA to anticipate and plan for shared threats to stability and to coordinate crisis response and management. Military cooperation structures are often already regionally organised and have been used in the past to address humanitarian needs for disaster-affected populations; e.g., after the 2004 Indian Ocean Tsunami and 2010 earthquake in Haiti. To facilitate cooperation in response to disaster events, the North Atlantic Treaty Organization's (NATO's) Euro-Atlantic Disaster Response Coordination Centre (EADRCC) serves as a clearing-house for both requests and offers of assistance among its member countries. Planning for and simulation of responses to such disaster events has also proven to be a useful application for regional field exercises. In addition, NATO, and other regional military organisations, also actively engage in planning for chemical, biological, radiological and nuclear risks as a result of war, terrorist activities, or industrial accident and thus have a strong overlap in interests with other regional stakeholders in DRR / CCA.

Conclusions

This chapter has taken a wide-ranging view of the definition of regional organisations to examine the variety of roles that regional organisations can play in promoting DRR, including CCA. While the term regional organisation has often been primarily applied to inter-governmental organisations organised around geographic regions, a wide range of government, business, academic, and civil society organisations increasingly are engaged in varying degrees of collaboration or partnership through regional associations or networks whose activities either influence or could be impacted by risks associated with all hazards and vulnerabilities.

Regional organisations representing these diverse sets of stakeholders are in many ways well positioned to respond constructively and creatively to evolving risks and trends that challenge or threaten global, national, and local goals for advancing and protecting development gains. The role of regions as an intermediary level between global cooperation and nation sovereignty is rapidly evolving with the rise of networks and connections that are no longer just based on geography or proximity but instead increasingly reflect a wider range of shared risk, affinities, or historical and cultural connections.

The lack of a single definition or model for what constitutes a regional organisation can provide a level of adaptability that enables regional organisations to fulfil diverse but essential roles in helping their stakeholders operate in complex and interdependent ecosystems of actors with opportunities for links at many levels. This diverse regional cooperation increases the solution space for DRR including addressing climate risks by providing additional channels to reach and engage not just national but also provincial and local stakeholders. This adaptability may be particularly advantageous with changing hazard patterns (e.g., due to climate change and river/ coastal engineering), hazards that might not before have been fully considered (e.g., pandemic), and trends exacerbating exposure and vulnerability (e.g., as urbanisation and evolving patterns of socialisation).

References

ADB (Asian Development Bank). (2014) *Operational Plan for Integrated Disaster Risk Management 2014–2020*, Mandaluyong City, the Philippines: ADB.

Ear, J. and Campbell, J. (2012) 'Regional Cooperation on Disaster Management and Health Security: APEC and Comprehensive Regional Strategy', in R. Azizian and A. Lukin (eds), *From APEC 2011 to APEC 2012: American and Russian Perspectives on Asia-Pacific Security and Cooperation*, Honolulu: Asia-Pacific Center for Security Studies and Vladivostok: Far Eastern Federal University Press, pp. 67–77.

Ferris, E. and Petz, D. (2013) *In the Neighborhood: The Growing Role of Regional Organizations in Disaster Risk Management*, London: The Brookings Institution and London School of Economics Project on Internal Displacement. Online https://www.brookings.edu/research/in-the-neighborhood-the-growing-role-of-regional-organizations-in-disaster-risk-management/ (accessed 10 September 2016).

Hettne, B. and Söderbaum, F. (2006) 'Regional Cooperation: A Tool for Addressing Regional and Global Challenges', in International Task Force on Global Public Goods (eds), *Achieving Global Public Goods*, Stockholm, Sweden: Swedish Foreign Ministry, pp. 179–244.

Hollis, S. (2015) *The Role of Regional Organisations in Disaster Risk Management: A Strategy for Global Resilience*, London: Palgrave Macmillan.

ICLEI (Local Governments for Sustainability). (2011) *Durban Adaptation Charter for Local Governments*. Online https://unfccc.int/files/meetings/durban_nov_2011/statements/application/pdf/111209_cop17_hls_iclei_charter.pdf (accessed 10 September 2016).

Linn, J. and Pidufala, O. (2008) *The Experience with Regional Economic Cooperation Organizations: Lessons for Central Asia*, Washington, DC: Wolfensohn Center for Development at the Brookings Institution. Online https://www.brookings.edu/research/the-experience-with-regional-economic-cooperation-organizations-lessons-for-central-asia/ (accessed 10 September 2016).

Peri-Peri U. (2015) *A Risk Education Network: Extending Connections, Enhancing Resilience*, Stellenbosch: Periperi U.

SAARC (South Asian Association for Regional Cooperation). (2013) *Regional Progress Report on the Implementation of the Hyogo Framework for Action (2011–2013)*, Kathmandu: SAARC.

SASDiR (Southern Africa Society for Disaster Reduction). (2014) *Position Statement on the Post-2015 HFA Dialogue in Africa*, Potchefstroom: SASDiR.

Schmeier, S. (2010) *The Organisational Structure of River Basin Organisations: Lessons Learned and Recommendations for the Mekong River Commission (MRC)*, Berlin: Hertie School of Governance.

UN-Habitat, CityNet and Seoul Metropolitan Government. (2015) *Safer Cities Communique*. Online http://citynet-ap.org/wp-content/uploads/2015/07/Safer_Cities_Communique_v2.pdf (accessed 10 September 2016).

UNISDR (United Nations International Strategy for Disaster Reduction). (2005) *UNISDR Highlights October 2005*, Geneva: UNISDR. Online http://www.unisdr.org/2006/highlights/2005/October2005-eng.htm (accessed 10 September 2016).

UNISDR (United Nations International Strategy for Disaster Reduction). (2011–2013) *Collection of Regional Progress Reports toward the Hyogo Framework for Action*, Geneva: UNISDR. Online www.preventionweb.net/english/hyogo/regional/reports (accessed 10 September 2016).

USAID (United States Agency for International Development). (2010) *Asia-Pacific Regional Climate Change Adaptation Assessment – Final Report: Findings and Recommendations*, Washington, DC: USAID.

White, S. (2015) *A Critical Disconnect: The Role of SAARC in Building the DRM Capacities of South Asian Countries*, London: The Brookings Institution and London School of Economics Project on Internal Displacement. Online https://www.brookings.edu/research/a-critical-disconnect-the-role-of-saarc-in-building-the-disaster-risk-management-capacities-of-south-asian-countries/ (accessed 10 September 2016).

World Bank. (2012) *Disaster Risk Management and Multilateral Development Banks: An Overview*, Washington, DC: The Global Facility for Disaster Reduction and Recovery, World Bank.

World Economic Forum. (2015) *Global Risk Report 2015*, Geneva: World Economic Forum.

31

NATIONAL AND SUB-NATIONAL LEVEL DOING DISASTER RISK REDUCTION INCLUDING CLIMATE CHANGE ADAPTATION

Livhuwani David Nemakonde, Dewald van Niekerk, and Gideon Wentink

Introduction

Governance of disaster risk reduction (DRR) and climate change adaptation (CCA) have until now evolved largely in isolation from each other – through different conceptual and institutional frameworks, response strategies, and plans, at both international and national levels. Research suggests that this situation permeates into sub-national level as well (Nemakonde 2016). Whilst the integration of DRR and CCA initiatives for policy, structures and budgetary processes is in its initial stages in many countries, it has only been during the last ten years that the agenda for bringing the two fields closer together has gained some momentum. However, progress has mostly been conceptually and on paper – rather than in practice. The majority of DRR and CCA initiatives at national level continue to function in parallel and isolation.

This chapter focuses on how CCA should be integrated within DRR at national and sub-national levels. The continued differences in application and understanding of these domains in national (and ultimately sub-national) government structures, policies, and strategies, will be highlighted. Through a number of case studies, the governance problematic and successes are emphasised, which leads us to propose recommendations on how governance of DRR would include CCA, thereby addressing both more effectively at national and sub-national level.

Governance of Disaster Risk Reduction Including Climate Change Adaptation

Governance is the 'rule of the rulers' (World Bank 2013). It can be seen as the process by which authority is given to those making the rules and how these rules are executed. It is the art of steering societies and organisations to achieve certain goals for the common good. Governance occurs through interactions among structures, processes, role-players and traditions that determine how power is exercised, how decisions are taken, and how citizens or other stakeholders have their say (or not!). Thus, governance is about power, relationships, and accountability: who has influence, who decides, and how decision-makers are held accountable (Hodgson 2006). It involves processes through which collective goals are defined and pursued wherein the state is not the only or most important actor. The development of a governance perspective involves recognising the roles of supranational and subnational states and non–state actors and their complex interactions in the process of governing. This holds true for disaster risk and climate change governance alike.

Addressing disaster risk, including the impacts of climate change, requires a move away from a hierarchical form of governing to an interactive one, based on polycentric organisations at multiple levels. The concept of multilevel governance extends its focus on multiple level interactions specifically at jurisdictional and spatial scales where government tasks are allocated to different levels with the involvement of civil society and/or private actors who need to coordinate their actions (Tai 2015). The adoption of the multilevel governance system is important because societal problem-solving mechanisms are also not designed, and have not evolved adequately, to cope with such interlinked problems of severity, scale and complexity such as disaster risks including climate change. DRR and CCA present examples of typical societal dilemmas in being characterised by complex causes and effects, uncertainty, interconnectedness, diversity of perspectives, and scepticism. They span different temporal, spatial and administrative domains and jurisdictions. Such complex problems force those addressing them to interact within networks and policy arenas to discuss the problem and the solutions with other affected and interest groups at different levels of governance. However, Becker *et al.* (2013) shows that the perceived interaction between DRR and CCA does not happen readily, nor spontaneously. Good governance of DRR and CCA will necessitate an integration of the two.

Multilevel Governance of Disaster Risk Reduction Including Climate Change Adaptation

As complex societal problems, DRR and CCA necessitate the involvement of actors located at different spatial and geographic scales, since they cannot be addressed through linear policy-making, nor fit into hierarchical decision-making (Raschky 2008). As such, the institutional architecture for DRR, which must include CCA, should encompass local through to national, as well as national to global arena institutions (UN 2015; Wilkinson 2015). Using a multilevel risk governance framework draws attention away from an understanding of the state as a single actor to better characterise the relationship between different levels of governance and between different types of actors within each of these levels.

The conundrum with multilevel governance is that there is no consensus on how it should be structured. In trying to address this challenge, Hooghe and Marks (2003) distinguish between two types of multilevel governance; Type I (which is the focus of this chapter) emphasises the multiple tiers at which governance takes place (national, provincial and local levels) where governments are the central authority; and Type II focuses on networks between public and private actors across levels of social organisation. In practice, the two types of governance coexist, or are more or less overlapping. What is important in multilevel governance is an understanding of the changing nature and role of the state. According to Peters and Pierre (2001), the emergence of multilevel governance challenges much of the traditional understanding of how the state operates, what determines its capacities, its contingencies, and ultimately the organisation of democratic and accountable government. The multilevel governance model does not reject the view that state executives and state arenas are important. Rather, the emphasis is on decision-making competencies that are shared by actors at different levels as opposed to decision-making being monopolised by state implementers. The justification is that dispersion of governance across multiple jurisdictions is more flexible than concentration of governance in one jurisdiction (Hooghe and Marks 2003).

The involvement of many different actors at different governance levels, as Mees *et al.* (2012) indicate, makes demarcation of responsibilities important. Unclear roles and responsibilities can lead to confusion when decisions have to be implemented and therefore can hinder the governance of disaster risk including from the impacts of climate change (Dovers and Hezri 2010; Storbjörk 2010). The next section outlines the different levels and respective roles at which DRR, including CCA, action takes place, as well as the respective roles of stakeholders.

Governance of Disaster Risk Reduction Including Climate Change Adaptation at National Level

Whereas international organisations and frameworks are important in DRR, including CCA, the state still holds ultimate responsibility due to its considerable political power and legal mandates to protect its citizens (Jones *et al.* 2014). Government has traditionally been seen as the locus of governing in terms of defining and orchestrating collective goals and actions within society. As such, due to its coercive power and its capacities for institution building and enforcement, the state is the main and most powerful actor in reducing the risk of disaster emanating from both climate change and other natural hazards and natural hazard drivers. An example is given in Box 31.1. States have the moral and legal duty to protect their citizens, thus making DRR including CCA a government responsibility, albeit through the involvement of private sector and civil society organisations (Wilkinson 2015).

Box 31.1 Vanuatu

Livhuwani Nemakonde[1], Dewald van Niekerk[1], and Gideon Wentink[1]

[1] North-West University, Potchefstroom, South Africa

Vanuatu is one of the most vulnerable countries in the world to disaster risks including those from climate change. The island state experiences cyclones, storm surges, landslides, flooding and droughts, which are being influenced by climate change as well as other human-induced actions such as urban development. The country published a Disaster Risk Reduction and Disaster Management National Action Plan 2006–2016, which mapped out priorities and a pathway for DRR and management, although it has not been revised to address climate change considerations. At the national level, disaster risk reduction and climate change agencies, activities and funding have been previously managed separately. Vanuatu started the process of integrating climate change and DRR initiatives with the establishment of the National Advisory Board on Climate Change and Disaster Risk Reduction in 2012. The government undertook a risk governance assessment to analyse Vanuatu's climate change and disaster risk governance capacity and needs at both national and local levels. The government of Vanuatu is committed to directing the country's climate change and DRR efforts. Vanuatu has had great success with mainstreaming climate change and DRR into sector policies. The Vanuatu Climate Change and Disaster Risk Reduction Policy (2016–2030) is an important advance in integrating work in these overlapping fields, placing Vanuatu at the forefront of innovative approaches in the Pacific and internationally. This policy's objective with regard to governance is to enhance strategic frameworks and institutional structures to deliver effective climate change and DRR initiatives in a coordinated, integrated and complementary manner. The effectiveness of the implementation of this policy is yet to be seen but it is heartening to see a small country like Vanuatu be so proactive in its approach to climate change and DRR, although still not going the full way by placing CCA within DRR.

Moreover, governments have a moral duty to provide and coordinate DRR including CCA goods and services such as early warning systems, seasonal forecasts and weather alerts, making sure that the infrastructure can withstand hazards, proper environmental management and land use planning, and ensuring societal safety. Multi-stakeholder involvement is important because hierarchical coordination of government action across sectors is said to be weak (the so-called silo effect) (Hanssen *et al.* 2013). Thus, coordination in this regard refers to the instruments and mechanisms that aim to enhance the voluntary or forced alignment of tasks and efforts of organisations. To achieve this coordination, national government must put incentives in place to facilitate individual and private sector actions that reduce disaster risk including supporting CCA, while avoiding actions that undermine it (Dixit *et al.* 2012), such as duplication and separation of DRR and CCA.

Specifically for DRR, the Sendai Framework for Disaster Risk Reduction 2015–2030 identifies eleven major activities that must be carried out at national level to address disaster risk. Amongst others it includes mainstreaming and integrating DRR within and across all sectors; adopting and implementing DRR strategies and plans across different time scales with targets, indicators and time frames; carrying out assessments of technical, financial and administrative disaster risk management capacities; and assigning appropriate, clear roles and tasks to community representatives within disaster risk management institutions and processes and decision making (UN 2015).

Over and above, risk reduction must be built into national development strategies, programmes and projects to protect development and to ensure that new development does not exacerbate disaster risk. DRR must be centrally placed within the structure of government in order to mainstream risk reduction into development policies and operations, including CCA. It is discussed in many countries that DRR should be anchored in a national ministry or office with perceived political authority to ensure policy coherence across development sectors (this, however, does not occur widely). Ishiwatari (2013) advocates for the creation of focal point agencies in all ministries and line functions at national level to play a leading role in promoting DRR. The agencies should have the authority to formulate the vision, develop national policies, allocate budgets for government organisations, demand compliance and define actions for government organisations. Another view is that separating DRR including CCA into a specific agency isolates it, rather than mainstreaming DRR into all activities as part of those activities; namely, DRR becomes a subset of development. In fact, the existence of a national DRR focal point, or several focal points, is by no means a prerequisite for the integration of DRR and CCA into various ministries and government sectors. Importantly, national governments are expected to establish and strengthen government coordination forums such as national platforms for DRR (UN 2015). These coordinating mechanisms are needed to properly design and implement risk reduction strategies. Therefore, national institutions and legislative frameworks must aim to support the creation and strengthening of national integrated DRR mechanisms such as multi-stakeholder national DRR platforms which include CCA.

As things stand, in adapting to the impacts of climate change, national governments tend to focus on strengthening climate risk assessments, raising awareness, providing general frameworks and guidance on how to adapt, and funding for adaptation projects (Dixit *et al.* 2012). Much of the adaptation work is policy focused and is concentrated at national level with less emphasis on addressing local level impacts. Keskitalo *et al.* (2013) argue that centralisation of adaptation policies at national level can be flawed as the impacts of climate change are mainly felt and differentiated at local scale. As such, macro level adaptation policy tends to be disconnected from the needs of marginalised communities where local specific adaptation needs exist independently from national and international policies. Urwin and Jordan (2008) argue that focusing on policies at national level constrains adaptation efforts at local level by limiting the ability of local government to respond to the challenges. As such, inappropriate national policy and institutional arrangements create barriers to adaptation or they can lead to maladaptation (Dixit *et al.* 2012).

Similarly, DRR activities using top-down government and institutional interventions at national level alone are often insufficient as they tend to have lower understanding of community dynamics, perceptions and needs, and thus ignore the potential of local knowledge and capacities. As such, national roles and responsibilities for DRR including CCA must be complemented by adequately decentralised and layered risk management functions, capacities and corresponding budgets. This decentralisation to lower levels of governance is important as it would enhance community participation, transparency, accountability and predictability, which determine the quality of governance. It is therefore important that national government strengthen linkages with local government to guide and support DRR including CCA.

The isolation of DRR and CCA and the resultant locations within separate government structures does not add to good governance of the complex problems shared by these two foci. For effective and efficient governance, CCA needs to be integrated into DRR owing to the long history and successes of DRR both at national and subnational levels. Particularly at community level, the focus of DRR is based upon experiences for which a full range of tools and methodologies has been developed. Using these tools, strategies and methodologies, many communities have for centuries coped with different hazards including climate trends, variabilities and extremes.

Governance of Disaster Risk Reduction Including Climate Change Adaptation at Sub-National Level

Local decisions are critically important to the design and implementation of risk reduction and adaptation goods and services. Disasters are often highly localised and therefore solutions to these disasters should be found at this level while also tackling wider-scale processes that create and maintain vulnerability, often forcing vulnerability onto the local level against local efforts, needs and desires. Local government is in a better position to develop and experiment with various new tools and techniques. DRR, including CCA measures, must be incorporated into development policies, strategies and investments at local level in order to strengthen the ability of communities to better respond to and cope with hazard events.

Local governments have the crucial roles of engaging communities and citizens in risk reduction and adaptation activities, and to link their concerns with government priorities. Moreover, local government has a central role in coordinating and sustaining multilevel, multi-stakeholder platforms to promote DRR, including CCA, in its region or for a specific hazard. Therefore, effective and integrated risk reduction and adaptation should occur through the dynamics of local governance, taking into account the complexities in other spheres which create and sustain vulnerability (e.g., international economic forces, wars and loss of biodiversity etc.). The dynamics of local governance involve engagement between citizens, civil society organisations, local authorities and the private sector. Government and other actors at local level help create an enabling environment around which households and communities can structure risk reduction including adaptation choices.

Ahrens and Rudolf (2006) are of the opinion that empowering local government through decentralisation of DRR has been shown to be beneficial to local communities, including marginalised groups. However, the local level of government must retain its autonomy from central government in order to define its own priorities and implement DRR measures without much interference (Wilkinson 2015).

Furthermore, local governments have the responsibility, which is often shared with national governments, of both identifying potential natural hazards, including those associated with climatic events, and of ensuring that consideration of such hazards is incorporated into statutory and non-statutory local government decision-making (Measham *et al.* 2011). Moreover, there is a 'duty-of-care' within local government to ensure that development decisions do not create the potential for significant, unmanaged exposure to hazards (Measham *et al.* 2011).

Baker *et al.* (2012) found in their study that limitations and deficiencies in local adaptation (planning) are the results of structural, procedural and contextual factors intrinsic to local governments. This, however, challenges the wisdom of devolving CCA responsibility to local government. For this reason, it is important to view local adaptation in the context of other levels and as an integrated part of DRR planning. The choice of specific adaptation practices is dependent on social and economic endowments of households and communities, and their location, networks of social and institutional relationships, institutional articulation and access, and access to resources and power (Agrawal 2010), all of which has a direct impact on the underlying drivers of disaster risks. Therefore, actors engaged at all levels and scales of governance are mostly best placed to develop and implement integrated DRR and adaptation policies, foster adaptive capacity and take adaptation action in order to reduce the impacts of climate change.

Considering that DRR encompasses CCA, it stands to reason that these two activities should not be treated in isolation. However, in practice this tends to be the case.

Integrating Parallel Structures for Disaster Risk Reduction and Climate Change Adaptation

DRR and CCA structures have largely developed in parallel, and as a result they mostly operate in isolation. Practitioners addressing DRR and CCA are affiliated to separate organisations and institutions both internationally and nationally. Particularly at national level, the traditional division of responsibilities into discrete areas has contributed to the location of DRR and CCA in different ministries and administrative units.

This fragmentation has been cited as the major government failure in tackling major problems (such as DRR including CCA) and cross-sectoral problems (Becker *et al.* 2013). A traditional bureaucracy, divided into vertical silos, in which most of the authority for resolving problems rests at the top of the organisation, is not well-adapted to support the kinds of process necessary for addressing the complexity and ambiguity of problems such as disaster risk including hazards influenced by climate change. Such problems are exposing the contradictions between the system of departments and cross-sectoral governance – including the difficulties imposed by horizontal, in addition to vertical, silos.

As a result of their location in different administrative units of government, DRR and CCA have taken different evolutionary paths, have a different conceptualisation of terms, and they use different methods, strategies, and institutional frameworks to achieve their goals (Lavell *et al.* 2012). Consequently, DRR and CCA are pitted against each other in both policy and practice as discrete issues with limited overlaps. Much of the literature that advocates for the integration of DRR and CCA falls short of achieving this objective because it does not spell out precise actions that must be taken. The literature only focuses on those elements that link the two and makes them compatible with much attention given to the similarities, differences, areas of convergence and the challenges for integration.

This calls for re-adjustments or re-orientation of the structures or organisations for effective, efficient and appropriate response. In this regard, integration of DRR and CCA is conceived to be capable of resolving multifaceted problems. An integration of CCA into DRR is recognised internationally as a robust way to address multiple simultaneous risks and hazards (Handmer *et al.* 2014). Integration in this regard is important because it is able to manage cross-cutting issues that do not correspond to the institutional responsibilities of individual departments. Whereas there are two forms of integration, vertical and horizontal, the focus here is on horizontal integration between different departments or ministries in the public sector.

The integration of structures for DRR and CCA can draw lessons from the Human Services field particularly in the UK that has been involved in service integration as an organising principle for the last four or so decades (see Ahrens 2011; Axelsson and Axelsson 2006). This is so because most problems facing society such as disaster risk are cross-cutting and as such, no single organisation can address the problem effectively. Such problems can be dealt with effectively through joint inter-agency planning and programming for the reason that resources and capabilities to cope with the problems are contained within autonomous organisations and vested interest groups.

One should remain mindful that separate DRR and CCA structures have already been established in many countries, and that the government machinery needed to integrate CCA into DRR turns exceedingly slowly. The overlapping domains of organisations addressing CCA and DRR present the need and opportunities for those organisations to initially engage in inter-organisational relations. Domain similarity is a qualitative indicator of the kind of organisations likely to become jointly involved in a web of inter-relation. Thus, the greater the domain similarity, the greater the interdependence, and therefore the greater the need for integration. As such, inter-organisational integration can take many different forms ranging from management hierarchy, which has to do with the top-down coordination of the organisation, to market competition, which deals with contractual relations between organisations and voluntary cooperation and collaboration between organisations that are not part of the common hierarchy (Axelsson and Axelsson 2006). An example is in Box 31.2.

Box 31.2   Mozambique

Livhuwani Nemakonde[1], Dewald van Niekerk[1], and Gideon Wentink[1]

[1] North-West University, Potchefstroom, South Africa

Mozambique has a well-functioning DRR system at various levels, which serves as the basis for the national strategy on DRR and CCA. This strategy outlines the vision and mission, guiding principle, objectives, actions and resources necessary to increase the resilience of citizens and communities to deal with the impacts resulting from hazards, including those influenced by climate change. In recognition of the fact that CCA and DRR planning has been largely centralised until now, there are actions to improve opportunities for involvement at provincial and district level. This is so because of the realisation that subnational governments have a key role to play in linking the strategy with the grassroots level. At this level, Institute National Gestao de Calemedates (INGC) already has an effective institutional architecture in place, coordinated by district disaster management officers who, in turn, play a role in facilitating and training community level DRR committees. Given these institutions and the coordination mechanism, INGC is ensuring that risk reduction is undertaken, and that credible early warning messages are issued. In Mozambique, there are also local DRR committees that support the process of local adaptation plans. Implementation of policies and strategies at local level is currently impeded by a breakdown in the chain of communication from national to local level, which emphasises the importance of outlining the roles and responsibilities between the national and subnational levels.

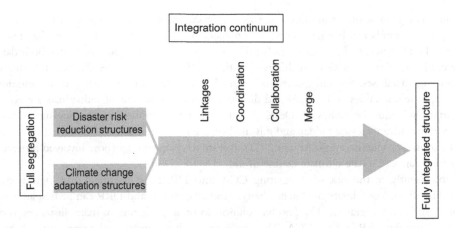

Figure 31.1   Continuum for Integrating CCA into DRR Structures

(By Authors)

Several authors have presented integration as a continuum (see Figure 31.1 with full segregation on the one extreme and a fully integrated structure on the other) (Ahrens 2011; Axelsson and Axelsson 2006). On one end of the continuum are organisations that hardly interact with each other when it comes to dealing with public problems that extend beyond their capabilities or mandates (Bryson *et al.* 2006; Page *et al.* 2015). This side of the continuum is characterised by a highly fragmented system with organisations working in isolation and wasting government resources. At the other end are organisations that have merged into a new entity meant to address the public problem through fully shared authority and capabilities (Page *et al.* 2015). Such organisations are completely integrated and there is no differentiation (Axelsson and Axelsson 2006). At the midrange of the integration continuum are organisations that share information, undertake coordination activities or develop shared power arrangement, through collaboration in order to pool their capabilities to address the challenge (Bryson *et al.* 2006; Page *et al.* 2015).

Whereas some cases of integration might require limited and loose contacts, coordination and/or collaboration between the organisations involved, this chapter is advocating for the merger between government departments or the sections thereof dealing with DRR and CCA, to form one entity with its own statutory status and appropriate authority. A merger is described as an arrangement, which brings together or transfers all or parts of different organisations or their authorities, jurisdiction, personnel and resources on a permanent basis to other organisations either as a new or existing department or agency. In this regard, integration will involve the organisational re-layout where a part or whole of two or more organisations are merged to create a new organisation. Specifically for the integration of DRR and CCA, a number of authors including Forino *et al.* (2015) and Kelman *et al.* (2015) advocate embedding climate change adaptation into core disaster risk reduction operations in order to attain simultaneous benefits for the social systems coping with climate extremes and change. In this sense climate change adaptation must be mainstreamed into disaster risk reduction owing to the long history and successes of disaster risk reduction both at national and subnational levels. Particularly at community level, the focus of disaster risk reduction is based upon experiences for which a full range of tools and methodologies have been developed. Using these tools, strategies and methodologies many communities have for centuries coped with different hazards including climate extremes.

The management of integration is not an easy task. In order for integration to succeed as anticipated, a number of barriers to integration need to be addressed. Most of these barriers are structural and conceptual and they involve the existence of different administrative boundaries, different laws, rules and regulation, different budgets and financial streams, different information streams and databases. Similarly, other soft issues like different professional and organisational cultures, different values and interest and differences in commitment of individuals are equally important and must be addressed. Developing functional integration can be devastating when collaborative advantages are hidden and missing because professionals are defending their territories in contra-productive ways (Ahrens 2011). Failure to cooperate by those involved protecting their turfs can lead to the ultimate demise of ties.

Subsequently, in the case of integrating CCA into DRR, the argument and logic is quite clear. The shared complexity of climate change (and adaptation) and DRR can be best addressed through such an integration. The possible solution to these problems, in many instances, is one and the same for DRR and for CCA. The integration will not only serve common issues but it is necessary to ensure good governance and accountability of governments to their citizens. In short, integrating CCA into DRR makes good governance sense.

Conclusion

This chapter has shown that both DRR and CCA can adequately be addressed by applying multilevel governance frameworks and by placing CCA within DRR. What is clear from the literature is that effectively integrating CCA into DRR will demand changes in the mind-sets of those involved in order to take up new challenges. Where feasible, this will require merging the different departments or sections within the departments that deal with DRR and CCA. Where the merger is not feasible, interrelations forums must be developed where common issues will be discussed and joint planning, budgeting and implementation will occur. Alternatively, departments should be arranged according to outcomes as opposed to functions. It should be noted that this integration of the structures is in no way a panacea for solving disaster risk problems, including those influenced by climate change, but will go a long way in contributing to good governance of DRR including CCA.

References

Agrawal, A. (2010) 'Local institutions and adaptation to climate change', in R. Mearns and A. Norton (eds), *Social Dimensions of Climate Change: Equity and Vulnerability in a Warming World*, Washington, DC: The World Bank, pp. 173–198.

Ahrens, J. (2011) 'Governance, development, and institutional change in times of globalisation', in J. Ahrens, R. Caspers and J. Weingarth (eds), *Good Governance in the 21st Century: Conflict, Institutional Change, and Development in the Era of Globalization*, Cheltenham: Edward Elgar Publishing, pp. 1–22.

Ahrens, J. and Rudolph, P.M. (2006) 'The importance of governance in risk reduction and disaster management', *Journal of Contingencies and Crisis Management* 14, 4: 207–220.

Axelsson, R. and Axelsson, S.B. (2006) 'Integration and collaboration in public health – a conceptual framework', *The International Journal of Health Planning and Management* 21, 1: 75–88.

Baker, I., Peterson, A., Brown, G. and McAlpine, C. (2012) 'Local government response to the impacts of climate change: An evaluation of local climate adaptation plans', *Landscape and urban planning* 107, 2: 127–136.

Becker, P., Abrahamsson, M. and Hagelsteen, M. (2013) 'Parallel structures for disaster risk reduction and climate change adaptation in Southern Africa', *Jàmbá: Journal of Disaster Risk Studies* 5, 2: 1–5.

Bryson, J.M., Crosby, B.C. and Stone, M.M. (2006) 'The design and implementation of cross sector collaborations: Propositions from the literature', *Public Administration Review* 66, s1: 44–55.

Dixit, A., McGray, H., Gonzales, J. and Desmond, M. (2012) *Ready or Not: Assessing Institutional Aspects of National Capacity for Climate Change Adaptation*, Washington, DC: World Resources Institute.

Dovers, S.R. and Hezri, A.A. (2010) 'Institutions and policy processes: The means to the ends of adaptation', *Wiley Interdisciplinary Reviews: Climate Change* 1, 2: 212–231.

Forino, G., Von Meding, J. and Brewer, G.J. (2015) 'A conceptual governance framework for climate change adaptation and disaster risk reduction integration', *International Journal of Disaster Risk Science* 6, 4: 372–384.

Handmer, J., Mustelin, J., Belzer, D., Dalesa, M., Edwards, J., Farmer, N., Foster, H., Greimel, B., Harper, M., Kauhiona, H., Pearce, S., Yates, L., Vines, K. and Welegtabit, S. (2014) *Integrated Adaptation and Disaster Risk Reduction in Practice*, Brisbane, Australia: RMIT University, Griffith University and the National Climate Change Adaptation Research Facility.

Hanssen, G.S., Mydske, P.K. and Dahle, E. (2013) 'Multilevel coordination of climate change adaptation: By national hierarchical steering or by regional network governance?', *Local Environment* 18, 8: 869–887.

Hodgson, G.M. (2006) 'What are institutions?', *Journal of Economic Issues* XI, 1: 1–25.

Hooghe, L. and Marks, G (2003) 'Unraveling the central state, but how? Types of multilevel governance', *American Political Science Review* 97, 2: 233–243.

Ishiwatari, M. (2013) *Disaster Risk Management at the National Level*, Tokyo, Japan: Asian Development Bank Institute.

Jones, S., Oven, K.J., Manyena, B. and Aryal, K. (2014) 'Governance struggles and policy processes in disaster risk reduction: A case study from Nepal', *Geoforum* 57: 78–90.

Kelman, I., Gaillard, JC, and Mercer, J. (2015) 'Climate change's role in disaster risk reduction's future: Beyond vulnerability and resilience', *International Journal of Disaster Risk Science* 6, 1: 21–27.

Keskitalo, C., Juhola, S., Westerhoff, L., Scholten, P. and Ashgate, L. (2013) 'Connecting multiple levels of governance for adaptation to climate change in advanced industrial states', in J. Edelenbos, N. Bressers and P. Scholten (eds), *Water Governance as Connective Capacity*, London: Routledge, pp. 69–88.

Lavell, A., Oppenheimer, M., Diop, C., Hess, J., Lempert, R., Li, J., Muir-Wood, R. and Myeong, S. (2012) 'Climate change: New dimensions in disaster risk, exposure, vulnerability, and resilience', in C.B. Field, V. Barros, T.F. Stocker, D. Qin, D.J. Dokken, K.L. Ebi, M.D. Mastrandrea, K.J. Mach, G.K. Plattner, S.K. Allen, M. Tignor, and P.M. Midgley (eds), *Managing the Risks of Extreme Events and Disasters to Advance Climate Change Adaptation*, New York: Cambridge University Press, pp. 25–64.

Measham, T.G., Preston, B.L., Smith, T.F., Brooke, C., Gorddard, R., Withycombe, G. and Morrison, C. (2011) 'Adapting to climate change through local municipal planning: barriers and challenges', *Mitigation and Adaptation Strategies for Global Change* 16, 8: 889–909.

Mees, H.L., Driessen, P.P. and Runhaar, H.A. (2012) 'Exploring the scope of public and private responsibilities for climate adaptation', *Journal of Environmental Policy and Planning* 14, 3: 305–330.

Nemakonde, L.D. (2016) *Integrating Parallel Structures for Disaster Risk Reduction and Climate Change Adaptation in the Southern African Development Community*, PhD Thesis, Unpublished.

Page, S.B., Stone, M.M., Bryson, J.M. and Crosby, B.C. (2015) 'Public value creation by cross sector collaborations: A framework and challenges of assessment', *Public Administration* 93, 3: 715–732.

Peters, B.G. and Pierre, J. (2001) 'Developments in intergovernmental relations: Towards multilevel governance', *Policy and Politics* 29, 2: 131–136.

Raschky, P.A. (2008) 'Institutions and the losses from natural disasters', *Natural Hazards and Earth System Science* 8, 4: 627–634.

Storbjörk, S. (2010) 'It takes more to get a ship to change course: Barriers for organisational learning and local climate adaptation in Sweden', *Journal of Environmental Policy and Planning* 12, 3: 235–254.

Tai, H.S. (2015) 'Cross-scale and cross-level dynamics: Governance and capacity for resilience in a social-ecological system in Taiwan', *Sustainability* 7, 2: 2045–2065.

UN (United Nations). (2015) *Sendai Framework for Disaster Risk Reduction 2015–2030, Adopted at the Third United Nations World Conference on Disaster Risk Reduction Sendai, Japan, 14–18 March 2015, UN General Assembly, A/CONF.224/L.2*, Geneva: UN.

Urwin, K. and Jordan, A. (2008) 'Does public policy support or undermine climate change adaptation? Exploring policy interplay across different scales of governance', *Global Environmental Change* 18, 1: 180–191.

Wilkinson, E. (2015) 'Beyond the volcanic crisis: co-governance of risk in Montserrat', *Journal of Applied Volcanology* 4, 1: 1–15.

World Bank. (2013) *Building Resilience: Integrating Climate and Disaster Risk into Development: Lessons from the World Bank Group*, Washington, DC: The World Bank.

32

COMMUNITIES DOING DISASTER RISK REDUCTION INCLUDING CLIMATE CHANGE ADAPTATION

Zenaida Delica-Willison, Loreine B. dela Cruz, and Fatima Gay J. Molina

Introduction

Development requires disaster risk reduction (DRR) including climate change adaptation (CCA) not only to ensure the safety of citizens and communities, but also to maintain the development gains so far achieved, especially those by less wealthy countries. The Sendai Framework for DRR (SFDRR) 2015–2030 ushered in last year through the United Nations, emphasises that for DRR including CCA to be a fundamental pillar of sustainable development, it requires all of society to be involved, in partnership with each other and engaged for shared responsibility, especially with regard to local communities. Experiences highlight the importance of the role of local communities, since whatever the scale of hazards, big or small, it is the local community that either suffers the brunt of or survives from hazards' devastating effects (Delica-Willison and Gaillard 2012; Maskrey 1989).

This is significant in relation to community governance. DRR including CCA should be an all-community and inclusive approach that is shared by all. Community governance is in essence applying community participation in decision-making on matters pertaining to public concerns at the local level affecting communities. Community governance requires that CCA be an integral part of reducing disaster risk and that DRR be framed within the wider backdrop of development.

Understanding Community, Community Participation, and Community Governance

Abarquez and Murshed (2004) define 'community' as a cluster of households, a small village, or a neighbourhood in a town. It can also be a shared experience, such as particular interest groups, professional groups, or hazard-exposed groups. It can likewise be groups that are both affected by and can assist in the mitigation of hazards and reduction of vulnerabilities. The concept of community can therefore have varied meanings, which is a challenge for practitioners (Cannon *et al.* 2014). Whatever the context or setting, a frequent but often-challenged definition emerging of a community is a group of people with diverse characteristics who are linked by social ties, who share common perspectives, and who engage in joint action in a geographical location or in common settings. There are at least five core elements of such communities: locus, which is a sense of place; sharing, which pertains to commonality in interests and perspectives; joint action, which serves as a source of cohesion and identity; social ties, which make the foundation of a community; and diversity, which captures the social complexity within communities (Green and Mercer 2001). The latter point is essential as community members may have varying perceptions of disaster risk and vulnerability depending on social class, gender, age, ethnicity, and disability (Wisner 1993). However, when affected by disasters, they are often homogenised as 'disaster victims', which leads to their diverse needs and capacities being overlooked (Oliver-Smith *et al.* 2016).

Community participation is essential to fully address the needs and capacities of those affected by disasters as no one better than them knows what these needs and capacities are (Delica-Willison and Gaillard 2012). Oakley and Marsden (1987) define community participation as the process by which individuals, families, or communities assume responsibility for their own welfare and develop a capacity to contribute to their own and the community's development. In the context of development, community participation refers to an active process whereby beneficiaries influence the direction and execution of development projects rather than merely receiving a share of project benefits (Paul 1987). There are five aspects of contributions in which community participation can be realised: share in project costs; increase in project efficiency; increase in project effectiveness; build beneficiary capacity; and increase in empowerment. Fostering community members' participation in DRR contributes in addressing the shortcomings of the top-down approach in development planning, including DRR (Victoria 2003). The top-down approach usually fails to address local needs and sidelines the potential of indigenous resources and capacities, hence contributes to increasing people's vulnerabilities. The new trend is towards vertical integration, learning from the top-down approach and now anchoring on the grassroots, bottom-up, with integration at varying levels (Gaillard and Mercer 2013).

Community participation contributes to achieving successful community governance. Bowles and Gintis (2002) state that communities are part of good governance for they address certain problems that cannot be handled either by individuals acting alone or by markets and governments. The interconnected elements of decision-making, engaging the community and capacity development are all important components of community governance. But community governance has no simple definition or description that can capture its essence, dynamics and nuances including certain complexities if put in contexts which involved diverse players and actors. McKinlay (2014) suggests that community governance is a collaborative approach to determining a community's preferred futures, and developing and implementing the means of realising them. In practice, it can include one or more of the different tiers of government institutions, civil society, and private sector interests and follow one of the following three approaches, depending on the different roles of these stakeholders:

- It can begin with the government. It may put focus on how government can utilise its formal decision-making power to provide more responsibility for decision-making back to the community. This is referred to as decentralisation, but with government imposing responsibility, is also a top-down approach.
- The second approach is looking at autonomous community organisations outside of the formal decision-making process; i.e., people forming decision-making forums on their own. This is popularly called the bottom-up approach.
- The third is extending how local institutions can facilitate independent community governance through pooling of resources with community trusts as example or nurturing the social economy. This is referred to as the collaborative approach.

Community governance therefore refers to processes for making all decisions and plans that affect life in the community, whether made by public or private organisations or by citizens. As such, for community governance to be effective, it should be more than a process but more so, it should be about getting things done in the community so that what has been done and achieved makes a positive difference in community life. Henceforth, measuring results is also crucial and important. Results may vary from one individual to another within a community; or from one community to another community. Citizen engagement in decision-making is vital and makes all the difference in action, targeting results and measuring performance goals.

Ultimately, community governance should draw upon a holistic and inclusive approach in targeting community-level issues (Totikidis *et al.* 2005). It is about community management and decision-making in addressing community needs and building community capacity including its well-being. Martin (2005) explained that governance and capacity building are seen as crucial precursors to addressing entrenched social and economic disadvantages. Empowerment is therefore a by-product of community governance that depends on changes that have happened and whether there are improvements in the well-being and quality of life of the people in the community.

Facilitating Empowerment and Development of Communities for DRR Including CCA

UNDP (2010) points out that supportive governance is necessary to foster DRR. Governance has an influence on the way national and sub-national actors (which include governments, parliamentarians, public servants, the media, the private sector, Civil Society Organisations (CSOs), and local communities) are willing and able to coordinate their actions to reduce disaster risk. DRR mainstreaming is a governance process that enables the systematic integration of DRR concerns in all the relevant development spheres. In short, the responsive, accountable, transparent and efficient governance structures underwrite the environment where DRR can be institutionalised as an underlying principle of sustainable development. DRR requires both the sustainable consideration of the underlying risk factors in all relevant sectors and the integration of risk reduction initiatives in planning and delivery of core development services and processes such as education, environment and health for the benefit of local communities (Twigg 2004).

Addressing the foregoing social issues is more than speaking the language of communities and transferring responsibilities to communities. Organising communities is needed with a structure capturing the shared vision (what the community wants), working out the vision through a mission (what they want to be), bringing the community as a whole in a cohesive way (how they treat each other and build unity), and working together to prioritise scarce resources and to find new resources for community development. Different communities have differing needs and their nature and areas of work surely depend on their own capacity to address their own various social issues and needs, and, in due course, take appropriate action. The success of community governance therefore highly depends on people's level of awareness and holistic understanding of local issues covering economic, social, cultural, political, and spiritual aspects and how they respond to these issues and needs. In the end, it is about how people live together and do their work as a community and aspire to improve and become better.

Partnership between local governments and CSOs is often encouraged in organising communities and fostering community governance in DRR including CCA (Delica-Willison 2006). Communities need to be supported in order to achieve their development goals. As such, local governments need to have a deeper orientation process to better understand their responsibilities in partnership with the communities. They remain the leading partner of communities in terms of support. Local governments play an important role in the development and empowerment of communities and in various collaborative engagements. They can facilitate access to resources and technical services from external sources.

Capability development of the local people is indeed a constant need to enable them to become effective actors in governing their own communities. There are unique and specific capabilities required that are not well within the experience of even the typically informed community members. Capability building may range from establishing and managing appropriate organisational structures for better understanding formal processes and responsibilities of various stakeholders to differentiating varied capability building needs in community governance that will address the many needs of the wider community members.

Local governments can serve as enablers and facilitators of the capability development process (Delica-Willison 2006). Their role may take the form of providing knowledge and skills to facilitate the necessary exchanges where communities can identify community needs and priorities. Bottom-up budgeting and monitoring, risk financing and insurance, drainage system improvement, embankments, early warning systems, land use regulations and policies, and ecosystem management are good examples of approaches and modalities where local governments are able to assist in community governance.

There are two paths through which local governments foster community governance and encourage capability development:

- Community members are elected to form a council for the community. But even as they are officially elected, council members may not always know and have the answers. In this situation, elected members need to initiate discussions with communities through community assemblies. They can also initiate community conversations or dialogues to gather the community's sentiments and inputs on community concerns that need to be decided upon by the council. This approach is called representative democracy.
- Community input may also be sought on a case-by-case basis. Indeed, extensive engagements with the community do not necessarily yield consensual understanding or decision-making. More often than not, community members need not want responsibility in decision-making, but they do value the opportunity to express their views and opinions to be heard. This ensures accountability and transparency when making decisions.

An example is given in Box 32.1.

Box 32.1   Fostering Participation of Communities in Disaster Risk Reduction: The Wellington Region Emergency Management Office Communicate and Collaborate Model

Mischa Hill[1]

[1] Wellington Region Emergency Management Office, New Zealand

Naku te rourou nau te rourou ka ora ai te iwi: 'With your basket and my basket the people will thrive'. Emergency management and disaster risk reduction (DRR) including climate change adaptation (CCA) is everybody's business: from local government and central government and the not-for-profit and private sectors, to communities and individuals. At the heart of emergency management are our diverse and capable communities.

The vision of the Wellington Region Emergency Management Group (WREMO) is: 'A resilient community, ready and capable'. Resilient communities are ready for emergencies and have the knowledge, skills, resources, and relationships to take action to reduce their risks, to respond to, and

---

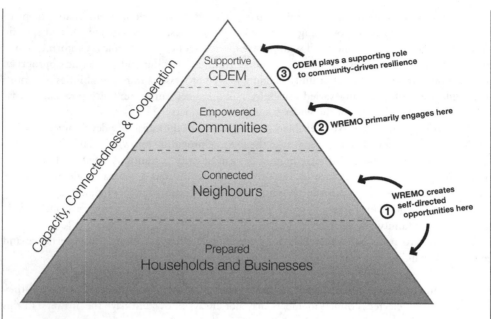

Figure 32.1   Model of Community-driven Emergency Management
(By Wellington Region Emergency Management Office)

to recover from, a disaster. To support this philosophy, WREMO has a team dedicated to working with communities to empower them to build resilience. The Community Resilience team employs a 'Communicate and Collaborate' approach by proactively engaging with diverse communities, supporting local ideas and existing structures, and facilitating various opportunities that lead to increased connectedness and preparedness. This approach is encapsulated by WREMO's Community-Driven Emergency Management (CDEM) model (Figure 32.1). This term emphasises engaged and empowered people and a supportive Emergency Management sector that collaborates with its communities as partners to create meaningful resilience outcomes.

The community resilience engagement is based on three strategic objectives: build capacity, increase connectedness, and foster cooperation. Through these, a range of participatory tools are used to facilitate wider community resilience.

Community response and resilience planning aims to bring a geographic community together to organise how they will self-activate in response to help one another after a disaster. The planning process dovetails with identifying opportunities to further strengthen the resilience of the community through projects they are interested in driving. This captures one of the three strategic objectives of fostering cooperation and is based on the social capital theory in which communities that work together day-to-day will ultimately work better together after a disaster.

The planning process is supported by the emergency management office, but owned by the local community. It rides on a whole of community approach; with participation from community leaders, organisations, and businesses in the area. This is based on the understanding that those who live and work in a community are best placed to know their local environment. Every community is unique with different interests. Emergency preparedness therefore must adapt to cater for diverse interests, priorities, and budgets.

This continuum of engagement draws on building resilience at different levels, creating as many opportunities for people to access, in ways that are most appropriate to them.

Fostering participation in DRR including CCA is underpinned by employing good community development principles. This enables trust and partnerships to form with those who can take meaningful actions to reduce their risk to the effects of disaster.

Community-based DRR Including CCA

In the specific context of DRR including CCA, community governance is usually encouraged through community-based DRR (or CBDRR with an example in Box 32.2). CBDRR encourages self-developed, context-specific, culturally and socially appropriate as well as economically and political doable approaches for dealing with disasters (Delica-Willison and Gaillard 2012). Molina (2016) elucidates that CBDRR involves (Figure 32.2): (1) initiating the process and establishing rapport amongst community members, led by local leaders in partnership with different stakeholders such as the vulnerable groups comprising children, women, elderly, persons with disabilities, and indigenous peoples; (2) conducting the participatory community risk assessment that includes the identification and analysis of hazards, vulnerabilities, and capacities of the community; (3) participatory development of the community DRR including CCA plan; (4) formation and/or strengthening of community organisation designed to tackle disaster risk; (5) community-managed implementation of the DRR including CCA plan; and (6) participatory monitoring and evaluation of the implementation of the plan that is foreseen to be progressing towards safer, less vulnerable, and more resilient communities.

Box 32.2   Integrated CBDRR Including CCA in Dagupan, Philippines

Zenaida Delica-Willison[1], Loreine B. dela Cruz[1], and Fatima Gay J. Molina[1]

[1] Center for Disaster Preparedness, Quezon City, Philippines

Prior to the legislation of the Philippine Disaster Risk Reduction and Management Act of 2010, also known as the Republic Act 10121 (RA 10121), Dagupan was already practising Community-Based Disaster Risk Reduction (CBDRR). Dagupan is a low-lying coastal city located in the province of Pangasinan, on the eastern edge of the delta of the Agno River. It is exposed to flooding, earthquake, fire, tsunami, and king tides. Milkfish farming, one of the main local resources, further contributes to flooding through siltation of local rivers. CBDRR stemmed from the partnership between the city, the Center for Disaster Preparedness, and the Asian Disaster Preparedness Center through funding from the United States Agency for International Development (USAID).

Through participatory risk assessment, the communities decided to engage in systematic DRR activities, notably in facing flooding. Through the project, the city organised a Technical Working Group (TWG) out of its City Disaster Coordinating Council (CDCC), now known as the City Disaster Risk Reduction and Management Council (CDRRMC), to implement DRR including CCA. TWG activities include planning, documentation, training, water quality monitoring, village/neighbourhood-level waste management, flood canal maintenance, and tree pruning. The villages/neighbourhoods were grouped into three teams and some members of the TWG were the team facilitators. Through this role, they were able to lead their team towards reducing and managing disaster risk at the community level. The TWG enabled the villages/neighbourhoods to participate in the monitoring of evacuation centres' readiness for disasters, survey the information on their vulnerability to floods, promote capacity building in DRR, and coordinate its DRR activities with other stakeholders.

The city has also established village/neighbourhood DRRMCs and community-based early warning systems (EWS). People participate in devising the EWS, manifested in the use of kanung-kong, an instrument made of bamboo that is used to inform people living in a zone, a smaller part of a village, of an impending flood event. The corresponding signal of the instrument for every alert level and number of strikes at designated time intervals that correspond to specific actions was agreed upon by the representatives of all the villages/neighbourhoods through a multi-stakeholder dialogue led by the city government in partnership with civil society organisations. The agreed warning codes have been integrated with EWS implemented by all villages/neighbourhoods in the city. People also monitor these codes based on the flood markers placed near rivers and flood-prone locations. The information comes from all DRR stakeholders. The information cascades from the regional level down to the community and then the community passes on the information they gather from their own experience to ensure that all stakeholders are aware of their situation.

The city government also made policy changes requiring all DRR programmes to include CCA. After RA 10121 was enacted, the city government crafted laws that support the institutionalisation of CBDRR within the City Disaster Risk Reduction and Management Plan, with corresponding budget appropriation and utilisation of the Local Disaster Risk Reduction and Management Fund (LDRRMF).

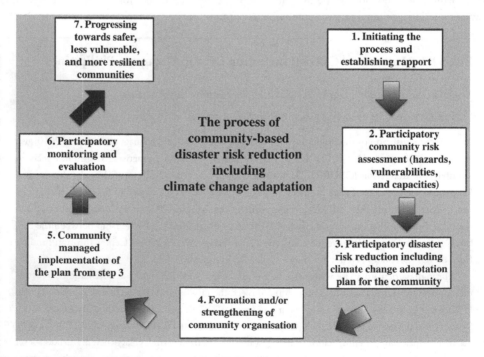

Figure 32.2 The Process of Community-based DRR Including CCA

(By Authors)

CBDRR aims for the empowerment of the members of local communities, particularly those who are most vulnerable, through genuine participation at all stages of the process. The community directly benefits from risk reduction and development as the process stems from the community members' felt and urgent needs. It is an integrated, comprehensive, and proactive process that considers all phases of DRR including CCA; i.e., pre-, during, and post-disaster measures are planned and implemented as necessary by the community. These include structural (infrastructure) and non-structural (health and literacy, economic resources, public awareness and training, community organising and advocacy, reforestation and environmental protection, etc.) initiatives.

The whole CBDRR process is owned by the community. Nonetheless, there is linkage of the community with other communities, organisations, and government agencies at various levels especially for vulnerabilities that the local community cannot address by itself. This is particularly important to combine indigenous/local knowledge with science and technology and foster support from outsiders. CBDRR thus addresses concerns of various stakeholders while upholding the basic interest of the most vulnerable sectors and groups in the community (Box 32.3).

Box 32.3   Fostering Community Governance for CBDRR Including CCA in Karachi, Pakistan

Zenaida Delica-Willison[1], Loreine B. dela Cruz[1], and Fatima Gay J. Molina[1]

[1] Center for Disaster Preparedness, Quezon City, Philippines

There are different motivations for and drivers of community participation. These vary depending on the culture and context of the communities involved. The case of the Orangi Pilot Project in Karachi, Pakistan, demonstrates the qualities of effective relationships between community participation and project effectiveness. The project's setting is the largest slum in Karachi, which faces an array of health hazards, including typhoid, malaria, diarrhoea, dysentery, and scabies.

The project managed to transform the community through its members organising into street committees to address waste disposal and participatory sanitation as the first priority. Community residents banded together to elect a project manager and contributed cash and voluntary labour to ensure that they own the sewers that were installed. Close to 100,000 households already enjoy the sewage facilities. Noteworthy is that community governance concerns had expanded to include housing, health, family planning, community-financed education, women's work centres, micro-enterprises, and reforestation.

What made all this possible and what are the lessons?

- The project was transformed into a community service for its community people. What has been key is the recruitment of organisers from within the community that have intimate knowledge of the locality that, in turn, facilitated the identification and design of effective services for the community people.
- The varied components of the project that were introduced were by-products of the community needs analysis and were evaluated and modified to be able to respond to continued changes happening in the community.
- The project opened windows of opportunity for community people to make improvements in their lives through community participation and collective action.

In summary, CBDRR aims to advance community members' empowerment through increasing people's options and capacities; more access to and control of resources and basic social services through concerted action; more meaningful participation in decision-making which affects their lives; and more control over the natural and physical environment. CBDRR develops the confidence of community members to participate in other development endeavours and contributes to addressing and reducing the complex relation of conditions, factors, and processes of vulnerabilities present in society, including poverty, social inequity, and environmental resources depletion and degradation (Victoria 2003).

Challenges to Community Governance for DRR Including CCA

Community governance for DRR including CCA faces many challenges. These may require different kinds of approaches, handling, and management.

Firstly, community governance involves tackling power relations and inequality. Power analysis cannot be constrained to the micro or macro level alone. Power permeates through internal boundaries at the country level but also externally at the global level. Power pervades domestic concerns in how knowledge is shared through networks and therefore exercises influence. It manifests in development policies and loan conditionalities that have impacts down to the grassroots and community level. Power and knowledge can influence poverty and inequality through the use of various strategies where some may be more appealing while the rest may be less desirable (Moncrieffe 2004).

Secondly, the increasing cultural diversity of communities within and across countries needs to be considered. Community risk assessments for informed DRR planning at the community level need to be participated in by diverse populations, including migrants, and all community members, including men and women, children and young people, persons with disabilities, indigenous peoples, and older persons. Such effort requires fair access to information and resources based on the fact that two parties may have different perceptions and socio-political power. As such, they may require equitable processes for negotiation.

Thirdly, there may be major institutional and social barriers to diverse community members' participation in DRR including CCA. It is, nonetheless, essential to ensure the inclusion of all groups at every phase of CBDRR, from risk assessment to DRR planning and implementation. CBDRR should also cover far-flung areas as much as easily accessible villages and neighbourhoods independently of political and/or social allegiances. Community assessment teams may be organised as a mechanism that promotes transparency and inclusivity to make sure that no one is left behind. Assessment teams may be composed of local officials and residents who are respected by the community.

Finally, fostering downward accountability is of paramount significance but often very difficult. This is due to the fact that there is a relationship factor between the people who make decisions and the people who are affected by the decisions. There is inherent transparency in the concept or idea of accountability. It is the capacity to provide explanations for carrying out responsibilities by an individual or organisations in particular concerns or services. There are three concepts embedded in downward accountability. These are transparency, trust, and people's participation. Local community governance can serve as an effective means to enhance downward accountability for particular services or cross-cutting community concerns. Enabling the community to go through the process of tackling and deciding its own particular services requirements and related concerns based on their unique context serves as a demonstration of community accountability to its constituents and a good example of governance and empowerment at this level (Box 32.4).

> ## Box 32.4 Overcoming the Challenges of Community Governance for DRR Including CCA in Japan
>
> *Zenaida Delica-Willison[1], Loreine B. dela Cruz[1], and Fatima Gay J. Molina[1]*
>
> [1] Center for Disaster Preparedness, Quezon City, Philippines
>
> The Women's Eye initiative is a good example of empowerment through genuine community governance in DRR including CCA, and post-disaster recovery. The Women's Eye initiative is a non-profit corporation in one of the prefectures in Japan. It seeks to engage and help women become active players in rebuilding their lives and their communities after a disaster like the Great East Japan Earthquake and Tsunami in March 2011. The organisation encourages local women's participation and empowerment by creating opportunities for dialogue, communication, and social change among them. The initiative puts focus on developing potential coordinators, organisers, and mediators rather than leaders as it realises and appreciates that there are more women who can become organisers, coordinators, and mediators than those that can become leaders.
>
> The initiative has achieved the following significant results:
>
> - It has provided space for the birthing of numerous small thematic communities that promote strong links and a sense of independence amongst members. A case in point is that participating women are now taking the lead in maintaining regular gatherings and planning new activities, displaying major shifts in their confidence and in community engagement.
> - Local women have become major actors in addressing issues in their locality. With the increase in the number of community organisers, the women's conditions have improved by developing them not only as leaders, but also more so as enablers and facilitators for varied needs of their communities.
>
> The initiative has helped in reducing communication barriers among women.
>
> Considering that the earthquake had broken up pre-existing communities, many individuals were left isolated and vulnerable. The initiative had harnessed the women's strength based on their daily focus in lives that made them equipped to identify and quickly help fellow community members who were socially vulnerable. The initiative had also helped them to participate actively in activities that discuss their conditions and issues that they face including their families and communities with fellow women in similar situations.
>
> In the long term, this initiative has strengthened DRR including CCA through creating safety nets as the situation called for withstanding social fragmentation, and forging mutual aid and support to one another. Utilising information sharing about living conditions through the newly established networks, many women became more active and concerned with their communities. In short, the many women whose social ties were weakened by the disaster have come out and taken on roles related to community organising and community engagement as an outcome of their participation in the Women's Eye initiative.

Conclusion

Lessons from community governance yield encouraging principles and guidelines that provide inspiration to further encourage CBDRR including for CCA. It draws upon the importance of people's participation in DRR for building a culture of safety and ensuring sustainable development for everyone, which is essential for reversing global and local trends of increases in disaster occurrence and losses from small-to medium-scale disasters and changing climate. The community further needs to develop and forge partnerships with various stakeholders, which include CSOs, community leaders, local and upper government agencies, NGOs, less vulnerable groups, the private sector, and donor agencies. Such a wide range of stakeholders is important to be taken into consideration to achieve the culture of safety and sustainable development.

There is no shortcut to DRR including CCA. Empowerment and development of communities is an ongoing process that needs to be undertaken towards the pursuit of more inclusive governance at the local level. It is a whole of community approach, engaging stakeholders so that they become involved in community concerns and set the agenda for sustainable community development and governance. Capability development is a constant need of the people in communities to enable them to become effective actors in governing their own communities. This is important and necessary for effective community governance as it depends highly on the level of awareness of people and their understanding of local issues whether economic, social, cultural, political, or spiritual, as well as on how they respond and take active roles on issues and needs of their community. Investing in capability building of communities and people contributes to realising the amplified voices that reverberate in various arenas and platforms at varying levels from local to national and global levels.

Community governance for DRR including CCA entails downward accountability. Transparency, trust, and people's participation need to be observed and applied. Accountable community governance therefore respects the rights, dignity, and safety of community members. All community members should participate in decisions that affect their lives, have access to information that guide their informed decision-making, and actively provide feedback for the continued improvement of the affairs of the community.

In the end, it is up to the communities themselves to make their community governance for DRR including CCA effective and to make it simple as they apply it in their own unique context. It lessens the burden of complexities and nuances of a governance set-up as they appropriately employ flexible leadership. Ultimately, it is truly about how people live together, do their work as a community, and aspire together to improve and become better as a community.

References

Abarquez, I. and Murshed, Z. (2004) *Community-based disaster risk management: field practitioners' handbook*, Bangkok: Asian Disaster Preparedness Center.

Bowles, S. and Gintis, H. (2002) 'Social capital and community governance', *The Economic Journal* 112, 483: F419–F436.

Cannon, T., Titz, A. and Krüger, F. (2014) 'The myth of community?', in International Federation of Red Cross and Red Crescent Societies (ed.) *World disaster report: focus on culture and risk*, Geneva: International Federation of Red Cross and Red Crescent Societies, pp. 92–119.

Delica-Willison, Z. (2006) *Integrating disaster risk management in local governance*, Bangkok: United Nations Development Programme.

Delica-Willison, Z. and Gaillard, JC (2012) 'Community action and disaster', in B. Wisner, JC Gaillard and I. Kelman (eds) *Handbook of hazards and disaster risk reduction*, London: Routledge, pp. 711–722.

Gaillard, JC and Mercer, J. (2013) 'From knowledge to action: bridging gaps in disaster risk reduction', *Progress in Human Geography* 37, 1: 93–114.

Green, L.W. and Mercer, S.L. (2001) 'Can public health researchers and agencies reconcile the push from funding bodies and the pull from communities?', *American Journal of Public Health* 91, 12: 1926–1943.

Martin, D. (2005) 'Rethinking aboriginal community governance', in P. Smyth, T. Reddel and A. Jones (eds) *Community and local governance in Australia*, Sydney: University of New South Wales Press, pp. 108–127.

Maskrey, A. (1989) *Disaster mitigation: a community based approach, Development Guidelines No. 3*, Oxford: Oxfam.

McKinlay, P. (2014) *An overview of developments and neighbourhood governance*, Tauranga: McKinlay Douglas Ltd.

Molina, F.G.J. (2016) 'Intergenerational transmission of local knowledge towards river flooding risk reduction and adaptation: the experience of Dagupan City, Philippines', in M.A. Miller and M. Douglass (eds) *Disaster governance in urbanising Asia*, Singapore: Springer, pp. 145–176.

Moncrieffe, J. (2004) *Power relations, inequality and poverty: a concept paper for the World Bank*, London: Overseas Development Institute.

Oakley, P. and Marsden, D. (1987) *Approaches to participation in rural development*, Geneva: International Labour Organization.

Oliver-Smith, A., Alcantara-Ayala, I., Burton, I. and Lavell, A. (2016) *Forensic Investigations of Disasters (FORIN): a conceptual framework and guide to research*, Beijing: Integrated Research on Disaster Risk.

Paul, S. (1987) *Community participation in development projects: the World Bank experience*, Washington DC: The World Bank.

Totikidis, V., Armstrong, A. and Francis, R. (2005) *The concept of community governance: a preliminary review*, Melbourne: Victoria University.

Twigg, J. (2004) *Disaster risk reduction: mitigation and preparedness in development and emergency programming, Good Practice Review No. 9*, London: Humanitarian Practice Network.

UNDP (2010) *Disaster risk reduction governance and mainstreaming*, New York: UNDP (United Nations Development Programme).

Victoria, L.P. (2003) 'Community based disaster management in the Philippines: making a difference in people's lives', *Philippine Sociological Review* 51: 65–80.

Wisner, B. (1993) 'Disaster vulnerability: scale, power and daily life', *GeoJournal* 30, 2: 127–140.

33

NGOS DOING DISASTER RISK REDUCTION INCLUDING CLIMATE CHANGE ADAPTATION

Terry Gibson

Introduction

Civil society organisations working at the local level can offer valuable insights into the relationship between Disaster Risk Reduction (DRR) and Climate Change Adaptation (CCA). They bring an understanding of how these thematic areas are perceived by risk-affected people in diverse contexts. However, as well as their potential, civil society organisations face many constraints. This chapter considers the recent development and roles of civil society organisations, recognising both the contributions they can make and the constraints they face. It focuses on the case of an international network of civil society organisations to draw out insights on the interface between DRR and CCA from the local perspective, ultimately arguing for CCA as part of DRR.

The Contribution and Constraints of Civil Society Organisations

A widely quoted statement regarding the relationship between the ideal and the practical reality of civil society is 'We dreamed of Civil Society and they gave us NGOs [non-governmental organisations]' (Misslevitz in Einhorn 2005: 12). It suggests disquiet about this relationship. To understand what lies behind this we need to consider the origins, development and context of civil society.

Whilst the term has held various meanings historically, it is commonly accepted (e.g., Banks and Hulme 2012; Fowler 2011) that civil society in its current form, particularly in the context of international development, emerged and expanded as a response to the humanitarian emergencies created by the Second World War (Duffield 2007). The subsequent decades saw growing programmes of humanitarian assistance and the emergence of civil society organisations (CSOs) at several scales. Most visibly, International Non Governmental Organisations (INGOs), a few of which (e.g., Save the Children Fund and the International Federation of the Red Cross and Red Crescent Societies) were already in operation, but many of which were established in the post-war period (including Christian Aid, CARE, Catholic Relief Services, Oxfam and World Vision) (Duffield 2007). Mainly headquartered in the north, representing the more wealthy world and thus relatively distant from local concerns, these represented a significant strand of civil society. As these agencies grew in scale and resources there was increasing awareness of the idea of 'doing development' as well as humanitarian response. This led to an emphasis in many cases on the importance of community development (Banks and Hulme 2012). A comparable transition from disaster response to preparedness led to the emergence of what the United Nations (UN) called 'Natural Disaster Reduction' in 1987, and the nomination of the 1990s as the 'International Decade of Natural Disaster Reduction'. The mushrooming development and humanitarian response industry was interpreted by some as a welcome means of substituting for the shrinkage of the State and its responsibilities, reflecting the dominant neo-liberal agenda of the 1980s and 1990s (Fowler 2011).

Alongside the large INGOs another growing layer of civil society consisted of smaller national and sub-national NGOs, often enlisted as 'partners' to the larger organisations to deliver their programmes at the local level. Distinct from these a further layer consisted of 'grassroots organisations'. These emphasised local legitimacy and involvement and often resisted becoming delivery agents for the larger NGOs and INGOs (Edwards 2011), instead maintaining action agendas defined and agreed locally.

To this burgeoning industry was added a further emergent phenomenon of transnational networks and movements including the Nestlé baby milk campaign (Keck and Sikkink 1998), the Jubilee 2000 campaign (Clark 2003) and the International Campaign to Ban Landmines (Cox 2011). Their existence became possible partly because of globalising trends, such as increasing impacts of multinational companies' operations and globalised economic flows. They were also particularly enabled by new communications technologies; initially email, followed by later technologies such as Skype, which enabled instantaneous and cheap communication (the asynchronous nature of email also helped to reduce timezone challenges) (Mawdsley et al. 2002; Waddell 2011).

Since the Second World War, a multi-layered and interdependent industry representing the practical embodiment of 'civil society' has therefore been constructed. Why does our opening quotation contrast the reality of NGOs with the ideal of civil society? A structuralist, actor-based perspective emphasising the different motivations and priorities of the range of actors making up organisations (Long 2001) suggests that the industry, like any institution, is partly configured by it and its participants' own need for survival (Clemens and Cook 1999). This can be seen at an organisational and individual level (Lister 2003). Organisations large and small have to secure funding, which in turn defines their behaviour. Projects have to be delivered according to the objectives set by the funding agencies, whether they are relevant to the ultimate goals of organisations and the communities they serve or not.

The methodologies through which projects and their objectives are delivered can become embedded as accepted practice and again can become self-serving rather than meeting the stated goals of the organisation. Participative methodologies are an example of this, becoming in some cases 'tyrannies' (Cooke and Kothari 2001). Choices made in development projects can be shaped by national and commercial priorities (e.g., see Baird and Shoemaker 2007). INGOs and NGOs dependent on state and institutional funding have to pursue programmes consistent with donor priorities and ideologies, leading one commentator with deep experience of the industry to argue that the development system maintains the dichotomy of social protection in the more wealthy world and self-reliance of the less affluent countries, rather than challenging it (Duffield 2007). It is critiques such as these, often emanating from actors within the industry, which lead to the heartfelt complaint 'we dreamed of Civil Society and we got NGOs'.

Evaluation of NGO impact has become increasingly important as the scale of investment leads governments and other donors to assess 'value for money'. An early survey of evaluations carried out by a number of governments found that a majority of programmes evaluated were 'successful' in terms of their immediate goals, but raised the question of whether they were achieving wider impact on societal change and progress (ODI 1996). This theme is echoed by other commentators including Edwards (2008) whose survey distinguished between substantial successes in 'service delivery' and more limited impact in addressing broader systemic factors such as the perpetuation of poverty and the abuse of human rights, reflecting his view that agencies have failed to innovate in their relationships with partners and in terms of downward accountability. Banks and Hulme (2012) argue that the growth of the development industry has led to a gradual separation of NGOs from their grassroots origins and closer liaison with and dependency on governments. Contrasting with this trend they give as an exemplar the case of 'shack/slum dwellers international', which has strong grassroots connections and uses processes of knowledge sharing through citizen-led surveying and through face-to-face meetings between communities to enable its participants to engage with structural issues. On a global scale NGO-led movements and consortia have achieved substantial impact through campaigns such as the Nestlé, anti-mine and Jubilee 2000 campaigns mentioned above, whereas other campaigns such as 'Make Poverty History' have been hampered by the constraints of the INGO consortia leading them (Cox 2011). The Global Network of Civil Society Organisations for Disaster Reduction (GNDR) allied its experience of a particular global action and campaign 'Views from the Front-line' to the concept of 'communities of praxis' (Gibson 2012), suggesting that cohering a range of organisations around a shared action reflecting the common goals of the organisations enabled it both to advance its cause and to achieve shared learning from reflection on its actions, echoing the Freirian concept of 'praxis' as an engine for change (Freire 1970).

Whether through restructuring to emphasise the local, through new collaborations or through new modes of action and learning there is a common view that NGOs at every scale should exercise critical reflection and consider radical reform to strengthen legitimacy at whatever level they work, avoiding co-option to political and commercial agendas (Banks and Hulme 2012; Edwards 2008; Mawdsley *et al.* 2002).

Within this diverse industry GNDR links several forms of civil society and is active at the meeting point between DRR and CCA. GNDR is a transnational network that draws together a large number of such organisations (over 1,000 in 2015) ranging from INGO scale down to grassroots organisations. With origins in the DRR thematic area its membership embraces organisations also concerned with CCA and Sustainable Development. This case highlights their practitioner and experiential perspective on links between DRR and CCA and why CCA should become embedded within DRR. The GNDR also demonstrates how the potential and constraints of NGOs have played out. In the following sections the term 'Civil Society Organisation (CSO)' will be used to refer to the wide range of GNDR member organisations, including INGOs, NGOs and grassroots organisations.

Learning from a Group of CSOs: The Local Perspective

The work of GNDR's membership, particularly their collaboration on the 'Views from the Front-line', 'Action at the Frontline' and 'Frontline' shared actions (see Box 33.1), has generated significant learning regarding local level experience of risk and resilience, and about the links between DRR and CCA. This has led to an identified need to embed CCA within DRR actions given the lack of distinction between the two at the local level (Wisner *et al.* 2014).

Box 33.1   The Global Network of Civil Society Organisations for Disaster Reduction (http://www.gndr.org)

Terry Gibson[1]

[1] Independent, Manchester, UK

The 'Global Network of Civil Society Organisations for Disaster Reduction' (GNDR) was formed in 2007 through the recognition that member organisations could do more together than apart. In

---

2015 the membership of Civil Society Organisations passed the one thousand mark. It initially coordinated its members to provide a complementary local level perspective on progress of the United Nations International Strategy on Disaster Reduction's (UNISDR's) 'Hyogo Framework for Action' on Disaster Risk Reduction (DRR). Its main action was the 'Views from the Frontline' (VFL) programme, conducted in 2009, 2011 and 2013 and reporting at the biennial UNISDR 'Global Platform for Disaster Risk Reduction'.

Consultation and learning from these actions led to a local level focus through the 'Action at the Frontline' programme. This was designed to support members in collaboration with local communities for action and learning, offering an alternative to short-term project-driven interventions. During 2013–2015 it was developed into 'Frontline', which provided the ability to gather together knowledge from local level to provide analysis and insights at national and global levels to inform better implementation and practice through understanding local needs and priorities. The work of the members has provided rich knowledge of local level risk and resilience as well as of the potential of and the problems faced by Civil Society, particularly working at local levels.

Initial work conducted in Views from the Frontline (2008–2013) suggested anecdotally that from the local perspective threats to prosperity, livelihoods and lives are not neatly segregated into different thematic areas. People consider the combination of factors that affect them: environmental, social, economic and political (see also Mercer 2010). It also suggested that many of these threats were not from large-scale, intensive, high visibility events but from multiple, small-scale, fast and slow onset events (GNDR 2013). The overall profile of these events seemed to be highly variable from locality to locality, rather than fitting a limited set of risk profiles. In understanding and addressing these the links between different scales were weak, and this observation encompassed governance, information and resources. It seemed that in many cases people at local level were the first, and often only, actors and responders (GNDR 2011a). Since many of the factors that affected them weren't classified as emergencies they weren't recorded or resourced beyond local level. Not only was understanding of these many factors often restricted to local level but information at other scales was often not accessible at local level (GNDR 2013).

More recently (2013–2015) GNDR's own reflection led it to develop 'Frontline' to explicitly gather this local level knowledge (GNDR 2015a). Findings from work in fifteen Latin American countries and now extending to a further fifteen countries in other regions provide an evidence base that largely confirms the anecdotal findings from VFL. In common with other investigations (e.g., UNISDR 'Global Assessment Review' (GAR) 2015) it shows that while large-scale disasters have a devastating impact, small-scale disasters account for over 40 per cent of losses and over 90 per cent of records at local level. It also confirms the finding that a mix of factors affect people; for example, in the Latin American data the third highest priority threat according to local level respondents is insecurity resulting from crime and violence. The data also displays high variability. For example, a dataset from thirty communities in Indonesia shows very limited correlation from community to community (GNDR 2015a). Respondents also cite poor governance, lack of information and lack of resources as major barriers to progress.

The findings from VFL and Frontline are summarised in Table 33.1.

A further body of work conducted by GNDR member organisations provides specific case study insights on the interplay between DRR and CCA. 'Action at the Frontline' (AFL) is the local component of the Frontline programme, in which participants take the Frontline data gathered in their locality and use it as a basis for discussion, action planning, partnership building and implementation. Not only does this meet the request of the members for the programme

Table 33.1 Observations on Local Level Risk and Resilience

Experience of risk and resilience at local level:
Multiple environmental, economic, social and political factors driving risks.
Predominantly small-scale 'everyday disasters'.
High variability in risk profiles from locality to locality.
Weak links to other scales of governance.
Weak flows of information to and from local level.
Limited resources.
Local people often first and only responders.

to be directly relevant at local level, but it provides case studies that complement the Frontline analysis, bring it to life, and test its conclusions against specific situations. These case studies reveal a complex interplay between different threats related to climate change and other causes; see, for example, the case from Kiribati (Box 33.2).

Box 33.2 Interconnected Hazards in a Small Island Developing State

Terry Gibson[1]

[1] Independent, Manchester, UK

Small Island Developing States face some of the starkest challenges resulting from hazards including climate change. GNDR member organisation 'Foundation for the Peoples of the South Pacific Kiribati' (FSPK) working on this Pacific island group right on the equator, has been conducting an Action at the Frontline programme in the most populous island of the group, South Tarawa (GNDR 2015b). People there face multiple threats including from increased flooding during king tides, saline incursion reducing access to fresh water, increased sea temperature and acidity reducing the haul from fishing, droughts damaging production of copra, and from social pressures resulting from the gradual loss of land for accommodation. These threats are amplified by migration to South Tarawa from other islands in the group as unemployment increases, driven by the reduced income from fisheries and agriculture. The increased population density also heightens environmental degradation. Particular groups such as those living with disabilities are neglected and marginalised as a result of the social and economic pressures. FSPK found that while people were extremely concerned about the decline in their livelihoods they did not understand the causes. For example, they blamed the increased salinity of drinking water from wells on the hot sun.

FSPK adopted the creative solution of making a disabled community, Tetoamatoa, champions to promote community understanding and action on the effects of climate change. This led to the development of a coalition of local people, organisations and government to identify and drive options for action taking account of all the threats facing the island's inhabitants.

The case illustrates the importance of embedding CCA within DRR activities and also that the effects of climate change are variable and very specific to local contexts. For example, Lavell (2015) found that Frontline data for Central America depicted high variability in risk profiles.

Acquiring useful knowledge and sharing community perspectives can be challenging. For example, on the North coast of Vietnam GNDR member organisation 'Development Workshop France', with a focus on the development of safe housing at community level, finds that communities have developed understanding of a wide range of threats which they have experienced for centuries – typhoons and floods, for example. What they find more difficult to understand is how to factor in the progressive changes in these phenomena resulting from climate change. The political context is one in which they neither have good access to technical knowledge from government institutions, or the ability to share their own experience and concerns. CCA adds a dimension of complexity to their DRR activities which is difficult for them, in the absence of this information, to understand (GNDR 2015b).

The threats that communities face can't be addressed separately but have to be considered holistically. For example, in Western Alexandria, Egypt, progressive reduction of agricultural productivity is believed to be a result of droughts potentially exacerbated by climate change. This triggers a pattern of migration in and out of the area, which in turn increases economic and environmental pressures. As the area is on the border of two administrative regions governance is poor. The work of the CSO in this area focused on strengthening local self-organisation to manage this range of interlocking pressures. Similarly, in another AFL programme on Kenya's Somalian border the CSO found that response to increasing droughts was impeded by local ethnic conflict. Ethnic tensions also led to poor local governance as people tended to travel from their homes to sympathetic ethnic areas to vote, resulting in them having no representation in their own locality. AFL activities facilitated by the CSO focused on critical self-reflection by the community to identify how they could take actions locally and also engage with local government. These case studies show the effects of hazards including climate change interacting with economic and political factors, social pressures, poor governance and conflict. They underpin the argument that at local level, silos (including that of DRR and CCA) must be broken down to achieve progress (GNDR 2015b).

Local level response alone is often insufficient. For example, participants in an AFL programme in Malawi saw engagement with government as critical to progress as without shifting from a situation where the government only responds to events classified as emergencies the majority of small-scale threats would continue to be neglected. A district government officer from the region involved in the discussion strongly supported this view (Gibson 2015). A further case study in Namibia showed a similar pattern, where repeated everyday disasters – parts of a provincial town flooding annually as a result of the topography and poor drainage – were ignored as they did not trigger emergency response (GNDR 2011b). Poor linkages of knowledge and governance create constraints that can't be addressed purely by local action.

This vein of rich qualitative information is drawn together in summary in Table 33.2, extending Table 33.1 and adding the additional factor of poor understanding of climate change, which has been highlighted in the AFL cases.

Overall, the cases support the idea that CCA represents a particular risk factor within people's integrated understanding of the risks they face. From a local perspective CCA should be situated within DRR. However, the cases show that climate change, as with other hazards and drivers, presents particular challenges, distinct from other threats understood and addressed locally. Its environmental effects are unclear and highly variable, the time frames and scale of changes are also unpredictable. Knowledge is often difficult to access at local level and integrating this threat into the complex multi-threat profiles experienced at local level demands an interplay between local experiential knowledge and other sources of technical knowledge which is also often lacking.

In all of the above, civil society, particularly where it combines engagement at local level with bridge-building capacities to other scales of knowledge and governance, is potentially

Table 33.2 Evidence for Observations on Local Level Risk and Resilience

Experience of risk and resilience at local level:	Evidence from GNDR, Frontline (FL) and Action at the Frontline (AFL):
Multiple environmental, economic, social and political factors driving risks.	FL data shows wide range of reported threats, for example 'insecurity' third highest priority in Central America. AFL cases show complex risk profiles.
Predominantly small-scale 'everyday disasters'.	Report on FL in Central America (Lavell 2015) shows emphasis on 'everyday disasters'. UNISDR GAR 2015 also highlights prevalence of 'extensive disasters'.
High variability in risk profiles from locality to locality.	FL database for Indonesia shows high variability between the 30 communities consulted. Report on FL in Central America (Lavell 2015) shows high variability in the region.
Weak links to other scales of governance.	FL data from Central America frequently reports this as a barrier. AFL cases (i.e., Kenya and Vietnam) highlight challenges of engaging with other scales.
Weak flows of information to and from local level.	FL data from Central America frequently reports this as a barrier. AFL case studies highlight constraint of limited technical information access and limited awareness of local information at other scales.
Limited resources.	Frequently cited by GNDR membership as a constraint. FL data from Central America frequently reports this as a barrier.
Local people often first and only responders.	AFL case studies (i.e., Malawi and Namibia) show that government response only triggered by emergencies.
Effects of climate change poorly understood.	AFL case studies (i.e., Vietnam, Egypt and Kiribati) cite this as a local level challenge.

well-equipped to build bridges of understanding, contextualised responses, and learning from action. The following section investigates this potential, and the associated constraints of civil society.

Civil Society: the Potential and the Challenge

Experience from GNDR working with member organisations since 2008 leads us to argue strongly on the basis of the work and evidence of the VFL, AFL and Frontline programmes that CSOs can make an important contribution to learning and action. This is rooted in an understanding of the interaction of many, often small-scale threats at local level and in the importance of linking local level action to other scales of knowledge, governance and resourcing. The case studies above reinforce the necessity for an integrated approach to multiple threats. They also demonstrate that local action can only achieve so much, and part of the CSO role is to build bridges to other organisations, sources of expertise and sources of political leverage and influence. In GNDR's case consultation with members has revealed that both the association with the GNDR's actions (such as VFL) and identity with a global network are factors that strengthen the influence of CSOs when attempting to influence government at local and national levels.

In exerting such influence CSOs work from a foundation of the integrated understanding of DRR and CCA outlined in Tables 33.1 and 33.2 above. They are natural advocates for this perspective. In practice CSOs also face constraints. To undertake this bridge-building role CSOs need to have local engagement, ability to operate across scales, and freedom to do so. However, in many cases CSOs become service delivery agents delivering programmes of work that are often externally defined, are not shaped to take account of specific local contexts, are often short term due to donor requirements, and allow limited learning and adaptation. See Box 33.3 for a case illustrating the constraints faced by many CSOs.

Box 33.3   Escaping the Project Treadmill

Terry Gibson[1]

[1] Independent, Manchester, UK

A member organisation in Malawi explained the value of the (Action at the Frontline) AFL programme, for which they receive a very small amount of funding, as being the only opportunity they have to work together with a community, develop common understanding, analyse local needs critically and fashion sustainable action plans (Gibson 2015). By contrast, they explained that the funded project work that fills the rest of their week may have an element of 'participation' but in reality the project goals are determined, the project is delivered and sustainability and learning are very limited.

Small activist CSOs have little time to be reflective or strategic. Without time to reflect and think critically they tend to reproduce standard approaches rather than learning from their experience. Without time to explore new options and partnerships they are unable to build the bridges which give them greater influence. All of this is driven by the project funding machine, which has the effect of enforcing the current view of development on organisations who know from their local experience that this view is often deeply flawed. Table 33.3 depicts both the potential and the constraints of CSOs identified in this discussion.

GNDR has provided one context to step out of these constraints through programmes focusing on local level learning and action, and building bridges to other scales. More generally enabling CSOs to fulfil a unique bridge-building role demands changes in their positioning and engagement including:

1   Recognition. Like the communities they serve, local CSOs are often the last who are heard in the fashioning of development priorities and programmes. Rather than treating them as passive service delivery agents their local experience and ability to engage cross-scale could be strengthened by involving them and resourcing them as active agents to build two-way flows of understanding and action. At least one large development fund – USAID's 'local-works' – is currently experimenting with a collaboration model to achieve this (https://www.usaid.gov/partnership-opportunities/ngo/localworks).

2   New resourcing models. Large-scale programmes and funding models exclude small-scale actors. This is inevitable as agencies try to manage transaction costs but models are needed that bridge between large-scale resource allocations and local level resourcing requirements. INGOs play a part in this in working through local partners, though the challenge here is to build genuine partnerships rather than top-down structures. Other novel models include community level grants. To work, these need to make monitoring and accountability requirements appropriate to the scale of funding rather than demanding the same heavy reporting requirements as larger funds. They also need to identify appropriate intermediaries to strengthen access from local level.

Table 33.3 Contributions and Constraints of Civil Society Organisations (CSOs)

Experience of risk and resilience at local level:	Evidence from GNDR, Frontline (FL) and Action at the Frontline (AFL):	Potential contribution of CSOs:	Constraints of CSOs:
Multiple environmental, economic, social and political factors driving risks.	FL data shows wide range of reported threats, for example 'insecurity' third highest priority in Central America. AFL cases show complex risk profiles.	Build bridges between thematic areas i.e., DRR/CCA/Sustainable Development Goals (SDGs).	Funding constraints, i.e., projectised funding, push CSOs back into 'silos'.
Predominantly small-scale 'everyday disasters'.	Report on FL in Central America (Lavell 2015) shows emphasis on 'everyday disasters'. UNISDR GAR 2015 also highlights prevalence of 'extensive disasters'.	Rich experience and understanding of the nature of everyday disasters encompassing DRR including CCA.	Lack of understanding and support from other scales such as national government and international institutions.
High variability in risk profiles from locality to locality.	FL database for Indonesia shows high variability between the 30 communities consulted. Report on FL in Central America (Lavell 2015) shows high variability in the region.	CSOs working at local level understand and contextualise their work to take account of high variability rather than taking a 'one size fits all' approach. Particularly important in understanding local level CCA impact and actions.	Limited capacity to act on their understanding.
Weak links to other scales of governance.	FL data from Central America frequently reports this as a barrier. AFL cases (i.e., Kenya and Vietnam) highlight challenges of engaging with other scales.	Able to engage with different scales including local, sub-national and national. Promoting CCA as part of DRR.	Challenge of securing political space and voice. Some members report this is becoming increasingly difficult.
Weak flows of information to and from local level.	FL data from Central America frequently reports this as a barrier. AFL case studies highlight constraint of limited technical information access and limited awareness of local information at other scales.	Able to access sources of knowledge and share local level knowledge through association with networks, platforms and institutions.	Focus on action tends to reduce preparedness and ability to secure information.
Limited resources.	Frequently cited by GNDR membership as a constraint. FL data from Central America frequently reports this as a barrier.	Able to promote dialogue and partnerships to strengthen access to resources.	Development funding and aid difficult to access directly at local level.
Local people often first and only responders.	AFL case studies (i.e., Malawi and Namibia) show that government response only triggered by emergencies.	Able to strengthen partnerships and collaborations to increase capacity.	Government structures and policies tend to focus on large-scale response rather than resilience building.
Effects of climate change poorly understood.	AFL case studies (i.e., Vietnam, Egypt, and Kiribati) cite this as a local level challenge.	Able to access specialist knowledge and contextualise it to inform understanding of climate change scenarios and action.	Limited ability to access knowledge. Practitioner/ academic divide.

3   Allow more time. Whilst donors often want to see rapid results, collaborations to act and
    learn at local level sustainably are often more effective where the scale is smaller and the
    timescale longer. Project cycles of three years are common (for example, the UK Department
    for International Development's (DFID) Building Resilience and Adaptation to Climate
    Extremes and Disasters (BRACED) programme (DFID 2015)) and when the project startup
    and the final reporting stages are included in this the period allowed to develop engaged and
    sustainable actions is extremely limited.
4   Strengthen networks which enable CSOs to benefit from peer to peer learning, mutual sup-
    port, shared actions and increased recognition. GNDR is one of a growing group of 'Global
    Action Networks' (Waddell 2011) that enable CSOs to maintain their local presence and
    legitimacy whilst strengthening their ability to act as bridge-builders.
5   Support joint civil society actions for learning, monitoring and accountability. Whilst there
    are many high level monitoring processes these are usually insensitive to local level impact,
    whereas civil society has the potential to monitor the local level effectiveness of national
    policies and international frameworks.

Conclusion

At the local level people face very specific combinations of threats to their prosperity and well-
being, which evidence in this chapter suggests are often small scale, driven by environmental, social,
economic and political factors and which interact in very specific ways dependent on local contexts.
This has led to an identified need to embed CCA actions within DRR. However, climate change
presents particular challenges of understanding and action as a new and unpredictable factor.

Local experience provides a valuable layer of knowledge, combined with other sources, to
inform effective and integrated action to reduce the impact of the many threats people face. This
layer is often neglected and CSOs working at local level have the potential to play an important
bridge-building role, enabling two way flows of understanding and targeting resources and action
to specific contexts. This potential is often constrained by the structures of development and
humanitarian action. To realise the potential of CSOs as bridge-builders, and make action more
targeted, effective and sustainable their potential and also their constraints should be recognised,
their role valued and strengthened and their voices and experience heard. Mechanisms for small-
scale, long-term local resourcing should be developed, collaborations and networks should be
strengthened and monitoring from the local level perspective should be supported.

References

Baird, I. and Shoemaker, B. (2007) 'Unsettling experiences: Internal resettlement and international aid agen-
    cies in Laos', *Development and Change* 38, 5: 865–888.
Banks, N. and Hulme, D. (2012) *The Role of NGOs and Civil Society in Development and Poverty Reduction*,
    Brooks World Poverty Institute Working Paper No. 171. Online http://papers.ssrn.com/sol3/papers.
    cfm?abstract_id=2072157 (accessed 9 February 2016).
Clark, J. (ed.) (2003) *Globalising Civic Engagement*, London: Earthscan.
Clemens, E.S. and Cook, J.M. (1999) 'Politics and institutionalism: Explaining durability and change',
    Annual Review of Sociology 25: 441–466.
Cooke, B. and Kothari, U. (eds) (2001) *Participation: The new Tyranny?* London: Zed Books.
Cox, B. (2011) *Campaigning for International Justice*, Bond for International Development. Online https://
    www.bond.org.uk/data/files/Campaigning_for_International_Justice_Brendan_Cox_May_2011.pdf
    (accessed 9 February 2016).
DFID (Department for International Development). (2015) *Building Resilience and Adaptation to Climate
    Extremes and Disasters (BRACED) Programme Project Development and Full Proposal Guidelines*, United

Kingdom's Department for International Development. Online https://www.gov.uk/government/uploads/system/uploads/attachment_data/file/286951/Application-Guidelines.pdf (accessed 9 February 2016).

Duffield, M. (2007) *Development, Security and Unending War*, Cambridge: Polity Press.

Edwards, M. (2008) *Have NGOs Made a Difference? From Manchester to Birmingham with an Elephant in the Room*, Global Poverty Research Group Working Paper No. 028. Online http://www.gprg.org/pubs/workingpapers/pdfs/gprg-wps-028.pdf (accessed 9 February 2016).

Edwards, M. (ed.) (2011) *The Oxford Handbook of Civil Society*, Oxford: Oxford University Press.

Einhorn, B. (2005) *Citizenship, Civil Society and Gender Mainstreaming: Contested Priorities in an Enlarging Europe*, Pan-European Conference on 'Gendering Democracy in an Enlarged Europe', 20 June, Prague, Czech Republic. Online www.qub.ac.uk/egg/PragueConference/BarbaraEinhorn.doc (accessed 9 February 2016).

Fowler, A. (2011) 'Development NGOs', in M. Edwards (ed.), *The Oxford Handbook of Civil Society*, Oxford: Oxford University Press, pp. 42–54.

Freire, P. (1970) *Pedagogy of the Oppressed*, London: Penguin.

Gibson, T. (2012) 'Building collaboration through shared actions: The experience of the Global Network for Disaster Reduction', *Jàmbá: Journal of Disaster Risk Studies* 4, 1: 48.

Gibson, T. (2015) *Conversations Held at Global Network of Civil Society Organisations for Disaster Reduction (GNDR) workshop, 21 July*, Johannesburg, RSA.

GNDR (Global Network of Civil Society Organisations for Disaster Reduction). (2011a) *Views from the Frontline (VFL) 2011*, UK: GNDR. Online http://www.gndr.org/programmes/views-from-the-frontline.html (accessed 9 February 2016).

GNDR (Global Network of Civil Society Organisations for Disaster Reduction). (2011b) *Action at the Frontline (AFL) 2011*, UK: GNDR. Online http://www.gndr.org/learning/resources/case-studies/case-studies-afl-2012.html (accessed 9 February 2016).

GNDR (Global Network of Civil Society Organisations for Disaster Reduction). (2013) *Views from the Frontline (VFL) 2013*, UK: GNDR. Online http://www.gndr.org/programmes/views-from-the-frontline.html (accessed 9 February 2016).

GNDR (Global Network of Civil Society Organisations for Disaster Reduction). (2015a) *Frontline*, UK: GNDR. Online http://www.gndr.org/programmes/frontline-programme.html (accessed 9 February 2016).

GNDR (Global Network of Civil Society Organisations for Disaster Reduction). (2015b) *Action at the Frontline (AFL)*, UK: GNDR. Online http://www.gndr.org/programmes/action-at-the-frontline.html (accessed 9 February 2016).

Keck, M. and Sikkink, K. (1998) *Activists beyond Borders*, Ithaca: Cornell University Press.

Lavell, A. (2015) *Summary report of Frontline Central America*, Teddington, UK: Global Network of Civil Society Organisations for Disaster Reduction (GNDR).

Lister, S. (2003) 'NGO legitimacy: Technical issue or social construct?', *Critique of Anthropology* 23, 2: 175–192.

Long, N. (2001) *Development Sociology: Actor Perspectives*, London: Routledge.

Mawdsley, E., Townsend, J., Porter, G. and Oakley, P. (2002) *Knowledge, Power and Development Agendas: NGOs North and South*, Oxford: International NGO Training and Resource Centre (INTRAC).

Mercer, J. (2010) 'Disaster risk reduction or climate change adaptation: Are we reinventing the wheel?', *Journal of International Development* 22: 247–264.

ODI (Overseas Development Institute). (1996) *ODI Briefing Paper 2: The Impact of NGO Development Projects*, London: ODI. Online http://www.odi.org/sites/odi.org.uk/files/odi-assets/publications-opinion-files/2636.pdf (accessed 23 February 2016).

UNISDR (United Nations International Strategy for Disaster Reduction). (2015) *Global Assessment Report (GAR) on Disaster Risk Reduction: Loss Data and Extensive Risk Analysis*, Geneva: UNISDR. Online http://www.preventionweb.net/english/hyogo/gar/2015/en/gar-pdf/Annex2-Loss_Data_and_Extensive_Risk_Analysis.pdf (accessed 9 February 20160.

Waddell, S. (2011) *Global Action Networks*, London: Palgrave.

Wisner. B., Oxley, M., Budihardjo, P.H., Copen, K., Castillo, G., Cannon, T., Mercer, J. and Bonduelle, S. (2014) '"Down home, it's all the same": Building synergisms between community-based disaster risk reduction and community-based climate change adaptation', in E.L.F. Schipper, J. Ayers, H. Reid, S. Huq and A. Rahman (eds), *Community-Based Adaptation to Climate Change: Scaling it up*, Oxford: Routledge.

34

PRIVATE SECTOR DOING DISASTER RISK REDUCTION INCLUDING CLIMATE CHANGE ADAPTATION

Joanne R. Stevenson and Erica Seville

Introduction

In many economies globally, the private sector creates significant employment and economic activity, and consequently has significant influence over the way disaster risk is borne and transacted in society. The private sector can refer generally to all organisations that are not under direct state control. The discussions in this chapter will refer primarily to for-profit organisations and revenue generating activities that are not owned or operated by government entities. In this chapter we explore the relationship between the private sector and disaster risk reduction (DRR) including climate change adaptation (CCA) through three lenses.

The first lens provides an examination of the way that private sector organisations assess and respond to the risks they face from natural hazards and hazard drivers, including climate change. The second lens examines the role of private sector investment in creating or exacerbating disaster risk for the communities with whom they interact. Finally, the third lens provides a view of the way the private sector can and does consciously work on DRR. This is then followed by a discussion of the various ways the public sector can manage and enhance the role the private sector plays in disaster risk reduction (DRR), including CCA.

This chapter only provides a preliminary survey of the issues relating to the private sector's role in DRR, including CCA. Ultimately, a consistent theme emerges: the private sector has a critical role in DRR, including the advancement of CCA, and collaboration between the public and private sectors is essential.

The Private Sector and Disaster: Context

Disasters have a range of potential impacts on private sector businesses and the global economy (Figure 34.1). The year 2015 saw some of the lowest recorded losses (reported losses include direct loss or damage of assets and stock and some indirect losses such as business disruption) in almost a decade, yet still an estimated USD 90 billion was lost as a result of disasters (Munich Re 2016). The wider impacts of disasters, such as the estimated 23,000 disaster related deaths in 2015, are much greater and longer-lasting. Such losses reverberate through societies and economies, but the associated financial losses are difficult to meaningfully quantify.

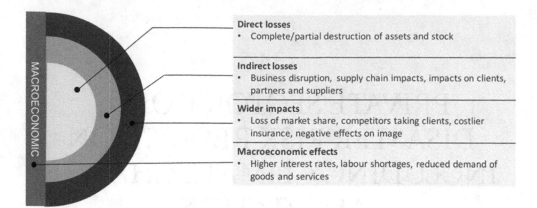

Figure 34.1 Levels of Economic Loss Caused by Disaster

(Adapted from UNISDR 2015a)

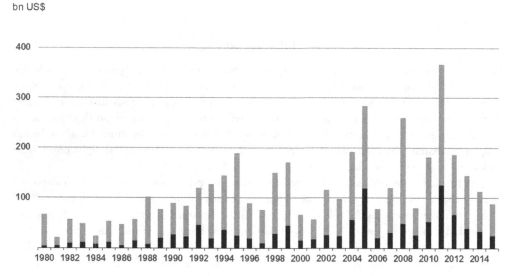

Figure 34.2 Number of Disaster Events Causing Financial Losses Worldwide 1980–2015

(By Munich Re 2016)

Direct, indirect, and the unquantified 'wider impacts' of disasters feed into the macroeconomic responses to disasters including those influenced by climate change. Responses can include decreased aggregate economic activity, lower levels of consumer spending, or a weakened trade balance and resulting downward pressure on exchange rates (Stevenson *et al.* 2016). The macroeconomic environment before the disaster may also affect the way impacts unfold. For example, Heger *et al.* (2008) suggest that small island nations, whose economies are often dominated by a few large sectors, are especially vulnerable to the short-term macroeconomic impacts of disasters.

There has been a distinct upward trend in the inflation-adjusted direct and indirect disaster losses over the last several decades (Munich Re 2016). This is true for both insured losses and overall economic losses. Figure 34.2 shows that over the last thirty years, both insured and overall financial losses have increased on average. There are several reasons for this increase, including growth of wealth in emerging economies (e.g., Brazil, China, and South Africa), concentrated urban development in areas exposed to river flooding and coastal disasters, and to some extent increasing frequency of reporting disasters and recording losses (Beilharz *et al.* 2013).

Within the private sector, there is a wide range of business structures, from individual traders to large multinational companies, and an equally wide range of risk profiles. Disasters, including those influenced by climate change, vary significantly across sectors of the economy, industry groups, and for different business structures.

Businesses will encounter industry-specific challenges in the face of climate change. For some industries, climate change is a clear and present risk, while others gain significant opportunities and yet others appear more insulated from the potential effects of climate change (Agrawala *et al.* 2013) (Table 34.1).

Table 34.1 Examples of Climate Change Related Risks Faced in Different Private Sector Industries (Agrawala *et al.* 2013)

Manufacturers	Physical risks	Disruption to operations due to extreme weather events; Damage to infrastructure; Restrictions to production due to rising temperature, variations in water quality and in water availability
Agriculture and mining businesses	Physical risks	Extreme weather events increase physical risks to business operations; Risk of overflow of storage due to increased rainfall; Resource extraction could be limited by sea level and water availability
	Supply chain and raw material risk	Water scarcity affects production
	Product demand risks	Changes in quality, quantity, and type of agricultural products
	Logistics risks	Risks to the transport corridors and transport hubs from where raw materials are processed and exported
Retailers and distributors	Physical risks	Damage to products during transportation due to extreme events
	Supply chain and raw material risks	Interruption, inefficiency or delays in supply chain; Difficulties with water scarcity and increased fuel prices
	Reputational risks	Decrease in product quality affecting reputation and consumers' satisfaction
Transportation	Physical risks	Extreme weather events causing delays, supply disruptions and losses of goods; Access to transport routes affected by flooding, permafrost thawing and mass movements, subsidence due to drought
Utilities	Physical risks	Disruptions of supply due to flooding or extreme events; Business interruption due to extreme weather
	Supply chain and raw materials risks	Reduced output due to water scarcity impacting hydropower and power plants using a thermal plant cooling system
	Product demand risks	Demand effects due to temperature changes
	Regulatory risks	Increasing pressure to conserve water in water scarce areas
Financial businesses	Financial risks	Risks in investment portfolio where investments are made in areas with climate vulnerabilities; Increased risk of customer default
Information businesses	Physical risks	Disruptions of operations due to extreme weather events; Difficulties in transportation

(Continued)

Table 34.1 Continued

Real estate businesses	Physical risks	Delays and disruptions in construction projects; Damage to buildings and drainage problems; Additional costs due to temperature changes increasing cooling loads
	Regulatory risks	Changes in building and design requirements
	Financial risks	Loss of value due to climate change impacts
Other service businesses	Product demand risks	Tourism industry affected in its infrastructure and by changes in tourism demands caused by different climatic conditions

Private sector opportunities also exist in the complex global hazard context. Business continuity planning is a rapidly growing industry and is increasingly being utilised by organisations of all sizes (Xiao and Peacock 2014). Insurers and reinsurers are developing innovative financial products that are financially beneficial to their industry and interested stakeholders. In areas where households and the government are able to invest adequately in reconstruction, and where resource limitations and debt do not hinder human capital investment, disasters themselves may bring some economic benefits (Stevenson *et al.* 2016). In areas damaged by disasters, consumer demand often shifts toward construction related industries, at least temporarily, creating opportunities for growth and expansion in these businesses (Xiao and Nilawar 2013; Zhang *et al.* 2009). This is not to say that disasters are ever to be considered social 'goods', but to illustrate that natural hazards generate both risks and opportunities and these are distributed unevenly across the economy and society.

Lens 1: Investing in Risk Reduction and Resilience in the Face of Uncertainty

The first lens in this chapter presents a view from inside the private sector. We examine how private organisations assess and respond to their risks in the face of natural hazards.

Self-interest is the most powerful driver for companies to assess and manage risk. For businesses to voluntarily undertake risk reduction activities, such as business continuity planning, investing in construction above minimum building codes, or making risk-sensitive land development choices, it must make financial sense.

Accurately evaluating the costs and benefits of risk reduction is difficult in complex environments where the risks and their potential consequences are unclear and, in some cases, entirely unknown. Both lack of information and a number of cognitive biases affect risk decision making in organisations. For example, discounting (i.e., determining the time value of money) often causes businesses to favour investments that have a clear return in the present rather than diverting resources into risk reduction activities with future returns that may or may not eventuate (Kunreuther and Useem 2010).

Those interested in stimulating DRR in the private sector need to find ways to increase the present value of risk reduction and resilience building investments. Hallegatte (2009) outlines five approaches to CCA investment that help manage uncertainty. The first method is to invest in win–win solutions, that increase net returns today and also in a crisis. Building financial safety nets into a business – e.g., reducing the debt-to-equity ratio or integrating electric or fuel efficient vehicles into a transportation fleet – increases the welfare of business regardless of climate outcomes. 'Co-benefits' that accrue from DRR initiatives are realised in real-time as well as increase resilience in a crisis event (Mcdermott 2016). The second approach is for businesses to choose to favour reversible or flexible DRR strategies to minimise sunk costs. Early warning systems, for example, can be adjusted as new information becomes available.

Hallegatte's (2009) third proposed method describes incorporating safety margins into investments to accommodate future change. Scenario modelling is now helping decision makers anticipate possible future changes and build margins into infrastructure, such as drainage systems that can accommodate increased precipitation and river flows (Wilby 2013).

The fourth approach moves away from technical solutions toward lower cost soft strategies that improve organisations' planning, response, and recovery capacities. The final private-sector led adaptation strategy addresses decision-making time horizons. Accepting that projections of disaster risk including those influenced by climate change are uncertain, some industries and investors reduce long-term commitments. For example, forestry businesses may choose species with a shorter rotation time, or people may build cheaper homes in areas that may have an increased flooding risk in the future (Hallegatte 2009). Such investments externalise risks to the environment or public.

In traditional economic theory, businesses optimise their investment decisions by weighing discounted benefits against anticipated costs. Such techniques require 'well-defined and well-understood risks', conditions that are not met in the face of a changing climate (Mcdermott 2016: 4). In the face of deep uncertainty about the likely impacts of climate change, DRR strategies including CCA should support investment decisions that are robust to a range of possible outcomes. Ultimately, investment strategies should maximise possible co-benefits of risk management investments, while minimising regrets (Hallegatte 2009; Mcdermott 2016).

Lens 2: The Private Sector and the Creation of Public Risk

Private sector organisations not only manage many of their own risks, but they also influence the risks of the surroundings where they reside and those with whom they interact. Investment and development activities can create or exacerbate public risk. Private sector actions that lead to environmental degradation, such as the loss of mangroves due to water diversion for agriculture, pollution from nearby industries, and other developments, increase exposure to hazards, and can exacerbate hazards, including flood and wind damage. In Bangladesh, for example, actions by large landowners focusing on export markets, such as for shrimp, have exposed the land of nearby smallholders to saltwater and coastal hazards (Fieldman 2011).

Globally, the private sector builds, owns, and operates the majority of infrastructure (UNISDR 2015a). The vulnerability of critical infrastructure (e.g., electricity, water, sewerage) is a particular concern, as the social costs of infrastructure system failures multiply as they flow through an economy. As market economies increasingly push for greater efficiency, more firms are outsourcing critical infrastructure services. Because disaster risk is difficult to quantify, such firms chronically under-invest in vulnerability reduction and resilience building (Auerswald *et al.* 2006).

Indeed, vulnerability in any industry where there is a lack of alternative suppliers, has severe consequences for a large number of stakeholders (Schneider 2014). For example, when flooding submerged a hard-drive manufacturing plant in Thailand in 2011, there were international ramifications, with a global shortage of hard disks and major companies posting significant losses due to decreased sales (Surminski 2013).

The creation of public risk can occur as the result of the myopic self-interest of private sector actors, short-sighted planning, and as a result of incomplete information about the consequences of decisions and actions. External risk neglect refers to the phenomenon of self-interested decision makers only considering the benefits and costs that accrue to them while ignoring benefits or costs imposed on others (Berger *et al.* 2010: 91). This can be particularly difficult to regulate, when the creators and recipients of risk are in different jurisdictions or even different generations as with climate change (Berger *et al.* 2010).

Short planning horizons can also facilitate decision making that increases disaster risk. Conscious and unconscious trade-offs are often made for shorter-term economic gains at the expense of longer-term environmental, human, and property losses. Locating manufacturing facilities close to a port in a coastal area, for example, gives a business the benefit of reduced transportation costs and faster market access, but exposes the company, its assets, and workers to greater risk of flooding (Surminski 2013).

Finally, in complex systems it can be difficult to predict and assess the consequences of all decisions and actions. Historic ignorance of the consequences of burning fossil fuels on global climate change or the influence of CFC-based (chlorofluorocarbons) industrial products on the stratospheric ozone layer has serious ongoing ramifications.

A lack of interest in accounting for the creation of public risk by private sector organisations can be seen as economically rational although potentially socially irrational. In competitive, resource scarce environments, myopic self-interest is a matter of business survival and the maintenance of jobs and standards of living. In cases where the benefits of the DRR initiatives do not accrue to those making the investments, it can be difficult to justify the costs. In situations where private decisions and actions have social impacts that are unaccounted for by the private sector it is more likely that outcomes will not be optimal for society (Auerswald *et al.* 2006). This is not a sustainable position for wider society affected by such risks.

Lens 3: The Private Sector and DRR

In the third lens, we focus on the conscious contribution of private sector entities to public sector risk reduction activities. Private sector organisations increasingly supplement the capacity of governments to prepare for and respond to hazards and disasters by providing access to labour, strategic capabilities, and efficient distribution systems. Private sector organisations contribute valuable expertise, products, and services to the development and delivery of risk reduction and resilience building activities in resource restricted environments. Additionally, private sector finance plays a large role in reconstruction after disasters. As a result, businesses are increasingly being treated as critical partners in the DRR, including CCA, policy space.

The most active businesses in this space are insurance companies, with interests in managing their own risk profile, and engineering, architecture, and construction firms interested in disseminating best practice in building construction (Twigg 2001). These organisations are well placed to bring together profit-driven business imperatives with wider societal interests. Businesses in other industry sectors have historically been less inclined to engage in DRR without government intervention, in part because of unclear returns (Twigg 2001).

Although it may not be standard practice across all industries there are a number of ways the private sector can reduce public risk. Most examples of private sector risk reduction activities that benefit the public can also demonstrate clear economic benefit to the investing firms (Box 34.1). There is a continuum of private sector involvement in managing public risk. The examples in Box 34.1 are at the more engaged end of the continuum. As a bare minimum all companies should comply with subnational and national regulations and assess their local and national operations against international laws, conventions, and standards (Nelson 2000; Twigg 2001).

Box 34.1 Private Sector Decisions and Actions that Reduce Public Disaster Risk

Joanne R. Stevenson[1] and Erica Seville[1]

[1] Resilient Organisations, Christchurch, New Zealand

Investing in construction above minimum building codes – SM Prime Holdings Inc., the largest shopping mall operator in the Philippines, build all of their malls above code and provided resilient housing to communities affected by Typhoon Haiyan in 2013. This has benefitted their corporate reputation and brand value (Johnson and Abe 2015).

Developing, adopting, and sharing technologies and practices that reduce greenhouse-gas emissions and other sources of pollution – Between 2000 and 2004 Thai-owned AT Biopower Co., Ltd built a number of small power plants using rice husks (an agricultural waste product) as an alternative form of energy. The investment has been profitable for AT Biopower, but also implements DRR for electricity generation in Thailand and reduces fuel imports (AT Biopower 2016).

Conducting transparent assessments, rating, and reporting on risk and resilience and making reports accessible to governments, investors, and customers – The Global Reporting Initiative (GRI) provides standards and tools for companies to produce sustainability reports outlining the economic, environmental, and social impacts caused by their daily operations. Companies can then publicly publish this information, although independent verification is not always easy. Rating and reporting enhances and protects company reputations, can help improve efficiencies, and can ease compliance activities. In theory, the public benefits through fewer adverse environmental and social impacts (Global Reporting 2016).

The next level of engagement is proactive risk minimisation. This reaches beyond compliance to companies' voluntarily monitoring their actual and potential social, economic, and environmental impacts, and developing policies to minimise damage from their operations and those of their partners (Nelson 2000; Twigg 2001). Finally, the most active companies engage in value creation, where they create 'positive societal value' through social investment, participating in or facilitating stakeholder and policy dialogue, and collaborations (Nelson 2000; Twigg 2001: 5).

Pathways for Improving Private Sector Risk Management

The public sector shares the risks and consequences of disasters borne and created by the private sector. Governments and inter-governmental bodies like the United Nations can incentivise, compel, or collaborate with businesses to implement DRR including CCA.

Incentivising Risk Reduction and Resilience Building

The portfolio of potential government-backed incentives includes (but is not limited to): tax-credits and grant programmes, subsidies to defray the immediate costs of DRR including CCA, reporting and accreditation programmes, and preferential contracting for companies demonstrating compliance. The Green Climate Fund (GCF) created by the United Nations Framework Convention on Climate Change (UNFCCC), for example, allocates resources to low-emission and climate-resilient projects and programmes in the private sector. The GCF employs a number of financial instruments including, grants, concessional loans, equity, and guarantees, which are available to incentivise private investors, developers, and enterprises to make climate-sensitive investments (Green Climate Fund 2016).

Similarly, the European Union finances DRR including CCA through a number of instruments including grants distributed through European Structural and Investment Funds (e.g., the European Agricultural Fund for Rural Development), Horizon 2020 for research, and the LIFE Instrument, which co-finances projects that contribute to the EU's environment and climate action goals (European Commission 2015; European Commission 2016ab).

DRR including CCA incentives also originate from private sector sources, such as the insurance industry. Major reinsurers (e.g., Box 34.2) commonly engage in international policy forums (e.g., the UNFCCC and the World Economic Forum) to promote financial tools that they argue improve the DRR capacity of cities and critical goods and service providers.

Box 34.2   Swiss Re Resilience Bonds

Joanne R. Stevenson[1] and Erica Seville[1]

[1] Resilient Organisations, Christchurch, New Zealand

Swiss Re, a major global reinsurance firm, has collaborated with a number of partners to develop an insurance-based product designed to generate capital for risk reduction projects. Resilience bonds (Figure 34.3) would be issued by insurance companies. Sponsors include large public or private asset holders (e.g., cities, hospitals, universities) that intend to purchase insurance and use investments to mitigate risk and build resilience. Investors purchase the bonds and receive coupon payments as systems become more resilient and increase the value of the bonds.

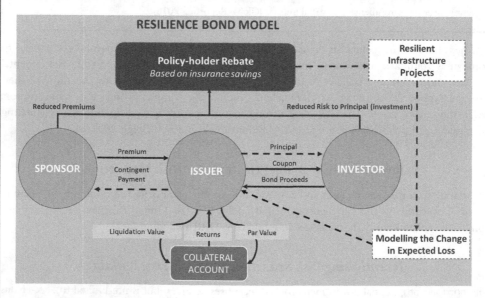

Figure 34.3   Swiss Re Resilience Bonds

(Adapted from Swiss Re 2015)

Resilience bonds are still unproven in their effectiveness. They offer a potentially innovative financial incentive for proactive investments in resilience building for major asset holdings (Swiss Re 2015).

Compelling Risk Reduction and Resilience Building

Governments can explicitly compel organisations through process regulations that mandate DRR actions that may create additional public risk. Alternatively, governments can use taxation as a means of compelling businesses to, at least partially, internalise the adverse externalities of their operations.

Process regulations that compel private sector organisations to address risk can be broken down into three broad categories: risk-avoidance, risk-reduction, and risk-sharing (Hutter 2010). Risk-avoidance strategies focus primarily on land-use and building regulations. Zoning and building codes reduce the exposure of people and property to hazard zones and create a minimum standard for resilience investment in the built environment. There is a range of implementation strategies for such approaches.

In Manizales, Colombia, for example, urban development policies are integrated with local risk management policies; analyses that identify risk micro-zones guide the type of private development allowed in various areas of the city (Hardoy et al. 2011). In California, property buyers must be shown a Natural Hazard Disclosure Statement prior to closing, which means that prices more appropriately reflect the cost of mandatory flood insurance (Troy and Romm 2004).

Risk-reduction regulations include strategies designed to decrease risk exposure in places where development has already occurred. The primary instrument for this strategy is building and infrastructure upgrades (Hutter 2010). Retrospective building upgrades are less economically efficient than constructing a new building to code. As a result, governments may choose to offer subsidies (Spence 2004). In Japan following the Great Hanshin-Awaji Earthquake of 1995, the national government created a subsidy for earthquake inspection and retrofitting of condominiums and offices in the interest of public safety (Suganuma 2006).

Risk-reduction regulations can also focus on approaches that help businesses build their capacity to respond and continue operating following a crisis. In the US, the Health Insurance Portability and Accountability Act (HIPAA) makes disaster recovery and business continuity planning mandatory for health care organisations and carries financial penalties for noncompliance (Brown 2005).

Finally, risk-sharing strategies, in this context, refer to regulatory mechanisms that allow the state and the market to share the burden of disaster risks through insurance or other financial hedging instruments (e.g., catastrophe bonds or weather derivatives). Governments may, for example, make it mandatory for commercial firms to provide insurance for disasters or enter partnerships with commercial insurers to make hazard insurance affordable for residential and commercial premises (Hutter 2010). Such schemes, including the Turkish Catastrophe Insurance Pool, the National Flood Insurance Programme (NFIP) in the US, and the New Zealand Earthquake Commission (EQC), are motivated by the need for specialised treatment of specific risks.

Taxation is another tool used to alter risk behaviour. This is a prominent strategy in the climate change mitigation arena, where regulation has focused on requiring carbon emitters to internalise the environmental and social costs of pollution through the sale of emission permits or through taxes. Conversely, governments may choose to incentivise DRR including CCA by offering tax deductions for certain activities. Officials in Manizales, Colombia, employ both approaches: reducing taxes for developers who work to reduce the vulnerability of housing in areas at risk from landslides and flooding, and levying an environmental tax on existing properties, which is then reinvested in DRR (Hardoy et al. 2011).

As capital becomes more mobile in the globalised economy, taxation has lost some efficacy as a tool for social protection. Corporations increasingly engage in tax-avoidance arbitrage, moving operations and capital to states with 'friendlier' tax policies (Fieldman 2011). Other jurisdictions may then lower tax rates or establish tax-exempt zones to appeal to corporations. As a result, such states lose an important tool to influence DRR including CCA (Fieldman 2011).

Collaborative Approaches

Government strategies to increase private sector engagement in DRR, including CCA, increasingly tend to focus on building a convincing business case. UNFCCC's Adaptation Private Sector Initiative (PSI) aims to promote private sector involvement in DRR including CCA, noting numerous co-benefits for industry partners (UNFCCC 2014; Table 34.2).

Policy is increasingly leaning toward more formal partnerships between the public and private sectors in the form of public–private partnerships (PPPs). PPPs offer practical opportunities and ventures for innovative research; information sharing; co-development of standards, policies and plans; awareness raising and education; and investment in resilient critical services, facilities and infrastructure (UNISDR 2015b). PPPs cover a broad spectrum of arrangements in which traditionally public activities are run partially or wholly by private sector entities. Such arrangements are common across the world for a range of important social functions, such as building and maintaining transport infrastructure, running waste management operations and facilities, and managing social support programmes. Chen *et al.* (2013) identified some of the most common cross-sectoral partnerships for disaster management. These are summarised in Table 34.3.

The Sendai Framework for Disaster Risk Reduction (2015–2030) recommends the establishment of PPPs to reduce the primary risk factors that cause or exacerbate the negative consequences of disasters. This framework endorses the further development of tools that enable new DRR, including CCA, ventures between private sector and public institutions at all levels (UNISDR 2015b).

Table 34.2 Benefits of Becoming a PSI Partner (UNFCCC 2014)

1   Associate with the United Nations process on climate change.
    The United Nations Framework Convention on Climate Change (UNFCCC) is the lead United Nations process for climate change where the direction of future policy and government regulation is decided. The UNFCCC Climate Change conferences have attracted an unprecedented degree of public attention, reaching more than 10,000 accredited climate change professionals.

2   Demonstrate that your business is already preparing for the impacts of climate change.
    Partnership with the PSI and regular reporting of your activities can document solid risk management and demonstrate to investors that their investments will be resilient to a changing climate.

3   Increase the visibility of your corporate citizenship work on adaptation.
    Climate change is the defining challenge of this century . . . Engaging with adaptation at this early stage demonstrates a forward-looking and sustainable commitment to corporate citizenship.

4   Network and share knowledge with leading international organisations and companies.
    Partnership with the initiative will enable you to draw on the extensive expertise of the United Nations on all aspects of adaptation. The database of PSI partner companies will further offer an important opportunity to share best practices and lessons learned.

5   Stake your claim to an emerging multi-billion-dollar market.
    Early involvement with adaptation will allow you to better understand and respond to the needs of future customers and stake your claim to this growing market.

Table 34.3 Common Types of PPPs by Disaster Management Phase

Building Resilience	1. Public–private contractual and non-contractual partnerships for critical infrastructure
	2. Government–community collaborative resilience building
Responding	1. For-profit, NGOs, and government partnerships
	2. Government–civil society partnerships
	3. Government as one of many actors in a 'many-to-many' network partnership
Recovering	1. Public–private partnerships for physical reconstruction
	2. Inter-sectoral partnerships for learning

Modified from Chen *et al.* (2013: 133)

While there are examples of successful working models of PPPs in a DRR context, there are also substantiated concerns about social equity, the risk of businesses capitalising on the suffering of others, and improper management of funds (Gunewardena 2008). Careful management of these relationships and a deeper understanding of the boundaries and flow-on effects of such arrangements are important as such partnerships proliferate globally.

There are constraints across all forms of risk regulation, from governmental willingness and capacity to implement and enforce regulations to the creation of moral hazard and the misinterpretation of tools. Creating a regulatory mix that includes communication, support, incentives, sanctions, and strong enforcement mechanisms can make these efforts more successful (Hutter 2010; Kunreuther 2010).

Remaining Questions and the Future of the Private Sector in DRR Including CCA

The private sector has a significant influence on disaster risk, including that influenced by climate change, across all levels of society. A growing body of case studies and best-practice guides exists on private-sector DRR including CCA; however, many companies still feel unsure how to address the topic or fail to see the value in extending their efforts beyond the bare-minimum of compliance (Surminski 2013).

Historically, a popular private-sector strategy has been to postpone CCA while waiting for more conclusive research, or to avoid direct DRR including CCA in order to redistribute risk through insurance or financial instruments like catastrophe bonds (Schneider 2014). Such approaches are no longer financially and socially sustainable. As a result, both private organisations and the public sector are finding ways to increase private sector participation in DRR, including CCA, as reported in this chapter. Their appropriateness and efficacy varies greatly by industry sector and local context.

Regardless of the specific tools employed, national and international actors feel that there needs to be an 'immediate and urgent shift' toward collaboration between the public and private sectors and cross-sector investments to address DRR including CCA (Johnson and Abe 2015: 25). The UNISDR's post-2015 private sector engagement blueprint presents five visions for enhanced private sector engagement for DRR (paraphrased below) (UNISDR 2015a):

1   Strong public–private partnerships: Form partnerships and other collective arrangements to combine private sector expertise, products, and services and public sector information, strategic oversight, and resources.

2   Resilience in the built environment: Use a mix of standards, certifications, regulations, incentives, compliance, and other measures to invest in, build, maintain, and enhance the built environment.

3   Risk-sensitive investments and accounting: Establish corporate policies, education programmes, and higher public expectations of supply chain and business continuity and investments that add to resilience.

4   Positive cycle of reinforcement for a resilient society: Influence public attitudes toward resilience through corporate policy and workforce education. Increased expectations will (ostensibly) enhance market demand for products that increase resilience.

5   Private sector risk disclosure: Promote enabling environments and increased demand for corporate social reporting and transparent risk assessment and disclosure.

The private sector has a critical role in DRR, including CCA, but many challenges remain. The structure of the global economy and increased mobility of capital has limited states' capacity to tax and regulate private sector investment decisions (Fieldman 2011). Disconnects between the public, private, and non-profit sectors hinder trust and effective partnerships (Twigg 2001). Short-term horizons still dominate business planning and reduce DRR including CCA incentives (Agrawala *et al.* 2013).

Therefore, important questions that researchers, policy makers, and practitioners need to consider going forward include: (1) What role does public sentiment and social expectation play in altering private sector behaviour, and what role should the state play in shaping the awareness of these issues in the public consciousness? (2) What is the optimal regulatory mix to contain the negative impacts of private sector activity and promote socially positive investment? and (3) How can we continue to build the economic case for DRR, including CCA?

References

Agrawala, S., Carraro, M., Kingsmill, N., Lanzi, E. and Prudent-richard, G. (2013) *Private Sector Engagement in Adaptation to Climate Change (Environment Working Paper No. 39)*, Paris: Organisation for Economic Co-operation and Development (OECD).

AT Biopower. (2016) *AT: Agricultural Waste to Energy*. Online http://www.atbiopower.co.th/about-us (accessed 2 May 2016).

Auerswald, P., Branscomb, L.M., La Porte, T.M. and Michel-Kerjan, E.O. (2006) *Seeds of Disaster, Roots of Response: How Private Action Can Reduce Public Vulnerability*, Cambridge: Cambridge University Press.

Beilharz, H.J., Rauch, B. and Wallner, C. (2013) *Economic Consequences of Natural Catastrophes: Emerging and Developing Economies Particularly Affected – Insurance Cover is Essential*, Munich, Germany: Munich Re.

Berger, A., Brown, C., Kousky, C. and Zeckhouser, R. (2010) 'The five neglects: Risks gone amiss', in H. Kunreuther and M. Useem (eds), *Learning from Catastrophes: Strategies for Reaction and Response*, Upper Saddle River, New Jersey: Wharton School Publishing, pp. 83–99.

Brown, C. (2005) 'HIPAA programs: Design and implementation', *Information Systems Security* 14, 1: 10–20.

Chen, J., Chen, T.H.Y., Vertinsky, I., Yumagulova, L. and Park, C. (2013) 'Public-private partnerships for the development of disaster resilient communities', *Journal of Contingencies and Crisis Management* 21, 3: 130–143.

European Commission. (2015) *European Structural and Investment Funds*, Brussels: European Commision. Online http://ec.europa.eu/contracts_grants/funds_en.htm (accessed 2 May 2015).

European Commission. (2016a) *Environment: LIFE Programme*, Brussels: European Commision. Online http://ec.europa.eu/environment/life/about/index.htm (accessed 2 May 2015).

European Commission. (2016b) *HORIZON 2020: The EU Framework Programme for Research and Innovation*, Brussels: European Commision. Online https://ec.europa.eu/programmes/horizon2020/en/what-horizon-2020 (accessed 2 May 2015).

Fieldman, G. (2011) 'Neoliberalism, the production of vulnerability and the hobbled state: Systemic barriers to climate adaptation', *Climate and Development* 3, 2: 159–174.

Global Reporting Initiative. (2016) *Global Reporting Initiative: Empowering Sustainable Decisions.* Online https://www.globalreporting.org/Pages/default.aspx (accessed 22 May 2016).

Green Climate Fund. (2016) *Private Sector Facility*, Incheon: Green Climate Fund. Online http://www.greenclimate.fund/home (accessed 4 May 2016).

Gunewardena, N. (2008) 'Human security versus neoliberal approaches to disaster recovery', in N. Gunewardena and M. Schuller (eds), *Capitalizing on Catastrophe*, Lanham: AltaMira Press, pp. 3–16.

Hallegatte, S. (2009) 'Strategies to adapt to an uncertain climate change', *Global Environmental Change* 19, 2: 240–247.

Hardoy, J., Pandiella, G. and Barrero, L.S.V. (2011) 'Local disaster risk reduction in Latin American urban areas', *Environment and Urbanization* 23, 2: 401–413.

Heger, M., Julca, A. and Paddison, O. (2008) *Analysing the Impact of Natural Hazards in Small Economies: The Caribbean Case*, Helsinki, Finland: United Nations University (UNU).

Hutter, B.M, (2010) 'The role of risk regulation in mitigating natural disasters', in H. Kunreuther and M. Useem (eds), *Learning from Catastrophes: Strategies for Reaction and Response*, Upper Saddle River, New Jersey: Wharton School Publishing, pp. 121–138.

Johnson, D.A.K. and Abe, Y. (2015) 'Global overview on the role of the private sector in disaster risk reduction: Scopes, challenges, and potentials', in T. Izumi and R. Shaw (eds), *Disaster Management and Private Sectors*, Tokyo, Japan: Springer, pp. 11–29.

Kunreuther, H. (2010) 'Long-term contracts for reducing losses from future catastrophes', in H. Kunreuther and M. Useem (eds), *Learning from Catastrophes: Strategies for Reaction and Response*, Upper Saddle River, New Jersey: Wharton School Publishing, pp. 235–248.

Kunreuther, H. and Useem, M. (eds) (2010) *Learning from Catastrophes: Strategies for Reaction and Response*, Upper Saddle River, New Jersey: Wharton School Publishing.

Mcdermott, T.K.J. (2016) *Investing in Disaster Risk Management in an Uncertain Climate*, London: World Bank Group.

Munich Re. (2016) *2015 Natural Catastrophe Losses Curbed by El Niño; Brutal North American Winter Caused Biggest Insured Losses*, Princeton, New Jersey: Munich Re Media Relations.

Nelson, J. (2000) *The Business of Peace. The Private Sector as a Partner in Conflict Prevention and Resolution*, London: International Alert, Council on Economic Priorities and Prince of Wales Business Leaders Forum.

Schneider, T. (2014) 'Responsibility for private sector adaptation to climate change', *Ecology and Society* 19, 2: 8.

Spence, R. (2004) 'Risk and regulation: Can improved government action reduce the impacts of natural disasters', *Building Research and Information* 35, 5: 391–402.

Stevenson, J., Noy, I., McDonald, G., Seville, E. and Vargo, J. (2016) 'Economic and Business Recovery', in S.L. Cutter, D. Benouar, S. Chang, F. Chen, M. de Castro, R. Djanlante, A.M. Esnard, B. Gerber, B. Glavovic, W. Krajewski, E. Michel-Kerjan, E. Penning-Roswell and G. Ziervogel (eds), *Research Encyclopedia, Natural Hazard Science*, Oxford: Oxford University Press USA.

Suganuma, K. (2006) 'Recent trends in earthquake disaster management in Japan', *Science and Technology Trends* 19: 91–106.

Surminski, S. (2013) 'Private-sector adaptation to climate risk', *Nature Climate Change* 3, 11: 943–945.

Swiss Re. (2015) *Leveraging Catastrophe Bonds as a Mechanism for Resilient Infrastructure Project Finance*, Zurich: Swiss Re. Online http://www.refocuspartners.com/reports/RE.bound-Program-Report-December-2015.pdf (accessed 2 May 2016).

Troy, A. and Romm, J. (2004) 'Assessing the price effects of flood hazard disclosure under the California natural hazard disclosure law (AB 1195)', *Journal of Environmental Planning and Management* 47, 1: 137–162.

Twigg, J. (2001) *Corporate Social Responsibility and Disaster Reduction: A Global Overview*, London: Benfield Greig Hazard Research Centre, University College London.

UNFCCC (United Nations Framework Convention on Climate Change). (2014) *Adaptation Private Sector Initiative (PSI)*. Online http://unfccc.int/adaptation/workstreams/nairobi_work_programme/items/4623.php (accessed 10 September 2016).

UNISDR (United Nations International Strategy for Disaster Reduction). (2015a) *Disaster Risk Reduction Private Sector Partnership: Post 2015 Framework – Private Sector Blueprint Five Private Sector Visions for a Resilient Future*. Geneva: UNISDR. Online https://www.unisdr.org/we/inform/publications/42926 (accessed 24 May 2016).

UNISDR (United Nations International Strategy for Disaster Reduction). (2015b) *Reading the Sendai Framework for Disaster Risk Reduction*, Geneva: UNISDR. Online https://royalsociety.org/~/media/policy/Publications/2015/300715-meeting-note-sendai-framework.pdf (accessed 24 May 2016).

Wilby, R.L. (2013) 'Decision- rather than scenario-centred downscaling: Towards smarter use of climate model outputs', *Geophysical Research Abstracts*, 15. Online http://meetingorganizer.copernicus.org/EGU2013/EGU2013-13593.pdf (accessed 24 May 2016).

Xiao, Y. and Nilawar, U. (2013) 'Winners and losers: Analysing postdisaster spatial economic demand shift', *Disasters* 37, 4: 646–668.

Xiao, Y. and Peacock, W.G. (2014) 'Do hazard mitigation and preparedness reduce physical damage to businesses in disasters? Critical role of business disaster planning', *Natural Hazards Review* 15, 3: 1–11.

Zhang, Y., Lindell, M.K. and Prater, C. S. (2009) 'Vulnerability of community businesses to environmental disasters', *Disasters* 33, 1: 38–57.

35

FROM POLICY TO ACTION AND BACK AGAIN FOR DISASTER RISK REDUCTION INCLUDING CLIMATE CHANGE ADAPTATION

Emily Wilkinson and Ilan Kelman

Introduction

Public policy is a system of principles to guide decision-making and action to deal with problems and opportunities. Disasters and human influences on hazards and hazard drivers such as climate change are complex social problems, so the sets of policy responses (and lack thereof) that they have promoted are tricky to untangle and evaluate. From regional agreements on drought early warning systems to municipal land-use plans, policies to galvanise action – and to avoid action – have been deployed in all countries and at multiple governance scales from global to local in one form or another.

This chapter examines some of the types of policies that can be and have been adopted by governments and institutions at all levels to yield action on disaster risks, including those with connections to climate change, as well as their limitations such as policies with disincentives to action. Incentives and disincentives behind policies and investments for DRR including CCA are explored.

Public Policy Options: Progress and Limitations

What is the Role of Government in DRR Including CCA?

Five clear roles for government, from global to local, can be identified with regards to DRR including CCA (Wilkinson 2012):

1   Governments as providers of goods and services for DRR including CCA.
    Examples include those measures that are often not well-provided by the private sector such as early warning systems; buildings, such as shelters and hospitals, to reduce loss of life and property during and after a disaster; and ecosystems such as mangrove belts and coral reefs, which in some circumstances might reduce the impact power from tsunamis and storm surges.

2 Governments as risk avoiders.

In order to reduce risk in society, governments not only have to provide public goods and services, but they also have to refrain from actions that generate risk. They are responsible for substantial investments in infrastructure, such as roads, hospitals, and schools. These need to be located, designed, built, and maintained in such a way as to minimise vulnerability to disasters.

3 Governments as regulators of private sector activity.

To prevent construction in high risk areas and ensure that infrastructure is safe from environmental hazards, governments can produce recommendations, standards, and regulations on building practices and land use and, in the case of regulations, invoke penalties for non-compliance.

4 Governments as promoters of collective action.

Not all DRR including CCA measures are within the public domain. Families, social groups, non-profits, and businesses also act. Government programmes, such as education and communication strategies, can help to raise awareness of disaster risk and encourage people to develop community plans for DRR including CCA, organising activities together so that no one is excluded.

5 Governments as coordinators of multi-stakeholder activities.

DRR including CCA requires coordinated action by public and private stakeholders. Flood risk management, for example, requires the participation of meteorological, hydrological, environmental, water, and sanitation authorities; public and private land users; community groups; planning departments; and civil protection departments. Governments can provide leadership and coordination, but community and private sector involvement is also needed.

For all the government roles identified above, a range of policy options or types, from legally binding to voluntary incentives or disincentives, can be adopted to deal with disaster risks including those linked to climate change. Vulnerability to natural hazards seems to be increasing, generating a steady rise in disaster losses and with serious implications for poverty reduction efforts and development progress (Wilkinson and Peters 2015).

Managing risk requires not only reducing those risks that already exist and threaten development but also taking action to manage development processes in such a way as to avoid risk generation and accumulation in the future (UNISDR 2015a). Policies therefore need to be understood in terms of three processes linked to development (UNISDR 2015a):

1 Avoiding the accumulation of new risk.
2 Reducing existing risk.
3 Building resilience of people and societies to residual risk; i.e., risk that cannot be effectively reduced or managed, such as living on the slopes of a volcano, which could generate pyroclastic flows, so the main options are pre-eruption evacuation or safe rooms in buildings.

Governments have a variety of policy options and tools to encourage actions in all these areas – although, as with all public policy approaches, there is no recipe for success – and national, sub-national, and supra-national government agencies around the world have done so in many different ways. Common approaches that have been adopted are discussed below, along with a commentary on some of the most critical factors limiting progress on managing risk.

How are Governments Doing in Practice?

In the case of preventing risk accumulation, the government's role as a risk avoider, regulator of private sector activity, and coordinator of multi-stakeholder activities is critical. In most contexts, DRR including CCA has traditionally been approached through a set of policies and practices to protect development against exogenous threats rather than to prevent or avoid the generation and accumulation of risks within development by focusing on vulnerability reduction. This interpretation has limited the effectiveness of DRR including CCA in achieving the policy objective of reducing detrimental impacts (UNISDR 2015a: 26). Economic losses from disasters are now reaching between USD 150 billion and USD 200 billion each year; with more people living in coastal areas and floodplains – partly due to them moving there and partly due to changing floodplains – and the continued degradation and loss of natural ecosystems, this trend looks set to continue (UNISDR 2015a). Projected future disaster losses in the built environment alone are estimated at USD 314 billion per year (UNISDR 2015a).

Risk Avoider

Some governments have explicit policies to avoid risk accumulation by incorporating risk assessments into public investment planning decisions. Peru, for example, has a national law on public investment that stipulates that disaster risk and the likely impact of climate change on it must be considered in project identification, formulation, and evaluation stages (UNISDR 2015b). These planning requirements are rare and most public infrastructure goes up without a disaster risk assessment, or even subsets such as a climate risk assessment, having ever been conducted.

Regulator of Private Sector Activity

A major concern can be the lack of private sector inclusion in dialogues and policy developments on land use and disaster risk assessment, even though the private sector is heavily involved in the design, construction, and maintenance of infrastructure. Governments are responsible for regulating private sector activity, but land use planning regulations and building codes controlling construction in high-risk areas have proven extremely difficult to implement (GFDRR 2015; Krimgold 2011). Consequently, compliance with and enforcement of these regulations tends to be low.

Coordinator of Multi-stakeholder Action

The role of global institutions and governments as coordinators of dialogue, negotiation, and action is also important. Strong institutions capable of implementing coercive measures are lacking in many countries, hence more cooperative arrangements are often adopted and can be more effective (May *et al.* 1996). The development of 'Geoparks' and other protected areas has proven to be effective in limiting risk accumulation when developed in partnership with local stakeholders (Fraga 2006), while participatory monitoring of building construction can offer a low-cost alternative to more formal and costly inspection processes in places where compliance with and enforcement of building codes is a problem (GFDRR 2015; Krimgold 2011).

Service and Goods Provider

Policies to reduce existing risks are varied and most are only relevant for particular hazards. These include, but are not limited to, public sector investments in: retrofitting of buildings (earthquakes), strengthening of roofs (volcanoes), dredging canals (floods), and building water storage (droughts). Sometimes, more than one hazard can be involved, such as building dams and reservoirs for floods and droughts, but this specific example also demonstrates how policies supporting actions to reduce risk end up exacerbating it or creating other risks. Large dams and reservoirs are frequently implicated in forced displacement of populations leading to poverty as well as generating a false sense of security about floods and droughts, so further risk reduction measures are deemed to be unnecessary (Etkin 1999; World Commission on Dams 2000).

Promoter of Collective Action

Awareness programmes (often for all hazards) and risk assessments are common components of policies for DRR including CCA aiming to improve collective action to reduce risk and manage residual risk, but by far the majority of resources has been focused on building protective infrastructure, with mixed success. There are many other examples of protective infrastructure dealing only partially with risks or creating new risks, thereby failing to address the root causes of vulnerability and the underlying drivers of risk (Box 35.1).

Box 35.1 The Co-benefits and Unintended Consequences of DRR Investment in Infrastructure in Tabasco, Mexico

Emily Wilkinson[1]

[1] Overseas Development Institute, London, UK
Based on Vorhies and Wilkinson (2016).

Between 2008 and 2010, the Mexican government invested USD 500 million in DRR, of which 84 per cent covered protective infrastructure. One of the largest investments was in flood protection for the state of Tabasco, which was affected by floods in 2007 with economic losses calculated at nearly USD 3 billion (World Bank unpublished). A joint World Bank and Mexican government study found that the cost–benefit ratio of these investments was 4:1, contributing to avoided damages and losses when floods occurred in 2010 equivalent to USD 3 billion, or 7 per cent of the GDP of Tabasco.

The investment also stimulated local actors to take greater care of the environment. Small-scale projects with environmental benefits have been initiated, including tree planting on riverbanks to prevent landslides. People are beginning to dispose of litter more responsibly, throwing less in the streets or into drains to avoid these becoming blocked during the rainy season. There have been unintended costs associated with Tabasco's flood project. Channelling water away from the capital Villahermosa has led to increased flooding elsewhere in the state of Tabasco, mainly in rural areas.

There have also been negative environmental impacts as a result of these large construction projects, especially in that it transpired that floods were not caused only by heavy rainfall but also by the way the dams operate. López-Méndez *et al.* (2007) stated that 'These results suggest that the suitable operation of the dams, based on better forecasts, would have reduced considerably the damages caused by the event'. Overall, the Tabasco flood protection case study suggests that better methodologies are needed to measure the full range of costs and benefits of investments in DRR including CCA, including unintentional consequences.

Controlling, Reducing Risk, and Managing Residual Risk

In terms of the different types of approaches to managing risk in development, the management of residual risk is by far the most common approach to DRR including CCA. It has been very effective in reducing loss of life. Bangladesh and Cuba are well-cited examples of this and demonstrate that with strong political backing, preparedness actions such as early warnings, evacuation training, and the construction of cyclone shelters, can succeed. In Bangladesh, the absolute number of deaths from tropical storms has fallen significantly, although not all the storms are comparable because they affect different areas and different numbers of people. A better way of calculating how effective preparedness policies are is to look at numbers of deaths relative to houses 'destroyed' by the wind and surge. Over a 40-year period in Bangladesh, there has been a 100-fold reduction in fatalities in housing destroyed by cyclones, indicating above all else that evacuation policies have been successful (Haque *et al.* 2011). Meanwhile, Cuba has rarely seen hurricane fatalities over past decades, although drought risk reduction has not been as successful (Aguirre 2005).

According to UNISDR (2015a), over 100 countries now have dedicated national institutional arrangements for DRR and, as of 2014, more than 120 countries had undergone legal or policy reforms. Much of this policy reform involves expanding government risk management roles and responsibilities beyond reducing residual risk to include measures aimed at reducing existing risk. The integration of DRR into development policy and practice, which is needed for controlling future risk, remains a challenge (UNISDR 2015a). In particular, disaster risk management systems, which are made up of different government agencies (and sometimes non-government organisations), have struggled to play a significant role in non-emergency situations.

Risk management policies tend to ignore the cumulative impacts of 'everyday hazards' such as small-scale floods, fires, and landslides, which result in small, isolated losses cumulatively totalling more than those of large disasters (Bull-Kamanga *et al.* 2003; Lewis 1984). These impacts are less visible and are rarely registered in international databases as disasters. They rarely capture the headlines and do not prompt public demand for policy reform or reflection by local or national authorities. Fires and spontaneous building collapses in Kenya over the past 20 years, for example, illustrate how systems of building code regulation can fail in dealing with chronic risks (GFDRR 2015). The construction of the disaster database Desinventar (http://www.desinventar.org) aims to include such disasters, no matter how small, thereby overcoming the limitations of the main disaster databases.

Another important shortcoming of many risk management policies is that they tend to ignore or pay insufficient attention to marginalised and vulnerable groups (Lovell and Le Masson 2014). Women are reportedly excluded from emergency preparedness and response programmes, while engagement with children directly in the design and delivery of policies and activities is not yet well-understood or implemented (Peek 2008). The exclusion of people with disabilities from post-disaster reconstruction processes was evident in Indonesia in areas affected by the 2004 tsunami, resulting in slow, ineffective and in some cases non-existent relief (Yeo *et al.* 2005).

What Drives National DRR Including CCA Policy Direction?

Despite escalating disaster-related losses, comparatively few resources have been committed to DRR over the years. From 1990 to 2010, over USD 3 trillion was spent on development aid but only USD 106.7 billion on disasters and, of this total, only 13 per cent (USD 13.5 billion) was spent on preventing and preparing for disasters, compared with USD 69.9 billion on emergency response and USD 23.3 billion on reconstruction (Kellett and Caravani 2013). Without a major increase in investment to reduce current and future risks, spending on relief and reconstruction is likely to become unsustainable. This section looks at some of the political and economic disincentives to investing in DRR including CCA, as well as some of the major drivers of change that help explain why there has been progress in implementing DRR measures in some countries and contexts.

Why Governments Rarely Prioritise DRR Including CCA

Compared with post-disaster response, levels of investment in DRR including CCA are low, particularly investment to avoid risk and reduce existing risk (Kellett and Caravani 2013). There are many reasons for this underinvestment including allocation of resources to other endeavours, a limited understanding of risks and impacts, greater political buy-in for more visible post-disaster support initiatives, and the ready availability of international post-disaster assistance.

In particular, DRR including CCA can suffer from a lack of salience with citizens where the benefits are hard to perceive or other priorities are highlighted. Policy makers tend to under-invest or not invest in projects to manage risk because the upfront costs of such investments are perceived to be high, visible, and immediate, whereas their direct benefits and distribution of benefits are perceived to be unclear, uncertain, and distant. Yet empirical evidence suggests that these perceptions are not always correct and that economic benefits can accrue irrespective of a hazard manifesting (Shreve and Kelman 2014). Existing methods of appraising investment decisions often fail to incentivise DRR including CCA because they undervalue the resulting benefits. Governments will often focus on short-term financial goals, rather than on the potential long-term benefits in the form of reduced risks.

Attracting investment by lowering the threat of losses from disasters is one potentially huge benefit of DRR including CCA. This 'background risk' can restrict long-term investments in income-generating assets, entrepreneurial enterprises, and other economic areas. For example, Mexico's Component for the Attention of Natural Disasters (CADENA) programme shows that weather-indexed insurance not only helps to compensate farmers for drought losses, but also enables farmers to overcome credit constraints and invest in tools and fertilisers, boosting their productivity.

Other 'co-benefits' of DRR measures can be produced through increases in productivity as well as positive social and environmental outcomes. DRR investments in farming and fishing communities can have important benefits in terms of avoiding disaster losses and providing more robust livelihoods. A floodgate rehabilitation project in Laos increased flood protection, reducing losses by USD 13,200 on average per gate, but the investment also resulted in farmers increasing their fish catch in the floodplain, with an average annual benefit of USD 3,600 per floodgate. Similarly, in Jamaica, public investments to reduce drought risk in farming, including a dedicated irrigation system, have increased productivity and output as well as reducing soil erosion and deforestation by optimising previously inefficient farming practices (Vorhies and Wilkinson 2016). These and other benefits produced by investments in DRR including CCA are not always well-understood, well-documented, or systematically measured. If the full benefits over all timescales of DRR measures could be captured, then they would help to improve the case for support from, and incentives for, private sector companies and governments to invest in DRR including CCA.

In terms of avoiding risk creation, there are particularly strong economic disincentives to implementing policies or applying regulations. Governments in the USA and many Caribbean islands are inclined to allow property developers to build on the coast in hurricane-prone areas, destroying the mangroves that offer natural hazard mitigation against storm surge, because of the high value of these properties, the tax revenues gained, and lobbying from the property owners. On the other hand, activities to reduce current risks, such as relocating settlements or retrofitting buildings, are of enormous value to the construction sector and can be lucrative for local politicians, despite the fact that housing solutions and sites offered to low-income families are often inappropriate (Jha *et al.* 2010). Too frequently, such decisions are made by a small minority, often external to the groups being affected, without fully considering how the entire population is impacted, especially with respect to social and cultural issues.

There appear to be fewer disincentives for investing in supporting communities to deal with residual risk, through early warning systems and preparedness planning. These activities are not only relatively low cost (compared with structural measures), but they also appear to be cost-effective, with studies suggesting that early warning systems have been extremely effective in reducing loss of life (Shreve and Kelman 2014). Nonetheless, dealing with residual risk still requires an impetus to enact the measures, which is not always forthcoming. Despite decades of efforts prior to 2004 to implement an Indian Ocean tsunami warning system, it was not enacted until after the catastrophe of 26 December 2004. Climate change seems to be paralleling this situation, whereby decades of warning and policy discussions have still not led to the level of concerted action needed.

Policy makers tend to be biased towards actions that are tangible and visible. Long-term projects such as nuclear power plants, airports, and train lines are often given the go-ahead because they can be monetised immediately while providing publicity opportunities and substantive foci (e.g., specific sites or large construction contracts). Conversely, the benefits of DRR including CCA can be hard to pinpoint, often being about changing attitudes or behaviour rather than building something focused and photographable.

Disasters as Drivers of Change

Disasters themselves can generate public pressure, reflection, and learning – as with the Indian Ocean tsunami warning system. Large-scale disasters, in particular, and to some extent recurrent events, have often provided a strong impetus for improvements in DRR including CCA in response to perceived policy or practice weaknesses. Prominent hazard drivers and major disasters affecting Latin America in the late 1990s – namely El Niño (1997–1998) followed by La Niña (1999–2000); the earthquake in Armenia, Colombia (1999); Hurricanes George across the Caribbean and Mitch across Central America (1998); and the landslides and flooding in Venezuela (1999) – all prompted criticism of existing DRR models (Lavell 2000: 5). In India, the Orissa cyclone (1999) and the Gujarat cyclone (1998) and earthquake (2001) led to a redesign of national legislative and institutional arrangements. In Pakistan, following the 2005 earthquake in Kashmir, a National Disaster Management Commission and National Disaster Management Authority were established, responsible for coordinating, implementing, and monitoring disaster risk reduction at national, provincial, and district levels (UNISDR 2007).

In the USA, too, major disasters have often been attributed to yielding legislative reform and policy change (Box 35.2) aiming to encourage DRR actions: Hurricanes Iniki and Andrew in 1992, for example, fed into two reports on the state of the emergency management system, contributing to important reforms including the creation of a mitigation directorate at the Federal Emergency Management Agency (FEMA) in 1993 and an amendment of the Robert T. Stafford Disaster Relief and Emergency Assistance Act in 1994. Emergency preparedness was improved, vesting responsibility jointly in the federal government and states and their subdivisions.

Box 35.2 Hurricane Sandy: A Catalyst for Climate Change Action in the USA?

Emily Wilkinson[1] and Ilan Kelman[2]

[1] Overseas Development Institute, London, UK
[2] University College London, London, UK and University of Agder, Kristiansand, Norway

The impact of Hurricane Sandy, which stands at nearly USD 80 billion worth of damage, has influenced climate change discourse in the United States in a way that science failed to in the past. A recent study by the Carbon Disclosure Project (CDP 2013) says that events like Hurricane Sandy are acting as a catalyst to get businesses to take action and develop contingency plans. The report shows that at least 70 per cent of major businesses are concerned about the threats that climate change poses to their companies. Just over 50 per cent of businesses contacted for the study said that climate change, specifically heavy rains and droughts, which they presumably feel are linked to climate change, has already had a very large impact on operations, always in a negative way.

This viewpoint is akin to what are termed 'focusing events', disasters that focus people on DRR, or 'windows of opportunity', the possibility of using the post-disaster attention to implement long-standing DRR policies and actions, as with the Indian Ocean tsunami warning system. Neither concept has been shown to appear consistently in reality with focusing events particularly being critiqued (Thompson 2016). In many circumstances, traits of focusing events and windows of opportunity are present, suggesting that a disaster has played a catalysing role, but it is hard to determine how unique this role is; i.e., whether it is the sole or dominating catalyst.

It is the same with Hurricane Sandy. Shortly after Hurricane Sandy, New York City mayor Michael Bloomberg and President Barack Obama spoke of the need to take action on climate change. While Obama did not link the storm directly to climate change, he told the press:

> When you combine stronger storms with rising seas, that's a recipe for more devastating floods. Climate change didn't cause Hurricane Sandy, but it might have made it stronger. The fact that the sea level in New York Harbor is about a foot higher than a century ago certainly made the storm surge worse.
>
> *(Obama 2015)*

This discourse matched what Obama was saying anyway in many other venues while Bloomberg was already promoting climate change action for New York City. Over the long term, it is unclear whether any wide-scale specific DRR policies, including those framed as CCA, emerged due to Hurricane Sandy only, even if individuals or organisations changed some of their actions. Hurricane Sandy likely had influence in supporting an already existing agenda, but the degree of that influence on DRR including CCA – like Hurricane Katrina before it – is inconclusive.

Despite the political attention given to large disaster events and the flurry of legislative and policy reforms that sometimes ensues, lessons drawn from disasters tend to focus on the symptoms not the cause. Most high-impact disasters lead to renewed policy interest in improving early warning systems and enhancing national disaster funds, so money is available to deal with the large-scale event that was just experienced. Much less attention is paid to the social and economic vulnerabilities and inequalities that exist and which explain why particular groups were so deeply affected and recovery so slow.

In the municipality of Lazaro Cardenas in southeast Mexico, for example, a hurricane that took three days to move further inland in October 2005 left many trapped in their houses and temporary shelters, without sufficient food and water. High winds caused extensive damage to housing and fishing boats, while flooding inland damaged crops, animals, and farming equipment (CENAPRED 2006). The emergency response was slow. Poor communication and non-consultative decisions about the type of building materials to be used in reconstruction led to social unrest and outbreaks of violence. Officials did not acknowledge the lack of DRR planning and minimised or bypassed concerns about underlying vulnerabilities, believing that the social issues could be easily fixed by anticipating more intense hurricanes and improving emergency response. Three years after the hurricane, the ruling party lost the local elections when citizens punished municipal authorities for their mismanagement of the disaster (Wilkinson 2011).

The Policy–Action Connection: Local ⇔ Global

Despite significant advances in international and national DRR policies and some progress in some governmental and intergovernmental policy and action, at multiple scales from global to local, the question remains regarding whether or not it is enough. When viewed from the top down, the question also remains on whether or not governments have the ability and interest to do enough regarding DRR including CCA. Should the responsibility fall on governments and international institutions to be leaders for DRR including CCA, perhaps even seeking global governance, or should governments lead by decentralising and using subsidiarity, which is the principle that policies and actions are enacted at the lowest governance level feasible, only to be scaled up if necessary? Even if the subsidiarity principle is robust, resources, skills, equipment, and personnel might be lacking.

Yet, in principle, the conclusion from disaster research and practitioner preferences from around the world is that local action is the starting point and the key for DRR including CCA. Instructions for the public in richer countries such as the USA and Australia emphasise that, after a major catastrophe, people will be on their own without outside assistance or emergency services for at least 72 hours. Training courses for the public suggest that 1–2 weeks might be more realistic.

The principles of local responsibility and local action are not just for the hazard that strikes quickly, such as an earthquake or tornado. A growing international body of community-based teams exists for emergency response in countries such as Australia, Japan, the Philippines, Turkey, and the USA. These teams are also being used to highlight DRR including CCA as well as long-term vulnerability reduction in their community, long before a disaster becomes prominent. Identifying people with disabilities living alone who will need help after a disaster leads to recognising their current needs and integrating them better into the community. Recognising the vulnerability of power lines in heat waves and ice storms can be an impetus towards developing more localised energy supplies as a DRR measure.

These examples are actions on the ground. Such actions should, in theory, then inform policy. Policies could involve developing guidelines for involving people with disabilities in DRR including CCA, as well as disaster response. Another example of a policy could be decentralising electricity supply systems, indicating the need to seek electricity substitutes where possible such as solar heating of water, and providing incentives for energy demand reduction. This policy would then encourage actions, so that policy and action become linked and mutually reinforcing.

Yet, local level processes can have drawbacks, which larger scales can assist in overcoming. In small, close communities, the lack of anonymity and confidentiality may hinder data collection, or leave the use of public data open to abuse. Examples are gender-based violence and populations of different gender and sexual identities, which are needed for comprehensive vulnerability assessments but which the people might not wish to have out in the open. Power relations might not be fully articulated because people fear the consequences of doing so. Furthermore, local or traditional knowledge can be recorded and misappropriated by outsiders for their own academic or governmental careers.

Such drawbacks from community-based actions can sometimes be overcome by combining different participatory processes or by having external oversight, such as from the national or international level. Additionally, despite the importance of community-based DRR including CCA, national and regional preparedness must continue for situations when communities become overwhelmed. This situation might occur for a local disaster, in which communities can cope on their own for several days, but nonetheless still need outside help to fully recover. For example, while every community could and should be actively involved in forest fire prevention, lightning can still lead to large fires. Forest fires require specialised equipment and skills to tackle. National and international support is frequently required to provide the appropriate equipment, personnel, and coordination.

Top-down policies and oversight starting at the global level can further assist in avoiding one community's actions causing problems for other communities. A community might decide that the infrastructure in current and, under climate change, future floodplains along a river is too expensive to move. The community prefers to engineer the river to increase the water flow rate through the city to avoid floodwaters overtopping the current barriers. This approach could cause flooding problems for other communities downstream. Instead, national level financial support and directed policy implementation might be useful to reduce the infrastructure's vulnerability. Alternatively, the two communities could collaborate to create and maintain a larger floodplain between their cities. Community-based preparedness does not mean carrying out actions in isolation for only one's own community, but means a mixture and blending of top-down and bottom-up.

Where does the initiative begin: from the top, from the bottom, or in the middle? Should the household and local level take the initiative on the premises of decentralisation and subsidiarity, then convincing higher governance levels to promote similar policies and actions? Or will it start with the international agreements negotiated under United Nations auspices, with the principles trickling down to top-down policies which push forward local action? Ultimately, it is a combination with both processes working in tandem to support each other. Combining the top-down and bottom-up approaches – while ensuring that the views and roles of individuals and groups across genders, ages, and other demographic differences are factored in – demonstrates how policies and actions can and should connect for DRR including CCA.

References

Aguirre, B.E. (2005) 'Cuba's disaster management model: Should it be emulated?', *International Journal of Mass Emergencies and Disasters* 23, 3: 55–71.

Bull-Kamanga, L., Diagne, K., Lavell, A., Leon, E., Lerise, F., MacGregor, H., Maskrey, A., Meshack, M., Pelling, M., Reid, H., Satterthwaite, D., Songsore, J., Westgate, K. and Yitambe, A. (2003) 'Everyday hazards to disasters: The accumulation of risk in urban areas', *Environment and Urbanization* 15, 1: 193–204.

CDP (Carbon Disclosure Project). (2013) *Reducing Risk and Driving Business Value, CDP Supply Chain Report 2012–13.* Online http://www.climatechangenews.com/2013/01/22/70-of-businesses-see-climate-change-as-a-major-risk (accessed 11 September 2016).

CENAPRED (Centro Nacional de Prevención de Desastres). (2006) *Características e impacto socioeconómico de los principales desastres ocurridos en la República Mexicana en el año 2005,* Mexico City: CENAPRED.

Etkin, D. (1999) 'Risk transference and related trends: Driving forces towards more mega disasters', *Environmental Hazards* 1, 2: 69–75.

Fraga, J. (2006) 'Local perspectives in conservation politics: The case of the Ría Lagartos Biosphere Reserve, Yucatán, Mexico', *Landscape and Urban Planning* 74, 3: 285–295.

GFDRR (Global Facility for Disaster Reduction and Recovery). (2015) *Building Regulation for Resilience: Managing Risks for Safer Cities*, Washington DC: The World Bank.

Haque, U., Hashizume, M., Kolivras, K.N., Overgaard, H.J., Das, B. and Yamamoto, T. (2011) 'Reduced death rates from cyclones in Bangladesh: What more needs to be done?', *Bulletin of the World Health Organization* 90: 150–156.

Jha, A.K., Barenstein, J.D., Phelps, P., Pittet, D. and Sena, S. (2010) *Safer Homes, Stronger Communities: A Handbook for Reconstructing after Natural Disasters*, Washington DC: The World Bank.

Kellett, J. and Caravani, A. (2013) *Financing Disaster Risk Reduction: A 20 Year Story of International Aid*, London and Washington DC: Overseas Development Institute (ODI) and the Global Facility for Disaster Reduction and Recovery at the World Bank.

Krimgold, F. (2011) 'Disaster risk reduction and the evolution of physical development regulation', *Environmental Hazards* 10, 1: 53–58.

Lavell, A. (2000) *Desastres durante una década: lecciones y avances conceptuales y prácticos en América Latina (1990–1999)*, San José: Facultad Latinoamericana de Ciencias Sociales.

Lewis, J. (1984) 'Environmental interpretations of natural disaster mitigation: The crucial need', *The Environmentalist* 4: 177–180.

López-Méndez, V., Zavala-Hidalgo, J., Romero-Centeno, R. and Fernández-Eguiarte, A. (2007) *Analysis of the Extreme Flooding during October 2007 in Tabasco, Mexico Using the WRF Model*, Mexico City: Centro de Ciencias de la Atmósfera, Universidad Nacional Autónoma de México (UNAM).

Lovell, E. and Le Masson, V. (2014) *Equity and Inclusion in Disaster Risk Reduction: Building Resilience for All*, London: Climate and Development Knowledge Network (CDKN).

May, P.J., Burby, R.J., Ericksen, N.J., Handmer, J.W., Dixon, J.E., Michaels, S. and Ingle Smith, D. (1996) *Environmental Management and Governance: Intergovernmental Approaches to Hazards and Sustainability*, London: Routledge.

Obama, B.H. (2015) *Remarks by the President at the Annual Hurricane Season Outlook and Preparedness Briefing*, The White House, Office of the Press Secretary, Washington DC. Online https://www.whitehouse.gov/the-press-office/2015/05/28/remarks-president-annual-hurricane-season-outlook-and-preparedness-brief (accessed 11 September 2016).

Peek, L. (2008) 'Children and disasters: Understanding vulnerability, developing capacities, and promoting resilience – an introduction', *Children, Youth and Environments* 18, 1: 1–29.

Shreve, C.M. and Kelman, I. (2014) 'Does mitigation save? Reviewing cost-benefit analyses of disaster risk reduction', *International Journal of Disaster Risk Reduction* 10, A: 213–235.

Thompson, D. (2016) 'Do catastrophes in poor countries lead to event-related policy change? The 2010 earthquake in Haiti', *Journal of Public Administration and Governance* 6, 2: 27–48.

United Nations International Strategy for Disaster Reduction (UNISDR). (2007) *Disaster Risk Reduction: 2007 Global Review*, Geneva: UNISDR.

United Nations International Strategy for Disaster Reduction (UNISDR). (2015a) *Global Assessment Report (GAR) on Disaster Risk Reduction: Loss Data and Extensive Risk Analysis*, Geneva: UNISDR.

United Nations International Strategy for Disaster Reduction (UNISDR). (2015b) *UNISDR Working Papers on Public Investment Planning and Financing Strategy for Disaster Risk Reduction: Review of Peru, Interim report*, Geneva: UNISDR.

Vorhies, F. and Wilkinson, E. (2016) 'Co-benefits of disaster risk management', in S. Surminski and T.M. Tanner (eds), *Realising the Triple Resilience Dividend: A New Business Case for Disaster Risk Management*, Dordrecht: Springer Nature.

Wilkinson, E. (2011) *Decentralised Disaster Management: Local Governance, Institutional Learning and Reducing Risk from Hurricanes in the Yucatán Peninsula, Mexico*, Unpublished PhD thesis.

Wilkinson, E. (2012) *Transforming Disaster Risk Management: A Political Economy Approach*, London: Overseas Development Institute.

Wilkinson, E. and Peters, K. (eds) (2015) *Climate Extremes and Resilient Poverty Reduction: Development Designed with Uncertainty in Mind*, London: Overseas Development Institute.

World Bank, unpublished. *Analisis de los impactos de las inversiones en prevención y reducción de riesgos: estudio de caso de Tabasco entre 2007 y 2010*, Washington DC, The World Bank.

World Commission on Dams. (2000) *The World Dams Report*, London and Sterling, VA: Earthscan.

Yeo, R., Kett, M., Stubbs, S., Deshpande, S. and Cordeiro, V. (2005) *Disability in Conflict and Emergency Situations: Focus on Tsunami-affected Areas*, London: Department for International Development (DFID).

PART V

Sectors and Implementation

36

EDITORIAL INTRODUCTION TO PART V: SECTORS AND IMPLEMENTATION

Do-it-ourselves Disaster Risk Reduction Including Climate Change Adaptation

Ilan Kelman, Jessica Mercer, and JC Gaillard

Disaster risk reduction (DRR) including climate change adaptation (CCA) could never remain isolated from other sectors, as outlined in Chapter 1. In implementing the process, many other sectors are involved, intersecting with core principles and turning them into policies and actions. This interaction demonstrates the fluidity of applying DRR including CCA in practice: it must apply to and support disparate areas, connecting directly to basic needs and livelihoods which cannot be fulfilled in the face of disasters.

Basic needs covered by chapters include water, food, health, and housing and settlements. Although they are presented as sectors, they interact, being interdependent on each other. Other topics are not covered by chapters, yet are as essential, including waste management, sanitation, energy including electricity, communications including telecommunications, and social services, which incorporates prisons, youth and elderly care, libraries, and welfare – amongst many others.

If basic needs are not covered, then immense vulnerability exists. To a large degree, providing for basic needs is DRR including CCA. Proper development means completing DRR by the very definition of 'proper development'. For implementation, two simultaneous directions are necessary.

First, tackle basic needs and use them as a springboard to connect to further DRR. For example, it is not just about providing housing and settlements, but also about ensuring that housing and settlements have completed a risk analysis so that technical and non-technical vulnerability reduction measures are incorporated during the construction. It is much cheaper and effective to include hazard-resistant design from the beginning than to add it in afterwards or through retrofitting.

The second direction is pushing the other way, promoting DRR including CCA as important for what it does – and then using this lobbying to impart the responsibility to provide basic needs. As an example, pointing out that a school is in a high risk landslide zone and could be relocated starts the ball rolling to highlight the importance of safe education more generally – for boys and girls. Safe education as a right means education as a right, so relocating a school to reduce landslide risk also means reducing vulnerabilities against other hazards, making it safe for boys and girls to attend, and ensuring access irrespective of socio-economic status.

Neither direction is more important that the other. Both need to be boosted side-by-side.

Connections with further implementation of DRR including CCA and with livelihoods can be made. Relocating a school and making it safe and inclusive involves planning. Non-technical vulnerability reduction measures must involve early warning systems, since no location could ever be entirely risk-free from all environmental hazards. Anyone using the school – pupils, staff, and community groups – needs to be comfortable with the site as a safe space as well as knowing the signs and signals of when to evacuate, where, and how. Part of an early warning system would ensure that parents can communicate with their children and feel comfortable that their children are in safe hands during a hazard.

Planning is one of many implementation tools for DRR including CCA, linking directly to different design approaches. Financial mechanisms are part of the scope, with funding approaches and insurance mechanisms having the power to nudge or force people, communities, and institutions in specific directions for DRR including CCA. Money is not the ultimate power, but is a substantial directive. One clear message for all sectors and to support implementation is that DRR including CCA is not a cost; it is an investment. Paybacks are generally quick, large, and continuing over long time scales, even in the absence of a significant environmental hazard.

One challenge is that some of the paybacks are invisible, being costs not incurred. A focus on only benefits received does not give the entire description of the paybacks. Even where calculations include all factors, day-to-day decision-making might tend to highlight what is gained directly rather than losses not experienced. DRR including CCA requires continual effort to communicate the full range of gains and to relate them to everyday experience.

Ultimately, successful DRR including CCA most often requires everyday investment in development to strengthen people's livelihoods, rather than specific and siloed initiatives labelled as DRR, CCA, or with other jargon. Hence, one may argue that DRR including CCA does not require much specific funding as long as the processes are being achieved through wider funding regimes.

The benefits from such investment ripple outwards, into the sectors not covered by *Handbook* chapters in addition to other implementation approaches. Standards and professional charters can go a long way to being an impetus for DRR including CCA. Economic and social policies are wide-ranging but major players in driving forwards the political environment to produce what is needed.

Because, ultimately, DRR including CCA cuts across all sectors and needs to be implemented by everyone. People have different opportunities, different resources, and different abilities, so the key is to provide opportunities, resources, and support for abilities to match with people's needs. Where some might not be able to contribute as much, others can pitch in more. 'We' and 'us' refer to the collective of society, indicating that everyone is affected by disasters and everyone has a role to play in dealing with them and their root causes.

No ephemeral authority will step in and save us like a deity does at the end of a Greek tragedy in the theatrical device called *Deus ex machina*. No 'others' will descend and instruct humanity on how to save itself from itself. With all the other abiotic and biotic elements, we are on this planet alone and together.

It is we who implement DRR including CCA.

37

FUNDING AND FINANCING FOR DISASTER RISK REDUCTION INCLUDING CLIMATE CHANGE ADAPTATION

Annika Dean

Introduction

This chapter provides an overview of funding and financing disaster risk reduction (DRR) including climate change adaptation (CCA). It examines the overlaps, connections and differences between the two types of funding, and seeks to critically evaluate the key challenges and opportunities for better integration of CCA funding mechanisms under broader DRR schemes.

First, this chapter provides a brief historical overview of mechanisms to fund DRR and CCA. It then visits the rationale and imperative for integrating CCA funding mechanisms into broader schemes to address DRR and sustainable development. Next, the chapter outlines five key challenges related to funding and financing DRR and CCA that are inhibiting effective implementation or integration. The first challenge is inequitable distribution of international finance for DRR. The second challenge is misalignment between institutional arrangements for DRR and CCA, and the budgetary systems in place to allocate, categorise and track resources at the national level. The third challenge relates to capacity constraints; some of the countries that are most vulnerable to disaster risks, including climate change, have the least capacity to access, use, report and acquit funding. The fourth challenge relates to inappropriate modalities. Much international finance for DRR and CCA is delivered on a project-by-project basis through modalities that hamper use of national systems, thereby threatening effective integration. The fifth and final challenge discussed in this chapter is the requirement for climate finance to be 'new and additional'. This requirement has perversely discouraged effective integration in some cases. Finally, the chapter explores opportunities for integration and makes a number of recommendations.

Brief Overview of Funding and Financing for DRR and CCA

Funding for DRR in one form or another has been around for a very long time. National governments have long taken measures to reduce the risk of disasters affecting their populations. International support for DRR has also been around for many decades. The United Nations Disaster Relief Office (UNDRO) was established in 1971 with a mandate for mobilising and coordinating disaster relief. A trust fund for direct relief assistance was established and managed by UNDRO, but the focus of the Office was on prediction and prevention of disasters, and assisting govern-ments in pre-disaster planning and establishment of early warning systems (Macalister-Smith 1985). Civil society has also been a major contributor to funding DRR over the past decades. The International Federation of the Red Cross/Red Crescent (IFRC) in particular has been a pioneer within civil society in actions to reduce disaster risks. Publishing the report 'Prevention is Better than Cure' in 1984, IFRC has not only been involved in delivering projects that reduce risks, but also in inventing new ways of doing DRR. For instance, in 2008, it launched the first ever pre-emptive appeal for flooding in West Africa, based on the forecasts for that season (IFRC 2009). Hundreds of other NGOs have been involved in activities to reduce the risk of disasters, mobilising funding from national government grants, from philanthropists or from public and private donations. Bilateral and multilateral disaster-related Official Development Assistance has also made a substantial contribution to international finance for DRR.

But despite increased recognition of the importance of DRR, in practice, insufficient fund-ing for DRR is persistent compared with funding for emergency response and recovery and rehabilitation. This is the case both for national governments allocating funding to reduce their own disaster risk, and for development partners allocating disaster related international finance (UNDP 2002; Kellett and Caravani 2013). The global cost of disasters far outstrips total Official Development Assistance and disasters cause serious setbacks in sustainable development. For instance, in 2011 the cost of disasters was USD 380 billion, roughly double aid flows in the same year (Munich RE 2012). Even so, only 0.4 per cent of Official Development Assistance over the past two decades has gone towards DRR (USD 13.5 billion). By contrast, emergency response has received USD 69.6 billion in disaster-related Official Development Assistance and reconstruction and rehabilitation has received USD 23.3 billion over the same time period. Admittedly, some of this funding for reconstruction and rehabilitation might effectively func-tion to reduce future disaster risks, if the practice of 'building back better' is adopted (Kellett and Caravani 2013).

The inadequacy of funding for DRR occurs at all scales: local, national, and international. The lack of disaster-related international finance for risk reduction is partially due to the fact that, hindered by media focus on disasters rather than DRR, few donors perceive DRR to be a priority. For this same reason, international finance for DRR is dominated by a small number of major donors. The key bilateral contributor of international finance for DRR is Japan, and the key multilateral contributor is the World Bank. Indeed, Japan and the World Bank combined have contributed 63 per cent of international finance for DRR over the past two decades (Kellett and Caravani 2013). Nevertheless, it is also important to acknowledge that addressing the root causes of disasters, as per DRR's definition, does not necessarily need dedicated funding per se. Any form of development funding – such as for health, education, or food security – can result in significant gains for DRR including CCA. As long as DRR including CCA considerations are part of development work – which is a big 'if' which is often not realised – DRR might not need to have its own independent, sectoral funding stream.

Since the early 1990s, climate change has gained increasing global attention. The United Nations Framework Convention on Climate Change (UNFCCC), which was drafted in 1992, emphasised the inequitable costs and burdens of climate change, and stated that the problem of climate change should be addressed in line with the 'common but differentiated responsibilities and respective capabilities' of the parties to the UNFCCC (1992, p. 1). This principle of common but differentiated responsibilities and respective capabilities is not only supposed to underpin the emission reduction commitments of parties to the UNFCCC, but is also intended to underpin commitments in relation to climate finance. Article 4.4 of the Convention states:

The developed country Parties and other developed Parties included in Annex II shall assist the developing country Parties that are particularly vulnerable to the adverse effects of climate change in meeting costs of adaptation to those adverse effects.

Various funds have been established under the UNFCCC over the years to facilitate transfers of climate finance for both climate change mitigation and CCA. Other multilateral and bilateral climate funds have been established outside of the UNFCCC process, and bilateral transfers of Official Development Assistance are now also tracked according to their relevance to climate change under the Rio Markers (although there are questions about whether climate relevant Official Development Assistance can be considered to be climate finance). The climate funds established under the UNFCCC have various objectives, various funding sources, and have been accessible via different access modalities (Table 37.1).

Table 37.1 UNFCCC Climate Funds

Name	Year	Access Modality	Focus	Funding Source
The Global Environment Facility Trust Fund	1991	Accredited implementing agencies	To fund the incremental costs of transforming projects with national environmental benefits into projects with global environmental benefits.	Voluntary contributions in replenishment rounds
The Global Environment Facility Strategic Priority on Adaptation	2001	Accredited implementing agencies	Temporary strategic priority under the GEF Trust Fund to safeguard the global environmental benefits achieved through the GEF Trust Fund under future climate change scenarios and pilot adaptation projects in less wealthy countries.	USD 50 million from the GEF Trust Fund
Least Developed Countries Fund	2001	Accredited implementing agencies	To fund adaptation projects in Least Developed Countries (starting with the formulation of National Adaptation Programmes of Action).	Voluntary contributions in replenishment rounds
Special Climate Change Fund	2001	Accredited implementing agencies	To implement medium – and long-term adaptation strategies in vulnerable sectors in less wealthy countries.	Voluntary contributions in replenishment rounds
Adaptation Fund	2001	Accredited national, regional or multilateral implementing entities	To fund tangible adaptation projects in less wealthy countries.	A 2 per cent levy on Clean Development Mechanism projects plus voluntary contributions
Green Climate Fund	2013	Accredited national, regional and international, public and private sector entities	To promote a paradigm shift towards low-emission and climate resilient development pathways.	Voluntary contributions

Despite the establishment of multiple climate funds under the UNFCCC, funding for CCA, like DRR, has not met demand. The Global Environment Facility (GEF) Trust Fund has been limited in its ability to fund CCA, because of a requirement that funded projects should deliver global environmental benefits, skewing funded projects towards climate change mitigation instead of CCA. The CCA-dedicated climate funds established in 2001 all experienced delays in becoming operational. When they did eventually become operational, funding was initially limited to 'pilot' projects and planning, not implementation. As a result, for most of the first decade after the UNFCCC was established, CCA funding for implementation of CCA projects was seriously inadequate. The fact that replenishment of the funds is voluntary also contributed to the persistent inadequacy of funding.

In 2009 at the UNFCCC negotiations in Copenhagen, some progress in relation to climate finance was made. The wealthiest countries committed to provide USD 10 billion each year from 2010 to 2012 in 'Fast Start Finance' to support urgent and immediate CCA and climate change mitigation priorities in less affluent countries. The wealthiest countries further agreed to jointly mobilise USD 100 billion per year by 2020 for CCA and climate change mitigation. The Copenhagen Accord specified that this finance should be 'scaled up, new and additional, predictable and adequate' (UNFCCC 2009, p. 3). It was also agreed that a new climate fund, the Green Climate Fund, would be established under the UNFCCC, which would initially be managed by the World Bank (UNFCCC 2009). The World Bank already manages a DRR-related fund, the Global Facility for Disaster Reduction and Recovery (Box 37.1). Although these commitments are not legally binding, they present an opportunity to address not only climate change mitigation and CCA, but also DRR and broader sustainable development.

Box 37.1   Global Facility for Disaster Reduction and Recovery (GFDRR)

Ilan Kelman[1]

[1] University College London, London, UK and University of Agder, Kristiansand, Norway
Based on GFDRR (2016).

The Global Facility for Disaster Reduction and Recovery (GFDRR; http://www.gfdrr.org) supports less affluent countries to implement DRR including CCA through a partnership of 34 countries including several donors and 9 international organisations. GFDRR is governed by its 'Partnership Charter', is administered by the World Bank, and has offices in Washington, D.C., Brussels, Geneva, and Tokyo.

In Fiscal Year 2015, running from 1 July 2014 to 30 June 2015, GFDRR provided grants in over 89 countries. USD 70 million was awarded for capacity building, knowledge sharing, and technical assistance. These added projects meant that GFDRR was managing a total of USD 216 million in grants, helping to leverage further funds. Specific activities included conducting post-disaster needs assessment, training over 11,000 people, incorporating disaster risk considerations into national planning, and producing over 220 products.

The Imperative for Integration

The rationale and imperative for integrating CCA into DRR and sustainable development is well established. Climate change will increase the exposure of populations by strengthening the intensity and frequency of fast and slow onset hazards (e.g., days of extreme heat, bushfires, cyclones, and sea level rise). More frequent disasters threaten sustainable development gains. At the national and sub-national level, many of the activities labelled as DRR, CCA, or sustainable development look the same. Taking a risk-focused approach, climate change is best viewed as one of a suite of risks that countries face. Integrating CCA into DRR and sustainable development planning is both a more theoretically rigorous approach, and is the most practical approach at the national and sub-national level (Mercer 2010). Especially in countries that are low on capacity, integration of CCA and DRR is a more efficient use of resources, allows better prioritisation of risks (by considering all risks alongside each other and in combination), reduces duplication of activities and reduces competition for the same resources.

This recognition has been at the heart of efforts to mainstream CCA into DRR and sustainable development. Mainstreaming requires that risk considerations be incorporated at every stage of development planning, including national, sectoral and ministerial policies and plans and legislation such as land use plans, building codes and social and environmental impact assessments.

Inequitable Distribution

International assistance for DRR should be delivered to those countries that have the highest risk profiles and the lowest capacity to adapt with their own resources. This is recognised in the Sendai Framework for Disaster Risk Reduction (2015–2030), which states that finance should be targeted towards Least Developed Countries, Small Island Developing States, African countries, landlocked countries and middle-income countries facing disaster risks (UNISDR 2015). In practice, the bulk of international support for DRR has focused on middle-income countries with high-risk profiles. Low-income countries with high-risk profiles have received negligible funding for DRR, but have received high levels of funding for emergency response and reconstruction and rehabilitation (Kellett and Caravani 2013). Countries in sub-Saharan Africa have also received negligible international finance for DRR. One possible explanation for this is that drought, which is the most common hazard affecting this region, is not included in the mortality risk index, which is often used to inform how funding is allocated (Kellett and Caravani 2013). Regional and bilateral relationships also play a significant role in influencing which countries receive funding. For instance, most of Japan's Official Development Assistance targeting DRR goes towards Southeast Asia, with Indonesia and the Philippines receiving roughly half of the total amount.

In general, more resources are allocated to DRR when effective mainstreaming is achieved, compared to funding standalone DRR projects. However, even with mainstreaming, resource allocation is not guaranteed. A review of the progress made to achieve the Hyogo Framework for Action (2005–2015) found that whilst there has been significant progress made to integrate DRR considerations into sustainable development policies and plans, insufficient resources to ensure effective implementation persist (UNISDR 2005, 2009). To understand why this is the case, it is crucial to understand the national budget process, which determines how resources are allocated (or not allocated) to various government plans and policies. Whilst resources can be mobilised from multiple actors and at multiple scales, effective systems of channelling resources at the national scale are paramount to effective integration. One of the main barriers to the effectiveness of disaster related international finance is that it can be unpredictable, ad hoc and reactive – perhaps explaining the greater emphasis on response and recovery compared to risk reduction. To effectively reduce disaster risks, it is essential to make national resources available for that purpose. Ideally, the bulk of DRR funding should come from national governments contributing to reducing risks facing their own people. Where these resources are not available, creating strong national strategies and systems for blending and accounting for finance from multiple sources is imperative.

Misalignment between Institutional Arrangements and Budgetary Systems

This leads to the second challenge discussed in this chapter: misalignment between the budget process and the institutional arrangements designed for DRR including CCA. In all countries, resource allocation for undertaking different activities is influenced by the budget process (Jackson 2011). There are three main ways to target public resources in the budget process. The first way is to target resources towards administrative units and sub-units, such as ministries, divisions and departments, through a departmental or administrative budget classification. The second is to allocate resources across different sectors and departments by classifying resources according to function with a specific budget code. The third is to target resources directly towards specific projects and programs (Jackson 2011). The various methods are not mutually exclusive, and each method has benefits and disadvantages. Crucially though, the way that DRR funding is allocated in the budget process must be matched with the institutional and governance arrangements for DRR within each respective country.

The first method involves the establishment of a specific department, agency or ministry responsible for DRR. Responsibilities of such an agency are to offer technical advice to ministries, ensure that DRR is effectively integrated into the national regulatory environment, coordinate the activities of ministries and provide monitoring and evaluation support. The advantage of establishing a specific department or agency responsible for DRR including CCA is that taxpayers and donors alike may be more willing to fund DRR through a named department. This can potentially increase the resources made available. Departments can also attract finance from both the recurrent and development funds. Once established, departments have an interest in defending their existence and interests in the budget process, potentially leading to greater sustainability. A potential downside is that the existence of a central agency may influence ministries to believe that DRR is being addressed, and is not part of their mandate or responsibility. For a cross-cutting issue such as DRR, which requires the cooperation of multiple sectors at various scales, this may turn out to be detrimental. Furthermore, if the department is not situated at a level of government that is higher than the ministries, it may not command the authority necessary to effectively perform a coordinating function (Jackson 2011). Either way, establishing a specific department or agency responsible for DRR requires that a budget classification is created to target sufficient resources to that department.

In the second method, expenditure for DRR is classified according to its specific function and can therefore occur across multiple departments and ministries. In this method, the government allocates funding to the function of DRR. This can help to avoid a common problem associated with the sectoral approach, which is the possibility of ministries redirecting resources away from DRR if it is not seen to be a priority. When resources are allocated for a specific cross-cutting function, departments and ministries have an incentive to find ways to build that function into their own mandate. The disadvantage of this method is that tracking funding according to function relies on a degree of complexity of budget codes. Most countries have budget systems that allow resources to be tracked according to function, but some countries with weak public finance management may require some strengthening of this capacity (Jackson 2011).

The third method is to fund DRR (including CCA) through projects and programmes. The main benefit of allocating resources to specific projects and programmes is that there is no burden on the recurrent budget. In addition, projects can be situated at the apex of organisational structures to give them clout and legitimacy. It can also be easier to cooperate with other non-government actors, such as civil society and the private sector, through projects. On the downside, project-based funding is frequently short-term and very poorly integrated into government systems, compromising the overall effectiveness and sustainability of outcomes and draining limited government capacity. It can also be unpredictable, reactive and ad hoc. This is discussed in more detail in the next section on modalities and is expanded upon in the case study of Kiribati (Box 37.2).

Box 37.2   Case Study of Kiribati

Annika Dean[1]

[1] University of New South Wales, Sydney, Australia

In the Republic of Kiribati there has been a long-standing focus on mainstreaming CCA and DRR into national development planning. This has been encouraged since 2003 as part of the national-scale Kiribati Adaptation Project (KAP) and more recently through the Kiribati Joint Implementation Plan on Climate Change and Disaster Risk Reduction (KJIP) (Government of Kiribati 2014). The KJIP Secretariat and the Kiribati National Expert Group (KNEG) were established in the formulation of the KJIP. The KNEG acts as a coordinating body for CCA and DRR activities, and theoretically consists of representatives from relevant ministries, civil-society organisations, faith-based organisations and the private sector. The intention of the KJIP is that it clearly links to the Kiribati Development Plan – the development master plan for Kiribati – which is updated every four years. The KDP has its own governance structure. Whilst there are no costed priority activities within the KDP, it is theoretically backed by medium-term Sector Strategic Plans, which are supported by Ministry Strategic Plans and Ministry Operational Plans and budgets.

Whilst in theory it seems that CCA and DRR have been effectively mainstreamed through the KJIP into development planning, in practice it is much more complicated. Official Development Assistance accounts for roughly 34 per cent of Kiribati's Gross Domestic Product (Sampson 2005). Nearly all of this assistance comes in the form of project-based grants channelled through the development fund and nearly all government activities addressing CCA and DRR have been funded in this way.

The Kiribati development fund is entirely separate to the recurrent budget. Transfers from the development fund to the recurrent budget are not permitted and ministry staff do not have access to the development fund. What this means in practice is that, despite a long-standing focus on mainstreaming CCA and DRR into development, the majority of ministries still do not incorporate donor-funded projects, which make up almost the entire development budget of Kiribati and account for almost all activities related to CCA and DRR, into their Ministry Operational Plans and budgets. Instead, ministries only include the reliable and predictable activities funded through the recurrent budget in their Ministry Operational Plans and budgets. The lack of predictability associated with project-based funding, and the lack of control and access to the development budget makes tracking these activities too cumbersome.

The fact that mainstreaming and integration efforts do not effectively reach ministry level, where activities are implemented, undermines their effectiveness. Mainstreaming is not something that can be done by donors in a one-off project – it is an ongoing process, requiring ongoing consideration and prioritisation of risks into national development plans, sectoral plans, ministry operational plans and budgets on a cyclical basis. The Kiribati experience shows how trying to achieve mainstreaming through unpredictable short-lived projects that are not directed through national systems inhibits the effectiveness of mainstreaming (and see also Gaillard 2012).

The most appropriate and effective institutional arrangements for DRR including CCA will differ between countries. Countries may choose to allocate resources to DRR via a combination of the three methods described above: establishing a department responsible for coordinating DRR activities but leaving implementation to the line ministries with a budget allocation for DRR activities classified according to function, and running projects and programmes, using either or both international and national finance. The key in all cases is ensuring that sufficient funding is allocated, that funding is legally mandated, and that the budget process is equipped to be able to allocate resources effectively and efficiently to match the chosen institutional arrangements.

Even when effective arrangements are put in place at the national level, it is rare that DRR is seen as a priority at the local level, and resources are frequently scarce. Fiscal grant systems and budget allocations to local level institutions can be introduced in cases where funding for DRR at the local level is insufficient. The provision of fiscal grants from central to local governments, and encouraging local participation in creating locally relevant projects and helping to implement national programmes, can help to ensure that the DRR agenda is taken seriously and reaches the local level (Jackson 2011).

Capacity Constraints

Another key challenge affecting CCA funding, and hindering effective integration of CCA funding into broader DRR schemes, is the complicated architecture that scaffolds climate finance. There are often high transaction costs for accessing climate finance. Huge time delays are common due to the cumbersome and bureaucratic processes of climate funds such as the GEF (Mace 2005). According to the Independent Evaluation Office of the GEF, the GEF's project cycle has become so complicated and elaborate over time, that it is burdening countries and resulting in higher transaction costs (GEF IEO 2014). Although the project cycles for the Least Developed Countries Fund and Special Climate Change Fund are slightly expedited compared to the GEF Trust Fund, they are still very complicated to navigate for small countries with limited capacity. Moreover, eligibility criteria and access requirements differ for different funds, adding to the complexity.

Accessing international finance for CCA and DRR relies on capacity. Because of the very complicated and protracted requirements that need to be met in order to access funding, manage projects and monitor and report on outcomes, lack of capacity can ironically prevent some of the most vulnerable countries from obtaining any funding at all. This partly explains why the middle-income countries with high-risk profiles have received comparatively greater amounts of international assistance for DRR than the least wealthy countries with high-risk profiles.

Lack of capacity, particularly in many low-income countries, is often cited by donor agencies as the main rationale for not channelling funding through recipient country systems. Instead, donor agencies often choose to deliver funding on a project-by-project basis. But project based funding can ironically undermine effective mainstreaming, and weaken the capacity of national systems, as discussed in more detail in the next section.

Inappropriate Modalities

The third challenge relates to the use of inappropriate modalities. There are a range of modalities for accessing international finance for DRR and CCA. These modalities differ between funds. The most common modality for accessing climate finance through the UNFCCC climate funds in the past has been in the form of project-based grants accessed via multilateral implementing agencies (otherwise called GEF agencies). Choosing a GEF agency is a requirement of accessing funds under the GEF Trust Fund, the Least Developed Countries Fund and the Special Climate Change Fund. The role of the GEF agencies is to help countries access their allocation from the GEF, design project concepts and navigate the GEF project cycle. There are currently eighteen accredited GEF agencies – all large development banks such as the World Bank, or United Nations organisations such as the UNDP. GEF agencies receive 10 per cent of the overall project grant in the form of agency fees. For some very small countries, working with such large, bureaucratic agencies with entrenched development paradigms and practices has added an additional layer of complexity and red tape to the process of CCA.

The Kyoto Protocol Adaptation Fund has a slightly different access modality enabling countries to gain direct access to climate finance through accredited National Implementing Entities (or regional access through Regional Implementing Entities). The Adaptation Fund Board has approved roughly two dozen agencies as National Implementing Entities in countries including Ethiopia, Morocco, Senegal and Kenya in Africa; Chile, Mexico, Peru, Belize and Argentina in Central and South America; Antigua and Barbuda, the Dominican Republic and Jamaica in the Caribbean; and India and Indonesia in South and Southeast Asia. In the Pacific, the Federated States of Micronesia and the Cook Islands have gained National Implementing Entity status and the Secretariat of the Pacific Regional Environment Programme (SPREP) has been accredited as a Regional Implementing Entity. Fortunately, the National and Regional Implementing Entities accredited by the Adaptation Fund Board are also recognised under the new Green Climate Fund. Though the ability to receive finance directly via National Implementing Entities is generally seen as an improvement from project-based delivery via Multilateral Implementing Entities, gaining National Implementing Entity status is very difficult, particularly for small countries with limited capacity.

Therefore, most climate finance and disaster related ODA is accessed in the form of project-based grants. This is the case for both DRR and CCA. For example, in the case of DRR, up until the year 2000, a small number of large-scale flood-protection infrastructure projects accounted for roughly half of all international finance for DRR (Kellett and Caravani 2013). The remainder of the funding was dispersed across hundreds of small projects, raising concerns about inefficiencies and high transaction costs for recipient governments (Kellett and Caravani 2013).

The project-based approach is burdensome in many small countries, as it increases the time and costs involved in project administration. There is also a tendency for projects to duplicate one another, so the establishment of coordination committees becomes a necessity. In small countries with limited government capacity, this can become a large burden. As projects often have their own parallel administration, financial management and reporting requirements, they are often poorly integrated into government plans and policies. This compromises sustainability of outcomes beyond the life cycle of the project. As well as draining limited capacity in the short term, reliance on parallel systems of reporting and financial management can also serve to weaken government systems (as government staff may become unfamiliar with national systems and more familiar with the systems of development partners).

Whilst the Paris Aid Effectiveness Principles and the Accra Agenda for Action (OECD 2005) and the Pacific Aid Effectiveness Principles (Pacific Islands Forum Secretariat 2007) encourage the use of local systems where possible, in practice, externally administered project-based grants are often a requirement of climate funds or the preferred modality of development partners, especially in countries deemed to have weak capacity. This is the case in the Small Island Developing States and many other countries most in need of international support.

Development partners often prefer the project-based approach because it obliges recipient countries to adhere to the processes of financial management and reporting requirements of the development partner. Project-based approaches also enable donors to maintain greater control over project objectives and procurement, allowing them to put out tenders and contracts for sub-components of the project. From the perspective of development partner governments, it can be difficult to make long-term predictable commitments, for instance in the form of direct budget support, because development partners also operate on an annual financial year basis, within an electoral cycle. Without strong bipartisan support, ongoing commitments of ODA can never be guaranteed.

Various new modalities are currently being trialled across the Pacific for accessing climate finance. General or sectoral budget support is one such modality, which is generally seen as preferable to many other approaches by recipient governments as it enables greater national ownership, and strengthens national systems by aligning with them (PIFS 2011). Budget support also generally reduces transaction costs, and can therefore be a more cost-efficient way of achieving outcomes. Budget support can come in the form of general budget support, or sectoral support. The European Union Global Climate Change Alliance (EU-GCCA) has provided some climate finance in the form of sectoral support to the Solomon Islands and Samoa in recent years. It remains to be seen how this is working. Mainstreaming of CCA and DRR into broader development strategies is a prerequisite for being able to receive funding from the EU-GCCA in the form of general or sectoral budget support, alongside a range of other criteria. Other donor agencies have their own requirements that need to be satisfied in order to qualify for direct budget support. It would be beneficial if donor agencies were able to coordinate in this respect and settle on a common set of standards or criteria.

National Trust Funds are a modality that can support both project-based approaches and budget support. National Trust Funds can also be used to save and accrue finance to act as a buffer in emergencies. National Development Banks have been shown to be a good modality for blending multiple types of finance from different sources such as the private sector, governments and donors. Tuvalu has established a national trust fund and the Federated States of Micronesia have established a sub-regional fund. Different modalities have different strengths and weaknesses, and the most appropriate modality for any given country depends on the context, preference, strengths and experience of the country (PIFS 2011).

Projects can help where innovative or one-off activities need to be conducted, but for ongoing critical functions, predictable, scaled up funding in the form of national budget commitments and general or sectoral budget support from development partners is vastly preferable. Accessing finance via general or sectoral budget support, or other modalities such as trust funds, generally allows for more flexibility and ownership, strengthening national systems and supporting the ability to blend different sources of finance, crucial for effective integration (PIFS 2011).

The Problem of Newness and Additionality

The fifth issue discussed in this chapter is both an opportunity and a challenge for integration. As already mentioned, the anticipated increase in climate finance over the coming years presents an opportunity to greatly increase the resources available not just for CCA, but for reducing the broad range of risks facing vulnerable populations, which underpin vulnerability to disasters including those exacerbated by climate change. Climate finance is supposed to be 'new and additional' to previous Official Development Assistance, predictable, adequate and scaled up over time. But the concept of newness and additionality has also had some adverse impacts that have discouraged effective integration of CCA into DRR.

Both donor and recipient countries have been eager to support the concept of newness and additionality, for different reasons. Donor countries have emphasised that climate finance should be used for addressing new and additional impacts – for the additional or incremental costs of climate-proofing baseline development projects. The requirement for demonstration of new and additional costs or impacts is an eligibility criterion under many climate funds. Donors are not prepared to pay new and additional finance for 'business-as-usual' development. Likewise, as climate change poses new and additional costs, less wealthy countries are keen to ensure that the distinction between Official Development Assistance and climate finance is recognised, to justify the mobilisation of new and additional resources beyond previous Official Development Assistance (Ayers and Huq 2009).

Ironically the preoccupation with newness and additionality, and the absence of an agreed baseline at the upstream policy level have fostered a 'dysfunctional adaptation discourse that excludes rather than incorporates fundamental underlying development objectives' and under-mines effective integration of CCA into DRR and development (Ayers and Dodman 2010, p. 167; Fankhauser and Burton 2011).

The preoccupation with climate finance being directly related to addressing climate change impacts has also favoured 'hard' CCA measures, such as construction of infrastruc-ture, rather than 'soft' measures such as improving adaptive capacity, or reducing vulnera-bility by addressing development factors such as healthcare, education and income poverty (Fankhauser and Burton 2011). These 'soft' CCA measures are 'central to efficient, effective and equitable adaptation' (Fankhauser and Burton 2011, p. 1037). As long recognised by DRR literature, without addressing the specific historical factors underpinning the vulner-abilities of groups, problems are likely to recur (Wisner *et al.* 2004, 2012). As noted above, addressing such deep vulnerability to achieve DRR including CCA does not necessarily need to be achieved through dedicated funding streams, because other development activ-ities should do so.

CCA should intensify efforts towards sustainable human development, which can be best achieved by placing it within DRR. Yet, in focusing on the incremental costs of CCA, climate finance discourse and policy has promoted standalone CCA approaches. Even where CCA has been mainstreamed into development in recognition of the close links between the two, incre-mental costs reasoning sometimes dictates that climate finance can only be spent on planning and not implementation; i.e., climate finance can be spent on the process of mainstreaming CCA into development planning, but not on actually implementing the baseline costs of priorities outlined in development plans. Under pressure to report on new and additional contributions, and in the absence of an agreed baseline from which to judge newness and additionality, donors also have incentives to promote impacts-driven CCA measures that can clearly and easily be linked to climate change impacts, rather than measures that address risks or broad socio-economic drivers underpinning vulnerability, which may be difficult to distinguish from other development meas-ures. That is, the climate financing structure to a large degree incentivises dissociation of CCA from DRR. On the other hand, there is work being done in some development organisations to embed concepts of risk and resilience to integrate and build coherence between multiple pro-grammes of work (Box 37.3).

Box 37.3   Financing Humanitarian Policy: Case Study of the Department for International Development, Government of the United Kingdom

Annika Dean[1]

[1] University of New South Wales, Sydney, Australia

The UK's Department for International Development (DFID) made a concerted effort over recent years to adopt a framework of resilience at the centre of its humanitarian policy 'Saving lives, preventing suffering and building resilience' (DFID 2011). This policy was developed in 2011 under the UK's coalition government (2010–2015), but at the time of writing this chapter, the UK government was in flux due to the end of the coalition in 2015 and, then in 2016, a new Prime Minister and Secretary of State for International Development. For the 2010–2015 time period, DFID's policy outlined a commitment to embed concepts of DRR and resilience into DFID's work on climate change and into all DFID programs by 2015. It is currently unclear whether or not their goals were reached.

In relation to funding and financing, the policy acknowledges that very few donors currently contribute towards DRR and resilience in humanitarian crises. It commits to advocating to other donors to expand the donor base for building resilience and risk reduction in humanitarian action. The policy also recognises the burdens that can be placed on the capacity of local systems and outlines a commitment to 'support . . . partners to strengthen their monitoring, evaluation and accountability systems and . . . avoid imposing additional bureaucracy on them' (DFID 2011, p. 8). The policy also outlines DFID's commitment to increase the range of funding mechanisms available to NGOs such as the Red Cross and the private sector. Expanded financing mechanisms might include contributing towards country-level or global-level pooled funds, committing to multi-year funding and making early pledges to appeals in an effort to increase the predictability and sustainability of funding and to avoid problems associated with short-term and unpredictable project-based funding.

Conclusions and Recommendations

Climate finance could potentially help to address shortfalls in funding for CCA, DRR and sustainable development, reducing the range of risks that vulnerable populations face. For this to occur, a number of challenges to integration need to be addressed. Most notable is that focusing on climate finance may mean that non-climate hazards and hazard drivers might be neglected, rather than tackling all risks and vulnerabilities simultaneously. This argument emphasises the need to ensure that climate-related topics are not avoided, but also do not dominate discussions, financing, and services.

At the international scale (donor level), activities to support effective integration should include directing resources towards the establishment of robust national systems, including through direct budget support where possible, rather than funding standalone projects. This may involve encouraging the strengthening of national systems of public finance management or other capacity-building activities. Donors should keep in mind that standards of public finance management should be appropriate to the context in order to facilitate effective implementation. In very small states, best-practice public finance management may not be appropriate and may in fact prove to be too burdensome, hindering implementation. Secondly, multilateral and bilateral funds (especially climate funds) should adjust their eligibility criteria and objectives to explicitly require or encourage activities that integrate CCA into DRR and sustainable development. Thirdly, more international finance for CCA and DRR should be directed to the least wealthy countries with high-risk profiles. If these countries do not have the capacity to apply for and absorb funding, capacity-building activities should be prioritised and access requirements should be simplified. Finally, DRR including CCA should be mainstreamed into all development activities funded through Official Development Assistance.

At the national level, effectively integrating CCA funding mechanisms into broader DRR and sustainable development funding, requires firstly that countries have mainstreamed risk considerations into strategies, policies, plans and regulations underpinning and supporting development. Mainstreaming needs to occur not only at the national level, but also ideally at regional, sectoral, ministerial and local government/community scales.

Mainstreaming can help to ensure the allocation of ongoing resources in the budget process. However, to ensure that sufficient resources are allocated to DRR including CCA, mechanisms should be developed at the national level that obligate governments to provide legally mandated resources to DRR including CCA within the budget process. This is important to overcome the pitfalls of intermittent project funding and to ensure that the mainstreaming of DRR is sustainable.

National governments should also think about the most appropriate institutional arrangements for DRR in their national context and strengthen national systems of budgeting and accounting to ensure that the budget system aligns with the chosen institutional arrangements. In addition, national governments should focus on pursuing a mix of modalities that are appropriate to the context and utilise strengths and experience. Modalities should, where possible, enable resources to be channelled through national systems and allow funding to be blended from multiple sources to satisfy multiple objectives (e.g., CCA, DRR and sustainable development). Strengthening systems of public finance management so that they are capable of efficiently tracking and reconciling funding from multiples sources is critical for effective integration. For cross-cutting issues like DRR, having the ability to categorise resources in the budget process according to function is useful in being able to resource ministries, local government levels, and even civil society groups to undertake activities to reduce risks.

Effective integration of CCA funding mechanisms into broader DRR funding schemes strengthens national ownership, enabling governments to assess risk levels (based on likelihood and severity), and to allocate and release resources where and when they are needed. For instance, national trust funds may enable countries to stockpile resources to address anticipated future emergencies, or for cases where slow-onset hazards do not cost much now but will cost enormous amounts of money in the future. This kind of future planning and resource allocation is difficult to achieve through project-based modalities.

Finally, national governments should focus on ensuring that resources for CCA and DRR do not only circulate at the national government level, but penetrate to the local level. Climate finance in particular has been very state-centric. National governments should make more effort to involve and include civil society actors in implementation and in coordination bodies.

References

Ayers, J. and Dodman, D. (2010) 'Climate change adaptation and development I: The state of the debate', *Progress in Development Studies* 10, 2: 161–168.

Ayers, J.M. and Huq, S. (2009) 'Supporting adaptation to climate change: What role for official development assistance?', *Development Policy Review* 27, 6: 675–692.

DFID (2011) *Saving Lives, Preventing Suffering and Building Resilience: The UK Government's Humanitarian Policy*, London: DFID (Department for International Development), Government of the United Kingdom.

Fankhauser, S. and Burton, I. (2011) 'Spending adaptation money wisely', *Climate Policy* 11, 3: 1037–1049.

Gaillard, JC (2012) 'The climate gap', *Climate and Development* 4, 4: 261–264.

GEF IEO (2014) *OPS5: At the Crossroads for Higher Impact*, Washington DC: Global Environment Facility Independent Evaluation Office.

GFDRR (2016) Global Facility for Disaster Reduction and Recovery Annual Report '15. Washington, D.C.: GFDRR (Global Facility for Disaster Reduction and Recovery).

Government of Kiribati (2014) *Kiribati Joint Implementation Plan for Climate Change and Disaster Risk Management (KJIP)*, Tarawa: Government of Kiribati.

IFRC (2009) *Disaster: How the Red Cross Red Crescent Reduces Risk*, Geneva: International Federation of Red Cross Red Crescent Societies.

Jackson, D., 2011. *Effective Financial Mechanisms at the National and Local Level for Disaster Risk Reduction*, Geneva, Switzerland: UNISDR.

Kellett, J. and Caravani, A. (2013) *Financing Disaster Risk Reduction: A 20-year Story of International Aid*, London and Washington DC: Overseas Development Institute and The Global Facility for Disaster Reduction and Recovery.

Macalister-Smith, P. (1985) *International Humanitarian Assistance: Disaster Relief Actions in International Law and Organization*, Dordrecht: Springer Science and Business Media.

Mace, M.J. (2005) 'Funding for adaptation to climate change: UNFCCC and GEF developments since COP-7', *Review of European Community and International Environmental Law* 14, 3: 225–246.

Mercer, J. (2010) 'Disaster risk reduction or climate change adaptation: Are we reinventing the wheel?', *Journal of International Development* 22, 2: 247–264.

Munich RE (2012) *Natural Catastrophes 2011 World Map*, Munich: Munich RE. Online http://semanticcommunity. info/AOL_Government/2011_Natural_Disasters_Costliest_on_Record#Preface (accessed 17 July 2016).

OECD (2005) *The Paris Declaration on Aid Effectiveness and the Accra Agenda for Action*, Paris: Organisation for Economic Cooperation and Development.

Pacific Islands Forum Secretariat (2007) *Pacific Aid Effectiveness Principles*, Suva: Pacific Islands Forum Secretariat.

PIFS (2011) *Pacific Experiences with Modalities relevant for Climate Change Financing*, Suva: PIFS (Pacific Island Forum Secretariat).

Sampson, T. (2005) *Toward a New Pacific Regionalism: Aid to the Pacific Past, Present and Future*. An Asian Development Bank – Commonwealth Secretariat Joint Report to the Pacific Islands Forum Secretariat, Suva: PIFS (Pacific Island Forum Secretariat).

UNDP (2002) *A Climate Risk Management Approach to Disaster Reduction and Adaptation to Climate Change: Integrating Disaster Reduction with Adaptation to Climate Change*, Havana: UNDP (United Nations Development Programme).

UNFCCC (1992) *United Nations Framework Convention on Climate Change*. New York: United Nations General Assembly.

UNFCCC (2009) *Report of the Conference of the Parties on its Fifteenth Session, held in Copenhagen from 7 to 19 December 2009*, Bonn: UNFCCC (United Nations Framework Convention on Climate Change).

UNISDR (2005) *Hyogo Framework for Action 2005–2015: Building the Resilience of Nations and Communities to Disasters*, Geneva: UNISDR (United Nations International Strategy for Disaster Reduction).

UNISDR (2009) *Global Assessment Report on Disaster Risk Reduction (2009)*, Geneva: UNISDR (United Nations International Strategy for Disaster Reduction).

UNISDR (2015) *Sendai Framework for Disaster Risk Reduction 2015–2030*, Geneva: UNISDR (United Nations International Strategy for Disaster Reduction).

Wisner, B., Blaikie, P., Cannon, T. and Davis, I. (2004) *At Risk: Natural Hazards, People's Vulnerability and Disasters*, 2nd ed., London and New York: Routledge.

Wisner, B., Gaillard, JC and Kelman, I. (2012) 'Framing disaster: Theories and stories seeking to understand hazards, vulnerability and risk', in B. Wisner, JC Gaillard and I. Kelman (eds), *The Routledge Handbook of Hazards and Disaster Risk Reduction*, Abingdon and New York: Routledge, pp. 18–33.

38

INSURANCE FOR DISASTER RISK REDUCTION INCLUDING CLIMATE CHANGE ADAPTATION

Freddy Vinet and David Bourguignon

Insurance has been part of the economy for a long time, always having taken long-term trends into consideration. Traditionally, non-life insurance products were supposed to be stable over a long-term period and insurers' tasks consisted of adjusting premiums and losses in order to save money so that they could face unexpected pay-outs due to huge disasters. Nevertheless, an increase in insurance losses has been observed over the last thirty years, mainly due to an increase in exposed assets. Therefore, the challenge addressing disaster risk reduction (DRR) including climate change adaptation (CCA) for insurers is both internal and external.

From an internal point of view, insurance companies must address DRR including CCA as hazard drivers continue to modify the environment and change one of the baselines on which insurers work. From an external point of view, the question is: How can insurance be a solution to enhance DRR including CCA? Key issues are:

- How to prepare for and warrantee solvency for insurance companies facing the consequences of climate change which, rightly or wrongly, are often articulated as being new risk, larger losses, and increased variability of losses.
- How to measure the products and capital savings needed to face the suggested, rightly or wrongly, expected rise of costs.
- Are the traditional rules of the insurance sector prepared to face challenges that are often described as being new?
- How useful and efficient is insurance as a tool to support and enhance DRR including CCA?
- How can insurers be involved more effectively in DRR including CCA to minimise any adverse impacts from climate change?
- What are the similarities and differences between more wealthy countries and less wealthy countries in terms of both disaster consequences and consequences of the hazard driver of climate change?
- How can disaster risk insurance be integrated as a solution into a comprehensive set of measures for DRR including CCA?

As Linnerooth-Bayer (2012) underlines, many financial tools exist to deal with disaster consequences – such as loans, micro-credit, and savings – with each having advantages and disadvantages. For example, credit can be a suitable solution, but after a disaster, local banks do not always have enough cash to meet demand nor might the banks or electronic transfers be accessible. Insurance, with its advantages and disadvantages, is one more possible tool, often assisting in tandem with others. This chapter explores insurance for DRR including CCA.

Different Forms of Disaster Insurance

Disaster insurance comes in different forms. Porrini and Schwarze (2014) list five models of disaster insurance systems basing their typology on Europe. The most common framework is the free market for insurance (cover is optional). In this case, each insurer is free to estimate and underwrite the risk, while people are free to purchase insurance they want and deem affordable. Many people located in exposed areas are not insured or pay high premiums, such as in Germany and Chile. At the other end, some countries such as Denmark, the Netherlands, and Thailand have a guarantee scheme organised by the state through an annual budget or fund. Private insurance plays a minimal role. Between these two extremes, there are intermediate solutions. In the case of Spain, France, and Norway, various solutions combine the extension of mandatory insurance coverage for fire and the state intervention as a reinsurer. In Europe, many other countries are trying to emulate this model, such as Austria, Belgium, Czech Republic, Italy, and Romania.

Parametric insurance is another system, which is increasingly used in less affluent and highly exposed countries, often at the national level. In this case, compensation is automatic when a parameter or a 'parametric threshold' is exceeded, which could be temperature, wind speed, precipitation, ground acceleration, or other metrics for hazard magnitude. Insurance tends to cover only large-magnitude events rather than frequent ones. It ensures a minimum of protection even in places where private insurance is poorly disseminated. Such a system also exists with the provision of immediate liquidity to states affected by a disaster, such as the Caribbean Catastrophe Risk Insurance Fund.

Thus, the rate of coverage for disaster insurance differs greatly depending on countries and hazards considered. It ranges from almost 100 per cent in some countries like Spain or France where insurance is compulsory to only several per cent in less wealthy countries. Yet, insurance is traditionally seen as being reserved for more affluent countries. In less affluent countries, it tends to remain as a privilege for wealthy people. For floods, Surminski (2013) estimates that in less affluent countries only 5 per cent of damage is covered by insurance, whereas the rate can be 40 per cent in wealthier countries.

Insurance, Development and DRR Including CCA: Is Insurance Only for Wealthy People?

Although insurance follows some general rules, its situation regarding DRR including CCA varies according to affluence levels. In low- and middle-income countries, the possibilities for insurance to enhance DRR including CCA are high, whereas in wealthier countries, the insurance market is sometimes considered to be saturated. In non-saturated markets, insurance can be a tool to quickly and efficiently reduce the economic impacts of disasters. Nevertheless, insurers tend to consider themselves or tend to be considered as risk-averse, so many are reluctant to develop activities in less affluent countries, for various reasons.

The first obstacle is that the risk might not be well-known. In wealthier countries, insurers have frequently been collecting data on losses for a long time, sometimes for more than a century. They can have long-standing historical information on some hazards such as fire and diseases. Meanwhile, in many less affluent countries, there is a lack of data on losses, on assets at risk and on hazards. Moreover, climate change is changing some of the weather-related data baselines around the world.

Another constraint comes from the potential customers. In low-income countries, and increasingly in traditionally wealthier countries, many families live on a day-to-day basis. Their time horizon is around several days while insurance implies long-term planning. Insurance is often considered as a luxury in a context of preoccupation with short timeframes. As Clarke and Grenham (2013) note, the need for insurance does not necessarily mean demand for insurance. The lack of national frameworks on insurance rules (financial, fiscal, and political) is also a constraint for insurance companies to cover new markets.

Insurers and DRR Including CCA: Limits and Gaps of Insurance

Are insurers DRR-averse? This question can be posed legitimately as, even if climate change and its impacts have been addressed for a long time in the insurance sector (see Keskitalo *et al.* 2014; Munich Re 1973), its involvement in CCA and wider DRR topics is comparatively recent. Historically, the insurance sector has been managed by economists whose goal was to adjust premiums on losses. To illustrate this disconnect between insurance and DRR including CCA, note that, until recently, the concept of vulnerability in the insurance sector was the opposite of 'social vulnerability'. For insurers, the richer an insured person, the more vulnerable that person is because wealth generates significant costs in case of disaster. From an insurance perspective, poor people without many assets are not vulnerable to large losses. Additionally, insurance mostly focuses on keeping equilibrium at national and international scales. Thus, insurers developed frameworks and studies to be consistent with international requirements in terms of climate change accountability. The key issue for insurers is to link broad frameworks addressing CCA and concrete local actions to account for the real impacts of climate change at the local level in terms of DRR.

At first sight, DRR and insurance may appear as opposite or incompatible. Indeed, the willingness to contract an insurance policy relies on the idea that losses can occur at any moment. If one suppresses disaster risk by efficient risk reduction measures, the need for insurance disappears. This contradiction may explain why insurers were not traditionally involved in DRR including CCA. Moreover, in a market context, insurers are reluctant to get involved in DRR actions – e.g., by accepting premium reductions in exchange for DRR actions – because they are not sure that they will benefit from those actions in terms of reduced damage and pay-outs. Indeed, customers are free to contract with another insurance company at the end of their contract, so the new company might reap the rewards from the DRR measures supported by the old company. This situation also explains why insurance-related DRR progress can be slow without consistent, comprehensive and sometimes coercive DRR policies at national levels.

Insurers sometimes estimate that it is too expensive for them to carry out local DRR actions. This must be a collective task led by authorities or an individual effort implemented by people at risk. Another obstacle is the lack of knowledge about the real effect of DRR on losses. For individual hazards such as fire or diseases, insurers have been promoting preventive measures for a long time because they are aware of the effectiveness of the measures (such as smoke detectors in every house, now enshrined in law in many jurisdictions around the world) and they can benefit quickly from loss reduction. But this relationship is less clear for many other hazards, especially those with low occurrence probability. For most such hazards, potential losses would not be scattered and diffuse within a large sample of customers, but could strike all insured people in a spatially delineated region.

Finally, insurance is a sector focused mainly on the economic dimension of risk, including actuaries, brokers, agents, modellers, underwriters, risk managers, asset managers, and regulators. The first and main reaction of insurers is to develop an immediate financial response by limiting their exposure to losses, tightening terms of contracts, and raising premiums. The increasing costs of disasters due to exposure alongside international mechanisms seeking a suite of DRR tools are opening opportunities for insurers to explore different approaches, especially in trying to understand how to reconcile DRR including CCA with insurance's traditional rules and aims.

Climate Change Consequences for the Insurance Sector

Insured Losses are Growing

For insurance companies, the economic impacts of weather-related events are increasing, so many suggest climate change as being the cause. This trend might actually be due to better reporting and recording; more financial assets existing; more assets covered by insurance; inflation; and a whole host of other vulnerability and exposure factors not linked to changing hazards. Nevertheless, since 1970, the 35 most expensive weather-related events for the insurance industry have cost a total of USD 400 billion (Swiss Re 2015). Seventy-five per cent of these costs occurred in the past decade. Although it took insurers a long time to acknowledge the reality of climate change (Mills 2009), recent events have been a wake-up call for them (Smolka 2006), legitimately or otherwise. This trend of increasing insured losses applies to all regions throughout the world (Hallegatte *et al.* 2007). In Asia, for example, the number of damaging weather events has increased by 30 per cent over the last 30 years (Munich Re 2014).

The Potential Impact of Climate Change on the Insurance Industry

Besides continuing to deal with frequent hazards, insurance and reinsurance sectors have led many prospective studies regarding preparing for the occurrence of extreme hazards (AXA 2012). Hurricane Andrew in Florida in 1992, which cost more than USD 16 billion and coupled with other weather extremes led to the bankruptcy of nine US insurance companies, was seen by some as a turning point that significantly increased insurance companies' concerns (Changnon *et al.* 1997).

Although disaster risk generates a need for insurance, there are several conditions for insurance to be effective, especially over the long term. First, risk pooling must be broad and extensive. The portfolios should therefore not be geographically concentrated in high-risk areas. Second, insurability can be guaranteed only when the probability of natural hazard occurrence has uncertainties, which is typically the case. However, many areas are subject to disasters, sometimes inexpensive, but more and more frequent. Under these conditions, and (rightly or wrongly) speculating that climate change could worsen the situation, insurers fear having to pay many more insurance claims. This perspective may create a fear of bankruptcy for some insurance companies which, in turn, could lead to a denial of insurance to people in certain vulnerable areas.

In the UK, ABI (2005) conducted several studies to measure the impacts of climate change on the cost of damage from 2005 to 2080. Notwithstanding expectations of increased assets and increased value of assets, the intensification of weather-related extreme events due to climate change is expected to result in an increase in the level of both premiums and capital needed, because the likelihood of insurance company bankruptcy will be higher (Nussbaum 2013).

For France, the most feared flood disaster for insurers remains the flood of the Seine in the Paris region (OECD 2014). A 100-year return period flood, as occurred in 1910, would have catastrophic social and economic implications across the country. The cost of direct damage could vary from EUR 3 billion to 30 billion, exceeding the maximum capacity of the French Natural Catastrophe insurance system's coverage and requiring the intervention of the national government. In this case, the major risk would not be the disaster itself but the potential bankruptcy of the country.

Insurance Sector Facing the Increasing Cost of Disasters in a Context of Climate Change

For decades, insurers have used traditional catastrophe models to estimate potential damage. Those tools were first developed by international reinsurers or reinsurance brokers (e.g., Swiss Re, Munich Re, AON Benfield, and Willis) and are now marketed by modelling agencies (e.g., AIR, RMS, and EQECAT). The models are widely criticised, including by the insurers themselves, because they consider the natural phenomena as the main driver of the loss variability; for example, windstorm cost models were calibrated on wind speed. Many analysts have highlighted the gap between the natural hazard parameters (e.g., water depth in the case of floods) and the observed cost (Black and Evans 1999).

Insurers are trying to improve their own knowledge on natural hazard impacts through analysis methods such as mapping natural hazards for different scenarios, exposures, and vulnerability assessments with tools including Geographic Information Systems (GIS). Furthermore, models increasingly include physical vulnerability variables such as a building's materials, age, roof, and number of storeys, amongst others (Penning-Rowsell *et al.* 2010). Non-structural variables and social vulnerability parameters, though, also influence damage; e.g., individual behaviour, collective and individual preparedness, crisis management effectiveness, and urban planning (Merz *et al.* 2012).

AFA (2015) calculated that the cost of climate-related disasters in France could rise annually from current values of EUR 8 billion to EUR 21 billion by 2040. Socioeconomic factors (e.g., the growth of asset value and changes in their locations) are expected to account for 40 per cent of this increase while climate change is labelled as accounting for 60 per cent of the increase. Using climate change only to explain the rising cost of losses would be a serious mistake. Insurers understand that to deal with climate change's consequences, it is necessary not only to act through financial tools, but also to take into account all drivers affecting damage.

Involvement of Reinsurance in DRR Including CCA

When insurers deal with DRR including CCA in more affluent countries, a principal question is: How can insurers get involved effectively to reduce the cost of disasters? In low-income countries, the question tends to be: How can insurance coverage be enhanced under appropriate conditions? For insurers, getting involved in DRR including CCA requires extensive energy and administration costs are high. The challenge is to target effective measures to control losses. As Surminski and Oramas-Dorta (2014, p. 154) underline, after examining 27 flood insurance schemes in low and middle income countries, 'the dearth of linkages between risk reduction and insurance is a missed opportunity in the efforts to address rising risk levels, particularly in the context of climate change'. They show that barely one third of the schemes directly or indirectly link insurance and flood reduction measures.

The involvement of insurers in DRR including CCA can take two main forms:

- Developing internal competencies by recruiting people with skills in DRR including CCA.
- Contracting private companies or public authorities to develop programmes in DRR including CCA.

Through these frameworks, insurers can address DRR including CCA before disasters by acting in advance of hazards, during disasters by being involved in crisis preparedness and management, and after disasters by promoting reconstruction with DRR including CCA measures.

Integrating New Skills and New Tools

In wealthier countries, insurers are more and more open to change because they are questioning their traditional way of handling insurance. Uncertainties surrounding climate change, while not necessarily completely new, have galvanised insurers to change their point of view, shifting from an historical actuarial knowledge of risk to exposure based models of risk estimates. This tendency concerns all hazards, but is particularly emphasised for climate-related hazards.

Many insurance companies such as AXA and Generali have integrated new skills by recruiting disaster risk managers to update their strategies and adapt them for climate change in ways other than through economical or financial tools. Insurers are particularly interested in research around individual beliefs and behavioural responses to disaster risk in order to identify new demand in insurance products and to fit existing products (Botzen and Van den Bergh 2012).

How Can Insurance Be Involved in Each Step of DRR Including CCA?

Insurers play a predominant role in promoting knowledge of risk through their own experience as well as through funding research. The AXA Research Fund has granted 492 research projects in risk evaluation or prevention between 2007 and 2018 across 33 countries. They also have a role in disseminating know-how on DRR including CCA. Insurance loss adjusters, i.e., employees who record damage after hazards, are highly competent in understanding risk reduction. Their expertise could be better shared with others involved in DRR including CCA, potentially leading to requirements for certain measures to have been implemented as a requirement of purchasing insurance or for a lower premium, such as with smoke detectors.

Insurers have further become involved in preparedness by developing their own competencies to manage crises or by contracting with other organisations (Box 38.1). The objective is to warn the population and local authorities in order to reduce damage, for instance by moving furniture and valuables out of harm's way.

Finally, insurers traditionally intervene after disasters. By compensating losses, they reduce the shock related to disasters and help restart economic and social life. But the post-disaster role of insurers does not stop there. Another way for insurers to reduce long-term losses is to be involved in preventative post-disaster reconstruction. The post-disaster recovery period can be considered as a window of opportunity (Christoplos 2006) to establish a set of rules that would account for DRR including CCA within a framework of sustainable reconstruction.

After a disaster, 'build back better' is frequently heard. Insurers could support the contractors and the population in rebuilding houses so that they are adapted to specific hazards. For example, for a house designed for freshwater floods, stone tiles could be used instead of wooden

---

Box 38.1   Insurers Involved in Early Warning

Freddy Vinet[1] and David Bourguignon[2]

[1] UMR GRED Université Paul-Valéry Montpellier / IRD
[2] MRN Mission Risques Naturels

Within the insurance sector, tools and measures for prevention, preparedness, and warning remain poorly developed, but continue to be explored. Since 2003, the PREDICT Services Company in Montpellier, France, has implemented prevention services combining meteorological technology, GIS, hydraulic and hydrological engineering, and innovative communication systems to anticipate and manage extreme weather. The concept was developed to address communities, companies, and citizens' needs when facing floods, storm surges, and storms in France, as well as in several Caribbean Islands such as Haiti.

The solution comprises three aspects of risk management:

- Elaborating contingency plans, to organise crisis management and to reduce risks after a vulnerability assessment of local communities.
- Implementing an early warning system and service, working 24/7, to assist and inform end users facing risks through helping them to anticipate, activate, and adapt their emergency plans and behaviour.
- Analysing hydro-meteorological data collected in real time. This information is provided through websites, emails, smartphone applications, SMS, and phone calls.

Feedback analysis is carried out to improve emergency plans.

The French insurance company Groupama was the first to offer PREDICT's prevention and early warning solution in its insurance services. In 2011, Groupama (and one year later, Gan) integrated the service into its contract, making it operational for 20,000 out of the 36,000 French municipalities, which include it in their safety plans and are informed in case of a crisis.

---

floorboards and electrical sockets could be placed higher, reducing flood damage and permitting quicker re-occupancy. Care would be needed to ensure that other problems are not exacerbated or created.

Insurance within a Set of DRR Including CCA Measures

Disasters and disaster-related insurance existed long before the question of climate change arose. DRR existed long before CCA and must continue without the unique reference to climate change. A key point in DRR is the level of knowledge of risk reduction measures. Information and communication by authorities often address the risk awareness, but it is unhelpful to encourage people to protect their home if they do not know what to do.

Insurance must be associated with other risk reduction measures, not only from the point of view of an insurance company that needs to reduce risks, but also from the point of view of end users who must not be too dependent on insurance. Arboriculturists in Western Europe learned this twenty years ago when huge hailstorms destroyed an important part of fruit crops. They diversified their mitigation tools to reduce losses as insurance became too expensive and sometimes ineffective. They put nets on part of their orchards to keep fruit free from losses and, in parallel, they covered another part of their orchards with a crop insurance contract to be able to reimburse loans and fees. They also kept some orchards free from protection; they referred to these as 'bonus' or 'lottery'.

Insurance must be connected to a comprehensive DRR including CCA strategy rather than being a stand-alone measure (Warner *et al.* 2009). Even with a highly effective risk reduction plan, insurance must remain a means to compensate residual risk when all other prevention measures fail. It requires at least a framework at national level and the willingness of policy makers to get involved. Insurance must not remain just a compensation tool. Insurance will contribute better to DRR including CCA if pay-outs promote investments instead of simply compensating losses.

State authorities can also develop tools to oblige insurance companies to contribute to risk reduction. In France, the Fund for Risk Prevention (Fonds de prevention des risques naturels majeurs, FPRNM) collects around EUR 150 million per year through a tax on insurance premiums. As insurance companies can be reluctant to get involved in DRR on a local scale this fund permits payment for DRR actions, such as land use plans, relocation of destroyed houses, and retrofitting.

Insurance in Less Affluent Countries

While insurers are changing their tools and practices to better handle climate change related risks, another challenge is insurance penetration in low-income countries. Examples of tools to address this challenge are microinsurance and weather index crop insurance.

Microinsurance: A Solution to Strengthen DRR Including CCA?

As Clarke and Grenham (2013) state, there is nothing 'micro' about disasters and insurance can be a solution to mitigate disaster risk. Churchill and Matul (2006, p. 12) define microinsurance as 'the protection of low-income people against specific perils in exchange for regular premium payments proportionate to the likelihood and cost of the risk involved'. There have been many experiences of microinsurance and countries like the Philippines and India already have specific rules for regulating microinsurance (IRDA 2005). The conditions to make microinsurance affordable and sustainable are summed up as follows (Churchill and Matul 2006):

- Products with simple rules and easily accessible claims documentation.
- Low premiums, paid at different times of the year depending on income.
- A large number of purchasers.

Warner *et al.* (2009) add that a regulatory system is necessary to give confidence to insurers and potential customers. Insurance premiums must be affordable, transparent, and stable. Insurers must be able to make provisions for quick pay-outs and to have solvency regulations. Microinsurance can be a solution when other traditional means for transferring or compensating for risk are less effective; e.g., when migrants do not provide enough remittances or when other sources of income are lacking. Mills (2009) introduces many cases where microinsurance is successful, such as in Uganda where AIG developed Climate-Friendly Insurance Products and garnered USD 45 million in premiums from 2.25 million microinsurance policies in 2007. One condition for microinsurance sustainability is trust in insurance companies to meet claims quickly under any circumstances; i.e., in major disasters and irrespective of climate change. Populations are also worried that purchasing microinsurance may reduce public or donor support for post-disaster relief and recovery (Clarke and Grenham 2013). Linnerooth-Bayer and Mechler (2006) suggested a strategy to strengthen the role of insurance in poor countries by linking donors, country administrations, and local stakeholders.

In 2015, AXA insurance company contracted with several African governments to disseminate insurance. The guarantee of the national government can give people confidence, although there is no guarantee that people will have confidence in the national government. The collaboration with microinsurers such as MicroEnsure allows a large number of people to be reached. In 2016, AXA increased its stake in MicroEnsure, a UK-based microinsurance company founded by the non-governmental organisation (NGO), Opportunity International. This will enable MicroEnsure to develop its activities in Africa and Asia. The participation of a big insurance company in microinsurance is seen as a means for AXA to start a business relationship with people who often have a negative opinion of insurance (e.g., too expensive, not swift to respond, or reserved for wealthy people). AXA's aim is to win customers through microinsurance and above all, in the long-term, move towards classic insurance products in less affluent countries. New ways of insuring (e.g., public – private partnerships) might be developed to ensure the affordability of insurance and to contribute to DRR including CCA. Clarke and Grenham (2013) report the success of the Horn of Africa Risk Transfer for Adaptation (HARITA) project supported by Oxfam in which farmers can pay insurance premiums by providing labour for CCA projects. This has the dual benefit of providing insurance and of reducing the risk of disasters involving hazards affected by climate change.

Developing Crop Insurance in Low-income Countries

Crop insurance was amongst the first major sectors in which insurance initiatives were undertaken in less affluent countries (Hazell 1992). Crops remain a principal subsistence sector and source of income in many regions of these countries, with agricultural products supporting a good proportion of the population. Weather conditions, naturally affect crops, including extremes such as floods, droughts, wind, and hail, as well as possible knock-on effects such as diseases and pests. Insurance can assist farmers in securing wages and minimising income variability.

Weather index insurance is now a common form of parametric insurance in less wealthy countries (Baarsch and Kelman 2016). For the payment of annual premiums, pay-outs are conditioned by a triggering threshold level (e.g., rainfall depth). Premiums are defined each year, avoiding moral hazard. Thus, weather index insurance is presented as a measure for DRR including CCA for farmers in less affluent countries. In Northern Peru, the El Niño Southern Oscillation (ENSO) index insurance uses the monthly sea surface temperature in November and December provided by the National Oceanic and Atmospheric Administration (NOAA). If temperatures reach a certain threshold determined in the contract, pay-outs are triggered, even before any floods manifest. Farmers can use the money to pay for the drainage systems to be cleaned, for example (Skees and Collier 2010). In this case, payments can have a direct and immediate effect on prevention. Under climate change, the development of weather indexed insurance may require deeper risk assessments looking farther into the future, subsidies for start-up costs, and a mechanism to transfer extreme disaster costs.

DRR including CCA, by definition, is about minimising adverse effects and grasping positive opportunities. What opportunities could be maximised from the point of view of insurers? To bring in more premiums? To find new customers? To develop new products? To support vulnerable people? To be involved in and to support long-term DRR including CCA?

Climate change has provided one impetus amongst many others for insurers to work with governments, scientists, people in communities, and other stakeholders to enact CCA. This impetus can also be an opportunity to reduce replication and overlap between DRR and CCA at all levels through better embedding CCA within DRR, making the development process more efficient and more effective.

Conclusion

Disaster insurance as a mode of loss compensation is old, at least in more wealthy countries, but still used extensively and with significant impact. The last thirty years have been marked by a growth of losses related to increasing assets, increasing values of assets, changing insurance penetration, and at some level influences from environmental changes.

Climate change, while not necessarily bringing anything immediately completely new, has nonetheless shaken up insurers' traditional skills and interests. Insurers have been learning the need to diversify their approach from focusing on an actuarial historic calculation of expected losses to more analysis of exposure and considering other techniques. Even in more affluent countries, efficient and effective disaster insurance requires significant government regulation and intervention. Insurers, governments, and local stakeholders are developing partnerships and tools for DRR including CCA to limit the exposure of portfolios, and hopefully to help people.

In low and middle income countries (and similar communities in more affluent countries), the question of insurance penetration can be more complex, because insurers may believe that they ought to be wary about facing the local consequences of climate change, even though it is not usually clear what climate change's local consequences will be. In many countries, historical and prospective knowledge of disaster risk is still required before developing insurance products and services.

Climate change certainly must not be a pretext for insurers to justify rising premium costs. First, climate change is not always negative, it also presents opportunities and it is the point of DRR including CCA to use those opportunities. Secondly, climate change is not the only factor influencing losses now and in the future. In many circumstances, individual choices could have a more significant impact on disaster losses than climate change. Moreover, it would be wrong to assume that insurance will regulate or solve all aspects of DRR including CCA. Low and middle income communities therefore have much to gain not only from microinsurance but also from improved risk management, and much wider and deeper development activities. Many forms of DRR including CCA exist already. Insurance is, and always has been, only one means amongst many to handle present and future consequences of disasters.

References

ABI (Association of British Insurers). (2005) *Financial Risks of Climate Change: Summary Report*, London: ABI.
AFA (Association française de l'assurance). (2015) *Impact du changement climatique sur l'assurance à l'horizon 2040*, France: AFA.
AXA. (2012) *Climate Risks*, AXA papers, No. 4, Paris: AXA.
Baarsch, F. and Kelman, I. (2016) 'Insurance mechanisms for tropical cyclones and droughts in Pacific Small Island Developing States (SIDS)', *Jàmbá: Journal of Disaster Risk Studies* 8, 1: 1–12.
Black, A. and Evans, S. (1999) *Flood Damage in the UK: New Insights for the Insurance Industry – A Report Presenting the Dundee Flood Loss Tables*, Dundee: Department of Geography, University of Dundee.
Botzen, W.J.W. and van den Bergh, J.C.J.M. (2012) 'Monetary valuation of insurance against flood risk under climate change', *International Economic Review* 53, 3: 1005–1025.
Changnon, S.A., Changnon, D., Fosse, E.R., Hoganson, D.C., Roth Sr, R.J. and Totsch, J.M. (1997) 'Effects of recent weather extremes on the insurance industry: Major Implications for the atmospheric sciences', *Bulletin of the American Meteorological Society* 78, 3: 425–435.

Christoplos, I. (2006) *The Elusive 'Window of Opportunity' for Risk Reduction in Post-disaster Recovery*, Bangkok: ProVention Consortium Forum.

Churchill, C. and Matul, M. (eds) (2006) *Protecting the Poor: A Microinsurance Compendium*, Geneva: International Labour Organization.

Clarke, D.J. and Grenham, D. (2013) 'Microinsurance and natural disasters: Challenges and options', *Environmental Science and Policy* 27, supplement 1: S89–S98.

Hallegatte, S., Hourcade, J.C. and Dumas, P. (2007) 'Why economic dynamics matter in assessing climate change damages: illustration on extreme events', *Ecological Economics* 62, 2: 330–340.

Hazell, P. (1992) 'The appropriate role of agricultural insurance in developing countries', *Journal of International Development* 4, 6: 567–581.

IRDA (Indian Regulatory and Development Authority). (2005) *Micro-insurance Regulations*, New Delhi, India: IRDA.

Keskitalo, E.C.H., Vulturius, G. and Scholten, P. (2014) 'Adaptation to climate change in the insurance sector: examples from the UK, Germany and the Netherlands', *Natural Hazards* 71, 1: 315–334.

Linnerooth-Bayer, J. (2012) 'Financial mechanisms for disaster risk', in B. Wisner, JC Gaillard and I. Kelman (eds) *The Routledge Handbook of Hazards and Disaster Risk Reduction*, Abingdon, Oxfordshire: Routledge, pp. 654–663.

Linnerooth-Bayer, J. and Mechler, R. (2006) 'Insurance for assisting adaptation to climate change in developing countries: a proposed strategy', *Climate Policy* 6, 6: 621–636.

Merz, B., Kreibich, H. and Lall, U. (2012) 'Multi-variate flood damage assessment: A tree based data-mining approach', *Natural Hazards and Earth System Sciences* 13: 53–64.

Mills, E. (2009) 'A Global review of insurance industry responses to climate change', *Geneva Papers on Risk and Insurance-Issues and Practice* 34, 3: 323–359.

Munich Re. (1973) *Hochwasser, Ueberschwemmung*, Munich: Muenchener Rueckversicherungsgesellschaft.

Munich Re. (2014) *Natural Catastrophes 2013 – Analyses, Assessments, Positions*, Munich, Germany: Munich Re.

Nussbaum, R. (2013) 'La résilience des sociétés vue au travers du prisme des assurances: une comparaison internationale', *Responsabilité and Environnement* 72: 42–47.

OECD (Organisation for Economic Co-operation and Development). (2014) *Seine Basin, Île-de-France: Resilience to Major Floods*, Paris, France: OECD Publishing.

Penning-Rowsell, E., Viavattene, C., Pardoe, J., Chatterton, J., Parker, D. and Morris, J. (2010) *The Benefits of Flood and Coastal Risk Management: A Handbook of Assessment Techniques (The Multi-Coloured Handbook)*, London: Flood Hazard Research Centre.

Porrini, D. and Schwarze, R. (2014) 'Insurance models and European climate change policies: An assessment', *European Journal of Law and Economics* 38, 1: 7–28.

Skees, J.R. and Collier, B. (2010) 'New approaches for index insurance: ENSO insurance in Peru international food policy research institute', in R. Kloeppinger and M. Sharma (eds) *Innovation in Rural and Agricultural Finance*, Washington, USA: International food research institute / The World Bank, Chapter 11.

Smolka, A. (2006) 'Natural disasters and the challenge of extreme events: Risk management from an insurance perspective', *Philosophical Transactions of the Royal Society A: Mathematical, Physical and Engineering Sciences* 364, 1845: 2147–2165.

Surminski, S. (2013) 'The role of insurance in reducing direct risk – The case of flood insurance', *International Review of Environmental and Resource Economics* 7: 241–278.

Surminski, S. and Oramas-Dorta, D. (2014) 'Flood insurance schemes and climate adaptation in developing countries', *International Journal of Disaster Risk Reduction* 7: 154–164.

Swiss Re. (2015) 'Natural catastrophes and man-made disasters in 2014', *Sigma* 2.

Warner, K., Ranger, N., Surminski, S., Arnold, M., Linnerooth-Bayer, J., Michel-Kerjan, E., Kovacs, P. and Herweijer, C. (2009) *Adaptation to Climate Change: Linking Disaster Risk Reduction and Insurance*, Geneva: UNISDR.

39

THE PLANNING NEXUS BETWEEN DISASTER RISK REDUCTION AND CLIMATE CHANGE ADAPTATION

Judy Lawrence and Wendy Saunders

Introduction

Disaster risk reduction (DRR) and climate change adaptation (CCA) have largely developed as parallel but separate discourses. What they have in common is a focus on reducing risk of natural hazards. The ever-changing character of risk is defined by the timing and rate of change in both hazards and vulnerabilities, including culture, values and capacities of those at risk. Some have suggested that by being separate discourses, CCA has merely reinvented policy approaches that were already known to the DRR community. In turn, this has entrenched the CCA discourse as a separate one from DRR, and has discouraged the development of much-needed new ideas and approaches.

Arguably, a sense of convergence between DRR and CCA has started to emerge. The Inter-governmental Panel on Climate Change (IPCC) undertook a Special Report on disaster risk management and climate change adaptation (IPCC 2012) which some have critiqued (Kelman *et al.* 2016). IPCC (2014) has more recently reported on the new developments in design of flexible pathways for adjusting to changing climate (IPCC 2014) acknowledging that such approaches need to be implemented within planning systems to deal with the uncertainty and dynamics of climate change considering disaster risk comprising extreme events, exposure and vulnerability. Some have critiqued whether there is convergence of DRR and CCA, suggesting further movement from a hazard-centric approach towards a focus on vulnerability is required (Kelman and Gaillard 2010).

This chapter critiques the institutional frameworks and practices of planning that seek to reduce risk to exposed and vulnerable populations. It does this by providing an overview of planning in the context of DRR and CCA, especially where connections and overlaps occur.

What does Planning Encompass?

We use 'planning' to mean the ability to plan in a democratically accountable way; and the activities of social and service sectors that have spatial or land use consequences in their wider social and environmental context (Wilson and Piper 2010). This includes spatial planning, land use planning, asset management, housing and economic development, transport, water, waste and storm water utilities, and in some jurisdictions, wider concerns of health and welfare. Planning is thus more than purely land use planning and spatial location decisions; it encompasses the processes that integrate land use planning decisions with other policy domains that also influence the nature and functioning of places, formal and informal.

Informal institutions provide integrative links to societal values and preferences that can support the formal institutions when operating in conditions of uncertainty and dynamic change. Haughton *et al.* (2010) characterise a spatial planning framework as comprising a focus on long-term planning; measures for integrated policy making; a central role for sustainable development; and focusing on inclusivity. Such a holistic framework can enable disaster risks including those linked to climate change to be fully understood and addressed within their societal contexts in an integrated way.

DRR and adaptive management are not new. The limitations of technical DRR measures such as structural barriers, reliance on building codes, insurance and emergency management responses, without adequate support through social interventions, have been recognised for some time. Path dependency and 'creating' disasters were consequences identified from use of these technical measures. More recently the onset of anthropogenic climate change, in an age where there are greater populations exposed to hazards, has intensified vulnerability and increased disaster consequences around the world. This has led to increased global attention on the potential role of planning for DRR, rather than just planning for hazard, grounded within the concept of sustainable development (Godschalk *et al.* 1998). Community consultation on policy choices dominated the discourse leading to the use of land use activity rules, as well as public education about inappropriate development in areas exposed to hazard (Burby 1998). Avoidance of risk became embedded within national planning statutes in many countries, with risk assessments developed as a way of identifying the risk, and associated planning instruments designed to reflect it; e.g., as risk increased, so did planning restrictions. At the same time, CCA was focusing on impacts, vulnerability and risk management, building on the DRR discourses, but applying them to a changing climate. Experience in the USA such as Hurricane Katrina has shown graphically how land use planning can place communities at risk (Burby 2006). The importance of strong social networks and institutions, along with natural, economic and political systems that have integrity, were also highlighted in these events. Accordingly, what has evolved are planning systems that strive not to infringe on property rights, and a reliance on market forces with the withdrawal of government regulations to provide certainty to some, great uncertainty to others, and risk transfers from private to public interests. The promise of natural hazard planning has not been realised (Glavovic 2010), in part due to the difficulties of moving from a hazard- to a risk-based framework that includes vulnerability.

In addition, it is far from clear that the planning measures used have the ability to address uncertainty and the changes arising from both physical and societal sources. The need for planning instruments that can support institutions inside and outside formal statutory processes, at the different levels of governance, has also been a challenge. The New Zealand statutory planning framework was designed with characteristics that can address uncertainty and attempts to integrate DRR and CCA, by having hazards and climate change considered in a planning context under the Resource Management Act 1991. However, fragmentation of governance has meant that practice struggles to deliver on the intent of the institutional framework (Lawrence 2015). The Netherlands through the Delta Plan is integrating layers of concern in a spatial context and tools are developing to address changing risk in an adaptive manner (Haasnoot *et al.* 2013). Most countries, however, have found the tensions between public and private interests, and short- and long-term focus, to be highly challenging (Glavovic and Smith 2014) due to power asymmetries and capacity deficits.

What are the Similarities and Differences between DRR and CCA?

DRR anticipates disasters as well as the smaller events that have an impact on the functionality of the affected area, but are not considered disasters per se. Wisner *et al.* (2004) clarify what disaster risk is with a mnemonic stating that disaster risk equates to hazard × [(vulnerability divided by capacity) minus mitigation measures]. Notwithstanding this, DRR as practised globally has a focus on event-specific disasters involving many different hazards, while recognising their causes are related to the degree of exposure and vulnerability of populations at risk; institutional traditions; human capacities; and access to resources. CCA focuses on both slow-onset climate-related hazards such as sea level rise, and those like storm surges and flooding, that can add to the effect of sea level rise in some locations. Differences between DRR and CCA relate primarily to the rate of change and the scale of likely climate impacts globally, as exacerbated by greater exposure and vulnerability of populations at risk. For example, we know that the sea level will continue to rise for many centuries, even if emissions reductions are dramatically reduced; and that more intense rainfall events are likely in many regions of the world, beyond what has hitherto been experienced by civilisation, while populations at risk have increased. These add scale and frequency dimensions to coping capacity for both DRR and CCA, thus also adding pressure on existing institutions of planning and their practice.

Both DRR and CCA operate under conditions of uncertainty and change. This draws attention to the limitations of planning measures that rely upon static assumptions of risk that entrench risk through lock-in and path dependency. DRR, through its attention to societal values and preferences, highlights the need for consideration of greater empowerment of communities in decision making. Similarly, CCA has also highlighted the importance of societal values and preferences in building adaptive capacity, but has also raised the spectre of limits to risk reduction through adaptation. Differences between DRR and CCA have been artificially constructed through functional scale challenges between organisations and their mandates, institutional and cultural norms of the actors, and differences in knowledge bases (Birkmann and Teichman 2010); these become particularly apparent in planning practice.

Planning Challenges and Opportunities

Planning systems are typified by a decision model based on adjustments that are rationally bounded, where decision makers initiate the decision process based on limited knowledge, and within constraints set by their institutional and social systems. This often delivers skewed outcomes that are highly influenced by special and individual interests. This has the potential to transfer costs to future generations and to the general public.

The strength of economic forces favouring unfettered property development, the amount of existing development, deeply rooted social values and legal precedents supporting individual property rights, define and constrain politically viable policy options for DRR and CCA. As a result, the public policy response to disaster risk does not always adhere to DRR and CCA principles. Rather, a loss reduction paradigm accepts development of land that is exposed to hazards, and attempts to reduce the impacts of hazard events through risk reduction initiatives. Risk reduction interventions then include land development regulations, engineered solutions, emergency preparedness and response, and a combination of market and government-subsidised insurance. Addressing the root causes of vulnerability is frequently absent from the discourses, policies, and practices.

The interdependency of natural and human systems means that planning must be able to respond in a flexible yet robust manner to changing risk profiles. This is often a challenge when planning frameworks are static and non-responsive to advances in understanding over the short term. Berke and Smith (2010) offer four criteria that developments can be evaluated against: 1) risk reduction functions of natural systems not disrupted; 2) land use decisions on risk reduction initiatives support economic vitality; 3) environmental and economic benefits of risk reduction initiatives equal across society; and 4) all stakeholders participate in the risk reduction planning process. Land use planning in reducing risks from natural hazards is necessary, but not sufficient; support of financial institutions, multi-level governance approaches, and processes empowering communities, are all necessary.

Responding to disasters, and implementing 'adaptive' measures in anticipation of future occurrences, can, however, entrench risk by creating path dependency; the desire to re-establish people's lives quickly after a disaster can dominate. This limits the ability to plan flexibly for changing risks driven by climate and vulnerability of communities. Planning has a critical role where governance and implementation interact, especially over long timeframes where decisions in the present affect outcomes in the future. Many infrastructure investments today will be around for decades and, as such, are 'locked-in' and difficult to change as communities and hazard regimes evolve.

Planning systems also operate at different scales: led by a national planning framework but implemented at three tiers of sub-national government (United Kingdom); implemented at lower state levels of governance (Australia); designed as an integrated system of national guidance and policy, with implementation devolved two levels of local government (New Zealand); or where decision making is based on community consensus under the leadership of the *matai* (chief) (Samoa). Agencies responsible for planning can be uni- or multi-functional, responsible to different constituencies, or can have community 'authority' and thus accountability. Closeness to communities of interest can create pressures on elected decision makers when responding to short-term special interests, while being responsible for long-term interests of future generations and the public interest.

Planning regimes seek certainty through legal systems, which require 'evidence' or 'predictions' on which to base decisions. They are applied at a point in time and space, often followed by review at prescribed future points in time for specific locations, thus entrenching societies' need for certainty over time. As a result, formal institutions of land use planning exhibit time and space constraints in their design and practice; decision makers may not consider the future consequences of decisions today and how they might address future change that appears uncertain at the present time, despite policies encouraging a long-term view.

In addition, use of measures like levees and fixed hazard zones sets up societal expectations of 'safety' within the bounds of those measures (Tobin 1995). Consideration of residual risk (i.e., risk that remains once mitigation measures are in place) under current conditions is often ignored in planning considerations, thus increasing residual risk relative to the static protection level as the climate changes, when development continues in areas at risk. Static measures can also lead to path dependency if they cannot be adapted to the change over time or spatially, creating wider societal implications which prevent access to resources and interfere with long-established livelihoods. Land use planning is based on policies and plans that are often static. When planning reallocates rights and responsibilities and costs and benefits, attempts to use planning to change them have been thwarted, often due to inadequate community discussion of the issues at stake.

Formal planning frameworks that are able to engender flexible outcomes for changing risk have been slow to develop and adjust to changing risk in societies. A widespread view dominates that science will deliver more certainty over time, and then the best courses of action will emerge. However, planning to any level will result in a range of residual damages that require further risk reduction. As Burton *et al.* (2002) observed, presenting risk from a 'top-down' perspective has led to a culture of reliance on 'evidence' for decision making, and institutional frameworks and practices designed to deliver 'certainty' for those affected. The practice of 'central tendency' (averages), 'just give me a number' (single expression of risk), using 'fixed' and 'static' protection measures (levees) and spatially fixed hazard lines (on planning maps) can all mischaracterise the extremes of risk, its changing character, and engender a false sense of security about the future ('the levee effect'), thereby increasing risk by increasing vulnerability.

How Have Problems Been Addressed?

Planning regimes have been largely ineffective in reducing disaster risks, including those linked to climate change. Populations have continued to grow in areas exposed to hazard risk and population vulnerability is still high.

A hazard involves an interaction between human life and property with natural phenomena that could cause damage that produces a level of risk. Historically in planning systems, hazard has typically been assessed on its likelihood only (e.g., the '1/100 year flood'). This is problematic when uncertainties as to frequencies, magnitude, and spatial location of the hazard mean the likelihood cannot be determined with any degree of precision. The vulnerability and consequences – and hence the risk – and their change over time across the physical and social environments-are the required planning focus for DRR and CCA. If vulnerability and consequences are not adequately addressed, or have delayed attention because of the uncertainties, maladaptation is possible. New measures and supporting institutions will be necessary to bridge the implementation gap between DRR and CCA.

Attempts to examine the intersection between DRR and CCA to date (Birkmann and Teichman 2010; Glavovic and Smith 2014; IPCC 2012) do not give much guidance on how the planning system can integrate DRR and CCA. Repeated calls for their integration (IPCC 2012) and suggestions that CCA be nested within DRR (Kelman *et al.* 2016; Kelman and Gaillard 2010) do not extend to the design of particular institutions and practice that might achieve integration. Solving the issue of integration of CCA and DRR at the planning practice level requires an understanding of: the characteristics of the hazard; how the hazard will change over time in frequency and magnitude; vulnerability and consequences; the conditions under which policies will fail in future; whether and to what extent the hazards and vulnerabilities differ from those already being addressed by DRR and CCA in the planning system; and how capacities can be addressed and how they change over time.

Whether current institutional frameworks and measures confer the capacity to plan purposefully for changing risk is an open question, suggesting that more radical institutional change may be required (Kates *et al.* 2012). The changes and 'surprises' that characterise disaster and climate risk challenge risk framing as one-off 'events', static measures and certainty; approaches that have become embedded in current practice. However, it is far from clear whether governance and planning institutions can anticipate change that has not been experienced before, and which is not seen as an urgent policy priority.

Typically, radical change in decision systems is the very sort that society is often reluctant to implement; for example, managed retreat at the coast as a planning policy to anticipate sea-level rise that is perceived as uncertain and distant, and as affecting property 'rights' and valuations in many locations. It is often shocks to human systems, falsely seen as being unexpected and unanticipated, that engender institutional change. However, this is usually after the fact and inappropriate rebuilding in the place at risk can result. Placing CCA within DRR contributes towards shifting the present and immediatebias in planning for and responding to disasters towards a more anticipatory approach that considers how future populations will be affected by all changes to risk including those induced by climate change (Kelman and Gaillard 2010).

A recent example of inertia in the planning system was illustrated in New Zealand in 2012 when the Hutt City Council notified a City Plan change allowing increased development in an area exposed to several natural hazards (fault rupture, subsidence, sea level rise, liquefaction, flooding, and tsunami). Only flooding had been previously experienced. The district plan for the area had limited rules to address the risks from natural hazards and no new rules were proposed to address the known risks, nor the increased risk from climate change. Expert submissions through the formal planning process became the means by which consideration of these risks was included within the Plan for ground rupture, subsidence, liquefaction, tsunami, flooding and sea-level rise.

Box 39.1 shows an example of how municipal foresight addresses CCA by having attended to DRR in the past. While this example had a positive outcome, in other cases, long-term anticipatory re-thinking has not occurred after disasters. In this case, the combination of a horrific disaster occurring at the same time as major governance changes enabled prevention measures for a similar flood disaster to be enfolded within the wider system changes that already had political and societal support. Box 39.2 shows an example of managed coastal realignment that has occurred over history in the United Kingdom.

**Box 39.1   Planning for Disaster Risk Reduction means Planning
          for Climate Change Adaptation**

Ilan Kelman[1]

[1] University College London, London, UK and University of Agder, Kristiansand, Norway

On 15 October 1954, Hurricane Hazel slashed through Toronto killing 81 people, 35 of them with addresses on one flooded street, Raymore Drive. Several more died elsewhere when rivers became raging torrents.

In the aftermath, Toronto and its surrounding region reorganised planning for and management of floods, within the context of wider changes to Toronto's governance, which had started in 1953 when Metropolitan Toronto was formed as an additional level of local government across several municipalities. Metro Toronto (now defunct) undertook urban planning functions, like construction of public transportation, motorways, and water and sewage utilities.

For flood risk reduction, coordination increased amongst the agencies responsible, focusing on river engineering and land use changes. The parts of Raymore Drive destroyed by Hurricane Hazel became a park and a floodplain. Recreational pathways were set up, preventing riverside urban development.

Toronto now has a lengthy greenway system, for walking, cycling, picnics, appreciating nature, and environmental education and which also serves as a floodplain. Subsequent hurricanes and storms affecting Toronto, like Hurricane Isabel in 2003, caused riverbanks to burst from the high flow rates. No widespread damage occurred due to the lack of property in the floodplain. DRR has been effected. Meanwhile, the railway tracks and a motorway built in the floodplain are frequently flooded causing transportation chaos when they are shut down. The frequency of this flooding is expected to increase under climate change and no manageable solution is in sight.

No matter how climate change influences Toronto's storms, the greenways serve as floodplains. By implementing planning for DRR in the 1950s, planning for CCA for the 2010s is already in place.

Box 39.2   Managed Realignment in England: A Long-term Planning Strategy for Disaster Risk Reduction Including Climate Change Adaptation

Ilan Kelman[1] and James Lewis[2]

[1] University College London, London, UK and University of Agder, Kristiansand, Norway
[2] Datum International, Marshfield, UK
Based on Lewis and Kelman (2009).

England's North Sea coast is sinking and eroding, as it has been doing for millennia. In the past, such as in medieval Dunwich, it was accepted that towns would need to continually realign and rebuild infrastructure as the sea encroaches through collapsing cliffs and disappearing beaches and lowlands. Today, expectations can be much more about holding the line and 'defending' or 'protecting' against the sea. Yet, no engineered coastline can last for ever, especially where the land sinks and the sea rises.

Consequently, the planning process of 'managed realignment' for coasts has been pursued, labelled with that phrase for at least two decades even while drawing on lessons from past centuries. Managed realignment can avoid costly walls and other engineered works, favouring instead ecosystems such as marshes, beaches, meadows, and wetlands. Being protected sanctuaries, they provide plant and animal habitats as well as nature-based recreational opportunities for people. Then, during sea storms, inland downpours, and continuing subsidence, irrespective of climate change, people and property have a buffer that reduces the hazard magnitude experienced and that potentially keeps them out of harm's way entirely. Planning for DRR has been implemented that supports planning for CCA.

A Way Forward?

Planning practice and decision making require new approaches to act as catalysts for better integration of CCA and DRR, with a focus on the dependency of long-term outcomes on near-term actions. Uncertainty and change for both hazards and vulnerabilities are at the nexus of the planning problem, creating delay in taking anticipatory actions.

Considering a range of scenarios of the future – as used in many other fields, including through DRR – is one way of priming the planning system for rapidly exploring alternative futures and thus supporting a wider range of response options at the exploratory stage of decision making. This avoids responses predicated only in historical experience, or on single numbers that ignore extremes or changing vulnerability baselines and enables consideration of land uses and assets that have different lifetimes (Kwadijk *et al.* 2010). Using scenarios iteratively and interactively encourages decision makers to think systematically, ahead of disasters occurring, about adaptive response measures that are robust over a range of possible futures. Such an approach has traction for use in planning, to enable a transition from static planning approaches currently used, and thus increase the ability to change over time within DRR and CCA.

Such thinking in CCA, which builds on earlier applications in DRR, is beginning to shift the focus of decision makers from protection to managing risk and uncertainty through flexible approaches and staged decision making, and to consider perspectives of vulnerability; e.g., community triggers for pathway change (e.g., Wise *et al.* 2014), dynamic adaptive policy pathways (Haasnoot *et al.* 2013), games for social learning (Van der Wal *et al.* 2016), and real options analysis (Ranger *et al.* 2013). Pathways processes are also being applied in Singapore, New Zealand, Bangladesh, and Thailand for example, where they incorporate many drivers of risk including vulnerability, tolerability, how long current planning strategies will be effective over a number of plausible outcomes, whether the path can be changed easily and with minimum disruption and cost, and what social triggers will lead to pathway change. Accordingly, decision makers are being motivated to consider planning over longer timeframes and to consider the lifetime of their decision, keeping options open, and thus avoiding the path dependency that sets up costly legacy effects. Countries plan for vulnerability differently – in New Zealand there is not a requirement for consideration in land use planning; in contrast, it is a requirement in Mexico and Norway.

Few scholars have addressed whether adaptive planning approaches can be given effect in a spatial planning context (Roggema 2009; Wilson and Piper 2010), and whether current planning institutions can address long timeframes (Wise *et al*. 2014). Studies have shown that current tools for spatial planning (static design and practice) are not adequate for addressing uncertain climate change (Lawrence 2015), nor for addressing changing natural hazard incidences, exposures, and vulnerabilities and their root causes; a legacy of DRR and CCA, practice deriving from different and separate discourses, provides a powerful argument to better integrate them through the design of flexible planning institutions and for empowering their practice.

The role of law in planning for DRR received much attention during the 1990s (e.g., Burby 1998), but the practice has been unsuccessful in reducing risk. More recently, for CCA, McDonald (2011) and Ruhl (2010) have revisited the role of law in the context of climate change, local contestation and increased exposure of populations to risk since the 1990s. The barriers arising from legal institutions, processes, and principles that constrain planning for DRR and CCA include: compensable property rights that impede new regulation; high levels of uncertainty; irreversibility; place-based impacts; and the inter-relationships between disasters and climate change impacts in the socio-economic context, within which they occur. For legal processes and instruments to be more responsive to change, they need to maintain legitimacy and legality, and robustness and flexibility at the same time. Institutions and measures need the ability to reconfigure cross-policy linkages and trade-offs, at all scales and across scales of governance; have greater regulatory variety and flexibility; and require multi-scale governance networks and conciliation.

It has been suggested that convergence of DRR and CCA may come through a 'window of opportunity' following disasters (Birkmann *et al*. 2010). However, the opportunity to take advantage of such windows is short, and the focus is on recovery, characterised by institutional inertia for more anticipatory change in planning frameworks and practice. Dovers and Hezri (2010) characterised three possible future climate conditions challenging decision makers that are equally relevant to DRR: those that are within our experience and ability to cope, those outside our experience but within our abilities to deal with them, and those we have never experienced and which we do not have the ability to deal with. All categories are likely to require responses by DRR and CCA.

The more problematic is unexperienced risk that stretches coping capacity. This type of risk has dominated CCA discourse, creating a perception that it is distant in time and space and not yet worthy of attention. It also fits the 'disaster' paradigm, reinforcing perceptions of disempowerment and making attempts to develop capacity to address such risks ahead of time difficult. Issues of legitimacy go well beyond traditionally practised risk assessment, risk management, and risk communication. The legal, institutional, social, and economic contexts in which a risk is evaluated, the complex relationships between actors, rules, conventions, processes, and mechanisms that govern how risk information is framed and collected, analysed and communicated, and how management decisions are taken, is central to building capacity to reduce risk of disasters and adapt to climate change. Central to this is inclusiveness of the actors that can reflect the different values and perspectives better than purely technocratic risk considerations.

The Role of International Agreements in DRR and CCA

International agreements can provide legitimacy, through regulatory frameworks and principles, for governments to take DRR and CCA action. However, such agreements tend to be single purpose or so broad to be weak drivers for planning. Parts of DRR are governed through land use and emergency management planning which provides directions for and restrictions on land use in areas exposed to natural hazards (Saunders *et al.* 2015). While each jurisdiction is culturally and environmentally different, underlying principles of planning are similar. For example, the Sendai Framework for Disaster Risk Reduction (SFDRR) embodies principles that hold for DRR and CCA, such as responsibility for risk reduction and risk sharing at multiple governance scales and sectors; full societal engagement and partnership; the need for coordinating mechanisms; empowerment of local authorities and communities; the need for a multiple hazards approach that is risk informed; coherence in the institutional measures used; and building back better after a disaster. The United Nations Framework Convention on Climate Change (UNFCCC) provides an international driver for anticipatory design of national level CCA. The roles of land use planning, codes and standards, environmental, resource management, and health and safety institutions, are important links for integrating CCA and DRR at national levels.

The real test for SFDRR and UNFCCC is how they are implemented at the national and local levels. For example, to aid post-earthquake reconstruction of Christchurch, New Zealand, the reference to the Hyogo Framework for Action (predecessor to SFDRR) to justify risk reduction was deemed not to be relevant at a local level. Rather, the national level of government was seen to be responsible. It is only when international agreements or their principles are embodied in law that they begin to drive planning decisions. This has been demonstrated in New Zealand through the Resource Management Act 1991 and its statutory National Coastal Policy Statement which guides practice and the Courts in their decision making about CCA, a function required to be considered by everyone operating under the RMA. However, even though this has been a statutory requirement since 2004, local governments struggle to deal with changing risk through the current planning instruments, due largely to perceptions of uncertainty of hazard or its distant threat, political reluctance at all levels of government to lead the issue, the static nature of plans, contestation, and the slow development of new planning practices.

Conclusions

Planning systems are typically driven by assumed certainty of outcome. This clashes with the notion of uncertainty and changing risk that is inherent in all environmental processes including drivers of some disasters and climate change, as well as socio-economic changes to exposure and vulnerability. Planning must be able to respond in a flexible yet robust manner to changing risk profiles for DRR and CCA. Current planning systems have proven inadequate for addressing such changes.

Areas relevant to planning for DRR and CCA that could address the inadequacies include: placing CCA within DRR across sectors and at multiple scales across national policies to local plans; development of standardised methods and criteria for impact assessments, which in turn provide assumed certainty in decision making, and an ability to monitor progress over time; and the provision of information – which is often a challenge for planning which creates static plans – including scientific data, local and external knowledge, and institutional (formal and informal) and personal memory. Fitting institutional frameworks and planning measures to the characteristics of the hazards and climate changes (uncertainty and change) is critical to enabling DRR and CCA to build on their respective experiences – to be better integrated, rather than maintaining their separation as a discourse, and in their practical implementation. Cross-integration of tools used for hazard and vulnerability assessments, and planning measures designed for spatial and temporal change will enable flexible and robust response options for whatever future eventuates. Focusing on vulnerability and incorporating uncertainties will be critical. A greater variety and flexibility of regulatory instruments will be required, using formal and informal institutions.

Significantly, embedding CCA within DRR will require deliberative processes that work with a diversity of public and private values and preferences. This marks out a world of hitherto unattended change in planning frameworks, institutions and their practice.

References

Berke, P. and Smith, G. (2010) 'Hazard mitigation, planning, and disaster resiliency: Challenges and strategic choices for the 21st century', in U. Fra (ed.), *Sustainable Development and Disaster Resiliency,* Amsterdam, The Netherlands: IOS Press.

Birkmann, J., Buckle, P., Jaeger, J., Pelling, M., Setiadi, N., Garschagen, M., Fernando, N. and Kropp, J. (2010) 'Extreme events and disasters: A window of opportunity for change? Analysis of organizational, institutional and political changes, formal and informal responses after mega-disasters', *Natural Hazards* 55: 637–655.

Birkmann, J. and Teichman, K. (2010) 'Integrating disaster risk reduction and climate change adaptation: Key challenges – scales, knowledge, and norms', *Sustainability Science,* 5: 171–184.

Burby, R.J. (1998) *Cooperating with nature: Confronting natural hazards with land-use planning for sustainable communities,* Washington DC: Joseph Henry Press.

Burby, R.J. (2006) 'Hurricane Katrina and the paradoxes of government disaster policy: Bringing about wise governmental decisions for hazardous areas', *The Annals of the American Academy of Political and Social Science* 604: 171–191.

Burton, I., Huq, S., Lim, B., Pilifosova, O. and Schipper, E.L. (2002) 'From impacts assessment to adaptation priorities: The shaping of adaptation policy', *Climate Policy* 2: 145–159.

Dovers, S. and Hezri, A. (2010) 'Institutions and policy processes: The means to the ends of adaptation', *Wiley Interdisciplinary Reviews: Climate Change* 1: 212–231.

Glavovic, B.C. (2010) 'The role of land-use planning in disaster risk reduction: An introduction to perspectives from Australasia', *The Australasian Journal of Disaster and Trauma Studies,* 1.

Glavovic, B.C. and Smith, G.E. (2014) *Adapting to Climate Change: Lessons from natural hazards planning,* Dordrecht: Springer.

Godschalk, D.R., Kaiser, E.J. and Berke, P.R. (1998) 'Integrating hazard mitigation and local land use planning', in R.J. Burby (ed.), *Cooperating with Nature – Confronting Natural Hazards with Land-use Planning for Sustainable Communities,* Washington DC: Joseph Henry Press.

Haasnoot, M., Kwakkel, J., Walker, W. and Ter Maat, J. (2013) 'Dynamic adaptive policy pathways: A method for crafting robust decisions for a deeply uncertain world', *Global Environmental Change* 23: 485–498.

Haughton, G., Allmendinger, P., Counsell, D. and Vigar, G. (2010) *The New Spatial Planning: Territorial Management with Soft Spaces and Fuzzy Boundaries,* London: Routledge.

IPCC (Intergovernmental Panel on Climate Change) (2012) *Managing the Risks of Extreme Events and Disasters to Advance Climate Change Adaptation: Summary for Policymakers,* Geneva, Switzerland: IPCC.

IPCC (Intergovernmental Panel on Climate Change) (2014) *Climate Change 2014: Impacts, Adaptation and Vulnerability: Summary for Policymakers,* Cambridge, UK: Cambridge University Press.

Kates, R.W., Travis, W.R. and Wilbanks, T.J. (2012). 'Transformational adaptation when incremental adaptations to climate change are insufficient', *Proceedings of the National Academy of Sciences (PNAS)* 109: 7156–7161.

Kelman, I. and Gaillard, JC (2010) 'Embedding climate change adaptation within disaster risk reduction', in R. Shaw, J.M. Pulhin and J.J. Pereira (eds), *Climate Change Adaptation and Disaster Risk Reduction: Issues and Challenges,* Bingley: Emerald, pp. 23–46.

Kelman, I., Gaillard, JC, Lewis, J. and Mercer, J. (2016) 'Learning from the history of disaster vulnerability and resilience research and practice for climate change', *Natural Hazards* 82, S1: S129–S143.

Kwadijk, J., Haasnoot, M., Mulder, J., Hoogvliet, M., Jeuken, A., Van Der Krogt, R., Van Oostrom, N., Schelfhout, H., Van Velzen, E., Van Waveren, H. and De Wit, M. (2010) 'Using adaptation tipping points to prepare for climate change and sea level rise: A case study in the Netherlands', *Wiley Interdisciplinary Reviews: Climate Change* 1: 729–740.

Lawrence, J. (2015) *The Adequacy of Institutional Frameworks and Practice for Climate Change Adaptation Decision Making*, Unpublished PhD, Victoria University of Wellington.

Lewis, J. and Kelman, I. (2009) 'Housing, flooding and risk-ecology: Thames Estuary South-Shoreland and North Kent', *Journal of Architectural and Planning Research* 26, 1: 14–29.

McDonald, J. (2011) 'The role of law in adapting to climate change', *Wiley Interdisciplinary Reviews: Climate Change* 2: 283–295.

Ranger, N., Reeder, T. and Lowe, J. (2013) 'Addressing "deep" uncertainty over long-term climate in major infrastructure projects: Four innovations of the Thames Estuary 2100 Project', *European Journal of Decision Process* 1: 233–262.

Roggema, R. (2009) *Adaptation to Climate Change: A Spatial Challenge*, Dordrecht: Springer.

Ruhl, J. (2010) 'Climate change adaptation and the structural transformation of environmental law', *Environmental Law* 40: 363–465.

Saunders, W.S.A., Grace, E., Beban, J.G. and Johnston, D. (2015) 'Evaluating land use and emergency management plans for natural hazards as a function of good governance: A case study from New Zealand', *International Journal of Disaster Risk Science* 6: 62–74.

Tobin, G.A. (1995) 'The levee love affair: A stormy relationship', *Journal of the American Water Resources Association* 31: 359–367.

Van Der Wal, M., De Kraker, J., Broeze, C., Kirschner, P. and Valkering, P. (2016) 'Can computer models be used for social learning? A serious game in water management', *Environmental Modelling and Software* 75: 119–132.

Wilson, E. and Piper, J. (2010) *Spatial Planning and Climate Change*, Abingdon: Routledge.

Wise, R.M., Fazey, I., Stafford Smith, M., Park, S.E., Eakin, H.C., Archer Van Garderen, E.R.M. and Campbell, B. (2014) 'Reconceptualising adaptation to climate change as part of pathways of change and response', *Global Environmental Change* 28: 325–336.

Wisner, B., Blaikie, P., Cannon, T. and Davis, I. (2004) *At Risk: Natural Hazards, People's Vulnerability and Disasters*, 2nd ed., Abingdon: Routledge.

40

EARLY WARNING SYSTEMS FOR DISASTER RISK REDUCTION INCLUDING CLIMATE CHANGE ADAPTATION

Zinta Zommers, Darren Lumbroso, Rachel Cowell,
Asha Sitati, and Elisabeth Vogel

> Just run! Run uphill! . . . Tell future generations that a Tsunami once reached this point.
> And those who survived were those who ran. Uphill. So run! Run uphill!
>
> *(Inscription at the bottom of Tsunami monument at Unosumai-chou*
> *near Kamaishi, commemorating the Great East Japan Earthquake.)*

On 11 March 2011, north-eastern Japan suffered a devastating tsunami, following a massive and comparatively shallow 9.0 magnitude earthquake. The loss of life could have been much worse if it were not for an earthquake warning system. Fifteen forecasts were issued within two minutes of the initial seismic detection, with the first public warning given fifteen seconds prior to the S-wave arrival in Sendai (Yamasaki 2012). As one witness reported,

> Our meeting was being held on the second floor of an office building near Sendai Station. Suddenly, during our discussions, our mobile phones began to ring with the distinctive earthquake alarm. This is part of the earthquake early warning system that we have here in Japan. This gave us a window of a few seconds to prepare before the shaking began.
>
> *(Tonks 2011)*

Four years later, in 2015, representatives of governments from around the world gathered in this same city. Creating the Sendai Framework for Disaster Risk Reduction (UNISDR 2015a), they committed to enhance 'multi-hazard warning systems'. Early warning systems (EWS) were also discussed that year during the twenty-first session of the Conference of Parties to the United Nations Framework Convention on Climate Change (UNFCCC) and are mentioned in the Paris Agreement on climate change, which emerged from this meeting (UNFCCC 2015).

Discussions in such diverse forums illustrate the broad relevance and appeal of EWS. Indeed, EWS can help countries, communities and households adapt to climate change and thereby reduce disaster risk. EWS illustrate how disaster risk reduction (DRR) encompasses climate change adaptation (CCA).

Yet, to ensure the success of EWS, greater focus must be paid to strengthening the capabilities of individuals through broader development processes. Levels of marginalisation – political, social, and economic – influence an individual's ability to understand and respond to warnings (Zommers and Singh 2014). While it has long been acknowledged in scientific literature that psychological and sociological processes influence the outcome of warning systems (e.g., McLuckie 1970), investment has predominantly focused on technical improvements to monitoring, prediction or dissemination.

EWS are no longer considered to be something based solely on technology, meteorology, or other hazard-centric sciences, but instead are products of multiple interactions among local knowledge, technology, meteorological and climate change science, as well as socio-economics. As expressed by Kelman *et al.* (2015), CCA sits within DRR, but both should be considered a subset of wider development and sustainability processes.

Defining EWS

According to UNISDR (2009) an EWS is,

> the set of capacities needed to generate and disseminate timely and meaningful warning information to enable individuals, communities and organisations threatened by a hazard to prepare and to act appropriately and in sufficient time to reduce the possibility of harm or loss.

The phrases 'timely' and 'meaningful information' are context-specific, depending on factors such as the hazard type, the needs of communities and their livelihood profiles, among others. Different sectors of society need different amounts of information and time to respond. What may be early for some populations is insufficient time for others. Further 'meaningful' information may range from a broad outlook to a specific alert including details of response options. Finally, EWS may seek to reduce 'harm or loss' in different ways. Some EWS may focus on reducing infrastructure damage and economic costs, while in others, the primary focus may be saving lives.

Regardless of such diversity, all EWS generally address five key questions (Glantz 2004):

- What is happening with respect to the hazard(s) of concern?
- Why is this a threat in the first place (i.e., what are the underlying causes for potential adverse impacts)?
- When is it likely to impact?
- Where are the regions most at risk?
- Who are the people most at risk (i.e., who needs to be warned)?

To be effective, EWS should integrate four key elements, or sub-systems (UNISDR 2006): knowledge of the risks (a warning sub-system); monitoring, analysis and forecasting of the hazards (a risk information sub-system); communication or dissemination of alerts and warnings (communication sub-system); and local capabilities to respond to the warnings received (preparedness sub-system). Failure in any one of these elements may cause failure of the entire system (Grasso 2014).

EWS in the Context of DRR Including CCA

EWS featured heavily in the Second World Conference on Disaster Reduction in 2005. Following this, the Hyogo Framework of Action (HFA) – the first detailed international plan to reduce disaster losses – was formally endorsed by the UN General Assembly. Adopted by 168 states, HFA outlines five priorities for action, and offers guiding principles and practical means to achieve these. In Priority 2, Core Indicator 2.3, governments acknowledge the need to enhance EWS.

Besides playing an important role within the DRR community, EWS are seen as a tool for CCA (IPCC 2013–2014). The Cancun Agreement, from the sixteenth Conference of Parties of UNFCCC, specifically invited governments to enhance action on adaptation through EWS (UNFCCC 2011).

Consequently, EWS for climate-related hazards, such as floods and droughts, are relevant to all aspects of DRR including CCA, especially given projected changes to the weather due to climate change. EWS may help increase preparedness for these changes and should be designed to do so in conjunction with people affected.

Progress

EWS have a long history of development. Informal systems based on indigenous knowledge may have existed in a community for generations (Ouma *et al.* 2013). Formal systems have been established in the last 100 years. For example, in 1949 a tsunami EWS – the Seismic Sea-Wave Warning System – was operationalised for the Pacific coastal communities of the United States (WMO 2015). Tsunami warning procedures helped reduce loss of life in Hilo, Hawaii, in 1964 (Anderson 1969). Warnings did not reach communities in other countries in 2004 when the Indian Ocean tsunami hit because the Pacific Tsunami Warning System was not mandated to warn outside its geographical remit. Those in charge tried nonetheless (Kelman 2006), but an EWS generally does not function well when implemented on an ad hoc basis after a hazard manifests. Instead, EWS is a long-term process that must begin long before it is needed. The devastation around the Indian Ocean highlighted these fundamental aspects of EWS along with the urgent need for greater investment in EWS as part of a broader strategy for DRR including CCA.

Since then, some advances in EWS have been made, as indicated in self-assessment reports prepared by governments for the Hyogo Framework of Action Review Cycle using a five-level assessment tool (Table 40.1; Figure 40.1). Improvements often result from investments in risk monitoring, forecasting, satellite data quality and increasing computer power (UNISDR 2015b). Box 40.1 provides an example.

Table 40.1 Hyogo Framework for Action (HFA) Five-Level Assessment Tool

Hyogo Framework for Action (HFA) level	Achievements
Level 1	Minor progress with few signs of forward action in plans or policy
Level 2	Some progress, but without systematic policy and/or institutional commitment
Level 3	Institutional commitment attained, but achievements are neither comprehensive nor substantial
Level 4	Substantial achievement attained but with recognised limitations in key aspects, such as financial resources and/or operational capacities
Level 5	Comprehensive achievement with sustained commitment and capacities at all levels

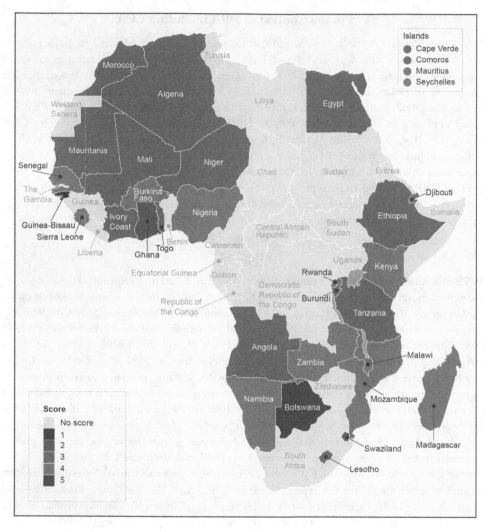

Figure 40.1 Map of HFA Self-Assessment Scores, Core Indicator 2.3 – EWS, for Africa. The means of verification are based on the evidence that: 1) early warnings are acted on effectively; 2) local level preparedness is in place; 3) communication systems and protocols are in place; 4) There is active involvement of media in early warning dissemination. The majority of countries report institutional commitment or substantial achievements

(By Lumbroso *et al.* 2014)

Box 40.1   Example of EWS Improvements

Zinta Zommers[1], Darren Lumbroso[2], Rachel Cowell[3], Asha Sitati[1], and Elisabeth Vogel[4]

[1] United Nations Environment Programme, Nairobi, Kenya
[2] HR Wallingford, Wallingford, UK
[3] iShamba Ltd, Nairobi, Kenya
[4] University of Melbourne, Melbourne, Australia
Based on UNEP (2013).

On the evening of 12 October 2013, Tropical Cyclone Phailin brought torrential downpours and damaging winds of up to 223 km/h to the eastern Indian states of Odisha and Andhra Pradesh. Twenty-one people died as a result of the cyclone and an additional 23 died due to severe flash flooding in its aftermath. When a comparable cyclone, Cyclone 05B, hit the same area in 1999 with winds of up to 260 km/h more than 10,000 lives were lost. Lessons learned from Cyclone 05B spurred investments in DRR. In total, 200 new cyclone shelters were constructed. These were used during Phailin, providing shelter to over 100,000 people. The India Meteorological Department (IMD) disseminated warnings four days before Phailin made landfall compared with two days before Cyclone 05B. This allowed for the evacuation of approximately 400,000 people on or before 11 October. Multiple forms of communication were used to warn people. Satellite phones were distributed to representatives in the 14 most vulnerable districts to ensure that warning communications continued during the storm.

Evidence of the rainfall is exhibited by the influx of Chilika Lake, Asia's largest brackish water lake, which is bordered by Khurdha, Puri and Ganjam districts and the Bay of Bengal (Figure 40.2). Approximately 40,000 villagers who live among the islands scattered in and around Chilika were able to evacuate prior to the cyclone's landfall. But the environment was not as fortunate. The cyclone breached the natural coastal barrier of Chilika, destroying kilometres of its delicate mangrove forests, which are favoured by some migratory species and several endangered plants and animals. A significant proportion of casuarina forests, which served as a protective barrier for residents of the area, were buried by sand. Arrows in Figure 40.2 indicate significant areas of change between the extent of the lake in October 2012 and the extent a few days after Phailin dissipated. Now that the barrier between Chilika and the Bay of Bengal has been breached, protection from future events is compromised, demonstrating the importance of continued early warning efforts as well as broader adaptation and DRR efforts. Activities could include the establishment of new damage-control mechanisms, restoration of forest ecosystems, and incorporation of ecosystem-based adaptation measures.

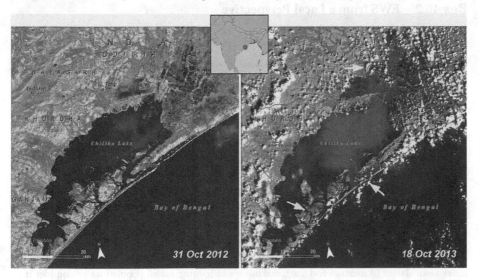

Figure 40.2   Landsat Satellite Images of Chilika Lake in 2012, When the Coastline was Intact, and Six Days After Phailin in 2013, When the Coastline Had Been Breached

Images: USGS/NASA; visualisation by UNEP/GRID-Sioux Falls (By UNEP 2013)

Identified Challenges

In 2009, the Global Network of Civil Society Organisations for Disaster Reduction (GNDR) carried out the first independent assessment of progress undertaken towards implementation of the HFA at the local level. The review covered 48 countries in Africa, Asia and the Americas. The work included 5,290 survey questionnaires (primarily through face-to-face interviews) with three stakeholder groups: local governments, civil society organisations and community representatives (GNDR 2009).

Results were significantly different from those presented by governments for the HFA assessment cycle. This review found a large gap between national and local level action (GNDR 2009) for both risk assessments and EWS. The reports of progress towards the HFA indicators were found to 'fade' as activities get closer to vulnerable people where the 'impact is at best limited and patchy and at worst not happening at all' (GNDR 2009: 36). The data collected by the GNDR showed significant differences between the level of perceived progress by the three groups: local government, civil society organisations and community groups. Communities indicated progress on EWS and risk assessments as being 'very limited', which was consistently lower than both local government and civil society scores.

Involving the Most Vulnerable

This gap in perception may result from a 'Last Mile' approach to EWS (Box 40.2). EWS need to start at the local level, integrate local knowledge and address local needs. Regional or national warning systems are not necessarily appropriate for, or applicable to, local communities or vulnerable groups within communities. Without contextualisation and engagement with the

---

Box 40.2   EWS from a Local Perspective

Ilan Kelman[1]

[1] University College London, London, UK and University of Agder, Kristiansand, Norway

Because EWS serve local needs, they frequently need to start with local needs indicating the information, technology, and dissemination required rather than the other way around. This approach is termed 'The First Mile' of EWS, rather than 'The Last Mile' approach, which sets up EWS technology and then, as the last step, connects the dissemination mechanism to communities.

The First Mile retains the Last Mile's ethos of reaching the right people at the right time with the right information for EWS. But it starts with the people who need and use the EWS and their knowledge, rather than starting with the remote operators of the technical system and assuming that external knowledge must be transferred to communities. Nonetheless, the EWS might end up being the same from technological and operational perspectives, but it has involved the affected people from the beginning, ensuring improved uptake and credibility.

The First Mile is particularly powerful for using EWS to support CCA. Climate change can at times seem distant in location and time, making it challenging for all populations to consider it in their day-to-day activities. A First Mile EWS using local observations of environmental changes to connect to possible futures might better support people in enacting DRR measures to deal with a range of possible futures, thus building up communities' CCA capabilities.

---

population, warnings may not prompt appropriate responses. For example, threshold levels of 'danger' will depend on people's livelihoods and will sometimes be different for diverse groups living in the same areas. Groups such as women, elderly, immigrants, and people with disabilities may be less likely to receive warnings, and are often unable to act on them due to differences in power relations and socio-economic inequalities (UNEP 2015), so that they are continually made to be more vulnerable.

Investing in Monitoring

In 2014, another independent survey was conducted, this time to identify specific barriers to EWS effectiveness (see Brown *et al.* 2014). Some 250 individuals who operate, carry out research on or use the outputs from EWS in Africa, the Caribbean and South Asia were asked about their perceptions. Effectiveness here was defined as the ability of EWS to reduce loss of life (see Brown *et al.* 2014; Lumbroso *et al.* 2014). The results of this survey are shown in Figure 40.3.

Particularly in Africa and South Asia, the main barrier to EWS effectiveness in terms of reducing loss of life was seen to be the lack of high quality data followed by the lack of technological capacity to generate forecasts (see Figure 40.3) (Brown *et al.* 2014). Massive underfunding of monitoring networks has led to their deterioration and this is coupled with a lack of modern equipment, poor quality services and a dearth of trained specialists (Rodgers and Tsirkunov 2013). The World Bank considers that the modernising of national hydrological and meteorological monitoring networks is a high value investment in low-income countries (Rodgers and Tsirkunov 2013).

Communicating Warnings

Stakeholders also emphasised that EWS effectiveness and risk assessments in the three regions depend upon communication of warnings from forecasters to end users and the accessibility of warnings, especially in Africa and the Caribbean (Brown *et al.* 2014). Many EWS examples were accurate and capable of providing useful information, but this was not effectively communicated or acted upon (e.g., IFRC 2014; UNEP 2013). The capacity to disseminate warning messages effectively at a local level often remains a significant challenge. While uncertainty remains about which dissemination methods are most appropriate and cost effective under which circumstances, decades ago Gruntfest *et al.* (1978) already noted that face-to-face communication is more effective than warnings delivered impersonally. Again, this points to a need for a broader application of the First Mile approach.

Focus group discussions with women, youth, the elderly and people with disabilities in Kenya, Burkina Faso and Ghana illustrate communication needs (UNEP 2015). In Kenya, 83.3 per cent of the focus groups indicated that they did not receive warnings for floods and 72 per cent did not receive warnings for droughts. People with disabilities and the elderly cited their lack of access to warnings: unsuitable information package (e.g., for those with visual or hearing impairments); inability to access information points due to limited mobility; lack of proper representation in committees; and low confidence or trust in existing institutions (UNEP 2015).

The same study found that people use and trust a wide range of information sources and providers (UNEP 2015). In urban areas in Kenya the national government enjoys the highest level of trust but in rural areas headmen and elders as well as religious leaders are the most trusted sources of information. Different groups should therefore be involved in information dissemination in different areas. To effectively address this diversity, it is critical to involve communities in the design and operation of all EWS sub-systems.

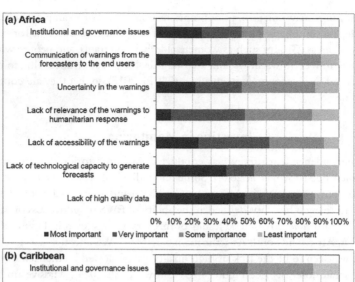

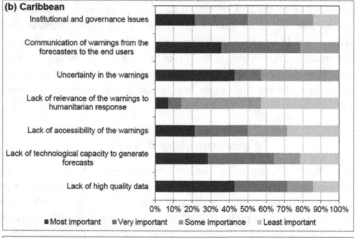

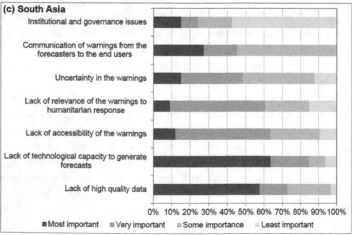

Figure 40.3   Stakeholders' Perceptions of the Barriers to the Effectiveness of Early Warnings for Weather-Related Hazards by Region

(By Brown *et al.* 2014)

Addressing Different and Multiple Hazards

There is significant variability in EWS needs, not only among regions and groups of people, but also by hazard type. There is a particular need to improve EWS for slow-onset hazards such as droughts (Grasso 2014). EWS for rapid/sudden-onset hazards have seen greater improvements, but prediction capabilities for hazards such as landslides and floods need to be enhanced and the area of coverage expanded systematically. For example, flood EWS are hugely variable in the Caribbean. In Jamaica, the first flood forecasting systems were set up in 1992. Today, the country has EWS for its larger catchments together with some community-based schemes (Haiduk 2004). Pilot flood forecasting systems have been implemented in some catchments in Barbados and Guyana (Boyce and Whitehall 2013), as well as a web-based system for Haiti. Yet, these do not appear to communicate effectively with vulnerable communities. Many Caribbean islands do not have the capacity or the resources to provide any flood EWS.

Ultimately, focus needs to extend to the creation of multi-hazard and multi-vulnerability EWS. In a survey of over 1,000 households in Kenya, Burkina Faso and Ghana, community members reported experiencing floods and droughts, but also disease outbreaks, crop pests, windstorms (Ghana, Kenya), bushfires (Ghana, Burkina Faso), and landslides (Kenya) (UNEP 2015). While sites within a country face different hazards, all face multiple hazards that may interact to increase disaster risk. A review of interactions among twenty-one natural hazards, drawn from six hazard groups (geophysical, hydrological, shallow Earth, atmospheric, bio-physical, and space hazards), indicates that hazards such as volcanic eruptions, earthquakes, and storms are likely to trigger other hazards (Gill and Malamud 2014). Landslides, floods, and even volcanic eruptions are often secondary results from other primary hazards (Gill and Malamud 2014). Separating warning systems by hazard is artificial, possibly even dishonest, and may undermine DRR including CCA.

Issues of scale also need to be addressed. While large disasters tend to receive the most humanitarian and media attention, small-scale and localised hazards can also be highly damaging, chipping away at the assets and coping capacity of households (Lewis 1984; Marulanda *et al.* 2010; UNEP 2015). Localised events are traditionally under-reported: rarely captured in global modelling or included in EWS (UNISDR 2015b). EWS should not only provide information about large-scale hazards, but must also be used to help communities reduce exposure to extensive risks.

Improving Coordination

Ineffective coordination can be a significant barrier to effective EWS. For example, in Sudan twenty organisations are involved in supporting different aspects of EWS. In the Horn of Africa, there are numerous government, non-governmental organisations and international agencies involved with forecasting droughts and famines. There is often little evidence of a systematic framework to direct, guide or coordinate these organisations. Where coordination of early warnings is weak, decision-making is often fragmented and accountability is diffused.

In India, different agencies issue warnings for different hazards, even though they are often interrelated. Stakeholders consulted as part of the study by Lumbroso *et al.* (2014) indicated that the dissemination of warnings by different Indian agencies often leads to ineffective communication and confusion amongst the intended beneficiaries.

Conflicting warning messages to stakeholders not only lead to misunderstandings, but can also result in the intended beneficiaries disengaging from EWS. In the long-term, the creation of multi-hazard and multi-vulnerability EWS could be used to better integrate the mandates of different hazard-specific agencies. In particular, adopting a 'First Mile' approach (Box 40.2) can help coordination in a specific locality, rather than assuming only top-down coordination is needed by the national or sub-national agencies.

Ensuring Early Action

In many cases, early warnings are disseminated but do not result in appropriate responses. Bailey (2013) reports that famine EWS in the Horn of Africa have a good track record of predicting food crises, but a poor track record of triggering early action. This is also sometimes the case with flood forecasts in large catchments, such as the 2000 floods in Mozambique when the government requested assistance before the floods but donors did not respond until after the floods.

Bailey (2013) predicts that continuing technological and methodological advances in forecasting systems may mean the gap between early warning and early action is only set to widen – another manifestation of taking a Last Mile approach. This may be because,

> the long lead times offered by famine early warning systems provide the opportunity for decisive early action, but also the opportunity for prevarication, delay and buck-passing. This disconnect persists despite major improvements in the sophistication and capabilities of modern systems.

> *(Bailey 2013: ix)*

Effective monitoring, evaluation, and the creation of bottom-up structures of EWS continue to be important for DRR including CCA.

Renewed Commitments?

In March 2015 in Sendai, government representatives committed to 'substantially increase the availability of and access to multi-hazard EWS and disaster risk information and assessments to the people by 2030' (UNISDR 2015a: 8). They also promised to 'Invest in, develop, maintain and strengthen people-centred multi-hazard, multi-sectoral forecasting and EWS, disaster risk and emergency communications mechanisms, social technologies and hazard-monitoring telecommunications systems', and to 'develop such systems through a participatory process. Tailor them to the needs of users, including social and cultural requirements, in particular gender' (UNISDR 2015a: 18).

Nine months later, in December 2015, EWS were a subject of discussion in climate change negotiations. In the Paris Agreement, 195 countries committed to keep global temperature rise to well below 2° Celsius above pre-industrial levels. EWS are explicitly mentioned in text related to CCA in Article 7 (UNFCCC 2015: 24):

Parties should strengthen their cooperation on enhancing action on adaptation, taking into account the Cancun Adaptation Framework, including with regard to: Strengthening scientific knowledge on climate, including research, systematic observation of the climate system and early warning systems, in a manner that informs climate services and supports decision-making.

EWS are also mentioned as a tool for the limits of CCA, when loss and damage occurs. Article 8, which relates to the Warsaw International Mechanism for Loss and Damage associated with Climate Change Impacts, states (UNFCCC 2015: 25):

Areas of cooperation and facilitation to enhance understanding, action and support may include:

(a) Early warning systems;
(b) Emergency preparedness;

(c) Slow onset events;

(d) Events that may involve irreversible and permanent loss and damage;

(e) Comprehensive risk assessment and management;

(f) Risk insurance facilities, climate risk pooling and other insurance solutions;

(g) Non-economic losses;

(h) Resilience of communities, livelihoods and ecosystems

Here, we see the convergence of DRR and CCA. Many of the areas of cooperation – EWS, emergency preparedness, risk assessment, and risk insurance – are essentially DRR tools and thus embrace CCA also.

EWS and Development

Despite these initiatives, gaps are already apparent in renewed commitments to act. SFDRR (UNISDR 2015a) does not specify important implementation issues and potentially creates both challenges and opportunities for complex, multilevel governance systems in coping with hazards and disastrous events (Zia and Wagner 2015). Zia and Wagner state that 'although the SFDRR includes people-centred, bottom-up rhetoric, it lacks explicit means of implementation for governance structures that incorporate people-centred, bottom-up design' (Zia and Wagner 2015: 197).

Further, while EWS are mentioned in both the DRR and climate change agreements, they are noticeably absent from the Sustainable Development Goals (SDGs) adopted by the UN General Assembly in September 2015 (UNGA 2015). This is a critical gap. As already concluded more than a decade ago (Glantz 2004: 7), 'An early warning system is an important tool in a government's programme to achieve sustainable development. In fact sustainable development prospects are very dependent on the effectiveness of the many early warning systems.'

At the same time, the success of early warning is determined by broader development patterns. Factors such as income levels, access to resources, the diversity of livelihood sources, education and the connectedness within a community can be key for DRR including CCA and so impact EWS. For instance, households with only one form of income may have limited options for alternative resources during disasters. Livelihoods in rural areas in Kenya often depend on pastoralism (UNEP 2015). During times of drought, focus has to be placed on the stabilisation or diversification of income sources of pastoralists (e.g., through governmental support or insurance schemes) as well as on livestock health.

Other important socio-economic factors influencing EWS success include health and age profiles of communities. This determines how individuals are affected by hazards (for example, elderly and people with health problems may be more affected by heat waves than young, healthy individuals) and how prepared they are to respond to hazards, such as if they need specific support. People with limited mobility, for example, may need specific forms of support during evacuation. Warning information has to be disseminated in a variety of formats and through multiple information channels (e.g., text, audio, and visual) so that illiteracy, deafness, blindness or other factors do not limit response.

Gender is important. There are varied gender-based limitations, specific needs and capacities that should be considered during the development and the operation of EWS. Generally, due to inequalities that exist in most, if not all, societies, women often have on average lower education, literacy and income levels than men, and thus lower access to resources and information. This affects their abilities to engage with developing and operating EWS suitable for their needs and making best use of their knowledge.

Figure 40.4   Word Cloud of Actions or Groups Needed to Prepare For and Respond to Hazards, Based on Answers from Community Surveys in Ghana (top) and Kenya (bottom) in Which Larger Words Are Mentioned More Often, Showing How Broader Development Needs Are Highlighted

(By Authors)

Achieving development goals – including gender equality, decent work, quality education, good health – is thus critical for improving EWS success (UNEP 2015). Addressing factors such as poverty, inequality, lack of education, and lack of employment opportunities can enable better understanding of warnings and better actions in response to warnings. EWS must thus not be considered a service provided solely by meteorological agencies, but must be linked to agencies and groups of people working to increase human capabilities. Indeed, when asked what was needed to improve response to early warnings, communities in Kenya, Ghana and Burkina Faso requested capacity building, training and education (Figure 40.4).

Steps Forward: EWS for DRR Including CCA

Going forward, to ensure success, it will be important to work closely with communities on all four subsystems of EWS – knowledge of risks, monitoring, communication, and capacity to respond. As Kelman (2006: 184) notes,

> warning systems should be planned as integrated components of the communities to be warned rather than as top-down impositions from governments or scientists. They should be incorporated into livelihood and development activities and should be continually relevant to the people who will be warned.

Communities and external agencies should both be involved in collecting data, interpreting data, and deciding when and how to issue warnings. There are several examples where indigenous knowledge has been combined successfully with newer technologies in EWS (Box 40.3). Ultimately, EWS should be a community-based process that can involve external technology, science, management, and operation rather than a technical process imposed on communities without their involvement.

Box 40.3   Nganyi Indigenous Knowledge of Rainfall Prediction

Zinta Zommers[1], Darren Lumbroso[2], Rachel Cowell[3], Asha Sitati[1], and Elisabeth Vogel[4]

[1] United Nations Environment Programme, Nairobi, Kenya
[2] HR Wallingford, Wallingford, UK
[3] iShamba Ltd, Nairobi, Kenya
[4] University of Melbourne, Melbourne, Australia
Based on Ouma *et al.* (2013).

The Nganyi community in Western Kenya uses indigenous knowledge to predict rainfall extremes including floods and droughts. Nganyi indigenous knowledge is grounded on a number of bio-physical, social and astronomical indicators combined with metaphysical and spiritual paradigms. Rain prediction runs in particular families who pass on the knowledge to the next generation through oral transmission and apprenticeship. In order to develop forecasts, rain prediction elders use ceremonial rituals involving prediction pots but also rely on rainfall shrines and natural forests with water sensitive plants, which respond to decreases and increases in atmospheric moisture. The flora and fauna of different shrines are monitored on a day-to-day basis and consensus forecasts are generated.

A recent pilot study combined the indigenous knowledge forecasts with outlooks generated by the Kenya Meteorological Department and downscaled to the Nganyi area. Modern scientists and Nganyi forecasters were brought together and asked to produce a harmonised consensus forecast. An evaluation was later conducted by asking community members if the forecast was accurate, received on time and helpful in reducing the impacts of the weather. There was general agreement that the forecast accurately predicted onset and distribution of rains; however, a large percentage of the community did not receive the advisory and others did not believe it. Further study and testing is needed.

It may also be valuable to further explore the use of public–private partnerships (PPP) for EWS. PPP may help spur innovations in monitoring and communication, develop information packages required by clients and provide additional services to remote areas. For example, iShamba (Box 40.4) is a mobile agriculture service designed to improve the livelihoods of Kenyan farmers. By providing farmers with information they demand, including weather-related information, in an easily accessible and actionable format, such as a highly targeted and free message to their mobile phones, iShamba aims to increase farming productivity. The push-pull mobile phone messaging system means that it can also serve as an important EWS, delivering messages in times of disaster. Alongside these messages, membership of iShamba provides farmers with access to a call centre of experts seven days a week who can be reached via mobile phone message or phone call. This is invaluable in a country with one agricultural extension officer to every thousand farmers. While PPP can complement government-run EWS, it should never replace government responsibility for and community involvement in warning.

Box 40.4   iShamba

Zinta Zommers[1], Darren Lumbroso[2], Rachel Cowell[3], Asha Sitati[1], and Elisabeth Vogel[4]

[1] United Nations Environment Programme, Nairobi, Kenya
[2] HR Wallingford, Wallingford, UK
[3] iShamba Ltd, Nairobi, Kenya
[4] University of Melbourne, Melbourne, Australia

In an average week, a subscriber receives a weekly weather forecast for their farm's location, the market prices for two crops from over 27 markets around Kenya, and a localised, time-based agri-tip about their chosen livestock or crop. The tips have been developed by local and industry farming experts for each micro-ecological zone in Kenya, as well as veterinarians and agronomists with many years of field experience. The mobile phone messages have been simplified and then further simplified, are designed to be actionable and accessible to farmers with low levels of literacy, and are delivered in a mixture of English and local terminology (e.g., hoho instead of capsicum and jambe instead of garden hoe). As an example, a farming instruction of 233 characters, 'Seed soaking: To supply the required moisture for germination, to shorten germination period and reduce seed rotting. During the soaking period change water to prevent accumulation of poisonous substances and allow entry of fresh air' was simplified to 149 characters, 'Soak maize seed for 1 week before planting. Put seed in bucket, keep covered with water that is always fresh. Plant seed to grow quickly and not rot'.

The impact of this m-Agri (mobile agriculture) product can already be seen, just one year into delivery. Research Guide Africa have shown that iShamba farmers had doubled the median potato yield of non-iShamba farmers in the 2015 growing season, resulting in twice the income (this internal research has yet to be published). In the same study 95 per cent of iShamba farmers found the information easy to understand and two thirds of iShamba farmers made a change to their farming practice as a result of the mobile phone messages.

The Mediae Company is the company behind iShamba Ltd. and is responsible for ensuring that the information is relevant and timely and that the product is evolving rapidly to meet the needs of a cost-sensitive user-base. As with Mediae's flagship agri edu-tainment TV and Radio programme, 'Shamba Shape Up', iShamba works with commercial partners to promote their products to farmers at the times most relevant to them.

For example, a message reinforcing that it is time to 'top dress' your maize (i.e., apply fertiliser) will be sent with a voucher code for a 10 per cent discount for a particular fertiliser product, benefiting both supplier and customer – keeping in mind the potential for abuse and conflict-of-interest. Private sector and government collaboration helps to ensure that agricultural information is delivered to farmers in formats in tune with farmers' evolving use habits. For Kenya in 2016 that format is mobile phone messaging via SMS, which reaches all mobile handsets, and areas where the signal is not strong enough for voice calls or where infrastructure is weak. As smartphone penetration increases and as 3G coverage expands, content delivered over the Internet could make business sense in terms of reducing running costs and in terms of meeting wider needs of their user base.

Ultimately EWS is one area where CCA and DRR are already closely linked. To be successful, a more networked and integrated approach is needed. This means integrating different hazards, different types of knowledge and information as well as different stakeholders – communities, governments, and the private sector. Information disseminated through the system could not only include details of hazards, but also broader adaptation options and material relevant to community development (Zommers and Singh 2014). By acknowledging that DRR including CCA must be rooted in sustainable development, early warning systems can be used to help build resilience, protecting livelihoods as well as lives.

References

Anderson, W. (1969) 'Disaster warning and communication process in two communities', *The Journal of Communication* 19: 92–104.

Bailey, R. (2013) *Managing Famine Risk: Linking Early Warning to Early Action*, A Chatham House Report, London: Royal Institute of International Affairs.

Boyce, S. and Whitehall, K. (2013) *Real-time Flood Forecasting for the Caribbean*. Online http://www.gwp.org/Global/GWP-C%20Files/IWRM%20Initiatives_CIMH_Flood%20Forecasting.pdf (accessed 24 September 2016).

Brown, E., Lumbroso, D. and Wade, S. (2014) *Science for Humanitarian Emergencies and Resilience (SHEAR) Scoping Study: Annex 1 – Results of a Stakeholder Questionnaire*. Online http://www.evidenceondemand.info/science-for-humanitarian-emergencies-and-resilience-scoping-study-annex-1 (accessed 31 March 2016).

Gill, J.C. and Malamud, B.D. (2014) 'Reviewing and visualizing the interactions of natural hazards', *Reviews of Geophysics* 52: 680–722.

Glantz, M. (2004) *Early Warning Systems: Do's and Don'ts*, Report of a Workshop 20–23 October 2003, Shanghai, China. Online http://www.riskred.org/fav/glantz2003.pdf (accessed 25 September 2016).

GNDR (Global Network of Civil Society Organisations for Disaster Reduction). (2009) *'Clouds but little rain . . .': Views from the Frontline: A Local Perspective of Progress towards Implementation of the Hyogo Framework for Action, June 2009*, Teddington UK: GNDR.

Grasso, V. (2014) 'The state of early warning systems', in Z. Zommers and A. Singh (eds) *Reducing Disaster: Early Warning Systems for Climate Change*, New York: Springer, pp. 109–125.

Gruntfest, E., Downing, T. and White, G. (1978) 'Big Thompson flood exposes need for better flood reaction system to save lives', *Civil Engineering: Magazine of American Society of Civil Engineers (ASCE)*: 72–73.

Haiduk, A. (2004). *Flood Forecasting and Hazard Mapping in Jamaica*, Thematic session on integrated flood risk management through appropriate knowledge sharing and capacity building systems, Kobe, Japan, 20 January, 2004. Online https://www.unisdr.org/2005/wcdr/thematic-sessions/presentations/session2-1/wraj-mr-haiduk.pdf (accessed 25 September 2016).

IFRC (International Federation of Red Cross and Red Crescent Societies). (2014) *Early Warning, Early Action: Mechanisms for Rapid Decision Making, Drought Preparedness and Response in the Arid and Semi-arid Lands of Ethiopia, Kenya and Uganda, and in the East Africa Region*, IFRC. Online http://www.droughtmanagement.info/literature/IFRC_Early_Warning_Early_Action_2014.pdf (accessed 25 September 2016).

IPCC (Intergovernmental Panel on Climate Change). (2013–2014) *Fifth Assessment Report*, New York: Cambridge University Press.

Kelman, I. (2006) 'Warning for the 26 December 2004 tsunamis', *Disaster Prevention and Management* 15, 1: 178–189.

Kelman, I., Gaillard, JC, and Mercer, J. (2015) 'Climate change's role in disaster risk reduction's future: Beyond vulnerability and resilience', *International Journal of Disaster Risk Science* 6: 21–27.

Lewis, J. (1984) 'Environmental interpretations of natural disaster mitigation: The crucial need', *The Environmentalist* 4: 177–180.

Lumbroso, D., Rance, J., Pearce, G. and Wade, S. (2014) *Science for Humanitarian Emergencies and Resilience (SHEAR) Scoping Study: Final Report.* Online http://dx.doi.org/10.12774/eod_cr.june2014.lumbrosoetal (accessed 25 September 2016).

Marulanda, M.C., Cardona, O.D. and Barbat, A.H. (2010) 'Revealing the socioeconomic impact of small disasters in Colombia using the DesInventar database', *Disasters* 34, 2: 552–570.

McLuckie, B. (1970) *The Warning System in Disaster Situations: A Selective Analysis*, Report Series 9, Delaware, USA: University of Delaware Disaster Research Center.

Ouma, G., Labal, O. and Onyango, M. (2013) *Coping with Local Disasters Using Indigenous Knowledge: Experiences from Nganyi Community of Western Kenya*, Saarbrücken, Germany: Lap Lambert Academic Publishing.

Rodgers, D. and Tsirkunov, V. (2013) *Effective Preparedness through National Meteorological and Hydrological Services*, Washington DC: World Bank.

Tonks, B. (2011) *Surviving Sendai, a First Person Account*. Online http://www.simcoe.com/news-story/2056940-surviving-sendai-a-first-person-account/ (accessed 7 February 2016).

UNEP (United Nations Environment Programme). (2013) *Cyclone Phailin in India: Early Warning and Timely Actions Saved Lives*, Global Environment Alert Service, UNEP. Online http://na.unep.net/geas/archive/pdfs/GEAS_Nov2013_Phailin.pdf (accessed 25 September 2016).

UNEP (United Nations Environment Programme). (2015) *Early Warning as a Human Right: Building Resilience to Climate Related Hazards*, Nairobi, Kenya: UNEP.

UNFCCC (United Nations Framework Convention on Climate Change). (2011) *The Cancun Agreements: Outcome of the work of the Ad Hoc Working Group on Long-term Cooperative Action under the Convention, FCCC/CP/2010/7/Add.1*, Bonn, Germany: UNFCCC.

UNFCCC (United Nations Framework Convention on Climate Change). (2015) *Conference of the Parties, Twenty-first session, Paris, 30 November to 11 December 2015, FCCC/CP/2015/L.9/Rev.1*, Bonn, Germany: UNFCCC.

UNGA (United Nations General Assembly). (2015) *Resolution Adopted by the General Assembly on 25 September 2015, A/RES/70/1*, New York: UNGA.

UNISDR (United Nations International Strategy for Disaster Reduction). (2006) *Developing Early Warning Systems: A Checklist*, Bonn, Germany: UNISDR. Online http://www.unisdr.org/2006/ppew/inforesources/ewc3/checklist/English.pdf (accessed 25 September 2016).

UNISDR (United Nations International Strategy for Disaster Reduction). (2009) *Terminology*, Geneva: UNISDR. Online http://www.unisdr.org/we/inform/terminology (accessed 25 September 2016).

UNISDR (United Nations International Strategy for Disaster Reduction). (2015a) *Sendai Framework for Disaster Risk Reduction 2015–2030*, Geneva: UNISDR.

UNISDR (United Nations International Strategy for Disaster Reduction). (2015b) *Global Assessment Report on Disaster Risk Reduction*, Geneva: UNISDR.

WMO (World Meteorological Organization). (2015) *Synthesis of the Status and Trends with the Development of Early Warning Systems. A Contribution to the Global Assessment Report 2015 (GAR15), Priority for Action (PFA) 2 – Core Indicator (CI) 3: Early Warning Systems are in Place for all Major Hazards with Outreach to Communities*, Background Paper prepared for the Global Assessment Report on Disaster Risk Reduction 2015. Geneva: UNISDR. Online http://www.preventionweb.net/english/hyogo/gar/2015/en/bgdocs/WMO,%202014a.pdf (accessed 25 September 2016).

Yamasaki, E. (2012) 'What we can learn from Japan's early earthquake warning system', *Momentum* 1: 1, Article 2.

Zia, A. and Wagner, C.H. (2015) 'Mainstreaming early warning systems in development and planning processes: Multilevel implementation of Sendai Framework in Indus and Sahel', *International Journal of Disaster Risk Science* 6: 189–199.

Zommers, Z. and Singh, A. (2014) 'Introduction', in Z. Zommers and A. Singh (eds) *Reducing Disaster: Early Warning Systems for Climate Change*, New York: Springer, pp. 1–19.

41

WATER FOR DISASTERS, WATER FOR DEVELOPMENT

Sarah Opitz-Stapleton

Introduction

Water. It is the source of life underpinning and linking core systems – food production, ecosystem functions and services, health and sanitation, energy and culture – and as such, cannot be considered in isolation from them. Water is one of the foundational resources required for sustainable development, a process articulated in the Brundtland Report (UNWCED 1987: 43) as meeting 'the needs of the present without compromising the ability of future generations to meet their needs'. Unequal and ecologically unsustainable development (mediated by institutional, social, cultural, political and economic processes) reduces or excludes particular groups' access to and use of physical and environmental resources (Bebbington 1999). This includes water and water-related services, like sanitation, sewerage and irrigation and the ability to live or earn a living in less hazard-exposed areas, particularly those prone to water-mediated hazards like flooding.

How water is developed and used, or exploited and degraded, and by whom has strong implications for differentiated vulnerability and the creation of historic and future hydrological risks and disasters. The nested nature of water cultures – e.g., national water laws imposed over 'local' water traditions, and transnational cooperation, power struggles and negotiation over major river systems – creates unequal rights and access to quality and quantity of water across space and time (Opitz-Stapleton 2009). Social and environmental water quality and quantity equity are important to realising sustainable development of water resources, and to achieving disaster risk reduction (DRR) including climate change adaptation (CCA) as part of that development. Inequities in water cultures conversely exacerbate immediate water-related disaster risk for some and may reinforce unsustainable development pathways for current and future generations. Imposed upon water cultures is climate change, which is altering the frequency, timing, spatial extent and intensity of many hydro-climatological hazards, and contributing to significant changes in local-to-regional water quantity and quality. The joint evolutions of water cultures and climate over space and time build the foundations of DRR including CCA, or lack thereof, upon which future generations will act.

This chapter explores water cultures and practices at three spatial scales – local, sub-national/national, and transnational – and the implications they have for short- and long-term vulnerability and risk. It draws loosely from political ecology traditions around water culture (Budds 2004) and constructions of vulnerability from the hazards and disasters scholarly tradition, while

445

acknowledging the embedded nature of CCA inside DRR and sustainable development (Wisner *et al.* 2012). As with the other chapters in this handbook, it does not present new research on water; it summarises some contemporary literature.

Water Vulnerabilities and Risks across Time

Water cultures – control over conceptualisations of access to and usage of water at a variety of spatial and temporal scales – have shaped and been shaped by the socio-economic, cultural and political processes and relationships with the environment. Water cultures – past current, and future – may increase vulnerability and exposure to hydrological hazards and give rise to differentiated disaster risks, including those influenced by climate change, today and in the future. Conversely, sustainable and flexible relationships with water and ecosystems may enhance abilities to deal with the challenges in the face of some hazards over space and time. This happens through both planned (e.g., promulgated by government and non-governmental organisations) and private (undertaken by business, community groups, households and individuals) water DRR measures, which incorporate CCA strategies and actions, and practices related to water allocation and environmental protection.

Social Vulnerability and Exposure: Local Water Traditions and Practice for Local Risks

Both proximity to water (exposure) and its use are mediated by human–environmental and social, cultural and political relationships. Local culture and social practices traditionally constructed (and construct) water rights governing how water can be used, how much, in what manner, by whom, and when within a community (Opitz-Stapleton 2009). They are complex, and frequently reflect asymmetric power relationships and cultural expectations tied to, for example, livelihood identities as farmers, herders, and landowners, among others; relationships with environmental and ecosystem services; and, exogenous influences like government actions around recognising, formalising or superseding traditional water rights (Hodgson 2004). Social structures related to gender, age and group identity also govern customary water rights, contributing to gender-differentiated hydrological vulnerability and risk, as elaborated later on in this section.

Human habitations were and are frequently located near a water source in order to grow food, raise livestock, run industries and businesses and assist in energy generation, among other activities. Proximity and access to a reliable water supply are crucial for lives, livelihoods and development, and for maintaining these during droughts and other water-related hazards. Access to a reliable water source contributes to societies' abilities to live with natural hazards, including those influenced by climate change (Brown *et al.* 2006).

In the Yellow River Basin, China, access to and use of irrigation water has been shown to reduce rural poverty and to increase crop productivity, contributing to food security (Zheng *et al.* 2015). The rivers of the Pacific Northwest in the United States and Canada provide important ecosystem services around salmon, which are in turn keystones to many Pacific Northwest indigenous tribes' unique cultures, subsistence and economic activities (Cozzetto *et al.* 2013). Historically, widespread yet locally constructed and managed irrigation systems throughout Iran, Jordan and Syria (*Qanat*); Pakistan and Afghanistan (*Karez*); and western China (*Kanerjing*) conveyed water from the mountains over large distances for irrigation. Such systems rarely exceeded the water supply as the amount withdrawn depended on gravity and was proportionate to the aquifer water tables; the systems also played a central role in many communities' cultures and traditional ecological knowledge (Balali *et al.* 2012). Where water is locally valued as inseparable from ecosystem wellbeing, and services and watersheds are protected, it can be managed in a sustainable manner contributing to DRR including CCA, through reducing vulnerability to the impacts of hydrologic hazards and through reducing the potential for people and communities to generate hazards from natural water processes.

Just as water cultures can contribute to a community's overall vulnerability or resilience, they can influence members' overall capacities, thereby supporting DRR, and through it CCA, or detract from their abilities to cope with and recover from the impacts of water hazards. An individual's or household's relationship with water, as mediated by the community in which they live, plays an important role in individual food security, hygiene and health, and workload – factors contributing to overall resilience or vulnerability in the face of natural hazards. Local water rights and practice do not necessarily grant individuals equal access to water or decisions about its management within the home or wider community (Brown *et al.* 2006). Income and social inequality that play out in localised water cultures can contribute to greater flood exposure, differentiated resilience capacities and flood risk to individual households and communities as seen in case studies in the United States (Zahran *et al.* 2008), and other locations.

Another such example of differentiated water cultures contributing to or reducing DRR including CCA is seen in gendered constructions of customary water rights and roles, leading to some women and girls possibly having lower resilience than men and boys in the face of water-climate hazards. Gendered roles around water are constructed and enacted differently among ethnic groups, communities and within households. Women and girls in many rural communities globally bear the primary responsibility for fetching water for domestic use and for smallholder plots growing the household's food supply (Ray 2007). Yet, men within the community frequently determine where women can draw water, how much, when and for what purposes. Women, particularly in female-headed households, may have to employ creative tactics for accessing irrigation water. Gendered customary water rights can also contribute to the overall disproportionate vulnerability that many women and girls face in daily life, and during and after natural hazards (Ahmed 2004). In times of meteorological drought (the reduction or absence of precipitation necessary for the functioning of a particular area's ecosystems and human systems, in contrast to droughts resulting from human overuse of water supplies), women and girls may face higher workloads to fetch water from farther distances, also potentially leading to higher heat stress and dehydration; greater health, personal safety and hygiene risks due to unmet water needs; greater food insecurity; and, unequal humanitarian response in receiving resources (Opitz-Stapleton 2014; Box 41.1).

While proximity to water resources is necessary for livelihoods and hence food security, it can also increase exposure to flooding and exacerbate health risks where there are inequalities in access to housing, safe domestic water supplies and sufficient sewage and solid waste management systems. Inequalities in linked local and state water cultures may lead to some parts of a community not being served or being underserved by wastewater and stormwater networks – or lead to poorer communities living in floodplains without adequate flood risk reduction measures being taken. This inadequate and unequal access to water services and drainage systems can compound flooding and damage while extending waterlogging after heavy rainfall or snowmelt events. It also contributes to contemporary water- and vector-borne disease and other health risks, including those exacerbated by climate change (Hunter 2003; Kouadio *et al.* 2012).

Child malnutrition has been shown to increase in routinely flood-exposed communities throughout Odisha, India (Rodriguez-Llanes *et al.* 2016). Lack of sanitation or treated drinking water can contribute to a variety of bacterial, viral or parasitic infections leading to everything from low-level fatigue and reduced work capacity to severe debilitation and death. Diarrhoeal diseases and leptospirosis outbreaks have been reported in low-lying, poorer communities and urban areas after heavy rainfall events and during flooding throughout Asia, Latin America and

---

Box 41.1   Gender and Water in Jinping County, Yunnan Province, China

Sarah Opitz-Stapleton[1]

[1] INTASAVE Asia-Pacific, Beijing, China
Based on Opitz-Stapleton (2014).

Men can be the primary water fetchers in some rural communities, depending on ownership of modes of transportation. In a village in Jinping County, Miao and Zhuang men are the predominant owners of motorbikes. The village relies on a single community water tank. With sufficient rain, there is enough water to be gravity-piped into houses. During times of drought, families are only allowed to withdraw two small buckets of water every three to four days. This is insufficient for domestic needs, and men and boys fetch water from nearby springs, lakes and streams on their motorbikes. Many women, not owning motorbikes, are excluded from water fetching. They and girls still face more hygiene, health and food insecurity during droughts as males' water needs are often met first. Women and girls from ethnic groups with particular gender roles mediating reduced access to water, information, transportation and financial resources are facing greater water risks during periods of water shortage. They also have fewer opportunities for addressing the difficulties than their counterparts in other ethnic groups, where females have higher social status and more access to information and transportation. DRR efforts – including CCA initiatives – to improve water quantity and quality security among rural communities in south China need to include considerations of gender constructions of water risk and access to socio-economic resources among different ethnic groups. Failure to do so may exacerbate water disaster risks for women and girls.

Africa (Kouadio *et al.* 2012). Not everyone within an affected area will face the same exposure and/or vulnerability to water-related health hazards. Those with sufficient access to water services and ability to live and/or work in less water-related hazard-prone areas may have greater resilience.

This section briefly touched upon how local water cultures can enhance or decrease an individual's, household's or community's vulnerabilities, resiliencies and risks around hydro-climatological hazards, indicating the need for DRR including CCA interventions. Risks evolve through linked processes of social and environmental change, and the abilities of water cultures to develop and adapt sustainably. Exogenous influences at a variety of spatial and temporal scales also strongly shape local water cultures, particularly through the imposition and overlay of state water cultures and transnational water management and DRR, including CCA. The next section examines some sub-national to national water issues influencing sustainable development and hydro-climatological hazards, including those influenced by climate change.

Sub-national to National Water Cultures

National and subnational jurisdictions often exercise their power through the formalisation of water cultures in order to control access and use of the resource, and promote particular socio-economic development ideals (Hodgson 2004). A number of factors influence vulnerabilities and resiliencies at a variety of spatial and temporal scales within a jurisdiction. These include the types of policies governments enact and enforce; water-related institutions and degrees of coordination between them (e.g., planning and co-management between an irrigation agency and an urban planning agency); and, determination of 'priority' uses by particular actors to the exclusion of other users and uses. States' water cultures create and tackle shorter-term (<5 years) and longer-term (>5 years) disaster risk, including those influenced by climate change, and ultimately socio-economic development pathways.

Local water cultures are embedded within subnational and national water cultures. Aspects of each may be complementary or in conflict. State water cultures were often formed through a hybridisation of local institutions, practices and laws (Getches 2001). Historically, many subnational and national water policies and management practices were enacted through local practice with several aims, including (Getches 2001; Hodgson 2004; Opitz-Stapleton 2009):

1. Conflict mitigation between individual water users and various local water cultures, including the power to exclude some users, uses and customary practices;
2. Formalisation of state power over all areas within state boundaries, often by formalising water rights as property rights;
3. Accommodating a growing diversity of water uses from municipal, flood and drought 'control', to environmental protection, energy generation and industry beyond localised domestic and agrarian use; and,
4. Determination of 'beneficial' use favouring particular economic activities and broader development trajectories by prioritising some water users and uses over others.

The formalisation of state water cultures over local traditional water practices has led to both increases and reductions in various risks associated with hydrologic hazards through large-scale water infrastructure and institutions. In doing so, state water cultures have simultaneously contributed to vulnerabilities and resiliencies with respect to water, supporting or inhibiting DRR including CCA. States are able to undertake large-scale water projects and moderate competing uses in a manner that local water cultures are frequently unable to do. They do so through policies and infrastructure designed to control water allocation that attempt to 'hazard proof' water management. State water cultures respond to water-climate hazards that include heavy rainfall contributing to flooding and meteorological drought, either of which can extend over wide geographical areas for lengthy periods. These hazards frequently did and still do exceed some local water cultures' abilities to manage and provide water for domestic, livestock and agricultural uses, contributing to food insecurity in rain-fed dominant agriculture/livestock production areas (Zheng *et al.* 2015). For meteorological drought, state sponsored and managed water storage reservoirs and irrigation projects currently may enable intensive agriculture and livestock production, and continued village and urban functioning during times of drought across larger areas (Olsson *et al.* 2010). State operated water infrastructure also permits large cities to live in semi-arid and arid regions, like Los Angeles or Dubai, where such population concentrations would not normally be able to live with locally available water resources.

Along many of the world's major river systems, such as the Yellow, Yangtze, Mekong, Limpopo, Nile and Mississippi Rivers, widespread flooding associated with heavy rainstorms, seasonal precipitation and human encroachment and alteration of riparian ecosystems historically contributed to loss of life and assets, reductions in socio-economic wellbeing, and displacement (e.g., Dun 2011). The same state-operated water storage systems that provide some buffer against drought or aridity also partially attenuate some flood depths and velocities along river reaches, potentially lessening flood damages and losses for low-to medium-severity flood events. Flood control is often a stated policy objective of state water cultures in constructing storage systems, like the Three Gorges Dam and other reservoirs along the Yangtze River in China (Wang *et al.* 2015), whether or not that objective is legitimate or is achieved. State water cultures also construct and maintain medium- and large-scale hydropower stations, providing electricity and greater energy access to communities, ostensibly for development. Formalised water cultures enable and maintain existing food-energy-water nexuses (whether or not these are currently socially and ecologically sustainable), and can contribute to DRR, including CCA, when sustainably developed.

Yet, the very state-related water cultures that have contributed to socio-economic gains for many and reduced some hazard risks have also contributed to widespread environmental degradation and water overexploitation that now threaten previous gains and risk reductions. Historically, as long as population densities were low and supply was 'adequate' for the needs and uses of the population, conflict over water was reduced and could be mediated by the state (Meinzen-Dick and Nkonya 2007). As population densities and water demand(s) increase and diversify, particularly in areas of unequal water distribution and/or access, some state water cultures favouring individual property rights as promulgated historically may actually contribute to long-term water insecurity for all within an area by promoting overconsumption.

Many water rights specify the amount, timing and appropriate use of a water resource; rights holders may not be able to deviate from the specification without losing their right or incurring a penalty. For instance, in Colorado, USA, water rights are allocated under a system of prior appropriation. The system developed in the late 1800s in response to local water cultures centred on agriculture. It grants individuals or entities (e.g., cities) specific water amounts for a stated purpose; with the rights that were registered first being granted seniority (precedence) over subsequent rights (Getches 2001). A farmer may own a right from 1882 to use 20 acre-feet (the USA still uses imperial measurements) of water for irrigation per annum. The farmer must use the entire amount, even though her or his irrigation efficiency has improved and only a portion is now needed, or risk forfeiting the entire right. Environmental flows – water required to sustain aquatic ecosystems – were not accounted for at all when Colorado's water rights system was developed. As a result, the system of prior appropriation contributes to water shortages and inequalities. More water has been allocated through paper rights than physically exists in the streams or aquifers of Colorado. Additionally, physical water availability has been shifting with increasing climate variability and environmental change; older water rights do not account for overall, longer-term decreases or increases in the amount of water available.

Systems of water allocation under state water cultures are often slow to respond to shifting demographic and water demand changes, and to the projected changing frequency, timing, intensity and location of water hazards under climate change. The impacts of climate change are likely to compound pre-existing inequalities in water cultures and water quality and quantity management issues, including those related to disaster risk such as flood and drought management. Examples of rigid subnational and national water culture conflicting with actual water availability due to shifting human and natural systems are also evident in the systems of water rights contributing to groundwater overdraft for predominantly agricultural use throughout the Ogallala aquifer in the United States and aquifers in the Gangetic basin in India (Sekhri 2013; Steward *et al.* 2013). Another is the overdraft of water resources and water pollution, coupled with numerous droughts since the 1950s, which threaten income, food security and DRR measures including CCA in some communities in China's arid north (Zheng *et al.* 2015).

Overdraft and socially unequal allocation are problematic in many countries and can reduce water security, impacting DRR including CCA, as highlighted in the case study in Box 41.2. Existing state-promulgated water rights systems may no longer support water allocation or serve as robust drought-risk mitigation strategies over the medium to long term in heavily depleted and or polluted aquifers – or in areas that may experience widespread decreases or variability in precipitation due to climate change. States can, however, promote more sustainable and resilient water use through systems of sharing or trading of water amounts and/or timing of allocations. Preferences leading to dedicated water resources for ecosystems, and allocating water rights for them, can introduce more flexible DRR including CCA into local to state water cultures.

Box 41.2   State Water Cultures and Water Security in Jaipur, India

Sarah Opitz-Stapleton[1]

[1] ISET, Boulder, Colorado, USA
Based on ISET and CEDSJ (2011).

The city of Jaipur in arid Rajasthan, India, encapsulates the challenges facing state water cultures in sustainably managing water resources in the face of dynamic human–natural systems interactions and shifting processes. Rural-to-urban migration rates are high and the city is growing rapidly. Jaipur relies heavily on groundwater resources for much of its needs. Water supplies are limited and often of low quality. Groundwater mining – both in the city by households due to inadequate services and in rural areas for irrigation – over many decades has heavily depleted local and district-level aquifers. Water quality and quantity are also impacted by pollution and loss of recharge zones due to urbanisation accommodating in-migration from the rural locations. The Rajasthan government constructed Bisalpur Dam, completed in 1999, to become Jaipur's principal water source from 2010 onward and to reduce city water shortages during times of meteorological drought and the dry season. As of 2011, the dam had only filled nine times since becoming operational in 1994. Increasing variability in the Indian Summer Monsoon, coupled with agricultural abstractions and competing demands on the Biswas River that supplies the reservoir, mean that Jaipur's water security is tenuous, particularly in dry years. Farmers and communities in the surrounding districts of Tonk, Ajmer and Bhilwara have been protesting that they were excluded from receiving reservoir water even as they face increasing water shortages. Lack of inclusion, albeit in an admittedly flawed system, decreases their disaster resilience and adaptive capacity during drought, heat waves and the dry season. The state government did not originally account for competing factors and risk drivers, including climate change, that may alter the timing, duration, amount and location of monsoon rainfall throughout Rajasthan, when designing the reservoir system and allocation policies. The slow nature of state institutions in incorporating climate change, accounting for shifting demographics and water demands, or continued groundwater depletion as part of their DRR strategies in water management have stressed inequalities in local water cultures. The state and district governments are now attempting to address these challenges at multiple administrative levels of formal water cultures.

Likewise, state water cultures relying heavily upon inflexible infrastructure for flood mitigation may find that such systems are inadequate at coping with shifting precipitation and/or snow-melt regimes. Rigid water infrastructure was designed to accommodate specific volumes of water demand and precipitation and/or flow thresholds, many of which were assumed to not vary greatly with time. Many reservoirs were designed to moderate specific flood thresholds. Exceeding these levels due to more intense rainfall over a shorter time period, for example, can overwhelm the reservoir's capacity to buffer the flood. Water infrastructure may not provide sufficient storage to cover existing demands during periods of prolonged meteorological drought and/or high temperatures. Infrastructure is also not adept at handling increasing development in and modification of waterways or evolving water priorities or demands. While constructing additional or retrofitting existing water infrastructure may play a role in future DRR strategies, including CCA, it may not provide as much 'climate-proofing' as desired or meet dynamic demand. In such instances, institutional inertia in state water cultures can slow the development of flexible, robust and sustainable non-structural responses to climate variability and change, thereby increasing the vulnerability of both people and ecosystems in the face of hydro-climatological hazards.

Additionally, widespread increases in vulnerability to natural hazards, including water-climate hazards, are partially due to highly sector-siloed approaches in water management between various state actors. This includes the unnecessary separation between DRR and CCA, whereas CCA should sit within DRR. While water is a foundational resource, its management is often split between multiple government agencies at the subnational levels and national levels according to sector. States moderate water use for urban consumption and wastewater discharge, agricultural and industrial diversions, energy generation, or transport and navigation of people and goods along waterways. Yet, co-management between various subnational and national agencies around water can be fragmented. In some instances, different agencies may conduct water planning and operations for DRR than those researching and making recommendations around water for CCA planning or economic development. Principles of DRR including CCA may not be mainstreamed into the planning and operations of various state institutions involved in or dependent upon water.

As a result, waterways can become clogged with silt, sewage and solid waste; additional diversions of water planned for a city or agricultural area; or new infrastructure constructed without considering dynamic social and climate processes. Levels of coordination and co-management of disaster risk, including risks influenced by climate change, across sectors with interests or influence on water management have direct and indirect water-climate risk repercussions. Siloisation of various aspects of water culture and DRR, including CCA, can enhance exposure to flood, drought or storm surge (Friend *et al.* 2014). As a result, construction and settlements may continue in floodplains or behind levees. Existing DRR strategies may become inadequate, particularly for CCA, due to reliance upon inflexible dams, flood levies and water structures constructed and managed by siloed state water cultures even as hydro-climatological regimes shift (ISET and CEDSJ 2011) – fundamentally also because they tend to assume implicitly that social and environmental baselines are not changing, which is clearly not the case.

In fact, isolated state institutions and water management approaches frequently ignore multiple human–ecological systems interactions and dependencies (Gober and Wheater 2014), can promote risky behaviours and contribute to poor DRR including maladaptation for climate change. Nowhere is this more evident than in the construction of large-scale water related infrastructure like reservoirs and irrigation systems, or planning and promoting unsustainable urban, agricultural or economic growth. Decisions and actions taken by states have long lifetimes. Reservoir and city design decisions, for example, impact water quantity, quality and hazard zones for decades. Ideally, CCA would be enfolded within DRR, and both would become an integral part of water planning and management across all agencies and embedded in more flexible and sustainable state water cultures. The aims and approaches of both are similar, but keeping CCA efforts separate from DRR within water cultures undermines and ignores the decades of research, experience and practice gained through DRR in creating and managing multiple hazard risks, and may magnify some water-climate risks through space and time. Sub-national and national water cultures not only have significant impacts on local water cultures, but also strongly influence water management, sustainable development and hydro-climatological risk (short and long term) at transnational scales, as explored in the next section.

Transnational Water Management and Risk

The water cultures at sub-national and national scales can have disproportionate impacts on transboundary water systems, strongly influencing water management and the hydro-climatological risk of upstream and downstream nations. States seek to maximise socio-economic development and gains for their populations and industries, drawing on water resources within their boundaries and often attempt to divert or store as much as possible for their own use. Multi-national interactions and competition over water resources along and in major surface water locations (e.g., the Nile, the Colorado, the Amazon, the Indus, and the Aral Sea Basin (e.g., Glantz 1999)) and aquifer systems have occurred for thousands of years. Today's water interactions – cooperation, conflict, and neutrality – are not new; only current national boundaries are new in comparison with humanity's long relationships with and interactions over such transboundary systems (Zeitoun and Mirumachi 2008). What is also new is the unprecedented exploitation of water resources within jurisdictions and across watersheds including large-scale ecosystem degradation, all of which collide with hydro-climatological variability and change at a pace which large human populations might not have experienced before at such widespread spatial scales. How governments and communities cooperate or not on management and usage of the transboundary water resources within their respective borders has implications for risks to all nations that share a particular watershed.

Many state water cultures continue to unilaterally pursue water management and use courses to promote their respective and interrelated food, energy and industrial development goals. In doing so, they may perpetuate and exacerbate water quality and quantity insecurity within both their own countries and for other countries that share the water resource. States continue to construct and operate large-scale reservoirs for hydropower, for water supply to meet growing demands (often irrespective of population changes) and for flood and drought risk management. They may undertake these actions in spite of or working within existing treaties with other riparian jurisdictions; most major transnational water systems, including the Columbia, the Amazon and the Zambezi rivers, are delicately managed through a balance of multiple state water cultures.

At the same time, climate change is beginning to have and will have a wide range of impacts on transboundary water systems, and a number of implications for the joint management of such systems. Dry season flows in snow-fed rivers like the Ganges (which spans several South Asian countries) or Amu Darya (flowing through multiple countries in Central Asia) may decrease as precipitation becomes more variable and/or glaciers and snowfields retreat. Intense, heavy rainfall events may increase over many areas, including those likely to experience an augmented incidence of meteorological drought, thereby also exacerbating flood risk. This projected greater climate variability poses significant challenges to transnational water system management for disaster risk while ensuring treaty-negotiated flows and uses for adjoining locations. Approaches to promoting multinational cooperation and coordination on watershed ecosystems protection, water use efficiency and DRR, including CCA, are needed to address uncertain and growing risks. One of the world's major transboundary rivers, the Colorado, provides a case study (Box 41.3) highlighting the challenges around reconciling subnational and national water cultures for DRR, including CCA, in coordination with different countries' water cultures.

As Box 41.3 highlights, local and state water cultures often conflict with transboundary water management in rapidly shifting socio-economic, environmental and climate contexts – especially when management aims for DRR including CCA, irrigation, urban water use and hydropower remain separated or poorly coordinated across agencies and organisations within a country. The legal and institutional frameworks at sub-national and national scales evolved to support particular economic development aims and sectors over others. Such water cultures often developed many decades previously, particularly when water demands were less than today. State water cultures can be inflexible and slow to respond to changing conditions, such as demographic shifts including urbanisation, increasing pollution and ecosystem degradation, engineering of the environment and climate change impacts. State water culture issues are magnified further in transboundary water management, particularly with in-country DRR and CCA efforts remaining separated. Siloed approaches to DRR and CCA in transnational water negotiations further hinder efforts toward sustainable and resilient water management, irrespective of multinational treaties, negotiations and management bodies, which exist for many transboundary rivers, such as the Mekong.

Box 41.3   The Colorado River: State Water Cultures Impacting Transnational Water Management

Sarah Opitz-Stapleton[1]

[1] Independent, Boulder, Colorado, USA

The Colorado River drains seven states and Native American lands in the southwestern USA and two states in northwest Mexico. The river is highly managed with a complex system of reservoirs (including the multi-year storage Lake Mead and Lake Powell) and diversions along its semi-arid to arid reaches – and is often described as one of the most over-allocated water systems in the world (Fuller and Harhay 2010). Its waters are abstracted to supply burgeoning urban populations, extensive agricultural irrigation and livestock production, hydropower, mining and shale gas exploitation, and growing recreational opportunities. A complex legal framework, the Law of the River, consisting of individual state water laws, interstate compacts and treaties amongst Mexico, the USA and indigenous peoples, effectively governs reserved water rights, apportionment, priorities and usage. The two pieces of seminal legislation within the Law are the Colorado River Compact (1922) and the 1944 treaty with Mexico that granted 10 per cent of the flows to Mexico, contingent on drought conditions along the river. The Compact was negotiated during a period of above-average flows in the river, and led to an apportionment of flows on paper (16.4 million acre-feet) between the upper and lower-basin states that did not exist as actual water in the river (long-term annual average flows are between 13.5 and 14 million acre-feet). This convoluted set of sub-national and transnational water cultures has created high degrees of vulnerability in the basin.

Human and natural systems within the Colorado River basin are highly exposed to prolonged, multi-year to multi-decadal droughts, which tree rings and other palaeoclimatic data proxies indicate have occurred numerous times over the past few millennia (Woodhouse *et al.* 2010). The southwestern USA also has a rapidly growing population, largely due to in-migration from other US states (Fuller and Harhay 2010). Cities like Los Angeles, Phoenix and Las Vegas are expanding rapidly, causing distributional conflicts with farmers and ranchers, among other historically prioritised uses. Due to the drought contingency clause in the 1944 treaty with Mexico, the Colorado River has flowed into its delta, the Sea of Cortez, only periodically over the past 60-odd years. A new bi-national pact, Minute 319, was signed in 2012 to provide additional deliveries of water to Mexico when reservoir elevations at Lake Mead exceed a certain level on an interim basis through 2017 (IBWC 2012).

The viability of this treaty and others with Mexico, along with the interstate and tribal treaties has been threatened by prolonged, widespread regional meteorological and socio-economic drought since about 2004, alongside growing demand. Climate change is likely to exacerbate meteorological and hydrological drought in the basin (Fuller and Harhay 2010). Some US states and tribes are exploring plans and options for water management along the Colorado River in the face of current and future stressors, although DRR and CCA considerations sometimes remain separated at different administrative levels and in planning and practice. Subnational, national and international water cultures pertaining to the Colorado River are slow to respond to the rapidly shifting human and ecosystem vulnerability and resilience processes plus changing hazards regimes, such as those influenced by climate changes within the basin.

Short-to Long-term Implications – Development through DRR including CCA

This chapter broadly summarised aspects of water cultures at a variety of spatial and temporal scales, and how the interactions amongst various water cultures simultaneously contribute to differentiated vulnerabilities, resiliencies and hydro-climatological risks.

Climate change is adding a layer to the continual changes affecting the frequency, intensity, duration and spatial extent of extremes (e.g., heavy rainfall or heat waves) and longer hazards (e.g., shifts in seasonality, multi-year oscillations such as El Niño and sea-level rise). The evolution in local to global climates is only one process contributing to dynamic water-climate risks. Local water cultures historically developed to take advantage of local water resources and climates; adopting DRR mechanisms, including for CCA, through practices ranging from irrigation to moving livestock herds to new water holes and fodder areas to trading water through virtual flows in commodities (although this point was not addressed in the chapter). Unequal access to water – in terms of quantity, quality, timing and location – created and creates differentiated vulnerabilities and capacities for coping with hydro-climatological hazards and their direct and indirect impacts. Depending on the local social and environmental context, some elements of various local water cultures may continue to serve as flexible and robust DRR measures capable of reducing the risk of water disasters, including those influenced by climate change.

At times, the severity and extent of hydro-climatological hazards have overwhelmed the abilities of local water cultures. State water cultures arose to consolidate state land and natural resource rights, as well as for DRR purposes. In doing so, state water cultures have simultaneously enhanced and aggravated local capacities and inequalities in water-climate risks. Additionally, state management and planning across sectors using significant amounts of water and/or impacted by shifts in water quantity and quality – agriculture, urbanisation, DRR and power generation – can be siloed between different agencies and departments. This jurisdictional separation can augment or create new hydro-climatological risks for local and state water cultures, and may hinder flexibility in responding to or being proactive about existing and emerging risks not only within local water cultures but also in transnational water system management.

While states are beginning to respond to climate change and to develop national adaptation strategies or plans of action, there is sometimes poor integration of such CCA approaches within existing DRR frameworks and strategies, especially with a frequent drive to keep CCA separate from DRR. DRR, including CCA, is not always effectively mainstreamed into policy and practice among different state actors involved in water management or water-related sectors like energy generation, health or agriculture. Additionally, state climate risk assessments and CCA strategies have not always accounted for shifting land use development, economic migration or changing economic structures, followed by the impacts of these factors on water quantity and quality, along with the creation of water risks and disasters. The failure to integrate CCA within DRR, and then to merge both as key elements of sustainable development, reduces flexibility and resilience within water cultures for responding to changing social and environmental conditions. Rather than treating them as separate concepts and practice, it is important that water cultures from the local to transnational levels be able to embrace DRR as including CCA.

References

Ahmed, S. (2004) *The Gendered Context of Vulnerability: Coping/Adapting to Floods in Eastern India*, Paper for Gender Equality and Disaster Risk Reduction Workshop 10–12 August 2004, Honolulu, Hawaii: East-West Center.

Balali, M., Keulartz, J. and Korthals, M. (2012) 'Reflexive water management in arid regions: The case of Iran', in S. Johnson (ed.) *Indigenous Knowledge,* Cambridge: The White Horse Press, pp. 137–154.

Bebbington, A. (1999) 'Capitals and capabilities: A framework for analyzing peasant viability, rural livelihoods and poverty', *World Development* 27, 12: 2021–2044.

Brown, O., Crawford, A. and Hammill, A. (2006) *Natural Disasters and Resource Rights: Building Resilience, Rebuilding Lives,* Manitoba, Canada: International Institute for Sustainable Development.

Budds, J. (2004) 'Power, nature and neoliberalism: The political ecology of water in Chile', *Singapore Journal of Tropical Geography* 25, 3: 322–342.

Cozzetto, K., Chief, K., Dittmer, K., Brubaker, M., Gough, R., Souza, K., Ettawageshik, F., Wotkyns, S., Opitz-Stapleton, S., Duren, S. and Chavan, P. (2013) 'Climate change impacts on the water resources of American Indians and Alaska Natives in the U.S.', *Climatic Change* 120: 569–584.

Dun, O. (2011) 'Migration and displacement triggered by floods in the Mekong delta', *International Migration* 49, supplement s1: e200–e223.

Friend, R., Thinphanga, P., MacClune, K., Henceroth, J., Tran, P. and Ngheim, T. (2014) 'Urban transformations and changing patterns of local risk: Lessons from the Mekong Region', *International Journal of Disaster Resilience in the Built Environment* 6, 1: 30–43.

Fuller, A. and Harhay, M. (2010) 'Population growth, climate change and water scarcity in the Southwestern United States', *American Journal of Environmental Science* 6, 3: 249–252.

Getches, D. (2001) *Water Law in a Nut Shell,* St Paul, USA: West Publishing Company.

Glantz, M.H. (ed.) (1999) *Creeping Environmental Problems and Sustainable Development in the Aral Sea Basin,* Cambridge: Cambridge University Press.

Gober, P. and Wheater, H. (2014) 'Socio-hydrology and the science-policy interface: A case study of the Saskatchewan River basin', *Hydrology and Earth Systems Science* 18: 1413–1422.

Hodgson, S. (2004) *Land and Water – the Rights Interface,* Food and Agriculture Organisation of the United Nations (FAO) Legislative Study 84, Rome: FAO.

Hunter, P. (2003) 'Climate change and waterborne and vector-borne disease', *Journal of Applied Microbiology* 94: 37S–46S.

IBWC (International Boundary and Water Commission). (2012) *Minute No. 319: Interim International Cooperative Measures in the Colorado River Basin through 2017 and Extension of Minute 318 Cooperative Measures to Address the Continued Effects of the April 2010 Earthquake in the Mexicali Valley, Baja California,* United States and Mexico: IBWC.

ISET (Institute for Social and Environmental Transition) and CEDSJ (Centre for Environment and Development Studies). (2011) *The Uncomfortable Nexus: Water, Urbanisation and Climate Change in Jaipur, India,* Boulder: ISET and Jaipur: CEDSJ.

Kouadio, I., Alijunid, S., Kamigaki, T., Hammad, K. and Oshitani, H. (2012) 'Infectious diseases following natural disasters: Prevention and control measures', *Expert Review of Anti-Infective Therapy* 10, 1: 95–104.

Meinzen-Dick, R. and Nkonya, L. (2007) 'Understanding legal pluralism in water and land rights: Lessons from Africa and Asia', in B. van Koppen, M. Giordano and J. Butterworth (eds) *Community-based Water Law and Water Resource Management Reform in Developing Countries,* Boston, MA: Centre for Agriculture and Biosciences International (CABI) Publishing, pp. 12–27.

Olsson, O., Ikramova, M., Bauer, M. and Froebrich, J. (2010) 'Applicability of adapted reservoir operation for water stress mitigation under dry year conditions', *Water Resource Management* 24: 277–297.

Opitz-Stapleton, S. (2009) *Political Ecology of Safe Drinking Water in the United States with a Case Study Focus on Puerto Rico,* Doctoral Thesis, Boulder, Colorado: University of Colorado at Boulder.

Opitz-Stapleton, S. (2014) *Understanding Gendered Climate Variability and Change Impacts in Jinping and Guangnan Counties, Yunnan Province, China,* Asia-Pacific: INTASAVE.

Ray, I. (2007) 'Women, water, and development', *Annual Review of Environment and Resources* 32: 421–449.

Rodriguez-Llanes, J., Ranjan-Dash, S., Mukhopadhyay, A. and Guha-Sapir, D. (2016) 'Flood-exposure is associated with higher prevalence of child undernutrition in rural eastern India', *International Journal of Environmental Research and Public Health* 13: 210–221.

Sekhri, S. (2013) *Missing Water. Agricultural Stress and Adaptation Strategies in Response to Groundwater Depletion in India, No. 406,* Virginia, USA: University of Virginia, Department of Economics.

Steward, D., Bruss, P., Yang, X., Staggenborg, S., Welch, S. and Apley, M. (2013) 'Tapping unsustainable groundwater stores for agricultural production in the High Plains Aquifer of Kansas, projections to 2010', *Proceedings of the National Academy of Sciences of the United States of America (PNAS)* 110, 37: E3477–E3486.

UNWCED (United Nations World Commission on Environment and Development). (1987) *Report of the World Commission on Environment and Development: Our Common Future,* Oxford: Oxford University Press.

Wang, G., Nadin, R. and Opitz-Stapleton, S. (2015) 'A balancing act: China's water resources and climate change', in R. Nadin, S. Opitz-Stapleton and Y. Xu (eds) *Climate Risk and Resilience in China*, London and New York: Routledge, pp. 96–128.

Wisner, B., Gaillard, JC, and Kelman, I. (2012) 'Framing disaster: Theories and stories seeking to understand hazards, vulnerability and risk', in B. Wisner, I. Kelman and JC Gaillard (eds.) *The Routledge Handbook of Hazards and Disaster Risk Reduction*, Abingdon: Routledge, pp. 18–33.

Woodhouse, C., Meko, D., MacDonald, G., Stahle, D. and Cook, E. (2010) 'A 1,200-year perspective of 21st century drought in southwestern North America', *Proceedings of the National Academy of Sciences of the United States of Amercia (PNAS)* 107, 50: 21283–21288.

Zahran, S., Brody, S., Peacock, W., Vedlitz, A. and Grover, H. (2008) 'Social vulnerability and the natural and built environment: a model of flood casualties in Texas', *Disasters* 32, 4: 537–560.

Zeitoun, M. and Mirumachi, N. (2008) 'Transboundary water interaction I: reconsidering conflict and cooperation', *International Environmental Agreements – Politics, Law and Economics* 8, 4: 297–316.

Zheng, Y., Meng, H., Zhang, X., Zhu, F., Zhanjun, W., Shuxing, F., Opitz-Stapleton, S., Jiahua, P., Zhongyu, M., Jianmin, F., Shangbai, S., Jianrong, F., Xinlu, X., Nadin, R. and Kierath, S. (2015) 'Ningxia', in R. Nadin, S. Opitz-Stapleton and Y. Xu (eds) *Climate Risk and Resilience in China*, London and New York: Routledge, pp. 213–241.

42

FOOD IN THE CONTEXT OF DISASTER RISK REDUCTION INCLUDING CLIMATE CHANGE ADAPTATION

John Campbell

Introduction

The second group of Sustainable Development Goals (SDGs) agreed upon by member states of the UN in 2015 has as its first target (2.1) to 'end hunger and ensure access by all people, in particular the poor and people in vulnerable situations, including infants, to safe, nutritious and sufficient food all year round' by 2030. This is a worthy goal and on the face of it an achievable one, given that global average daily food production is currently 5,000 Kcals per capita, almost triple the daily requirements of 1,800 Kcals per person (Mucke 2015). There are, however, major obstacles to enabling equitable access to global food resources and these lie largely in social, political and economic processes rather than in the environmental sphere. Moreover, the underlying factors that cause much of the global food insecurity are also the causes of vulnerability to natural hazards and hazard drivers such as climate change. Significantly, the sustainable development goals also include the following target (13.1): 'Strengthen resilience and adaptive capacity to climate-related hazards and natural disasters in all countries.' Herein lies the key conundrum facing those seeking to improve food security, reduce disaster risk and achieve equitable ways of adapting to climate change. All three are fundamentally about social, political and economic rather than environmental processes: while scientific research and technical innovation (e.g., modified seeds, new fertilisers and new tools) are important, the main issues relate to equity and fairness at both local and global scales.

This chapter is about food in the context of disaster risk reduction (DRR) including climate change adaptation (CCA). Food and food security are very complex issues and have their own specialised literature. But in this chapter the focus will be on how disaster risk and climate change vulnerability can be affected by existing conditions of food security, how food security can be affected by climate change and natural hazards and how in turn DRR including CCA can improve food security especially in those places deemed most at risk. The three elements are tightly connected and these linkages are very complex. Accordingly, focussing on climate variability and change allows only a partial analysis of the causes of food stress. Undoubtedly, preserving food security in the contexts of DRR including CCA will require social, political and economic responses.

Before embarking on this discussion of DRR including CCA it is important to acknowledge that climate change, with the exception of some high latitude areas that may experience increased productivity, is very likely to set in motion environmental changes that may well threaten sustained food production at levels needed to provide the nutritional, social, economic and cultural needs of the world's populations (Porter *et al.* 2014). It follows that substantial reductions of greenhouse gas emissions are a matter of considerable urgency, if the worst effects of climate change are to be averted, and many of these worst effects will have serious implications for food security. It is also clear that while there have been some recent steps to reduce greenhouse gas emissions (e.g., the climate change COP21 meeting in Paris in 2015) they have been very much too little, too late.

Disasters often lead to food crises, but in many cases vulnerable food systems are among the very causes of disaster. This chapter will begin with the assumption that sustaining food security is a critical goal of DRR including CCA. It will examine traditional systems of food security and illustrate how histories of colonialism, independence and globalisation have caused food security decline in many countries over the past century or longer. At its very basic manifestation, the most dangerous effects of climate change are likely to be on food production. Yet, the complexity of the relationship is such that many contemporary food production systems (re)produce the conditions of vulnerability to natural hazards and climate change that many communities around the globe are experiencing.

The World Food Summit Plan of Action (1996) provided a definition of food security that has become widely adopted. 'Food security exists when all people, at all times, have physical and economic access to sufficient, safe and nutritious food to meet their dietary needs and food preferences for an active and healthy life.' The FAO (2006) outlines the four dimensions of food security that support this definition. They are 1) food availability, which is largely related to local agricultural and fisheries production or imports, 2) access to food (entitlements), 3) utilisation of safe and healthy food and 4) temporal stability (protection against environmental, economic, political or other crises) of access to food. Christoplos (2012) noted that these four main pillars of food security (availability, access, stability and safety) are all in decline. Climate change has been implicated in the decline in food availability and is expected to have significant impacts in coming years and decades but its links to the other three elements are much less direct. Moreover, these processes may be exacerbating the effects of deteriorating food production. Table 42.1 summarises the links between the four pillars of food security, climate change and natural hazards (shown in the third column as changing patterns of climate extremes).

As many writers (e.g., Wisner *et al.* 2004) point out famines are often the tip of the iceberg in the context of food security and in periods when famines are not occurring millions around the world suffer from malnutrition and there are millions of fatalities that can be attributed to chronic or persistent food insecurity. One may ask, what is the point of achieving DRR or CCA that sustains the status quo for these people in their everyday lives (Lewis 1999)? As chronic and persistent food insecurity are symptoms of the underlying causes of vulnerability, if they are not adequately addressed, neither will DRR including CCA be successfully achieved. This is an important area of contention for those supporters of DRR who express concern over definitions of CCA which focus on maintaining social, economic and environmental conditions as if they were unmodified by a changing climate.

Table 42.1 Climate Change (Mean Conditions and Variability and Examples of its Effects on Food Security)

	Changing Means	*Changing Variability (Extremes: Magnitude and/ or Frequency)*
	Examples:	Examples:
	Increased atmospheric carbon	Droughts
	Increased atmospheric temperatures	Floods
	Sea level rise (SLR)	Tropical Cyclones
	Increased sea surface temperatures (SST)	Heat waves
	Increased ocean acidity	
Availability	Increased availability of carbon dioxide may increase productivity of some crops and weeds but may be negated by other effects of CC. Heat stress may reduce productivity of crops and animals. Sea level rise may cause the loss of agricultural land and crops and SLR, and increasing SST and acidity may degrade coral reef ecosystems (critical marine food resource areas).	All types of extreme event may reduce food productivity. Some impacts on crops and livestock may extend for several years and across wide areas. Where events occur more frequently and/or with greater magnitude the recovery periods for food production may be reduced. Some extreme events are likely to exacerbate the effects of secular change outlined to the left such as storm surge during tropical cyclones increasing loss of land, soil salinisation and crop destruction.
Access	Secular changes may reduce livelihoods as a result of reductions in productivity and also contribute to increasing food prices.	Increasing extreme events are likely to produce food price volatility and increased unreliability of livelihoods in both food and non-food sectors. Extreme events may disrupt transport networks (e.g., bridges washed out and droughts making waterways unnavigable).
Stability	Possibility that some thresholds may be crossed which will lead to severe ruptures in food availability.	Increasing frequency and/or magnitude of extreme events will increase the periodic disruption of food availability. Food crises may occur more often and/or with greater severity.
Utilisation (Safety)	Warmer temperatures and increased humidity may reduce the safe lifetimes of foods and increase the incidence of bacterial, viral, parasitic and fungal contamination of crops and animal food products.	Flood waters may contaminate foods. Crops and livestock under drought conditions may have reduced nutritional value.

Food and Disasters: Cause and/or Effect?

Food is often a central issue in many, though not necessarily all, disasters. Most directly, disasters impact upon food production and in turn food availability. This is especially the case with meteorological extremes such as droughts, tropical cyclones, frosts, floods and the like. The type of disaster most commonly associated with disruption of food systems is famine. But, famine has a number of explanations and they do not necessarily include the occurrence of an extreme environmental event and even where food production is disrupted by extremes (such as droughts, floods and locust plagues), other social, political and economic factors also have a significant role (Box 42.1). Such events can have direct impacts upon the agriculturalists whose crops are affected and who

Box 42.1   The Multiple Causes of Famine

John Campbell[1]

[1] University of Waikato, Hamilton, New Zealand

Famines are often considered to be among the worst forms of disaster with millions of people facing food shortages and starvation. But, famines are both notoriously difficult to define and to explain. The term is not included in the UNISDR 'Terminology on Disaster Risk Reduction' (UNISDR 2009) and is also excluded from the glossary for the IPCC (2007). According to the IPCC (2014) a famine is:

> Scarcity of food over an extended period and over a large geographical area, such as a country, or lack of access to food for socioeconomic, political, or cultural reasons. Famines may be caused by climate-related extreme events such as droughts or floods and by disease, war, or other factors.

This definition is an expanded version of the one used in IPCC (2012) which excluded 'or a lack of access to food for socioeconomic, political or cultural reasons'. The expanded version acknowledges the multiple causes of famine including those which are underlying rather than just the proximate.

Early theories of famine causation tended to focus on the natural processes that were believed to give rise to widespread starvation. In particular, famine occurrence was attributed to droughts (and less commonly to other natural extremes such as floods and insect infestations) and resulting declines in food productivity. Such theories were sometimes referred to as meteorological and in this sense famines were constructed simply as 'natural' disasters. But, seeking to understand famine led to significant changes in thinking about disasters. Because rich people never die from famines, purely environmental theories of causation could not be sustained. There had to be social, economic or political processes at work as well (see Devereux 1993 for a comprehensive outline of famine theories).

Indeed, understanding of the complex social causality of famines grew during the Sahelian famine of the 1970s. A number of social scientists from different disciplines showed that the reasons for the famine were rooted in colonial history, class politics and development practices amongst other socio-political processes (e.g., Copans 1975). Development strategies that promoted commodity production for export often forced food production onto increasingly marginal lands setting in motion a process that Glantz (1994b) described as droughts following the plough. This played a major role in disaster theorising about 'Taking the naturalness out of natural disasters' (O'Keefe *et al.* 1976) that underpins DRR to the present time and could be more incorporated into CCA, highlighting the importance of DRR informing CCA. While food availability decline cannot be dismissed as a factor contributing to famines, Sen (1981) showed in his work that famines could occur when food production had not declined but people's access to food was diminished. He introduced the concept of food entitlement decline (FED), which is sometimes now described as failure of exchange entitlements (FEE). So, to understand the complex processes of famine requires examination of the political ecology of food systems around the world (Glantz 1976).

are likely to sustain reduced subsistence and/or cash livelihoods. Other disaster events may have a relatively minor direct impact on food production. For example, crop damage is not necessarily a consequence of geological hazards like earthquakes and tsunami (though landslides triggered by earthquakes may destroy crops and tsunami may destroy coastal agricultural systems). But, even where crops, livestock or fisheries are unaffected, many people may lose their livelihoods and/or have other financial pressures that restrict their access to food.

As with other types of 'natural' disasters droughts are only likely to become famines when drought conditions affect a vulnerable population (Wisner *et al.* 2004). This vulnerability may be manifested in agricultural systems (crops grown – or not grown–soil conditions, land tenure, etc.) or social, political or economic conditions. Typically, actions to reduce famine revolve around the physical or technical aspects of the agricultural system (drought resistant crops, irrigation, etc.) or the provision of food relief (often not before conditions have become alarming), but rarely tackle the causes that underlie the vulnerability.

As noted, natural extremes do not only affect food production. They may affect access to food by disrupting storage (e.g., disrupting livelihoods, damage to stockpiles, electricity shut down and damage to freezers and refrigerated storage) and transport systems (road and rail damage, destruction of bridges etc.) and causing food prices to escalate. These occurrences may also impinge upon food safety. For non-food producers, the impacts of reduced food production are likely to be felt through increases in prices. In all cases the outcomes are likely to be increasing levels of widespread malnutrition and in the worst cases starvation.

Food, Vulnerability, and Resilience: A Brief History

Little has changed since Baker (1987), Copans (1975) and Glantz (1976) in their work on famine in Africa, considered the social, economic and political processes that dated back to early colonialism on the continent through to the more recent roles of experts from multi-lateral organisations and the drive to produce agricultural commodities for export as having played an important role in breaking down existing structures of independence and relative food security. In the Pacific Islands a similar process can be identified (Box 42.2), which has seen a trend emerge from pre-colonial resilience and interdependent self-sufficiency of island communities to one of contemporary food insecurity and vulnerability to environmental hazards such as tropical cyclones and droughts and also to the longer-term effects of climate change.

Box 42.2   From Resilience to Vulnerability: Community Food Systems in Oceania

John Campbell[1]

[1] University of Waikato, Hamilton, New Zealand

Figure 42.1 shows how there have been changes in the food systems on Mota Lava, a small island in northern Vanuatu that is often exposed to tropical cyclones and droughts. Prior to early European contact the population of around 2,000 had a robust agro-ecosystem with resilient crops (especially to strong winds), a diverse range of cultivars and a range of famine foods (e.g., sago and wild yams) that were mostly consumed only when the main crops were damaged. Because agricultural production was seasonal, methods were developed that enabled crops to be stored and where necessary preserved. These stocks took on an important role after major weather extremes. Food storage was based on surplus production which also underpinned complex networks of exchange and cooperation. Links were established and maintained through ceremonial exchanges that included feasting. These links, across many islands in northern Vanuatu, could be called upon for assistance when disasters occurred.

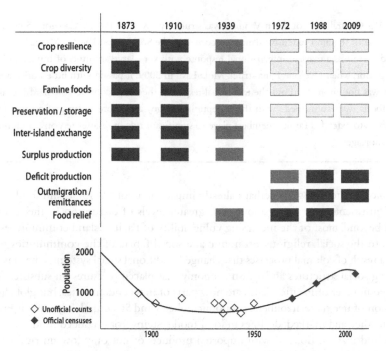

Figure 42.1 Food Security and DRR in Mota Lava, a Small Island in Northern Vanuatu

(After Campbell 2006)

Note: From the original citation, dark is high food security, grey is medium food security, and light is low food security

    With colonisation came a number of changes. People were encouraged to enter the cash economy mostly by expanding coconut plantations to enable the production and export of copra. This occurred especially in the mid-to late twentieth century. Missionaries restricted the cultural practices that were associated with the systems of inter-island exchange. Capitalism saw increasing imports into the island including rice which together with the decline in traditional exchange reduced the need for surplus production and storage. As demand for cash incomes grew, coconut plantations spread onto fertile land that was until then used for food crops. As a result, the land devoted to food production declined while commodity production increased. This was made possible without major impacts on food security because of population decline (mainly through introduced diseases), which was not halted until the mid-twentieth century. In 1939, disaster relief was received for the first time and people no longer used famine foods and no sago, for example, was prepared for consumption. By 1972 when Cyclone Wendy occurred, there was a strong expectation of relief from the government, which was met, and traditional means of coping with disaster were no longer turned to.

    With population growth in the second half of the century there has emerged a food security crisis with insufficient land available for food production. Shortened fallow periods have reduced soil fertility and traditional agro-diversity and crop resilience have declined as the less resilient introduced crop, cassava, has come to dominate subsistence production, almost as a monoculture. In recent decades the returns on copra have declined considerably so that the island has a major deficit in its balance of trade. There is insufficient subsistence food produced, creating increased demand for imported food, but returns from copra which are often close to nil, are insufficient to cover the costs. This has resulted in increasing outmigration and remittances from migrants support the island economy.

In 2015 Central and Southern Vanuatu was struck by Cyclone Pam, a Category 5 cyclone that caused some of the most extensive damage recorded in the South West Pacific. Media attention was high and there was a massive international response with very large amounts of humanitarian food assistance. But, when Cyclone Funa struck Mota Lava in 2009, it passed with little notice and disaster relief was not given. Without the traditional responses there was considerable hardship, although famine foods were consumed (with the exception of sago) and more outmigration was initiated. Remittances to extended family members played a significant role in disaster response and mitigation of food shortages.

But while climate change is perhaps already impacting upon Pacific Island food systems and, without adjustments, may bring about even greater levels of disruption in the relatively near future and beyond, most of the increasing vulnerability of Pacific Island communities is a result of changes to the social, religious, economic and social fabric of the communities themselves, initially as a result of colonial processes that changed traditional social systems, religions that modified existing social structures and a cash economy that placed pressures on subsistence production. A large number of Pacific islands are now parts of independent states, but globalisation and the expansion of free trade, neoliberal economic policies and Structural Adjustment Programmes have seen further, and marked, declines in local food systems from reduced food production and availability, and markets flooded with imported products of not only low nutritional value but which are also otherwise unhealthy. Many outer island communities throughout Oceania have, over the past century or so, devoted large areas to coconut plantations for the production of copra, a product with little contemporary global demand. This has often been at the expense of arable land. The livelihoods of mixed subsistence–cash farmers have been steadily reduced over a number of decades (Campbell 2006; 2014).

Crudely stated, precolonial food systems in many parts of the world appear to have been relatively sustainable. This may have been particularly the case for subsistence food systems which often included built-in mechanisms that could help offset disruptions to food production systems caused by natural extremes. So, while such systems were exposed to climatic hazards such as droughts and tropical cyclones, they often were not necessarily vulnerable. There has been a considerable decline in the role of subsistence food production throughout the world, not only as a result of urbanisation, but also in rural areas where farmers have sought to expand their involvement in commercial agriculture. However, the notion of subsistence agriculture forming a basis for sustainable agriculture has been 'debunked' (Christoplos 2012: 549) as throughout the world farmers have increased the commercial component of their agricultural activities (though in many small outer island communities in the Pacific islands subsistence agriculture and marine resource exploitation remain critically important).

There can be little doubt that food systems have undergone considerable transformation through time. Some of these changes have reflected technological innovation while others have resulted from colonisation and the subsequent conversion of more and more land to commercial production (often of single crops) at the expense of local food supply (be it for subsistence or for sale). These changes have included replacing traditional systems of agro-ecological diversity, reducing fallow periods, reductions in food preservation and/or storage, neglect of famine foods and losing traditional environmental knowledge (TEK) about such strategies to cope with shortages of food supply. In many Pacific islands prior to colonisation communities produced large surpluses much of which was stored but equally much of which was used for ceremonial feasting that helped establish and cement linkages among different groups that could be called upon in times of shortage. As Box 42.2 illustrates, these surpluses have declined as a significant component of subsistence production has been replaced by single crops of commodities.

Food relief has also played a role in these processes. As humanitarian responses to disasters have grown, particularly in the second half of the twentieth century, the need to preserve and store food has declined, and TEK related to these processes and the use of famine foods is also disappearing.

Contemporary Disasters and Food (In)security

One of the most common forms of disaster response has been the provision of food relief to those who have suffered damage to their subsistence food productions systems (both land and sea based) or whose livelihoods have been disrupted so that they have insufficient income to purchase food. Between 25 and 30 per cent of humanitarian assistance is in the form of food aid but by no means all of this is in response to the manifestations of environmental extremes or degradation in the context of vulnerable populations but in response to food price spikes and complex humanitarian emergencies (Harvey *et al.* 2010).

Often, when there is a 'natural' disaster there is an immediate call for food assistance even though it is not always necessary. In some disasters agricultural damage is limited and in others, it is possible to salvage crops in the short term. Longer-term problems may, on the other hand, become very significant both in terms of replacing destroyed or damaged food sources and in terms of rehabilitating food systems for future security. Where crops are seasonal, there may be a long wait before replanting is possible and where tree crops are damaged or destroyed recovery or regrowth may take several years.

If, as indicated by the IPCC (2013–2014), climate change manifests through increasing magnitude and/or frequency of some weather extremes (which are among the types of hazard events that cause most damage to food production), the demand for food relief may grow. This is especially likely to be the case where other adaptive actions have not been applied, fail to be effective or are too expensive or technically infeasible. If, at the same time food prices increase through both secular reductions in availability and instability caused by disasters, the costs of food relief are likely to steadily increase.

An important issue surrounds the concepts food relief, aid or assistance all of which imply that the provision of essential items for food security is a form of aid and/or charity. Article 4(4) of the UNFCCC clearly states that developed countries (historically the main polluters) have an obligation to assist those that are most vulnerable to climate change to meet the costs of adaptation. Accordingly, where access is limited and the costs of food aid increase as a result of climate change, provision of food should be seen as a form of compensation, rather than simply charity. This has the potential to create a dilemma where food assistance is deemed necessary and appropriate, but may at the same time serve to increase vulnerability. Food relief is not the only, nor the preferable, form of DRR in the context of food security. Agricultural and fisheries rehabilitation after disasters is also important and in the long term the development of food systems that are less vulnerable to the effects of climate change or climatic extremes is important. Equally, reduction of current imbalances in food access is also critically important.

Food in the Context of Climate Change

As Table 42.1 shows, climate change may affect food security through changes in average conditions through time. It may also be manifested in the form of changing patterns of extreme events. Under both scenarios food production may be significantly reduced either by gradual environmental degradation, namely creeping environmental changes (Glantz 1994a), or increasing magnitude and/or frequency of periods of extreme degradation (triggering disasters in vulnerable communities).

Much of what is written about the relationship between climate change and food has focused on climate change effects on agricultural and freshwater and marine productivity. This is understandable as the effects of increased heat and water stress and increasing intensity of extreme events are likely to result in significant declines in production. Accordingly, much of the focus on adaptation has been around finding ways to ameliorate Food Availability Decline, particularly by seeking to avert waning levels of food production. These include for example addressing crop productivity by altering cultivars and/or seasonal routines, and improving efficiencies in irrigation and fertiliser use. There can be little doubt that such approaches will be necessary.

As Wisner *et al.* (2004) note, although famine is often a result of political economic processes and failure of food entitlement systems, many are preceded by decline in the availability of food. These adaptations address the availability pillar of food security but do not address the other pillars, the improvement of which will also be critical in order to reduce food insecurity. Accordingly, it can be anticipated that under future scenarios of climate change patterns of food insecurity may well reflect inequalities of access as is the case today, although the inequalities may become more marked or the numbers facing reduced access to food may be greater, population growth notwithstanding. As Table 42.1 also indicates, stability and utilisation patterns may also be adversely affected by climate change. It follows then that DRR including CCA activities will need to address all of the pillars of food security. While climate change is likely to have an overall negative effect on agriculture some high latitude areas may experience increases in food productivity. DRR including CCA in the context of these areas will include measures that enable local populations to benefit from these changes. It may also include strategies that enable surpluses in these areas to be transferred to those places where food insecurity is likely to increase. Given that current food insecurity takes place in a context of global food surpluses successfully achieving such an end may be difficult.

Building Food Security, DRR and CCA: A Challenging Future

There are numerous synergies between DRR and CCA as they relate to food and in particular reducing food insecurity or food vulnerability. As many of the effects of climate change on food security will be played out through changing patterns of extreme events it follows that DRR measures will be an extremely important and appropriate response. In particular, a key set of activities include those that reduce food and livelihood vulnerability of both rural and urban communities. Such activities are likely to be challenging if the underlying (political economy) causes of vulnerability are to be addressed.

Other activities may involve learning from the past and applying, in a contemporary context, measures that increase the resilience of agricultural (crops, animals, wild foods) and food (distribution, storage, preservation) systems (Box 42.2). That many of these elements of resilience have fallen away perhaps indicates the persistence and power of the processes that have led to this decline. Nevertheless, there is potential for adapting traditional ecological knowledge, as it is applied to disaster reduction, in contemporary settings (e.g., Campbell 2014; McMillen *et al.* 2014).

If communities were to become increasingly reliant upon food relief to enable them to cope with food insecurity and food crises that occur with greater frequency, the implications are likely to be serious. Increased dependence on relief may create vicious cycles of more frequent and intense food crises occasioning increased food relief and further erosion of the remnants of food security that remain. As the longitudinal changes in food availability caused by climate change worsen, relief providers may find it difficult to find sufficient appropriate food supplies, and the costs of food relief are likely to also grow. This implies a bleak future for some communities if climate change projections, as outlined by IPCC (2013–2014), are to be played out.

Conclusion: Sustainable Food Futures?

Given that there have been many obstacles to the achievement of food security for all global citizens up to now the prospects for sustainable food security may seem somewhat glum. There will obviously be numerous challenges, not just in terms of enabling a growing population to be food secure, but also in reducing the effects of disasters on food security, in a context of changing environmental conditions that are likely to impinge on food availability and access. The challenge will not only be for those who are already chronically food vulnerable, but for those who currently consume more than their fair share of the globe's food resources. There will have to be changes in nutritional practices throughout the world and improvements in distribution systems to ensure that all people have an equitable share.

References

Baker, R. (1987) 'Linking and sinking: Economic externalities and the persistence of destitution and famine in Africa', in M. Glantz (ed.) *Drought and Hunger in Africa. Denying Famine a Future*, Cambridge: Cambridge University Press, pp. 149–168.

Campbell, J.R. (2006) *Traditional Disaster Reduction in Pacific Island Communities*, GNS Science Report, New Zealand: GNS Science.

Campbell, J.R. (2014) 'Development, global change and traditional food security in Pacific Island Countries', *Regional Environmental Change* 15, 7: 1313–1324.

Christoplos, I. (2012) 'Food security and disaster', in B. Wisner, JC Gaillard, and I. Kelman (eds) *The Routledge Handbook of Hazards and Disaster Risk Reduction*, London: Routledge, pp. 543–552.

Copans, J. (1975) *Sécheresses et famines du Sahel, Volume 1*, Paris: Maspero.

Devereux, S. (1993) *Theories of Famine: From Malthus to Sen*, Hemel Hempstead: Harvester Wheatsheaf.

FAO (Food and Agriculture Organisation of the United Nations). (2006) *Food Security, FAO Policy Brief, Issue 2*, Rome: FAO.

Glantz, M.H. (ed.) (1976) *The Politics of Natural Disaster: The Case of the Sahel Drought*, New York: Praeger.

Glantz, M.H. (ed.) (1994a) 'Creeping environmental problems', *World and I*, June, 218–225.

Glantz, M.H. (1994b) *Drought Follows the Plough: Cultivating Marginal Areas*, Cambridge: Cambridge University Press.

Harvey, P., Proudlock, K., Clay, E., Riley, B. and Jaspars, S. (2010) *Food Aid and Food Assistance in Emergency and Transitional Contexts: A Review of Current Thinking*, London: Humanitarian Policy Group, Overseas Development Institute.

IPCC (Intergovernmental Panel on Climate Change). (2007) *Climate Change 2007: Contribution of Working Group II to the Fourth Assessment Report of the Intergovernmental Panel on Climate Change*, Cambridge: Cambridge University Press.

IPCC (Intergovernmental Panel on Climate Change). (2012) *Managing the Risks of Extreme Events and Disasters to Advance Climate Change Adaptation*, Cambridge: Cambridge University Press.

IPCC (Intergovernmental Panel on Climate Change). (2013) *Climate Change 2013: The Physical Science Basis*, Cambridge: Cambridge University Press.

IPCC (Intergovernmental Panel on Climate Change). (2014) *Climate Change 2014: Impacts, Adaptation and Vulnerability*, Cambridge: Cambridge University Press.

Lewis, J. (1999) *Development in Disaster-Prone Places: Studies of Vulnerability*, London: Intermediate Technology Publications.

McMillen, H.L., Ticktin, T., Friedlander, A., Jupiter, S.D., Thaman, R., Campbell, J., Veitayaki, J., Giambelluca, T., Nihmei, S., Rupeni, E., Apis-Overhoff, L., Aalbersberg, W. and Orcherton, D.F. (2014) 'Small islands, valuable insights: Systems of customary resource use and resilience to climate change in the Pacific', *Ecology and Society* 19, 4: 44.

Mucke, P. (2015) 'Food insecurity and risk assessment', in L. Jeschonnek, P. Mucke, K. Radtke and J. Walter (eds) *World Risk Report, 2015*, Berlin and Bonn: Bündnis Entwicklung Hilft and United Nations University Institute for Environment and Human Security (UNU-EHS), pp. 5–11.

O'Keefe, P., Westgate, K. and Wisner, B. (1976) 'Taking the naturalness out of natural disasters', *Nature* 260, 5552: 566–567.

Porter, J.R., Xie, L., Challinor, A.J., Cochrane, K., Howden, S.M., Iqbal, M.M., Lobell, D.B. and Travasso, M.I. (2014) 'Food security and food production systems', in C.B. Field, V.R. Barros, D.J. Dokken, K.J. Mach, M.D. Mastrandrea, T.E. Bilir, M. Chatterjee, K.L. Ebi, Y.O. Estrada, R.C. Genova, B. Girma, E.S. Kissel, A.N. Levy, S. MacCracken, P.R. Mastrandrea and L.L. White (eds) *Climate Change 2014: Impacts, Adaptation, and Vulnerability. Part A: Global and Sectoral Aspects. Contribution of Working Group II to the Fifth Assessment Report of the Intergovernmental Panel on Climate Change*, Cambridge: Cambridge University Press, pp. 485–533.

Sen, A. (1981) *Poverty and Famines: An Essay on Entitlement and Deprivation*, Oxford: Clarendon Press.

UNISDR (United Nations International Strategy for Disaster Reduction). (2009) *2009 UNISDR Terminology on Disaster Risk Reduction*, Geneva: UNISDR.

Wisner, B., Blaikie, P., Cannon, T. and Davis, I. (2004) *At Risk. Natural Hazards, People's Vulnerability and Disasters. Second edition*, London: Routledge.

World Food Summit. (1996) *World Food Summit Plan of Action*, Rome: Food and Agriculture Organisation (FAO).

43

HEALTH SUPPORTING DISASTER RISK REDUCTION INCLUDING CLIMATE CHANGE ADAPTATION

Amina Aitsi-Selmi, Chadia Wannous, and Virginia Murray

Introduction

The conceptual framework used in this chapter places climate change within disaster risk reduction (DRR) as a subset of multiple disaster risk drivers; and climate change adaptation (CCA) as one of multiple DRR processes. In turn, DRR is placed under the umbrella of sustainable development. Health is treated as a cross-cutting theme and the health effects of disasters (both climate and non-climate related) are reviewed. Clearly, the causal relationships between the different areas of policy activity discussed in this chapter are bi- if not multi-directional.

For example, increasing access to health services (during and outside disasters) and improving living conditions (in alignment with the objectives of the Sustainable Development Goals (SDGs)) will reduce disaster losses in lives, livelihoods and health (as aimed for by the Sendai Framework (UN 2015)). At the same time, sustainable development through better food and transport systems meets objectives for health (by reducing the burden of non-communicable diseases and their risk factors such as obesity) but also for climate change mitigation and DRR including CCA (see Box 43.1).

The year 2015 presents a window of opportunity in terms of policy. Three landmark UN agreements coincide in time with health featuring in all of them. These are: The Sendai Framework for Disaster Risk Reduction 2015–2030 (UN 2015), which puts health at its core and gives it much more prominence than its predecessor the Hyogo Framework for Action 2005 (agreed in March in Sendai, Japan, by 187 countries) (Aitsi-Selmi and Murray 2015); the SDGs which are the successors to the Millennium Development Goals and expand beyond immediate health outcomes to include the upstream factors that influence health such as the physical environment and social justice (agreed in September in New York, USA, by 193 countries); and the Paris climate change agreement (UNFCCC 2015) aiming to strengthen and support the global response to climate change (agreed in December in Paris, France).

The coincidence of three such agreements is an opportunity of global significance for building coherence across these policy streams. In this chapter, the health links across these agreements are reviewed and specific policy opportunities to frame DRR including CCA as health issues are highlighted. A real possibility exists for improving people's health and preserving their environment if these frameworks are implemented synergistically. The health dimension could support the implementation of DRR including CCA through capacity development and joint policy initiatives between all relevant sectors including the health sector (Aitsi-Selmi *et al.* 2015).

Box 43.1   Obesity, Health Inequality, and Co-benefits with Climate Change

Amina Aitsi-Selmi[1], Chadia Wannous[2], and Virginia Murray[1]

[1] Public Health England, London, UK
[2] UN Office for Disaster Risk Reduction (UNISDR)

The 2011 UN High Level Summit gave global prominence to the growing problem of non-communicable diseases (NCDs), particularly heart disease and diabetes for which obesity is a key risk factor. Obesity is a growing and complex problem globally that cannot be solved by the health sector alone. Yet, it strains health systems, social systems and the economy. Obesity has become a matter of concern for policy-makers, for finance ministers seeking to curb spiralling health-care costs, for emergency services called upon to evacuate severely obese people, and even for armed forces – some of which now struggle to recruit physically fit personnel.

Mission Readiness – a coalition of retired US military leaders – released a landmark report in 2010 concluding that childhood obesity is 'a potential threat to national security'.

Contrary to some perceptions, the problem is not confined to affluent Western countries. Countries in the Middle East (e.g., Egypt) and the Pacific Islands are among those with the highest female obesity levels in the world (Aitsi-Selmi *et al.* 2014). Low- and middle-income countries are at a particular disadvantage owing to the fact that obesity-related diseases can coexist with undernutrition and infections. This mixed burden of disease reflects the uneven effects of rapid globalisation, economic development and accompanying cultural shifts and unregulated commercial activity in the food and beverage sector that local public health systems cannot address.

Furthermore, the unequal distribution or social gradient of obesity changes, as countries develop, from affecting mostly rich, educated women to affecting poor, uneducated women over time. Cross-country studies (Aitsi-Selmi *et al.* 2014) have shown that as countries begin to get richer, wealthy women are at greater risk of obesity but that education begins to protect them as economic development proceeds. This suggests that, like for DRR including CCA, education (of women in particular) can help to promote health and well-being.

This may call for strengthening of global and national governance systems overseeing markets and public health infrastructure alongside promoting access to formal education for women. From this perspective, investment in women's education may be viewed as an intervention with co-benefits for DRR including CCA. Other obesity-reducing policies that change consumption behaviour (particularly in relation to food and travel) promise co-benefits for climate change mitigation. Walking, cycling and improved diets (increased vegetable and fruit consumption, and reduced red and processed meat consumption) can reduce greenhouse gas emissions as well as reducing the risk of obesity and therefore diabetes, heart disease and cancer.

Health as a Synergistic Element across Disaster Risk Reduction and Climate Change Adaptation

The World Health Organization defines health broadly as 'a complete state of physical, mental and social well-being, and not merely the absence of disease or infirmity' (WHO 1948). Health puts a human face on what can sometimes seem to be a distant threat and can achieve greater public resonance. For example, public concerns about the health effects of climate change, such as heatwaves, infections, undernutrition and food insecurity, have the potential to accelerate political action in ways that attention to carbon dioxide emissions alone does not. It has been argued that when climate change is framed as a health issue, rather than purely as an environmental, economic, or technological challenge, it becomes clear that we are facing a predicament that strikes at the heart of humanity (Wang and Horton 2015). Below we review areas of synergy or overlap, in terms of scientific evidence and policy initiatives between health and each of CCA and DRR, followed by a brief discussion of the links with the SDGs to highlight the overall coherence under a broad sustainable development agenda.

Health Impacts of Climate Change

Human health is sensitive to the effects of climate change. Climate change influences disease, food, water, sanitation and weather, tending to have the greatest negative effects on the poorest populations (Preston *et al.* 2014). The WHO has estimated that 5.5 million healthy life years are lost to climate change globally in one year (WHO 2008).

In the UK, heat-related mortality is responsible for 2,000 premature deaths per year. Projections indicate that heat-related mortality will increase by 540 per cent compared to current levels by 2080. The risk of river and coastal flooding will also increase significantly, increasing the incidence of the physical and psychological burden of disease resulting from the impact of floods. As a result of rising temperatures and increased flooding, the UK will see the introduction and increased incidence of vector-borne diseases that are currently non-existent (Vardoulakis and Heaviside 2012).

The health impacts of climate change and the related scientific evidence have been discussed systematically by the Intergovernmental Panel on Climate Change (IPCC 2014). It is clear that while the worldwide burden of human ill-health from climate change is relatively small compared with other causes, it is not well quantified at present and the impacts may increase as climate change and its underlying drivers continue to evolve. These impacts can be negative (such as increased mortality due to heatwaves or rising levels of respiratory illness due to air pollution) or positive (such as reduced mortality from cold weather and reduced transmissibility of certain vector-borne diseases) (IPCC 2014).

Impacts from recent climate-related extremes, such as heat waves, droughts, floods, cyclones and wildfires – all forms of hazard that can lead to disasters – include alteration of ecosystems, effects on agriculture and food production and water supply, damage to infrastructure and settlements, morbidity and mortality, consequences for mental health and human well-being and vector-borne diseases. The effects can be direct or indirect:

- Direct effects include the greater likelihood of injury, disease, and death due to more intense heatwaves and fires. In 2014, the WHO documented that more than 7 million deaths every year are attributable to air pollution, making it one of the most important health risk factors globally and comparable to tobacco smoking (WHO 2014a). Mental health impacts from extreme events are also substantial; they can last for long periods of time and affect a large proportion of the population (Berry *et al.* 2010).
- Indirect effects result from impacts on infrastructure and the wider economy and society. They include the increased likelihood of malnutrition resulting from diminished food production; risks from lost work capacity and reduced labour productivity in vulnerable populations; and increased risks from food- and water-borne diseases and vector-borne diseases (IPCC 2014).

Indirect effects are difficult to estimate and monitor and are often temporally separated from the event leading to under-examination and, thus, the substantial underestimation of the total health burden (IPCC 2014). Until the mid-century, it is expected that climate change will impact human health mainly by exacerbating health problems that already exist, especially in low-income countries. However, globally, over the 21st century, the magnitude and severity of negative impacts on health are projected to increasingly outweigh the positive impacts mentioned above (IPCC 2014). Therefore, while a proportion of these impacts may be inevitable, DRR including CCA interventions may help to minimise them.

Health and Climate Change Adaptation

Significant co-benefits and synergies (as well as trade-offs) exist between CCA and human health. Examples of actions with co-benefits include (i) improved energy efficiency and cleaner energy sources, leading to reduced emissions of health-damaging climate-altering air pollutants; (ii) reduced energy and water consumption in urban areas through greening cities and recycling water; (iii) sustainable agriculture and forestry; and (iv) protection of ecosystems for carbon storage and other ecosystem services (IPCC 2014).

Worldwide, motorised transport is responsible for approximately 25 per cent of carbon dioxide emissions and emissions from motorised transport are rising faster than those from other sources (Woodcock *et al.* 2009). Motorised vehicles are a source of several pollutants, including nitrogen oxides, particulates, carbon monoxide and hydrocarbons, which have a direct impact on health. The health co-benefits of reducing motorised transport use are well documented and substantial: physical activity through walking and cycling can improve physical condition and fitness, mental well-being and social cohesion (IPCC 2014). Significant reductions in the incidence of obesity, ischaemic heart disease, cerebrovascular disease, high blood pressure, dementia, diabetes, breast cancer, colon cancer and depression can be achieved through measures to simultaneously reduce air pollution and increase physical activity (Woodcock *et al.* 2009).

Measures to improve home insulation, energy efficiency and energy expenditure could have appreciable benefits to health. Such benefits arise from improved indoor air quality and control of winter indoor temperatures and protection from respiratory illness, the effects of cold temperature and rising damp. Improved comfort, expanded use of space, increased privacy and improved social interaction are additional benefits that can boost both mental and physical health. It is estimated that combining insulation, ventilation control and switching to renewable energy sources in UK households alone could save 850 disability-adjusted life-years and avoid 5,400 premature deaths annually (Wilkinson *et al.* 2009).

The current global food system is a major contributor to human-made greenhouse gas (GHG) emissions. GHGs are produced at all stages in the system from farming to distribution and waste disposal. Agriculture accounts for approximately 10–12 per cent of GHG emissions, with the livestock sector contributing 80 per cent of these emissions (Vardoulakis and Heaviside 2012). An additional 6–17 per cent of GHG emissions result from agricultural changes in land use, such as deforestation (Friel *et al.* 2009). The impacts of climatic and environmental change, which the food system contributes to, can negatively affect food production and make food supplies unpredictable across the world.

Reducing the consumption of animal related products, alongside purchasing food and water which is locally produced reduces the distance products need to travel. Co-benefits of this approach are in addressing diets rich in saturated fat that contribute to the risk of non-communicable diseases including cardiovascular disease (heart attacks and strokes), type 2 diabetes and some types of cancer. For the UK population, a 15 per cent reduction in the burden of ischaemic heart disease and cerebrovascular disease could be achieved following a 30 per cent reduction in animal product consumption (Friel *et al.* 2009).

Taking these synergies and potential co-benefits into account, the central message of the second Lancet Commission on Health and Climate Change (Watts *et al.* 2015) is that 'Tackling climate change could be the greatest global health opportunity of the 21st century' (Wang and Horton 2015: 1798). The commission puts forward nine recommendations to governments and commits to supporting monitoring of their implementation:

1   Invest in climate change and public health research, monitoring, and surveillance.
2   Scale-up financing for climate resilient health systems world-wide.
3   Protect cardiovascular and respiratory health by ensuring a rapid phase out of coal from the global energy mix.
4   Encourage a transition to cities that support and promote lifestyles that are healthy for the individual and for the planet.
5   Establish the framework for a strong, predictable, and international carbon pricing mechanism.
6   Rapidly expand access to renewable energy in low-income and middle-income countries.
7   Support accurate quantification of the avoided burden of disease, reduced health-care costs, and enhanced economic productivity associated with a low-carbon economy.
8   Facilitate collaboration between Ministries of Health and other government departments, empowering health professionals and ensuring that health and climate considerations are thoroughly integrated in government-wide strategies.
9   Agree and implement an international agreement that supports countries in transitioning to a low-carbon economy.
10 To help drive this transition, the 2015 Lancet Commission will develop an independent Countdown to 2030: Global Health and Climate Action, designed to monitor progress on the implementation of climate change policies that promote health over the next fifteen years.

Implementing these strategies would build on the health-related synergies between sustainable development and climate change mitigation but also between DRR including CCA and sustainable development.

Health Impacts of Disasters

The consequences of disasters on human health and well-being are varied and, like climate change, include direct impacts on lives and livelihoods and indirect impacts on the wider economy and society.

Direct health impacts – whether climate related or not – can be acute such as those resulting from physical injury (including blunt trauma, drowning and air pollution) or chronic by leaving those affected with short- and long-term mental health consequences including general distress, post-traumatic stress disorder, anxiety, excessive alcohol consumption, and other psychiatric disorders (Malilay *et al.* 2013; Neria and Shultz 2012).

The short- to medium-term ramifications of disaster impacts can be extensive. The US Centers for Disease Control and Prevention link hazards to the transmission of infectious diseases, since disasters can damage vital infrastructure such as water supplies and sewerage systems; and these can be further compromised by population displacement and overcrowding (Malilay *et al.* 2013). Sexually transmitted diseases and gender-based violence may increase following disasters in situations of high stress, low sanitation and overcrowded temporary relief where privacy and access to reproductive health services are limited.

People with chronic diseases such as diabetes, heart disease and cancer have ongoing medical needs that can easily be affected when health services are disrupted in disaster situations

(Aitsi-Selmi *et al.* 2015). During post-disaster evacuations many people lose essential medicines and many do not even have a record of their prescriptions with them when evacuated (Murray *et al.* 2015).

Epidemics and pandemics are also of concern as disasters in themselves. The Ebola outbreak of 2014–2015 in West Africa is an example of how an infectious disease epidemic can have consequences on the scale of a disaster, on lives, livelihoods and the entire country. As a result of the pressure on the fragile health system capabilities in the region, more people are estimated to have died from lack of access to services that were disrupted – including deaths from childbirth, malaria and AIDS as well as other diseases – than died from Ebola (Walker *et al.* 2015).

Framing health risks arising from poor health system infrastructure, including weak monitoring and surveillance systems, as slowly emerging disasters could help to bring the different communities of policy, research and practice together (Viens and Littmann 2015). Parallels have been drawn between the objectives of those working in the infectious disease surveillance and related sectors such as antimicrobial resistance (AMR), and those working on DRR (Viens and Littmann 2015): while AMR presents unique and novel challenges, components of a comprehensive emergency and longer-term response will be comparable to the actions taken in the face of other public health disasters such as Ebola or more often accepted forms of disaster such as the Indian Ocean tsunami of 2004.

Health and Disaster Risk Reduction

The health sector through its public health arm is increasingly concerned with the total health system and societal factors that affect population health parameters (such as life expectancy and mortality) and not solely the eradication of a particular disease affecting an individual patient (Murray *et al.* 2015). This allows for greater synergy to be explored between the different sectors. The three main public health functions are all of importance for DRR as outlined by WHO (http://www.who.int/trade/glossary/story076/en):

1   The assessment and monitoring of the health of communities and populations at risk, to identify health problems and priorities.
2   The formulation of public policies designed to solve identified local and national health problems and priorities.
3   To ensure that all populations have access to appropriate and cost-effective care, including health promotion and disease prevention services.

Conversely, the continued delivery of these public health functions relies on adequately addressing disaster risk from prevention to response. WHO Member States have made high-level policy commitments to DRR and adopted a resolution at the 2011 World Health Assembly to strengthen national health emergency and disaster management capacities and the resilience of health systems (WHA 2011). In addition to the synergistic functions above, policy areas that are an opportunity for cooperation between health and DRR were presented at the 2011 Global Platform for DRR, including: health systems strengthening, risk assessments, the management of chronic diseases and mental health impacts during and after disasters (Murray *et al.* 2015). Through participation in the Sendai Framework policy process and negotiations, health actors worked to ensure that some of these approaches and policy areas, that are so important to people's health, were included in the Sendai Framework and that health was considered as an explicit outcome (WHO 2011).

These efforts are reflected in the Sendai Framework's emphasis on health (Aitsi-Selmi *et al.* 2015). Voluntary commitments to be delivered with agreement from health actors and their partners include:

[E]nhancing the resilience of national health systems through training and capacity development; strengthening the design and implementation of inclusive policies and social safety-net mechanisms, including access to basic health care services towards the eradication of poverty; finding durable solutions in the post-disaster phase to empower and assist people disproportionately affected by disasters, including those with life threatening and chronic disease; enhancing cooperation between health authorities and other relevant stakeholders to strengthen country capacity for disaster risk management for health; the implementation of the International Health Regulations (2005) and the building of resilient health systems; improving the resilience of new and existing critical infrastructure, including hospitals, to ensure that they remain safe, effective and operational during and after disasters, to provide live-saving and essential services; establishing a mechanism of case registry and a database of mortality caused by disaster to improve the prevention of morbidity and mortality and enhancing recovery schemes to provide psychosocial support and mental health services for all people in need.

Looking to the future, Member States and the WHO Secretariat have set a course that goes beyond disaster response into risk management and reduction. Following the Sendai Framework adoption, the WHO committed to designing a new policy framework on 'Reducing Health Consequences of Emergencies and Disasters: A Risk Management Policy Guide' to help countries to manage emergency risks and reduce their health consequences effectively (WHO 2015). Other commitments include providing greater input and participation from the health sector in DRR through national, regional and global DRR platforms.

Examples exist of coordinated action between the health and other sectors with a move away from emergency response to prevention and greater community participation including at local level (see Box 43.2). These novel initiatives need to be evaluated and scaled up where results are promising.

Box 43.2   Evolution of Health and Nutrition Services in Ethiopia through the Health Extension Programme

Bob Alexander[1]

[1] Independent, World Citizen
Based on UNICEF (2014).

Nutrition services in Ethiopia previously followed a strategy of reacting to emergencies. Interventions were generally short term, dependent on humanitarian funding and provided by the international community only during acute food insecurity and seasonal hunger crises. Many health services were also provided in a reactive curative approach either through such emergency interventions or through health centres (HCs) and hospitals that were often too removed from communities both literally and figuratively to be responsive to their needs. The government Health Extension Programme (HEP) aims to enable more adaptive service delivery by enabling a transition from such externally driven emergency and distant curative approaches to a community-based development approach of local diagnosis, treatment and referral that optimally addresses communities' needs and can adapt to crises.

The HEP is adaptive to evolving risks by definition: designed to be able to dynamically add new elements and drop or modify others over time as capacities and conditions change. How the

---

system can progress in this transition depends on the location-specific evolution of Health Extension Workers (HEWs) and volunteer Health Development Army (HAD) members through different approaches to manage these responsibilities.

Some generalisations can be made about the steps in this progression. The first transition occurs when HEWs are first deployed. The focus in the resulting initial HEW phase is on adopting a proactive preventive rather than reactive curative paradigm by using HEWs to do sensitisation, mobilisation, monitoring and referrals to the health centre for any curative needs. The second transition occurs when the knowledge and skills of HEWs have expanded enough that they can begin to organise, train and involve the community through an active HDA system responsible for more of this sensitisation, mobilisation and monitoring and the HEWs can initially expand into other coordination and curative roles. The amount of time to reach this transition depends on local conditions and achievement, so some areas in Ethiopia have already achieved this transition while others have yet to do so. The final transition is to an optimal state of complete local responsibility for surveillance, diagnosis, and either treatment or referral as appropriate, with no more district-level supervision, and in which HEWs and their HDAs have reached a point of optimally adapting roles and capacity based on conditions that arise.

The Social Determinants of Health and Their Relationship to Disaster Risk Reduction Including Climate Change Adaptation

Traditionally, society has looked to the health sector to deal with its concerns about health and disease through health care, notwithstanding the richness of traditional knowledge in dealing with some health concerns at the time (Box 43.3). However, the high burden of illness responsible for premature loss of life and the significant inequalities in health outcomes by age, gender, ethnicity, geography, socioeconomic group and so on, arise in large part because of the conditions in which people are born, grow, live, work and age, known in the health sector as the social determinants of health. Poor and unequal living conditions are the consequence of poor social policies and programmes, unfair economic arrangements and careless political choices.

The 2008 WHO Commission on the Social Determinants of Health recommended that 'action on the social determinants of health must involve the whole of government, civil society and local communities, business, global fora, and international agencies' and that '[P]olicies and programmes must embrace all the key sectors of society not just the health sector' (WHO 2008). In 2015, these statements were echoed in the Sendai Framework (paragraph 17, p. 12) which calls for the significant reduction of disaster losses in lives, livelihoods and health:

> through the implementation of integrated and inclusive economic, structural, legal, social, health, cultural, educational, environmental, technological, political and institutional measures that prevent and reduce hazard exposure and vulnerability to disaster, increase preparedness for response and recovery, and thus strengthen resilience.

The urban environment has a major impact on health equity through increasing exposure to Western diets, alcohol and cigarettes as well as motorised transport, poor air quality, greater electricity consumption and potentially unsafe buildings. As a result, urbanisation is reshaping population health problems, particularly among the urban poor, towards non-communicable diseases, accidental and violent injuries, as well as deaths and impacts from disasters, while infectious diseases and undernutrition will continue in particular regions and groups around the world (UNISDR 2015; WHO 2008).

Box 43.3   Traditional Health Practices and Climate Change

Soledad Natalia M. Dalisay[1]

[1] University of the Philippines Diliman, Quezon City, the Philippines

Communities have exhibited traditional measures or local knowledge that have helped them surmount stresses from climate change impacts. Three specific cases drawn from the Philippines looking into the role of traditional health practices in DRR including CCA are provided below.

First, it is not uncommon that groups of people draw their 'materia medica' from their immediate environment. For instance, among Aeta indigenous communities of Zambales, the abundant endemic tawa-tawa (scientific name: *Euphorbia hirta*) plant is traditionally used in the management of dengue, which has been identified as one of the critical health impacts of relatively higher levels of precipitation and humidity associated with climate change. Aeta usually refrain from telling non-Aeta how the plant is prepared prior to ingestion. They believe that such knowledge is sacred and divulging its preparation procedures would cause the plant to lose its efficacy as a medicine. Nonetheless, tawa-tawa is also used against dengue by non-Aetas. This spurred the interest of health organisations, which are now investigating its medicinal properties. The use of a locally available, accessible and abundant 'cure' for dengue such as the tawa-tawa is vital in light of the current surge in dengue cases.

Second, traditional food preparation practices as well as food prescriptions and prohibitions reflect societies' knowledge on how to deal with environmental change successfully and maintain food security. This could be gleaned in how some island and coastal communities have been addressing red tide occurrences. Harmful algal bloom or red tide has been associated with El Niño. A red tide raises the toxicity levels in shellfish and other seafood and its consumption can be debilitating or fatal. The red tide is not new to some coastal dwellers in the Eastern Visayas region who refer to it as the 'sea turning red' in their local language. They have traditional food proscriptions that cover bivalves and other shellfish, but allow the consumption of fish provided it is thoroughly washed in clean water and the intestines are discarded. Thus, they are able to consume an important food resource like fish even during an environmental crisis such as the red tide.

Third, fishers follow a schedule for their fishing expeditions that is based on their local knowledge of the seas, celestial bodies and the behaviour of the fish that they catch. Such knowledge has been the basis of their successful fishing expeditions in the past. Recently, fishers in Bohol started to complain of extremely hot days, especially in the summer months. Such temperatures were not experienced in the past and they felt that this contributed to elevated blood pressure that they now suffer from. Thus, while traditionally fishers adhered to a specific fishing calendar, which specifies even the time they should go out to fish, they have adjusted this to limit their exposure to the sun especially during the times of the day when the heat is at its peak. With their adjusted fishing schedules, they are able to continue fishing and at the same time counter the adverse effects of extreme heat on their health. In this instance, the dynamism of local knowledge allowed them to address the challenges posed by higher temperatures.

The poor and socially marginalised often live in places more exposed to hazards; have less ability to cope with and recover from disaster impacts; are less able to advocate and represent their issues as they have less influence on decision-making; often depend on informal safety nets that become stretched after major shocks; and are adversely affected by delays in, or lack of access to, relief/early recovery responses (WHO 2008). In general, low socioeconomic position means poor education, lack of amenities, unemployment and job insecurity, poor working conditions and unsafe neighbourhoods, all creating and maintaining vulnerability to hazards.

This WHO Commission took the view that a new approach to development is needed. For example, economic growth is treated as important, particularly for poor countries, as it gives the opportunity to provide resources to invest in improvement of the lives of their population. However, growth by itself, without appropriate social policies to ensure reasonable fairness in the way its benefits are distributed is seen to bring little benefit to health equity. Of course, while health and health equity may not be the aim of all social policies they will be a fundamental result.

Four distinct and largely independent research and policy communities – DRR, CCA, environmental management and poverty reduction (in which health is involved) – have been working to improve people's lives but face challenges in terms of facilitating learning and information exchange as well as overcoming misaligned financial structures. However, there are good reasons to work together, leading health to support the placement of CCA within DRR, which in turn would be part of sustainable development.

Traditionally separate communities of research, policy and practice can be brought together under place-based initiatives that address the needs of large populations through social movements. For example, the Healthy Cities movement (http://www.euro.who.int/en/health-topics/environment-and-health/urban-health/activities/healthy-cities) is a global programme with initiatives in all regions of the WHO. Health 2020, the European policy for health and well-being, has given the WHO Healthy Cities programme a mandate for action and leadership backed up by solid evidence on the social determinants of health.

Healthy Cities Networks optimise limited local resources by providing local governments with direct support through training, opportunities to share good practice and access to national and international expertise. National networks, through their partnership-based approaches, have shown their great potential to support countries towards the goals of Health 2020 to improve health for all, reduce inequalities and improve leadership and participatory governance. This includes cooperation with other networks focusing on, for example, health in schools, age-friendly cities and climate change (WHO 2014b).

In summary, the conceptual frameworks informing health indicate how DRR includes CCA. Therefore, it is critical that health is at the centre of whole-of-society efforts to reduce vulnerabilities through DRR including CCA. Evidence-informed guidance is increasingly available on what healthy urban design looks like as well as how urban planning and public health can work together to achieve common goals using policy and regulatory frameworks (WHO 2014c).

Conclusions

This chapter argued that health – in its broadest sense of human well-being rather than as a sector – is an element of synergy and point of convergence for the post-2015 agenda, especially for DRR including CCA. Recognising and articulating synergies between the different policy agendas through a health lens can offer a clearer vision and narrative for concerted action and funding reform to achieve common goals. The high level of interdisciplinary work and inter-sectoral cooperation needed requires an integrated vision of how the scientific community may interact efficiently with policy-makers. Lessons from health can inform that vision.

References

Aitsi-Selmi, A., Bell, R., Shipley, M.J. and Marmot, M.G. (2014) 'Education modifies the association of wealth with obesity in women in middle-income but not low-income countries: An interaction study using seven national datasets, 2005–2010', *PLOS ONE* 9, 3: e90403.

Aitsi-Selmi, A., Egawa, S., Sasaki, H., Wannous, C. and Murray, V. (2015) 'The Sendai framework for disaster risk reduction: Renewing the global commitment to people's resilience, health, and well-being', *International Journal of Disaster Risk Science* 6, 2: 164–176.

Aitsi-Selmi, A. and Murray, V. (2015) 'The Sendai framework: Disaster risk reduction through a health lens', Bulletin of the World Health Organization 93, 6: 362–362.

Berry, H.L., Bowen, K. and Kjellstrom, T. (2010) 'Climate change and mental health: A causal pathways framework', *International Journal of Public Health* 55, 2: 123–132.

Friel, S., Dangour, A.D., Garnett, T., Lock, K., Chalabi, Z., Roberts, I., Butler, A., Butler, C.D., Waage, J., McMichael, A.J. and Haines, A. (2009) 'Public health benefits of strategies to reduce greenhouse-gas emissions: Food and agriculture', *The Lancet* 374, 9706: 2016–2025.

IPCC (Intergovernmental Panel on Climate Change). (2014) 'Summary for policymakers', in C.B. Field, V.R. Barros, D.J. Dokken, K.J. Mach, M.D. Mastrandrea, T.E. Bilir, M. Chatterjee, K.L. Ebi, Y.O. Estrada, R.C. Genova, B. Girma, E.S. Kissel, A.N. Levy, S. MacCracken, P.R. Mastrandrea and L.L. White (eds) Climate Change 2014: Impacts, Adaptation, and Vulnerability. Part A: Global and Sectoral Aspects. Contribution of Working Group II to the Fifth Assessment Report of the Intergovernmental Panel on Climate Change, Cambridge: Cambridge University Press, pp. 1–32.

Malilay, J., Batts, D., Ansari, A., Miller, C.W. and Brown, C.M. (2013) 'Natural disasters and environmental hazards', in Centers for Disease Control and Prevention (CDC) (ed.) *Health Information for International Travel 2014: The Yellow Book*, Oxford: Oxford University Press, p. 121.

Murray, V., Aitsi-Selmi, A. and Blanchard, K. (2015) 'The role of public health within the United Nations post-2015 framework for disaster risk reduction', *International Journal of Disaster Risk Science* 6, 1: pp. 28–37.

Neria, Y. and Shultz, J.M. (2012) 'Mental health effects of Hurricane Sandy: Characteristics, potential aftermath, and response', *JAMA* 308, 24: 2571–2572.

Preston, I., Banks, N., Hargreaves, K., Kazmierczak, A., Lucas, K., Mayne, R., Downing, C. and Street, R. (2014) Climate Change and Social Justice: An Evidence Review, London: Joseph Rowntree Foundation.

UN (United Nations). (2015) The Sendai Framework for Disaster Risk Reduction 2015–2030, Geneva: UNISDR. Online http://www.who.int/hac/techguidance/preparedness/sendai_2015/en/ (accessed 23 December 2015).

UNFCCC (United Nations Framework Convention on Climate Change). (2015) Conference of the Parties, Twenty-first session, Paris, 30 November to 11 December 2015, FCCC/CP/2015/L.9/Rev.1, Bonn, Germany: UNFCCC.

UNICEF (United Nations Children's Fund). (2014) Review of Adaptive Basic Social Services Provision to Reduce Disaster Risk of Populations Especially Children in Selected Horn of Africa Countries, New York: UNICEF.

UNISDR (United Nations International Strategy for Disaster Reduction). (2015) UNISDR STAG 2015 Report: Science is Used for Disaster Risk Reduction, Geneva: UNISDR. Online http://preventionweb.net/go/42848 (accessed 23 December 2015).

Vardoulakis, S. and Heaviside, C. (eds) (2012) Health Effects of Climate Change in the UK 2012 – Current Evidence, Public Health Recommendations and Research Gaps, London: Health Protection Agency, Centre for Radiation, Chemical and Environmental Hazards.

Viens, A.M. and Littmann, J. (2015) 'Is antimicrobial resistance a slowly emerging disaster?', *Public Health Ethics* 8, 3: 255–265.

Walker, P.G., White, M.T., Griffin, J.T., Reynolds, A., Ferguson, N.M. and Ghani, A.C. (2015) 'Malaria morbidity and mortality in Ebola-affected countries caused by decreased health-care capacity, and the potential effect of mitigation strategies: A modelling analysis', *The Lancet Infectious Disease* 15, 7: 825–832.

Wang, H. and Horton, R. (2015) 'Tackling climate change: The greatest opportunity for global health', *The Lancet* 386, 10006: 1798–1799.

Watts, N., Adger, W.N., Agnolucci, P., Blackstock, J., Byass, P., Cai, W., Chaytor, S., Colbourn, T., Collins, M., Cooper, A., Cox, P.M., Depledge, J., Drummond, P., Ekins, P., Galaz, V., Grace, D., Graham, H., Grubb, M., Haines, A., Hamilton, I., Hunter, A., Jiang, X., Li M., Kelman, I., Liang, L., Lott, M., Lowe, R., Luo, Y., Mace, G., Maslin, M., Nilsson, M., Oreszczyn, Y., Pye, S., Quinn, T., Svensdotter, M., Venevsky, S., Warner, K., Xu, B., Yang, J., Yin, Y., Yu, C., Zhang, Q., Gong, P., Montgomery, H. and Costello, A. (2015) 'Health and climate change: Policy responses to protect public health', *The Lancet* 386, 10006: 1861–1914.

WHA (World Health Assembly). (2011) Resolution 64.10: Strengthening National Health Emergency and Disaster Management Capacities and Resilience of Health Systems, Geneva: WHO. Online http://apps.who.int/gb/ebwha/pdf_files/WHA64/A64_R10-en.pdf (accessed 23 December 2015).

WHO (World Health Organization). (1948) Preamble to the Constitution of the World Health Organization as Adopted by the International Health Conference, New York, 19–22 June 1946, and entered into force on 7 April 1948, Geneva: WHO.

WHO (World Health Organization). (2008) Closing the Gap in a Generation: Health Equity through Action on the Social Determinants of Health, Final Report of the Commission on Social Determinants of Health, Geneva: WHO. Online http://www.who.int/social_determinants/thecommission/finalreport/en/ (accessed 23 December 2015).

WHO (World Health Organization). (2011) WHO's Interdepartmental Mass Gatherings Group Best Practice, Geneva: WHO. Online http://www.who.int/csr/resources/publications/MassGatheringflyer_EN.pdf?ua=1 (accessed 23 December 2015).

WHO (World Health Organization). (2014a) Burden of Disease from the Joint Effects of Household and Ambient Air Pollution for 2012, Geneva: WHO. Online http://www.who.int/phe/health_topics/outdoorair/databases/FINAL_HAP_AAP_BoD_24March2014.pdf?ua=1 (accessed 23 December 2015).

WHO (World Health Organization). (2014b) Healthy Cities Promoting Health and Equity – Evidence for Local Policy and Practice: Summary Evaluation of Phase V of the WHO European Healthy Cities Network, Geneva: WHO. Online http://www.euro.who.int/__data/assets/pdf_file/0007/262492/Healthy-Cities-promoting-health-and-equity.pdf?ua=1 (accessed 23 December 2015).

WHO (World Health Organization). (2014c) WHO Report on Addressing Inequities in Obesity, Geneva: WHO. Online http://www.euro.who.int/en/publications/abstracts/obesity-and-inequities.-guidance-for-addressing-inequities-in-overweight-and-obesity-2014 (accessed 23 December 2015).

WHO (World Health Organization). (2015) Protecting People's Health from the Risks of Disasters, Geneva: WHO. Online http://www.who.int/hac/techguidance/preparedness/protecting_peoples_health_march2015.pdf (accessed 23 December 2015).

Wilkinson, P., Smith, K.R., Davies, M., Adair, H., Armstrong, B.G., Barrett, M., Bruce, N., Haines, A., Hamilton, I., Oreszczyn, T., Ridley, I., Tonne, C. and Chalabi, Z. (2009) 'Public health benefits of strategies to reduce greenhouse-gas emissions: household energy', *The Lancet* 374, 9705: 1917–29.

Woodcock, J., Edwards, P., Tonne, C., Armstrong, B.G., Ashiru, O., Banister, D., Beevers, S., Chalabi, Z., Chowdhury, Z., Cohen, A. and Franco, O.H. (2009) 'Public health benefits of strategies to reduce greenhouse-gas emissions: urban land transport', *The Lancet* 374, 9705: 1930–1943.

44

HOUSING AND SETTLEMENTS IN THE CONTEXT OF DISASTER RISK REDUCTION INCLUDING CLIMATE CHANGE ADAPTATION

Elizabeth Wagemann and Camillo Boano

Introduction

Housing is essential to the well-being and development of most societies. It is a complex asset, with links to livelihoods, health, education, security, and social and family stability. Housing acts as a social centre for families, a source of pride and cultural identity, and a resource of both political and economic importance. In 1972, John Turner pointed out that the word 'housing' can be used as a noun or as a verb. The noun 'housing' describes a 'commodity or product', while the verb 'to house' describes the 'process or *activity* of housing' (Turner and Fichter 1972, p. 151). According to this distinction, housing must be understood as what 'it is' and what 'it does' in people's lives, and therefore, people's experience in the way houses are promoted, built or used becomes crucial.

Housing is one of the most affected areas of the built environment when a hazard strikes and has an influence on the development of future risks. The destruction of homes or their loss through displacement or dispossession is one of the most visible effects of disaster. Furthermore, the result of disasters in housing has economic and social impacts because it affects many aspects of daily life, such as local businesses, employment, health, school attendance, and transportation among others. Moreover, the loss of housing stock has implications that go beyond the loss of a building. Houses provide more than shelter but human dignity, security, personal safety, and protection from the climate and diseases as well as cultural identity. Therefore, when houses are affected by any type of disaster, the consequences have repercussions on other aspects of life. Consequently, the reduction of risks and adaptation of housing to future environmental changes, short-term and long-term, is relevant, because it influences many interlinked aspects of daily life and also other sectors.

Usually houses are not isolated objects. From individual dwellings to neighbourhoods, villages, cities and megacities, human settlements are complex systems that interlink both the natural and the built environment with aspects of society, culture, economy, politics and technology. Human settlements are classified as rural and urban, although there are different opinions on how to differentiate them. Some characteristics used to make the distinction are population size, economic activities, and land use. Population size is the most used; however, definitions vary from country to country, and there is not a unified concept (see Satterthwaite 2000, p. 1144).

At different scales (between countries, within countries and within settlements), the risks from disasters are unevenly distributed and the economically poor, as well as the socially and politically excluded, often become the most vulnerable (UN-HABITAT 2008). Some researchers suggest that increases in the number of disasters triggered by climatological, hydrological and meteorological events are affected by the increase in the number of people living in vulnerable areas, and the use of lower-cost design and materials in buildings (McDonald 2003). The climate change-related risks that inhabitants face are a function of exposure to extreme events and the quality of housing, infrastructure and services. The lack of quality is related to the production of housing and settlements, which exacerbate inequality, poverty and vulnerability patterns.

The production of housing reflects socio-economic disparities and affects the exposure to different hazards. Inadequate housing solutions, as well as unsafe and poorly constructed houses, are known as the main causes of risks connected with climate threats. Moreover, disasters are known to be triggered by vulnerable situations due to socio-economic, political and physical factors. These different factors do not affect the population in the same way, and households face different levels of housing vulnerability. In this context, the urban poor have been recognised as most vulnerable due to sub-standard housing in hazardous areas (Tran *et al.* 2012).

Initiatives, merely focusing on hazard control infrastructure projects and post-disaster response and recovery, cannot effectively reduce risks associated with housing and settlements, without fully acknowledging the complex interplay between rapid urbanisation, changing climate and urban development (IFRC 2010). Also, in making their homes, communities use their strengths and resources, which can be applied to disaster mitigation, resilience and collective action. Therefore, it is necessary to consider approaches to mitigating hazards and reducing vulnerability that incorporate social justice and participatory approaches to research and planning (Phillips *et al.* 2010).

Housing and settlements lie at the intersection of DRR and CCA, and should be approached holistically as they are multi-scalar and multi-sectorial in essence. Collaborative approaches between different sectors (e.g., transport, land and infrastructures) and different disciplines (e.g., architecture, engineering, land economy, law and sociology) should occur. Whether focusing on cities, villages, or other type of settlements, the chapter suggests that incorporating DRR and CCA in housing and settlements is in essence avoiding the reductionist implementation of material (only)-oriented housing and delivering a holistic work around socio-spatial integrated strata of economic, cultural, political and social elements.

Dichotomy in the Production of Housing

One of the key drivers behind a population's exposure to hazards is the production of housing, especially in urban settlements, a process that reflects inequality and dualism (Boano *et al.* 2013). Those who are better-off have better tools to recover from and cope with any kind of threats, creating an unequal exposure to hazards (Box 44.1).

On the one hand, the public and private sectors produce formal housing for the ones that can afford it, settled in areas of low hazard, following building codes, with provision of infrastructure and services, within city planning, and insurance to cover rebuilding costs in case of disaster. The design and construction processes are supervised by professionals who ensure that certain quality and codes are accomplished. On the other hand, low income populations, without capital or access to credit, settle in informal areas, often in cheap hazard-prone land that is unstable, without compliance with building codes, though unsupervised self-built processes. Although housing in the formal sector also shows different levels of quality because high income households can achieve better quality houses than medium income households, the poor are the most exposed to risks.

Box 44.1   Social Differences before and after Hurricane Katrina

Elizabeth Wagemann[1] and Camillo Boano[2]

[1] University of Cambridge, Cambridge, UK
[2] University College London, London, UK

Different levels of exposure to hazards, vulnerability and capacity to recover of different groups living in the same affected area were evident after Hurricane Katrina hit the Gulf Coast of the United States in 2005. In the city of New Orleans, class and race were linked to housing quality and access to recovery. While economic resources were essential for getting housing, social networks were crucial for determining the extent of support. After the hurricane, access to jobs and security of income determined the difficulty in coping with the cost of life in the aftermath (Elliott and Pais 2006; Lavelle and Feagin 2006). Although illustrative, the effect of different socio-economic statuses on disaster preparedness, emergency response and disaster recovery is not exclusive to this particular disaster and has been observed in other cases in the United States (Fothergill and Peek 2004).

Inadequate shelter for low-income groups in informal settlements is the result of poorly planned and uncontrolled urbanisation. Urban population is expected to keep growing from 3.4 billion (in 2009) to 6.3 billion by 2050, with small cities projected to absorb over 40 per cent of the predicted increase (World Bank 2011, p. 56). As urban population increases, one of the priorities for city authorities is to prevent precarious land uses and invest in infrastructure. However, homes in informal settlements keep being built incrementally with cheap materials, without robust structures and in unsafe lands. IFRC argues that this reflects the failure of governments and authorities, which have been unable to ensure land with infrastructure for housing in appropriate locations (IFRC 2010, p. 142). Housing officials at the city level face many challenges in regulating informal settlements, such as insufficient capacity and lack of enforcement authority (World Bank 2011, p. 59). Moreover, disasters can have an impact on poverty by increasing the proportion of the poor in a low-income region creating a vicious circle of vulnerability (Chhibber and Laajaj 2013; Guha-Sapir and Santos 2013). Therefore, there is a reciprocal and complex relationship between development and disasters, which can have an effect on people's vulnerability (Wisner *et al.* 2004).

These practices of housing production across the blurred lines of the formal and informal processes also create a clear division between those that can afford to protect themselves from the risks and to adapt their houses, settlements and standards, and those without access to formal tools to cope with risks. There is a 'vulnerability gap' comprised by the lack of knowledge, financial capacity or unwillingness of urban authorities to reduce vulnerabilities, and the poor communities who try to reduce their vulnerabilities but are limited by financial and political capacities (IFRC 2010, p. 45).

Land and housing are certainly difficult to access in urban areas due to restricted land availability, high land and housing prices, complicated land acquisition processes and the presence of different tenure arrangements, including ownership, lease, rental, informal rental and squatting. In this context, communities with limited land rights or informal tenure arrangements are often disproportionately impacted by displacement and damage to shelter caused by disasters. This group generally includes informal dwellers and squatters on public and private land who lack formalised or legal rights; tenants who are unable to return to their homes or land; and households headed by women whose Housing, Land and Property (HLP) rights are not recognised (IFRC 2010).

Vulnerability and Resilience in the Context of Housing and Settlements

Vulnerability refers to the conditions that increase the susceptibility of a community to hazard impacts and are determined by physical, social, economic and environmental factors (UNISDR 2004, p. 16). The physical factors refer to susceptibilities of location and the built environment, and may be described as 'exposure' (UNISDR 2004, pp. 41–42). This factor is influenced by the site, quality of the design and materials used for buildings, housing and infrastructure. The social factors are linked to the well-being of individuals, communities and society, including levels of education, security, peace, human rights, governance, social equity, values, beliefs and overall collective organisational systems (UNISDR 2004, p. 42). In this context, some groups are more vulnerable than others due to different privileges, classes, marginalisation and power relations that may have an influence on the physical features of a settlement or a community, and create more exposure to risks. The economic factors depend on the levels of individual, community, and national economic savings, debts and access to credits, loans and insurance (UNISDR 2004, p. 42). Finally, environmental factors are linked to the extent of natural resource depletion and degradation (UNISDR 2004, pp. 42–43). All these four factors are interconnected, and influence the level of vulnerability of houses, communities and settlements.

The concept of resilience is interpreted in different ways from many different fields. Alexander (2013) summarises resilience as a multi-faceted concept adaptable to various uses and contexts in different ways. UNISDR describe resilience as 'the capacity of a system, community or society potentially exposed to hazards to adapt, by resisting or changing in order to reach and maintain an acceptable level of functioning and structure' (UNISDR 2004, p. 17), determined by the capacity for learning from past disasters and to improve risk reduction measures. Alexander positions resilience for DRR linked with social and psychological aspects, and for CCA linked with social and technical aspects (Alexander 2013, p. 2714). The use of resilience as an entry point to include CCA in DRR has advantages, such as use of the CCA status to reach a wider agenda and ensure long-term policies (Mercer 2010), to avoid the hazard approach and support reverting the root causes of vulnerability (Kelman and Gaillard 2010), and to understand impacts of unequal distribution and unmanaged development to better cope with natural hazards.

A disaster-resilient community is one that has the capacity to resist or adapt, manage or maintain basic functions, and recover from a disastrous event with specific behaviour, strategies and measures (Twigg 2009, p. 8). The emphasis on resilience means stressing what communities can do for themselves through strengthening their capacities and minimising their vulnerabilities while maximising the opportunities offered by risk reduction measures (Twigg 2009, pp. 8–9; Box 44.2). However, communities are complex social realities. Often, they are viewed in spatial terms despite the need to consider other dimensions such as economic differences, social status, religion and interests. The level of a community's resilience is influenced by a wider context, such as public infrastructure, political linkages, and social and administrative services (Twigg 2009). To recognise the marginalised and vulnerable groups is crucial to identify where resilience building is needed. The capacity of poor communities to reduce risk, then, does not only depend on the physical attributes of the settlements but also the complex institutional relationships with political and economic actors (Miles *et al.* 2012). Understanding power relations and the disproportionate effect of disasters on disadvantaged communities is required to understand the ability of residents to reduce risk (Boano *et al.* 2013).

The political economy approach to social vulnerability and resilience emphasises the distribution of power and wealth within cities which perpetuates ever-widening risk between different populations (Boano *et al.* 2013). The marginalisation of low income populations creates an even more notorious division between formal and informal, where poor communities are most

---

Box 44.2   Resilience as a Consequence of External Inputs

James Lewis[1]

[1] Datum International, Marshfield, UK
Based on Lewis (2013).

Chiswell village, once of 145 people, is situated within a trough between 196 metres of cliff face and a 14-metre-high shingle bank, beyond which is the English Channel and the Atlantic Ocean. The shingle bank normally protects the village from wind, sea spray and minor storms; the converse of Chiswell's access to the sea being that, from historical evidence, the sea has access to Chiswell.

Towards the end of 1978 and again in early 1979, Chiswell experienced two unusually violent storms in which many houses were damaged by wind and severe flooding. The Chiswell Residents Action Group was founded to secure help for the many people who could not continue to occupy properties on which they were committed to mortgage payments. Many people left the village, which appeared to be in danger of physical and social disintegration.

In 1986, an extension of an existing sea wall was completed, together with other improvements to better protect Chiswell. Some new houses were built, flats/apartments created and existing properties were maintained and improved. The population grew, new businesses opened and a new confidence prevailed. Portents of a new resilience?

In December 1989, a five-year-return-periodstorm caused waves to overtop the new sea wall, damaging eight properties. Sea waves do not always comply with design parameters. Will additional engineering works serve to protect Chiswell from an increased incidence and severity of storms now resulting from a changing climate? Will improved sea defences, a product of resources from beyond the community, bring about further additions to an increased village population in a historically vulnerable location? Will a newly restored resilience be undiminished, apply to all and be adequate against a historically established and increasing vulnerability?

---

exposed and vulnerable due to the location of their houses and their quality, among other factors (Boano *et al.* 2013). Vulnerability and resilience are therefore closely connected with the quality and location of housing. Lack of control of urban growth and oversight in housing production from local authorities mean that frequently low income populations resolve their housing needs through the informal market, often in a risk zone, without compliance with building codes, and inadequate materials. Furthermore, the marginalisation of low income populations is further deepened when territorial plans do not consider the existing conditions of informal construction in risk zones, increasing the gap between the formal and informal sectors (Boano *et al.* 2013). As a result, the poorest and socially marginalised are most likely to suffer from disasters associated with natural hazards (Bosher and Dainty 2011; Wisner *et al.* 2004), either related to climate change or not.

From Technocracy to Community

DRR has been focused on supporting the most vulnerable through community participation, while CCA has been centred on a top-down perspective, based on technocracy and global policy agendas rather than practical implementation (Mercer 2010). The difference in focus can be explained by the development of DRR, which historically has been focused on prevention, mitigation and preparedness based in development research and policy perspectives, while CCA is rooted in scientific theories of the changing climate; i.e., a hazard-orientated approach (World Bank 2011, p. 21). Although they both share an interest in community-based processes, DRR does it from current and historical risks in order to implement future planning while CCA emerges from future risks and policy agenda. Therefore, there is an opportunity in the convergence of them, with all elements of CCA already part of DRR, to create community-based measures that tackle current as well as projected risks in the housing sector and settlements.

The tools of CCA are mitigation and adaptation, two complementary strategies and inseparable elements for reducing and managing the risks of climate change and contributing to climate-resilient pathways for sustainable development, while raising issues of equity, justice and fairness (IPCC 2014). While mitigation addresses the root causes by reducing energy use and greenhouse gas emissions and enhancing carbon sinks, adaptation seeks to lower the risks posed by the consequences of climate change and to exploit beneficial opportunities (IPCC 2014, p. 76). Mitigation reduces the hazards and adaptation reduces the vulnerability to the hazards. Mitigation and adaptation are underpinned by common enabling factors – such as sound technologies and infrastructure, effective institutions and governance, sustainable livelihoods and behavioural and lifestyle choices – and synergies between them can increase the cost-effectiveness of actions (IPCC 2014, p. 26). Despite the uneven distribution across the world, what is important to note is that mitigation contributes to adaptation because its success on a global scale can reduce the need for adaptation in the long-term.

Mitigation at the scale of buildings, including housing, has been focused on a technocratic approach, through the rational use of energy in buildings associated to the reduction of emissions. Legal measures, regulation, and assessments for best practice have been promoted – mainly in wealthy countries, such as BREEAM (Building Research Establishment Environmental Assessment Methodology) in the UK, LEED (Leadership in Energy & Environmental Design) in the US, and CASBEE-UD (Comprehensive Assessment System for Built Environment Efficiency) in Japan. Under this umbrella, mitigation to climate change is seen as the promotion of best practices based on energy efficiency, sustainable urban development and the creation of building codes and energy strategies (ETC/SCP 2013).

On the other hand, adaptation strategies to climate change at the scale of settlement and housing have been divided into non-structural and structural approaches. The non-structural have been focused on regulating housing directly through building code and zoning, indirectly influencing transportation, infrastructure and investment (World Bank 2011, p. 58). Structural adaptation strategies include building elevation, resilient design, retrofit of homes and buildings, additions of green roofs or sun shading, smart ventilation and water storage, among others (World Bank 2011, p. 60). But codes, standards, assessments and tools either related to mitigation or adaptation are currently available to only a few, because safe construction is costly and the most vulnerable keep producing their houses outside any planning and control while the rich often ignore, circumvent, or use their financial and political clout to flout planning and control mechanisms and regulations. Petal *et al.* (2008) argue that 'building code enforcement' is crucial and must not be resolved by punishment but 'building-code compliance' where communities voluntarily accept codes.

The enforcement of high standards and codes has found difficulties being applied not only in less affluent countries but also in wealthier countries, as the cases of Turkey before the 1999 earthquake and Miami before Hurricane Andrew have shown. The main problem with implementing strict building codes has been the difficulty of enforcing and monitoring them. The cost of complying with building codes is prohibitive for some, and therefore in many cases regulations have failed to reflect what is possible to afford and achieve locally, considering that a high proportion of housing construction is an incremental process done by inhabitants themselves, in many cases, without knowledge of the codes or strategies for building structurally safe houses. Also, homeowners may be unaware of the extent of their houses' vulnerability or they may lack appropriate financial resources to be involved in any type of improvement (World Bank 2011).

Institutions involved in DRR are poorly connected to those involved in producing housing, and therefore they fail to coordinate their actions (Boano *et al.* 2013). Therefore, better indicators of coping capacity at the micro level are required in order to reinforce home-grown measures and progressively reduce dependence on and expectations of aid, as well as better assessing the short- and long-term impacts of hazards on communities (Guha-Sapir and Santos 2013). Better connections, good management, governance and leadership can promote reform and improve performance (Baker 2011). Other mechanisms, such as public awareness and community involvement can play a role in limiting vulnerabilities. Also, engaging vulnerable communities, as well as other stakeholders, such as local government, the private sector, academia, NGOs and community-based organisations can be the first step to making changes. Programmes designed to reduce risks and support adaptation can include social funds, community-driven development and slum upgrading programmes (Baker 2011, p. 18).

CCA as Part of DRR through Community Participation

Although researchers and practitioners of Community Based Disaster Risk Reduction (CBDRR) seek to integrate CCA and DRR at the community level (Mercer 2010; Twigg 2009), some have continued to work independently due to different funding mechanisms, concepts and institutional frameworks. However, overlapping areas between CCA and DRR are being acknowledged, with the straightforward approach being CCA as a subset within DRR (Mercer 2010). This would provide increased attention to community focused vulnerability assessments, including measurement and improvement of social resilience (World Bank 2011, p. 21). Knowledge accumulated over the years on DRR at community levels based on preparedness and resilience can teach plenty to CCA.

At the scale of settlements and housing, adaptive capacity of the most vulnerable can be integrated with DRR (World Bank 2011, p. 47). Placing CCA in DRR in community involvement can be crucial to identify diverse and unknown risks. Hazards identification discovered through local knowledge can help to understand diverse pre-existing risks, but weather-related hazards connected to climate change along with urban growth and other factors combined can change the nature of vulnerabilities. When communities are involved in the process, they can be more aware and get engaged in the design and planning of community adaptation. In the process, it is crucial that the relevant information is translated in a way that is understood by communities, and allows opportunities to share experiences, failures and successes. Also, by giving the most vulnerable a voice, persistent political disadvantage can be reduced through promoting more inclusive planning processes (World Bank 2011, p. 49).

Community participation in all phases of the design of safe houses can increase mutual understanding between stakeholders, allow sharing knowledge on safe housing, provide options that are adequate to local contexts, and ensure user satisfaction (Tran *et al.* 2012, p. 14). Community participation in the design and construction processes of housing is widely encouraged by NGOs, policy makers and scholars involved in post-disaster recovery and reconstruction, with the aim of building community resilience. Safe building construction can help to build resilience in pre- and post-disaster periods, can improve pre-disaster fragilities, and is seen as an opportunity to meet short- and long-term needs of disaster-prone communities (Tran *et al.* 2012, p. 14).

In relation to the involvement of communities in CCA, community-based adaptation (CBA) is a bottom-up approach based on experiences, knowledge and capacities of local people. The approach draws on participatory approaches developed in both DRR and community development work and can include structural and non-structural measures (McDevitt 2012, p. 29). Local and international NGOs have helped to develop positive models for CBA, as well as strategies to develop adaptive capacity at the local level (World Bank 2011, p. 49).

Comprehensive and 'at scale' participation mechanisms with all different kinds of actors e.g., local authorities, service deliverers, landowners, as well as NGOs and academia, are useful to develop innovations in community planning. Cities, instead of being a vertical 'unit of control', are provided with 'smaller people-based and local smaller units' (Boonyabancha 2005). These units 'can be a system of self-control for a more creative, more meaningful development' (Boonyabancha 2005, p. 23). They can contribute to DRR including CCA, conceived and practised as a vehicle for social change, maintaining a twofold function of improving the material reality of the urban poor and fostering confidence in marginalised groups concerning their skills and capacities, individually and collectively. Such visible actions illustrate and suggest alternative possibilities and transformative potentials, encouraging those in similar situations to follow in order to expand the resilience of the whole system. This is an iterative process in which, over time, material improvements reinforce the terms of engagement with different actors and vice versa, building up strength and power in and of the communities.

In many cases, individuals and communities living in settlements exposed to hazards do take measures to reduce risks by themselves, such as the protection of their houses using resistant materials, raising houses on plinths, coordinating temporary sites to move to and contingency plans in case of flood. These adaptations are influenced by the priorities and available resources, and they can be supported by the local community. Community-based organisations have shown to be resourceful in post-disaster work, supporting the reconstruction of housing and infrastructure, helping the affected to get control back of their lives, and strengthening their own social cohesion.

However, community participation can be developed in many ways, and not all types of participation ensure the best results (Davidson *et al.* 2007). CBA can contribute to vulnerability reduction but many risks are the products of deficiencies in infrastructure that have a different scale, and cannot be tackled by communities themselves. Also, in many cases true participation does not occur, suggesting a gap between theory and practice (Davidson *et al.* 2007). Moreover, the concept of community is not always clear. For example, in some rural settings, dispersed dwellings do not create a unified group that is recognised as a community, and in urban contexts the heterogeneity of its population has developed individualism and a low degree of involvement in community participation. Also, some adaptation strategies can produce some conflicting issues, bringing benefits to some inhabitants and problems to others. For example, labelling lands in risk maps may be seen as an excuse for displacing low-income communities and giving access to the land to other stakeholders for other uses, or the diversion of a river course to prevent flooding in one area may increase the risk of flooding in other areas.

The factors that guide the selection of adaptation strategies are specific to each settlement due to financial, technical, political and social issues, and there could be discrepancies between political leaders and communities (World Bank 2011). In order to balance the differences, assessments and participatory processes can benefit from an integration of top-down and bottom-up community based perspectives (World Bank 2011, p. 42), embedding CCA into DRR.

Conclusion

Housing and human settlements are complex systems that connect society, economy, politics, culture and technology with the natural and the built environment. They reflect socio-economic disparities, and therefore different groups are affected by hazards in different ways, and they face different levels of vulnerability. Housing provides more than a shelter, acting as a source of identity, pride, security, and as a social centre for families, but it is also one of the most affected areas when disasters occur, with an effect on the development of future risks and the production of vulnerabilities. Bad quality housing settled in unsafe urban or rural locations as the result of lack of planning and resources or wrongly conducted resettlement and reconstruction plans, makes low-income and disadvantaged groups the most vulnerable. Frequently low-cost housing does not meet safety standards, due to unaffordability to the poor. Also, lands exposed to hazards can be cheaper, and therefore the only option for less advantaged groups. Therefore, buildings in these areas are affordable to those less well-off but exposed to future risks, creating a dichotomy between those that can afford safe houses and land, and those that cannot. However, high income groups can be exposed to risks too; i.e., rich people wanting beach-side, river-side, or forest-side properties leave themselves exposed to the potential hazardous nature of these locations, as is seen in Canberra and Miami. Nevertheless, although exposed to risks, more advantaged groups can access savings, credit, loans and insurance, and have the capacity and tools to recover and adapt for future events.

Due to the complexity behind the production of houses and settlements, CCA and DRR strategies should be approached holistically, avoiding focusing on the material-built aspects only and incorporating the social, cultural, political and economic elements of the process. In this context, there are opportunities to embed CCA within DRR, where both contribute to a development agenda from complementary approaches. They both share an interest in community-based processes, and although they have a different emphasis can tackle current and future risks linked to climate related events. Better connections between institutions involved with DRR and CCA and those producing housing, urban infrastructures, good governance, public awareness and community involvement can play a role in limiting vulnerabilities. Embedding CCA in DRR through community participation can help affected groups to understand weather-related hazards connected to climate change alongside pre-existing risks and to engage in designing and planning for the future.

Integrating DRR and CCA in all disaster phases including reconstruction and recovery should be interpreted as a political process, through which the scale of urban governance itself is being re-defined (Mercer 2010). Embedding CCA within DRR across different scales of governance, especially at local and community level, would enable local residents to take the initiative on their own. In this context, initiatives doing just this would present an opportunity to critically question the scale of urban and rural governance through productive contradictions between local government and communities.

References

Alexander, D.E. (2013) 'Resilience and disaster risk reduction: An etymological journey', *Natural Hazards and Earth System Sciences* 13: 2707–2716.

Baker, J.L. (2011) *Climate Change, Disaster Risk, and the Urban Poor: Cities Building Resilience for a Changing World*, Washington DC: The World Bank.

Boano, C., Frediani, A.A., Aston, T., Chacon, M. and Salamanca Mazuelo, L.A. (2013) *Exploring Oxfam's Room for Manoeuvre to Address the Production of Urban Risk. A Research Report Prepared by the DPU London for Oxfam Latin America and the Caribbean, with Case Studies from Guatemala City and La Paz*, London: Development Planning Unit, University College London.

Boonyabancha, S. (2005) 'Unlocking people energy', *Our Planet Magazine* 16: 22–23.

Bosher, L. and Dainty, A. (2011) 'Disaster risk reduction and 'built-in' resilience: Towards overarching principles for construction practice', *Disasters* 35: 1–18.

Chhibber, A. and Laajaj, R. (2013) 'The interlinkages between natural disasters and economic development', in D. Guha-Sapir and I. Santos (eds) *The Economic Impacts of Natural Disasters*, New York: Oxford University Press, pp. 29–54.

Davidson, C.H., Johnson, C., Lizarralde, G., Dikmen, N. and Sliwinski, A. (2007) 'Truths and myths about community participation in post-disaster housing projects', *Habitat International* 31: 100–115.

Elliott, J.R. and Pais, J. (2006) 'Race, class, and Hurricane Katrina: Social differences in human responses to disaster', *Social Science Research* 35: 295–321.

ETC/SCP (European Topic Centre on Sustainable Consumption and Production). (2013) *Housing Assessment, Final Report,* Copenhagen: European Topic Centre on Sustainable Consumption and Production.

Fothergill, A. and Peek, L.A. (2004) 'Poverty and disasters in the United States: A review of recent sociological findings', *Natural Hazards* 32: 89–110.

Guha-Sapir, D. and Santos, I. (2013) 'Conclusion', in D. Guha-Sapir and I. Santos (eds) *The Economic Impacts of Natural Disasters,* New York: Oxford University Press, pp. 311–313.

IFRC (International Federation of Red Cross and Red Crescent Societies). (2010) *World Disasters Report: Focus on Urban Risk,* Geneva: IFRC.

IPCC (Intergovernmental Panel on Climate Change). (2014) *Climate Change 2014: Synthesis Report. Contribution of Working Groups I, II and III to the Fifth Assessment Report of the Intergovernmental Panel on Climate Change,* Cambridge: Cambridge University Press.

Kelman, I. and Gaillard, JC (2010) 'Embedding climate change adaptation within disaster risk reduction', in R. Shaw, J.M. Pulhin and J.J. Pereira (eds) *Climate Change Adaptation and Disaster Risk Reduction: Issues and Challenges,* Bingley, UK: Emerald Group Publishing Limited, pp. 23–46.

Lavelle, K. and Feagin, J. (2006) 'Hurricane Katrina: The race and class debate', *Monthly Review: ProQuest Social Sciences Premium Collection* 58: 52–66.

Lewis, J. (2013) 'Some realities of resilience: An updated case study of storms and flooding at Chiswell, Dorset', *Disaster Prevention and Management* 22, 4: 300–311.

McDevitt, A. (2012) *Climate Change Adaptation Topic Guide, 2nd edition,* Birmingham: Governance and Social Development Resource Centre (GSDRC), International Development Department, College of Social Sciences, University of Birmingham.

McDonald, R. (2003) *Introduction to Natural and Man-made Disasters and their Effects on Buildings: Recovery and Prevention,* Oxford: Architectural Press.

Mercer, J. (2010) 'Disaster risk reduction or climate change adaptation: Are we reinventing the wheel?', *Journal of International Development* 22: 247–264.

Miles, S.B., Green, R.A. and Svekla, W. (2012) 'Disaster risk reduction capacity assessment for precarious settlements in Guatemala City', *Disasters* 36: 365–381.

Petal, M., Green, R., Kelman, I., Shaw, R. and Dixit, A. (2008) 'Community-based construction for disaster risk reduction', in L. Bosher (ed.) *Hazards and the Built Environment: Attaining Built-in Resilience,* London: Taylor & Francis, pp. 191–217.

Phillips, B.D., Thomas, D.S.K., Fothergill, A. and Blinn-Pike, L. (eds) (2010) *Social Vulnerability to Disasters,* Boca Raton: CRC Press.

Satterthwaite, D. (2000) 'Will most people live in cities?', *British Medical Journal* 321: 1143–1145.

Tran, T.A., Tran, P., Tuan, T.H. and Hawley, K. (2012) *Review of Housing Vulnerability,* Hanoi: Institute for Social and Environmental Transition.

Turner, J.F.C. and Fichter, R. (1972) *Freedom to Build: Dweller Control of the Housing Process,* New York: The Macmillan Company.

Twigg, J. (2009) *Characteristics of a Disaster-resilient Community: A Guidance Note, 2nd edition,* London: UK Department for International Development (DFID).

UN-HABITAT (United Nations Human Settlements Programme). (2008) *Mitigating the Impact of Disasters: Policy Directions – Enhancing Urban Safety and Security, Global Report on Human Settlements 2007 volume 3,* London: Earthscan.

UNISDR (United Nations International Strategy for Disaster Reduction). (2004) *Living with Risk: A Global Review of Disaster Reduction Initiatives,* Geneva: UNISDR.

Wisner, B., Blakie, P., Cannon, T. and Davis, I. (2004) *At Risk: Natural Hazards, People's Vulnerability and Disasters, 2nd edition,* London and New York: Routledge.

World Bank. (2011) *Guide to Climate Change Adaptation in Cities,* Washington DC: The World Bank.

45

HUMAN MOBILITY AND DISASTER RISK REDUCTION INCLUDING CLIMATE CHANGE ADAPTATION

Lorenzo Guadagno and Daria Mokhnacheva

Introduction

Population movements and their implications have been the object of growing attention within research, policy and operational work focusing on the linkages between people and the environment for at least three decades. Topics such as migration, displacement, relocation and remittance transfers are becoming increasingly central to Disaster Risk Reduction (DRR) including Climate Change Adaptation (CCA) research and policy.

Over the few last years, the mobility and environment discourse has mainly been approached as part of a focus on the effects of global environmental change on communities and societies, and in particular around the notion of 'environmental migration', any movement of people who are forced or choose to leave their habitual homes predominantly due to sudden or progressive change in the environment (IOM 2011). As a subset of broader research, policy and practical efforts on sustainable development, and in particular on the interactions between people and the environment (of which mobility is a fundamental yet not unique dynamic), the domain is concerned with a series of issues that have been elaborated in a variety of other fields. Despite this fact, the mobility and environment discourse has only partly and slowly leveraged perspectives that have been fundamental to at least three decades of DRR thinking.

On the other hand, the attention given to population movements within the DRR domain has been more fragmentary and may yet result in the establishment of a dedicated body of research and practice. As this chapter will argue, adopting an integrated risk reduction perspective to look at mobility and environment dynamics could further support the theoretical evolution of the field – and contribute to the development of a more comprehensive framework to underpin research, policy and operational work.

A Theoretical Look at Mobility and the Environment

Environmentally Induced Mobility

Over the last few decades, the mobility and environment domain has been dominated by attempts to quantify, identify and define the legal status of 'environmental migrants'. Forecasts of hundreds of millions of people moving out of less wealthy countries as a consequence of climate change have shaped much of the mainstream perceptions on the topic (Box 45.1). These discourses have contributed to deepening the reflection on the environmental determinants of people's mobility and have brought mobility and environment issues to the attention of policy makers and of the general public (Ionesco *et al.* 2016). However, they have often been based on hazard-centric, deterministic interpretations, which have been refuted in academic debates (for an overview of this evolution see Morrissey 2012). Most of this work has developed in parallel with, yet isolated from, the evolution of DRR thinking in the 80s and 90s – however, it is integrating many elements that are key to the interpretation of the interactions between people and the environment from a DRR perspective.

The research on the environmental drivers of population movements, in particular, has highlighted how people's decisions to move (or not to move) are always multi-causal: they are based on the distribution of local and distant resources, opportunities and hazards, on people's capacities, preferences and perceptions, and on the conditions enabling or constraining their choices. The environment underpins all these elements, which are co-determined by the economic, social and political features and processes of any human context (Foresight 2011).

No matter how severe the environmental pressures on a given community are, the need, capacity and willingness to move of different people are always determined by their individual and collective aspirations, access to local and distant livelihood options, the availability of social and material resources to undertake the movement, and the presence of barriers and incentives to movement. This holds true even in the case of displacement induced by sudden and slow-onset hazards. Global estimates of tens of millions of people displaced every year in disasters (IDMC 2015), for instance, really aggregate a variety of patterns of movement – ranging from short-term, pre-disaster evacuations, to international migration and protracted displacement (Ionesco *et al.* 2016).

The human, economic, social and environmental impacts (including the mobility effects) of environmental hazards and changes are mediated by socially constructed factors of vulnerability and individual and collective capacities and responses. Hence, the relationship between environmental pressures and population movements is neither simple, nor univocal. Environmental hazards can reduce the availability of resources that are needed to move, thus reducing people's mobility (Black *et al.* 2012), while preventing or mitigating hazards can help protect and multiply resources that can be invested in migration projects, ultimately resulting in increased outward movements (Sakdapolrak *et al.* 2014).

This calls into question the existence of identifiable, specific categories of 'environmentally induced' (and even less of 'climate-induced') movements and migrants/displaced (Faist and Schade 2013) and requires broadening the focus away from the sole 'nature' to look at inextricable socio-environmental processes that influence, and are influenced by, mobility patterns – retracing the core DRR tenet of 'taking the naturalness out' of what are essentially social processes. While environmental features and dynamics underpin all local and global patterns of movement and immobility, they have to be interpreted in the light of social, political, economic and cultural processes determining people's capacities and conditions of vulnerability.

Box 45.1   Refugees and Trapped Populations

Lorenzo Guadagno[1] and Daria Mokhnacheva[1]

[1] International Organization for Migration, Geneva, Switzerland

The use of the 'refugee' terminology with reference to those forced to move due to environmental causes represents a stretch of the 1951 Convention and 1967 Protocol (UNHCR 2010). While clear and potentially useful to frame climate-induced movements as an environmental justice issue, it has instead been instrumental to promoting a security-based approach to mobility and environment issues in more affluent countries, based on the (unfounded) assumption that massive population inflows from less affluent countries would result in increased insecurity and instability in host countries (Bettini 2013).

Moreover, it creates an exclusive focus on forced cross-border movements. Instead, evidence shows that even in the face of environmental changes or disasters population movements overwhelmingly take place over short distances, and within national or regional borders (Faist and Schade 2013). These movements also remain intrinsically diverse: some movements follow patterns and routes typical of economic migration, and others look more like rush escapes to safe places. The most vulnerable people might actually not be able to move out of at risk or affected areas at all (so-called 'trapped populations') (Black *et al.* 2012). Therefore, a refugee-like regime might ultimately not serve all those suffering the most severe impacts of disasters and environmental change.

Migration and Risk Reduction

More recently, research has focused on the way moving contributes to people's and households' resilience in the face of environmental hazards. Drawing from the literature on migration and development, moving has been looked at not only as a symptom of vulnerability, but also as a potential option for households to anticipate, or cope with, the impacts of hazards, change and uncertainty (migration as adaptation), as well as an integral component of people's wellbeing strategies (migration for adaptation) (Scheffran *et al.* 2012). By moving, the members of a household can diversify their exposure to hazards, reduce pressures on local resources, and access opportunities and services across places, all of which expand their networks, resources and capacities (Greiner and Sakdapolrak 2012).

Other work has looked at the environmental impacts of population movements in areas of origin and destination, examining changing use of land and natural resources and occurrence of hazards and disasters as a consequence of circulation of people, skills and assets linked with emigration and immigration (Foresight 2011; for a critique of its recommendations, see Felli and Castree 2012). Emphasis has also been placed on planned relocations as a possible strategy to be adopted by states to protect communities exposed to the negative impacts of climate change (Yamamoto and Esteban 2014). Both elements have been central to decades of literature focusing on people's (re)settlement as a determinant of disaster risk (amongst others, see Cernea 1993; Davis 1981), which has exposed the potentially negative effects of these processes on hazard exposure and on the social and economic fabric of communities.

While still largely rooted in the idea of 'environmentally induced' movements, these studies have also focused on the consequences of moving (i.e., the implications of people's movement on their – and others' – ability to withstand, adapt to, and harness changing environmental conditions). This has resulted in the mobility and environment domain increasingly looking at movements of all kinds (migration for employment, education or family reunification, or conflict-induced displacement) for the wellbeing and risk impacts they produce through population distribution and availability and circulation of resources – a particularly significant shift as our societies are increasingly shaped by flows, unprecedented in size, of internal and international migrants, refugees and asylum seekers, students, tourists and other people on the move.

Mobility as a Dynamic of Risk

The evolution sketched above has progressively aligned the mobility and environment domain with previous concepts rooted in the migration and development literature, which shares a common approach with DRR perspectives. Looking at mobility within the framework of household-level livelihood decisions, this work has stressed the potential benefits of population circulation and associated transfer of remittances, knowledge and social resources for the wellbeing of individuals, communities and societies. It has also showed that moving can be the expression of limited options, and result in exploitation and marginalisation of those on the move, as well as impoverishment and increased vulnerability of people in their communities of origin and of destination (Gasper and Truong 2010).

All mobility decisions (including the decision not to move) result in a variety of positive and negative, intertwined wellbeing outcomes, for different people, across locales and timescales. Such outcomes include the creation and reduction of risk associated with natural hazards and environmental change: in areas of origin, transit and destination, mobility decisions contribute to shaping population distribution, pressures on ecosystems and occurrence of and exposure to natural hazards, as well as availability of and access to material and immaterial resources that can be leveraged for preventing, mitigating and recovering from their impacts. It is impossible to understand mobility's risk and risk reduction outcomes looking only at specific locales, actors and times (Guadagno 2016).

Who benefits from and who bears the cost of (im)mobility decisions is largely determined by underlying social, economic and political structures and dynamics – intra-household gender and age roles; class structures and relations in communities of origin and destination; national, regional and global political systems. These factors shape the variety of hazards people face (both 'everyday' ones, such as unemployment, impoverishment and poor health, and 'extreme' ones, such as earthquakes, cyclones and war) and the opportunities they can access to pursue security and wellbeing (Wisner *et al.* 2004).

These factors further determine how costly and risky moving and not moving are to different people, and therefore their need and ability to move, generating a spectrum of rather voluntary and rather constrained (im)mobility decisions for different individuals. This shapes people's capacity to freely choose a place to live within spaces that present unequal distribution of hazards and opportunities – arguably the most fundamental element of risk reduction.

Mobility (understood as the capacity to move, including in anticipation of and in response to both normal-life events and environmental shocks and stresses) has an intrinsic resilience value for people and households. Conversely, obstacles to people's choices for movement or non-movement compound their vulnerability. Whether in situations of migration, displacement, evacuations, relocations or immobility, different people's needs and abilities to move; the conditions in which they (do not) move; the resources and opportunities they and others acquire and lose in the process; and the ways in which the affected systems change are all determined by a context's structural conditions – expressed, crucially, in the legal and operational frameworks regulating the various kinds of movement (De Haas 2014).

Population movements and environmental change are two intertwined dynamics of risk, taking place in the context of these underlying social, economic and political dynamics that have to be adequately factored into all reflections on (im)mobility decisions and patterns, and their outcomes. For the mobility and environment field, in particular, this requires expanding the analysis to look at how policies and measures, investments and cultural stances hindering or facilitating people's movement (and the transfers of material and immaterial resources they induce) reflect on risk creation and reduction processes across groups, locales and timescales.

The Policy Picture

The main international policy instruments on climate change and DRR have, to some extent, reflected this evolution of the perspectives on mobility and the environment. The idea that population movements influence people's and communities' risk levels, in particular, has progressively been integrated within relevant policy discourses, alongside better established discussions on reducing and managing movements induced by environmental factors. However, this has yet to result in an explicit recognition of mobility decisions and patterns as one of the fundamental dynamics of risk.

Population Movements in the SFDRR

The Sendai Framework for DRR (UNISDR 2015) recognises migrants' potential to contribute to the resilience of communities and societies of origin and destination through their knowledge, skills and capacities (paragraphs 7 and 36.a.vi), and recommends engaging them in the design and implementation of DRM efforts (paragraph 27.h). It also includes references to the planned movement of settlements from risky sites through relocations (paragraph 27.k), as well as to the need to prepare for and manage disaster-induced evacuations and displacement (paragraphs 33.h and 33.m), including in the case of transboundary movements (paragraph 28.d). Lastly, it calls for efforts to address disaster-induced mobility in ways that build the resilience of mobile people and their host communities (paragraph 30.l).

Despite looking at various facets of the mobility and risk nexus, the SFDRR does not list migration status as one of the potential elements that shape vulnerability to disasters and leaves largely implicit the issue of human (im)mobility as one of the drivers of risk at the global and local level (only under the umbrella of 'demographic change' in paragraphs 6 and 30.f). Furthermore, it does not recognise the legal and practical conditions under which population movements take place as a determinant of mobility's risk and risk reduction outcomes – confirming how, despite some attention to global risk drivers, it largely steered away from politically contentious issues.

Mobility in the UNFCCC

Mobility-related topics have been more solidly integrated within the climate change policy process. In 2010, paragraph 14.f of the United Nations Framework Convention on Climate Change Cancun Adaptation Framework called for enhancing understanding, coordination and cooperation with regard to climate change induced displacement, migration and planned relocation. The impacts of climate change on mobility have subsequently been covered under the loss and damage programme in Doha in 2012 (decision 3.CP/18, paragraph 7.a.vi) and at the 2013 and 2014 Conferences of Parties (COP).

The Paris Agreement reached at the COP21 in December 2015 acknowledged migrants as a specific group to be taken into account in actions undertaken by the UNFCCC Parties to address climate change (UN 2015, paragraph 50), and led to the establishment of a task force to develop recommendations for integrated approaches to avert, minimise and address displacement related to the adverse effects of climate change. The scope of the work of the task force could be flexible enough to cover broader mobility patterns and their impacts, and the UNFCCC's call for an active involvement of such actors as IOM, UNHCR and the Platform on Disaster Displacement in the task force is significant in this regard.

Mobility in National Policies

National adaptation strategies could constitute one of the channels to work on mobility and environment issues. In fact, mobility is highlighted as a key element in nearly fifty National Adaptation Programmes of Action (NAPAs). Most countries look at environmentally induced displacement or rural outmigration as sources of additional pressure on local and urban infrastructures, identifying mobility as a driver of vulnerability. Some support community stabilisation and risk reduction strategies to prevent environmental pressures from resulting in migration and displacement (Ionesco *et al.* 2016).

Very few countries (e.g., Afghanistan and Samoa) note migration's role in supporting coping strategies for individuals and households at risk of or affected by hazards. Many Small Island Developing States and a few other countries see government-led resettlement as a potential measure to help communities seemingly threatened by rising sea levels and other climate change impacts. Twenty-four States have mentioned human mobility in their Intended Nationally Determined Contributions (INDCs) submitted to the UNFCCC ahead of the COP21, including by recognising its potential benefits for adaptation (Mach and Traore Chazalnoel 2015).

Few national DRR or DRM policies look explicitly at population movements, and only focus on preparing for and addressing disaster-induced displacement. Conversely, migration management policies rarely consider environmental change and disasters. Regional or bilateral migration or circulation schemes and agreements have been looked at as potential options to open up opportunities for reducing vulnerability through movements; yet, while seasonal or circular labour migration schemes might allow migrants to diversify their livelihoods and improve their incomes, they have also received harsh criticism, as they often have detrimental effects on migrants' rights and living conditions – largely to the benefit of employers and host communities (Gasper and Truong 2010).

There have been few attempts at the national level to comprehensively address population movements, environment, climate change and disaster risk as part of broader development planning. Efforts to develop national strategies in this domain can build on decades of experience and thinking in risk reduction, sustainable ecosystem management, management of migration, displacement and relocations. Some countries, such as the Federated States of Micronesia and the Cook Islands, have developed joint adaptation and DRR policies, which recognise migration and planned relocations as elements of the continuum of measures that might be needed in order to reduce risk for vulnerable communities (Ionesco *et al.* 2016), but such examples remain limited, and largely disconnected from broader national migration management or poverty reduction strategies.

Finally, the implementation of any of these policies – let alone a combination of them all – often remains challenging due to limited funding and technical capacity, as well as changing priorities and conditions. Many countries in need of urgent risk reduction actions are also subject to political instability and conflicts, with population movements representing a key dynamic of both these processes. This is yet another reason to take a comprehensive approach to sustainable development that integrates mobility issues within interventions aiming to address other fundamental environmental and socio-economic issues, including disasters, and climate change's impacts on hazards.

Practical Implications

Working on mobility in the context of risk reduction requires looking at a variety of areas, taking into account both risk and risk reduction outcomes of population movements and the mobility outcomes of environmental processes and events.

Integrating Migrants in Risk Reduction and Management

Given the ubiquity of population movements in today's world, addressing the specific conditions of vulnerability to disasters linked to mobility status is a priority for DRR and DRM actors all over the world.

Foreign services and other institutions in people's areas of origin have worked to reduce migrants' vulnerability through targeted awareness raising and preparedness activities, as well as by developing insurance schemes to buffer against the negative impacts of disasters on migrants and their families. Many examples exist of national and local authorities in destination areas adapting preparedness, planning, response and recovery to account for the specificities of mobile groups (Guadagno 2015). Some have worked to recruit them as staff or volunteers in civil protection or disaster management organisations, or have leveraged their own disaster prevention and preparedness efforts to strengthen broader institutional systems. This work is particularly relevant to ageing societies, in which migrants are disproportionately represented in the working-age groups.

Institutions in countries of origin and destination have also attempted to facilitate and channel remittance and skill transfers in support of risk reduction activities in at-risk areas, and/or recovery for disaster-affected sites (Sall 2005).

Addressing Environmentally Induced Mobility

Population displacement is a frequent outcome of environmental changes at all timesscales, which if left unaddressed, can extend their adverse impacts beyond the geographic boundaries of the directly affected areas and can make those impacts last longer (Esnard and Sapat 2014). Preparing for and addressing displacement in disasters is a key element of the work of many national and local DRM actors, while a number of bi-lateral or regional legal and operational arrangements target the management of environmentally induced cross-border displacement.

The mobility consequences of disasters and environmental processes include a variety of patterns of movement of directly and indirectly affected people and households (including the affected people's migrant relatives, and the members of the displaced people's host communities), ranging from short-term evacuations to long-term, long-distance movements, which often mirror normal-time mobility (Wang and Taylor 2016). Efforts to address disaster-induced mobility may therefore include: improving people's capacity to move out of affected or at-risk areas; protecting the newcomers who move into affected areas to support reconstruction and recovery; and facilitating movements and exchanges of resources and information among affected persons and their distant relatives. Logistical support for evacuations and longer-distance movements, flexible work arrangements and immigration regulations for migrants working far from the affected areas, and restoration and facilitation of communication channels and remittance-transfer systems to affected areas can all be more routinely integrated in DRM and disaster recovery work (Box 45.2).

Box 45.2   Disaster-Induced (Im)mobility After the 2015 Nepal Earthquake

Lorenzo Guadagno[1] and Daria Mokhnacheva[1]

[1] International Organization for Migration, Geneva, Switzerland

In April 2015, a large, shallow earthquake shook parts of Nepal, affecting the mobility patterns of people within the country and abroad. An estimated 2.8 million people were displaced. Most of the affected found shelter with relatives and friends or in spontaneous outdoor locations close to homes, workplaces and communities. Only about 37,500 moved into planned displacement sites – a majority of them Nepalese nationals who had migrated internally before the earthquake and who had little access to social safety nets in their location (Khazai *et al.* 2015).

Other internal migrants moved back to their areas of origin. About 300,000 people headed for rural areas in the days after the earthquake in order to find assistance or provide aid. At the same time, rural dwellers tried to move out of isolated, affected areas looking for support. Damaged and interrupted roads represented a major challenge for the mobility of both these groups.

Many of Nepal's estimated 4 million international migrants tried to return home to provide material and social support to affected families. Many were not able to do so due to their employers' unwillingness to provide salary advances and paid leave. In other cases, their families' level of indebtedness or their dependence on remittances discouraged them from returning, and put pressure on them to use savings and assume risky behaviour to remit even more (Sijapati 2015).

Moving was a key element of people's responses to the earthquake, and efforts by governmental institutions, other response actors and the private sector to address physical, legal and economic constraints to population movements might have been a key element of successful relief. As it is expected that internal and international migration will increase as part of affected families' medium- and long-term efforts to rebuild and recover, such efforts will remain key to the wellbeing of the Nepalese.

As mentioned above, specific efforts have been carried out to support the permanent relocation of communities out of areas at high risk of, or severely affected by, environmental changes. While relocations may reduce the exposure to hazards of at-risk communities, they are also costly processes that have the potential to deplete the human, social and economic capital of relocated persons, their hosts and the communities left behind. They rarely succeed unless they support people's long-term access to basic services, opportunities and resources, leverage their agency and safeguard the cohesion of all affected communities (Cernea 1993; Yamamoto and Esteban 2014).

All these movements, in particular when massive and sudden and at least in the short term, can have negative impacts on the communities that people come from and move to. They can increase the number of people living in high-risk areas, can result in reduced availability of opportunities and services, and can spike prices for basic goods and resources. Such changes are likely to affect more severely people (whether or not on the move) who are already vulnerable (e.g., low-income families, low-skilled workers, marginalised groups, and women-headed or single-income households). Responding to and recovering from environmental changes at all timescales in order to minimise their impacts also requires addressing the wellbeing and risk implications of these movements.

Mobility as a Demographic Dynamic

All population movements shape all the components of risk: hazard exposure and occurrence (through population distribution and ecosystem pressures), as well as vulnerability and capacities (through circulation of and access to resources, opportunities and services). Consequently, attention to the potential risk and risk reduction outcomes of population movements should not be limited to environmentally induced migration, displacement and relocations.

Instead, population movements, and the demographic changes and circulation of resources they induce, should be one of the dynamics to consider when assessing risk creation and reduction processes linked with productive investments, land-use planning choices, and social and economic development at large. This is particularly important for urban areas that often represent the main destination for migration flows. Urban systems that can accommodate growing pressures on the delivery of basic services and the availability of fundamental wellbeing opportunities by an increasingly diverse population will be key to human security and wellbeing over the coming decades (Foresight 2011; Box 45.3).

Box 45.3   Migration and Risk in Goma, Democratic Republic of Congo

Lorenzo Guadagno[1] and Daria Mokhnacheva[1]

[1] International Organization for Migration, Geneva, Switzerland

Goma, an economic hub and part of a conurbation straddling the border between Rwanda and the Democratic Republic of the Congo (DRC), is located in a highly seismic region, between two active volcanoes and a lake at risk of a limnic eruption. Over the last few decades, and in particular following the 1994 Rwandan Genocide, it has represented the destination of large flows of refugees fleeing armed conflict in the region, which have fuelled the uncontrolled growth of precarious, underserved neighbourhoods. In 2002, an eruption of the Nyiragongo volcano destroyed 13 per cent of the city. In total, 130,000 people were displaced and 200 died, mostly due to the absence of early warning mechanisms, evacuation plans and preparedness measures.

Since then, the city has continued to attract newcomers looking for economic opportunities or fleeing conflicts and hosts today an estimated 1.1 million inhabitants, three times as many as in 2002. People have resettled on the land covered by the 2002 lava flows, and increasingly concentrate in areas highly exposed to several hazards. Many of them are recent immigrants who did not experience the 2002 disaster and are little aware of the risks they face.

The city's vulnerability is underpinned by its institutions' weakness, compounded by decades-long cycles of armed struggles among various factions and the government. In particular, Goma's North Kivu region has been the theatre of fighting between the M23 movement, government forces and other militias, culminating in M23 seizing the city in November 2012. The conflict resulted in widespread abuses to, and displacement of, civilians in the area. Despite the December 2013 ceasefire, the area has still been affected by episodes of violence by armed militias.

Under these conditions, the local authorities' capacity and willingness to plan and manage the city's development in view of future disasters has been reduced. Strengthening local governance, and in particular the capacity of the city's government to manage incoming population flows, will be key to reducing risk in Goma (IOM 2014).

Levels of risk for different groups in rural and urban areas of origin can also be affected by outwards population movements through loss of working-age population, household split and community disruption, land abandonment and de-intensification of land uses, and degradation of housing and infrastructure stocks. Planning, enforcement of existing regulations and targeted support to communities of origin will also be needed in order to prevent and address these impacts of population movements.

The mobility impacts of DRR including CCA policies and measures are another element to consider in this regard. By transforming incidence of hazards and availability of opportunities, such efforts influence mobility decisions of local and distant communities. The displacement or relocation of communities to directly make space for risk reduction infrastructure or buffer zones presents great similarities with development-induced relocations. These measures are likely to produce additional, indirect and longer-term mobility effects, such as through gentrification processes (Sherwood *et al.* 2014), or increased movements facilitated by the improved protection of material and intangible assets that are necessary for moving (Sakdapolrak *et al.* 2014).

Lastly, it is necessary to consider the impacts of policies and measures that facilitate or constrain migration and other population movements on the vulnerability and exposure of those moving, as well as on other people and groups in home and host communities. The conditions in which moving takes place, the conditions in which those moving live and work in their destination, and the manner in which population movements affect the lives of households and societies of origin and destination, reflect decisions (primarily) made by states about who can move, where to, for how long and with what rights, benefits and obligations (Geddes *et al.* 2012). The way these and other relevant policies (e.g., labour regulations, access to land and services) are shaped is, and will be, key for distributing the costs and benefits of migration across places, groups and individuals (de Haas 2014) – including its risk reduction and risk creation outcomes.

Conclusions

Population movements are a fundamental dynamic of human communities and societies, at once the product and a determinant of their social and environmental features. (Im)mobility decisions and patterns are embedded in local, distant and global socio-environmental structures and relations that determine access to resources and opportunities for different groups and individuals. In turn, they produce a variety of risk and risk reduction outcomes, for different people, in different locations and across various timescales which shape the highly contextual, localised conditions in which risk associated with natural hazards and environmental change is manifested. DRR, including CCA, and sustainable development efforts, which all look at risk as a pivotal concept, cannot overlook (im)mobility as one of its key determinants.

In order to adequately capture the changes in hazard occurrence and exposure, vulnerability and capacities occurring as a result of (im)mobility patterns, it is important to think translocally and diachronically about processes that create and reduce risks (linked both with environmental shocks and longer-term changes). This means looking in particular at the risk implications of (im)mobility decisions and patterns, as they result from planning choices and investments, including for risk reduction. It also means recognising that all measures that facilitate or hinder population movements (or the transfers of material and immaterial resources they underpin) influence current and future risk levels of those moving, those staying behind and those living in host communities.

Planned and unplanned, forced and voluntary, local and cross-border movements (and not just those that are immediately induced by environmental factors) are relevant to this perspective: they all transform people, sites, communities and systems. Different stakeholders, procedures and priorities might however be relevant to different kinds of movements and contexts. Making risk reduction and adaptation efforts better tailored to migrants' needs, capacities and contributions will be key to reducing overall losses to disasters, in particular in increasingly mobile and diverse communities. Measures to anticipate, manage and address disaster-induced displacement are likely to become even more important to preparedness, response and recovery work as displacement risk grows around the world. Relocations might become a more cost-effective option in the face of increasing risks. All these measures share the common objective to reduce risks to those involved in, or affected by, mobility decisions.

As we head into an era of increasingly interconnected locales and systems, creating the practical and legal conditions for people's (im)mobility decisions not to result in further risk, in particular for those who are already more marginalised and vulnerable, is essential to promote sustainable development in home and host societies. Integrating DRR including CCA and human mobility perspectives will help to better understand and address the whole palette of these decisions and their implications on the wellbeing and risk of those moving and those staying behind, as well as the members of their host communities.

References

Black, R., Arnell, N.W., Adger, W.N., Thomas, D. and Geddes, A. (2012) Migration, immobility and displacement outcomes following extreme events', *Environment, Science and Policy* 27, S1: S32–S43.

Cernea, M.M. (1993) *The Urban Environment and Population Relocation*, Washington DC: The World Bank.

Davis, I. (1981) 'Disasters and settlements – Towards an understanding of the key issues', *Habitat International* 5, 5/6: 723–740.

de Haas, H. (2014) *Migration Theory: Quo Vadis?*, Oxford: International Migration Institute, University of Oxford.

Esnard, A.M. and Sapat, A. (2014) *Displaced by Disaster: Recovery and Resilience in a Globalizing World*, New York: Routledge.

Faist, T. and Schade, J. (2013) 'The climate-migration nexus: A reorientation', in T. Faist and J. Schade (eds), *Disentangling Migration and Climate Change*, Dordrecht: Springer, pp. 3–25.

Felli, R. and Castree, N. (2012) 'Neoliberalising adaptation to environmental change: Foresight or foreclosure?', *Environment and Planning A* 44, 1: 1–4.

Foresight: Migration and Global Environmental Change. (2011) *Final Project Report*, London: The Government Office for Science.

Gasper, D. and Truong, T-D. (2010) 'Movements of the 'We': International and transnational migration and the capabilities approach', *Journal of Human Development and Capabilities* 11, 2: 339–357.

Geddes, A., Adger, N., Arnell, W.N., Black, R. and Thomas, D. (2012) 'Migration, environmental change, and the "challenges of governance"', *Environment and Planning C: Government and Policy* 30: 951–967.

Greiner, C. and Sakdapolrak, P. (2012) 'Rural-urban migration, agrarian change and the environment in Kenya. A critical review of the literature', *Population and Environment* 34, 4: 524–553.

Guadagno, L. (2015) *Reducing Migrants' Vulnerability to Natural Disasters through Disaster Risk Reduction Measures*, Geneva: Migrants in Countries in Crisis Initiative.

Guadagno, L. (2016) 'Human mobility in a socio-environmental context: Complex effects on disaster risk', in K. Sudmeier-Rieux, M. Fernandez, I. Penna, M. Jaboyedoff and JC Gaillard (eds) *Identifying Emerging Issues in Disaster Risk Reduction, Migration, Climate Change and Sustainable Development: Shaping Debates And Policies*, Cham: Springer, pp. 13–31.

IDMC (Internal Displacement Monitoring Centre). (2015) *Global Estimates 2015: People Displaced by Disasters*, Geneva: IDMC.

IOM (International Organization for Migration). (2011) *Glossary on Migration*, Geneva: IOM.

IOM (International Organization for Migration). (2014) *IOM's Disaster Risk Reduction in North Kivu Update, 1/10/14*, Kinshasa: IOM.

Ionesco, D., Mokhnacheva, D. and Gemenne, F. (2016) *The Atlas of Environmental Migration*, Abingdon: Routledge.

Khazai, B., Anhorn, J., Girard, T., Brink, S., Daniell, J., Bessel, T., Mühr, B., Flörchinger, V. and Kunz-Plapp, T. (2015) *Shelter Response and Vulnerability of Displaced Populations in the April 25, 2015 Nepal Earthquake,*

CEDIM Report No. 2, Eggenstein-Leopoldshafen: Centre for Disaster Management and Risk Reduction Technology (CEDIM).

Mach, E. and Traore Chazalnoel, M. (2015) *Ahead of COP21 Intended Nationally Determined Contributions Take Stock of Human Mobility Questions*. Online https://weblog.iom.int/ahead-cop21-intended-nationally-determined-contributions-take-stock-human-mobility-questions#sthash.FtUGvqeL.dpuf (accessed 11 September 2016).

Morrissey, J. (2012) 'Rethinking the "debate on environmental refugees": from "maximilists and minimalists" to "proponents and critics"', *Journal of Political Ecology* 19: 36–49.

Sakdapolrak, P., Promburom, P. and Reif, A. (2014) 'Why successful in situ adaptation with environmental stress does not prevent people from migrating? Empirical evidence from Northern Thailand', *Climate and Development* 6, 1: 38–45.

Sall, B. (2005) 'Remittances and economic initiatives in sub-Saharan Africa', in Organisation for Economic Co-operation and Development (OECD) (ed.), *Migration, Remittances and Development*, Paris: OECD.

Scheffran, J., Marmer, E. and Sow, P. (2012) 'Migration as a contribution to resilience and innovation in climate adaptation: Social networks and co-development in Northwest Africa', *Applied Geography* 33: 119–127.

Sherwood, A., Bradley, M., Rossi, L., Gitau, R. and Mellicker, B. (2014) *Supporting Durable Solutions to Urban, Post-disaster Displacement: Challenges and Opportunities in Haiti*, Washington and Geneva: Brookings Institution and International Organization for Migration.

Sijapati, B. (2015) *Migration and Resilience, Experiences from the Nepal 2015 Earthquake*, Kathmandu: Centre for the Study of Labour and Mobility.

UN (United Nations). (2015) *Adoption of the Paris Agreement, FCCC/CP/2015/L.9/Rev.1*. Online http://unfccc.int/resource/docs/2015/cop21/eng/l09.pdf (accessed 11 September 2016).

UNHCR (United Nations High Commissioner for Refugees). (2010) *Convention and Protocol Relating to the Status of Refugees*, Geneva: UNHCR.

UNISDR (United Nations International Strategy for Disaster Reduction). (2015) *Sendai Framework for Disaster Risk Reduction 2015–2030*, Geneva: UNISDR.

Wang, Q. and Taylor, J.E. (2016) 'Patterns and limitations of urban human mobility resilience under the influence of multiple types of natural disaster', *PLoS ONE* 11, 1: 1–14.

Wisner, B., Blaikie, P., Cannon, T. and Davis, I. (2004) *At Risk: Natural Hazards, People's Vulnerability and Disasters, 2nd edition*, London: Routledge.

Yamamoto, L. and Esteban, M. (2014) *Atoll Island States and International Law: Climate Change Displacement and Sovereignty*, Berlin, Germany: Springer.

46

EDITORIAL CONCLUSION TO THIS HANDBOOK

From Action to Principles for Disaster Risk Reduction Including Climate Change Adaptation

Ilan Kelman, Jessica Mercer, and JC Gaillard

Within this volume, we have a handbook of ideas and actions for bringing together disaster risk reduction (DRR) and climate change adaptation (CCA) through placing CCA in DRR. No reason exists to separate them. No reason exists to be territorial. No reason exists to create silos and cliques, labelling each other as being different and searching for separation.

Those committed to CCA, DRR, and development generally share the same fundamental ethos that we have one planet and that one species is making a mess of it. There is so much that we can do about this situation, so let's work together in order to do so.

Many actions for DRR including CCA are well known and well tested, along with actions for climate change mitigation. We have been witnessing a subtle shift in such actions to focus on climate change as the main topic to address – sometimes exclusively, even if that occurs at the expense of other DRR and development issues. Could climate change nonetheless be a suitable starting point?

The Value of Climate Change

Would implementing actions solely for climate change, both mitigation and adaptation, tackle the root causes? Would it address the principles which cause climate change in the first place?

Even if human greenhouse gas emissions were to total zero as of tomorrow, we would still be facing climate change's legacy for decades, possibly even centuries. But if human-caused climate change could be miraculously halted tomorrow, would that solve the identified threats?

Climate change is causing major disruptions to ecosystems, to the point that major biome shifts appear to be occurring and species appear to be going extinct. Other human actions cause this level of environmental damage anyway.

Deforestation and overfishing have destroyed many ecosystems, even inside protected areas. Human-caused land use changes have exacerbated many environmental hazards, leading to widespread flooding and landslides. Human-caused impacts on water-based ecosystems have poisoned fish – persistent organic pollutants and mercury appear in Arctic marine life, far away from the pollutants' sources – and have led to toxic algal blooms.

Powerful interests behind overfishing and large-scale logging have even argued that climate change will ruin these resources, so humanity might as well exploit them now. Of course, without climate change, these interests would still be involved in these destructive activities and would still be ignoring the consequences. They would find another argument to justify their destructive actions.

So if human-influenced (or entirely natural) climate change were not of concern, then humanity – more to the point, a highly powerful but small minority within humanity – would still sport impressive results in wrecking for everyone the planet's environment. This value of immediate exploitation irrespective of the long-term costs is the same value that led to human-influenced climate change: fossil fuels are cheap and easy to use now, so some believe that we must use as much as possible regardless of the consequences.

This means short-term gain for long-term pain. A slogan summarises this value: 'Earth First! We can strip-mine the other planets later'.

Compared with the important, needed, and inspiring work ongoing for tackling climate change, how much effort is being put into tackling the root causes of these values? The need goes far beyond greenhouse gases and CCA. Instead, it is about changing the fundamental values which lead some within humanity to live completely out of balance with the environment, from the local level to the global level.

Many people have few options other than to exploit their local environment. A small global group actively selects their leadership of resource (mis)management so that it is designed to ruin the planet long into the future. Human-influenced climate change is one manifestation amongst many of unsustainable environmental values.

The real crisis is how this small, powerful minority thinks and acts. The crisis is values, not climate change. Climate change has value in bringing the situation to our attention. It has mobilised and galvanised to action many pro-humanity, pro-environment forces. But climate change is not the problem per se. It is one symptom amongst many of the manifestation of unsustainable values.

Climate Change and Disasters

Another symptom of the manifestation of unsustainable values is the presence of vulnerability to all hazards – especially when some groups are made to be more vulnerable than others based on characteristics such as gender, disability, age, sexual orientation, race, and economic status. Climate change can be a powerful motivator away from identifying and targeting the root causes of this vulnerability creation.

Consider environmental phenomena which are often related to climate, examples of which are floods, droughts, wildfires, storms, tornadoes, landslides, temperature extremes, lightning, and fog. They are not always connected to the climate. For instance, droughts can occur due to water overconsumption, lightning can occur due to volcanic eruptions, landslides can occur due to earthquakes, wildfires can occur due to arson, and floods can occur due to tsunamis. Irrespective of the origin of an environmental phenomenon or process, including when it displays hazardousness to society, the fundamental definition applies that disaster risk is a combination of hazard and vulnerability.

In considering climate's and climate change's influences on environmental hazards, few represent an unusual or atypical environmental phenomenon or process. Floods, wildfires, and droughts are abiotic ecosystem components with many species adapted to their regular occurrence. They are not even necessarily hazardous unless unprepared people or infrastructure are affected; i.e., unless vulnerability exists. That is, climate or climate change by itself cannot be a disaster because a disaster cannot happen without vulnerability; it cannot happen without society.

People can choose to live in harm's way; for instance, by purchasing a house in the floodplain to enjoy beautiful views of the river without taking adequate measures to reduce flood damage. More often, people are forced into vulnerable situations such as when women might choose not to evacuate because they fear sexual violence in the public storm shelter or their culture does not permit them to be outside without an accompanying man.

Even in more affluent countries, such as in Europe, many people cannot afford to heat their homes during winter or cannot afford fans or air conditioning for summer heat waves. Normal, seasonal weather can kill thousands of people each year in the heat or the cold due to the people's vulnerability rather than due to the temperature extreme per se. Poverty arises for numerous, complex reasons ranging from government ideology to individual choices. Understanding vulnerability and reasons for it is not straightforward and can be individualised.

So vulnerability, not hazards, causes disasters. Climate change's most significant influences are on potential environmental hazards, not on vulnerabilities. Climate change will make some environmental hazards worse, will reduce potential problems from other hazards, and will have no influence on further sets of hazards.

One example is that many storms seem to be decreasing in frequency but increasing in intensity due to climate change. This observation applies to tropical cyclones – referring to hurricanes, typhoons, and cyclones, depending on the location where they form – and to Arctic polar lows – which are some of the storms forming in some northern waters.

Another example is that, as average air temperature rises around the world under climate change, disease-carrying vectors seem to be expanding their range, moving into higher latitudes and higher altitudes. Examples might be malaria, dengue, and chikungunya. Higher temperatures accelerate vector and microorganism life cycles, meaning that they can breed faster and more frequently.

Yet, more intense storms have more chance of washing away vector eggs and larvae, preventing disease transmission. As sea level rises, coastal areas become salinated, which can also inhibit some insect breeding.

The influence of climate change on storm- and disease-related hazards is complex, as it is with many other environmental hazards. All these influences are only hazard-related, not considering vulnerability. Steps for reducing vulnerability to these hazards, including storms and vector-borne disease, have long been practised.

Consequently, climate change by itself cannot cause or increase disasters. Vulnerability must exist. Even where climate change makes hazards worse, we can choose to address vulnerability in order to avert disasters. DRR can and should be effected, and can be effective, irrespective of climate change. If poor people were supported for home heating and cooling costs, then in-home temperature deaths could be avoided no matter what climate change does to temperature extremes.

Subsidising poor people for basic living expenses is an ideological choice. In being for or against it, an active choice is made on vulnerability – and on more or fewer people dying in temperature extremes. Climate change neither creates nor destroys ideological stances.

So, future disasters under climate change, and independent of climate change, display few instances where society lacks choices to stop the hazards from becoming disasters through reducing vulnerability – effectively, the point of enacting DRR including CCA. If only CCA is considered, then only some climate-related environmental hazards could be considered, meaning that some vulnerabilities (such as to earthquakes and volcanic eruptions) might not be fully addressed. If DRR is considered, then all environmental hazards, including those which are climate-related, could be part of the endeavours.

Consequently, implementing CCA only cannot address all DRR concerns. Implementing DRR will, by definition, solve all CCA concerns.

Beyond DRR

DRR does not and cannot cover all development-related activities. Implementing DRR only cannot address all development concerns while implementing development will, by definition, solve all DRR concerns.

DRR as a principle has also been challenged with respect to the 'reduction' component, in contrast to disaster risk management (DRM). DRM is defined by the UN Office for Disaster Risk Reduction (UNISDR, formerly the secretariat of the United Nations International Strategy for Disaster Reduction) as 'The systematic process of using administrative directives, organisations, and operational skills and capacities to implement strategies, policies and improved coping capacities in order to lessen the adverse impacts of hazards and the possibility of disaster' (http://www. unisdr.org/we/inform/terminology).

Major overlaps are seen with DRR's definition from UNISDR:

> The concept and practice of reducing disaster risks through systematic efforts to ana-
> lyse and manage the causal factors of disasters, including through reduced exposure to
> hazards, lessened vulnerability of people and property, wise management of land and the
> environment, and improved preparedness for adverse events.
> *(http://www.unisdr.org/we/inform/terminology)*

It seems that DRM might potentially be a subset of DRR, since DRM highlights reducing 'adverse impacts of hazards and the possibility of disaster', while DRR includes this component and much more.

Consequently, not everyone in disaster and development uses both DRR and DRM, instead highlighting DRR. Others take different approaches to the differentiation between DRR and DRM. While we were preparing this handbook, Allan Lavell, who is responsible for many of the successes in Latin America's DRR, wrote to us (quoted with permission):

> DRM for us is the frame, strategy, and approach that allows, amongst other things,
> DRR to occur. DRR is one part, only one part, of what DRM attempts to achieve in
> an integrated fashion. This is how it is used throughout Latin America and in the GAR
> [Global Assessment Report] and it may be wise to accept from the outset that the way a
> 'Handbook' sees it is not necessarily the way other parts of the world see it!!

This view places DRR as a subset within DRM and has been institutionalised across Latin America.

To a large extent, as long as definitions are clear, could it be that either approach suffices? Or do we risk becoming buried in jargon and acronyms? Consider that UNISDR's terminology also includes 'Corrective disaster risk management' and 'Prospective disaster risk management' alongside 'Extensive risk' and 'Intensive risk'. Does understanding and implementation improve with increasing numbers of phrases and with increasing lengths of phrases? Any improvement is likely to be contextual. Many of the phrases and the discussions of their meanings might have use in professional, scientific, and political venues.

For people outside these venues, who are the vast majority and who are usually those with the highest vulnerability, what meaningful (to them) phrases exist for the actions we espouse? To inspire everyone to own DRR for themselves while supporting those who need support, what disaster-related phrases could be used for publicity to grab attention and to engage interest? 'Stopping disasters', 'preventing disasters', 'disaster-related activities', and 'making communities safer' are examples of phrases that have been proposed and applied. None is necessarily a good communicator across many boundaries. A simple, engaging, attention-grabbing idea is still missing.

Or, perhaps the question itself is undesirable to answer because it would trivialise the work needed, because a solid and universal answer does not exist, or because it would be too dependent on the whims of public opinion and public perception which are easily swayed by the latest fad or by media channels preferring their own profit or ideology over public safety. It might be better to retain the myriad of phrases, with selection each time based on the audience addressed, the specific interests of the speaker, and the ultimate goal of communicating and exchanging with that audience within that venue. At times, we could use, 'Extensive and intensive disaster risk reduction and management including climate change adaptation through holistic vulnerability reduction plus climate change mitigation for development and sustainability'. At other times, we could use, 'Be safer'.

This varied and targeted approach could achieve the most overall given our world of multiple and competing interests, short attention spans, and political games in which disaster risk topics are often sidelined. The typical attitude is that the event has not happened, so why worry? Then, a catastrophe strikes followed by demands about why nothing was done before the disaster. Action follows, preparing perfectly for the disaster that just happened.

Moving beyond this attitude of thinking and doing only when it is too late is, in fact, DRR – but it is also much more than DRR. It is the value shift championed above. It is the ethos of pooling minds and resources to bring everyone together that underlies and underlines the point of completing this handbook. Separation – such as DRR and CCA, rather than DRR including CCA – fuels partitioning of responsibility, as does doing strictly DRR rather than placing DRR firmly within development.

CCA in DRR in Development

To accept the actions of DRR including CCA (in addition to DRM, climate change mitigation, and other terms) as being development, is it possible to outline basic principles that embrace the vocabulary and definitions without unduly technocratic prose? Here, we attempt to do so, distilling many of the concepts and decades (if not centuries) of literature and actions into five key points which we suggest provide principles for DRR being part of development. No originality exists in these phrases or explanations; they are taken from long-standing discussions, scientific publications, and policy and practice reports which display the courage to dig deeply towards the defining root causes of disasters.

1 Disasters are Societal, Not Environmental, Processes

By definition, disasters are about society. Environmental hazards are frequently involved in disasters, although many environmental hazards and hazard influencers are affected by human activities. But without societal interaction, an environmental phenomenon or process is just an environmental phenomenon or process. When an environmental phenomenon or process interacts with society, then it can be a potential resource, a potential hazard, a combination of hazard and resource, or effectively neutral. Where hazardousness emerges, a disaster might result if vulnerability to the hazard exists. Because the interaction with society and the presence of vulnerability are both necessary baselines for a disaster to occur, the disaster is caused by societal, not environmental, aspects.

2 Natural Disasters Do Not Exist

Because disasters are caused by societal aspects, disasters are not natural; they are all human-caused. Environmental phenomena and processes occur and they are natural (although sometimes modified by human activities). Society's choice to create, perpetuate, and accept vulnerability to those environmental phenomena and processes is sadly typical, but it should not be the natural state of affairs. Since the societal process of vulnerability, not environmental phenomena and processes, leads to disasters, disasters are created by humanity and are not 'natural'.

3 Disaster Risk and Disasters Cannot Be Understood without Focusing on Vulnerability

Vulnerability dictates how society could be adversely impacted by environmental phenomena and processes and the reasons permitting the development and continuation of the situation that permits those negative impacts. Vulnerability is about what humanity does to itself to allow harm from environmental phenomena and processes – more notably, what some societal sectors do to other societal sectors. Individuals are rarely fully responsible for all of their own vulnerability. Instead, much vulnerability to environmental phenomena and processes is created by those not directly affected by the consequences of this vulnerability. Without understanding these processes and why they exist, we cannot understand disaster risk or disasters.

4 Disasters Are Slow-onset

Environmental phenomena and processes such as tornadoes and tsunamis can manifest rapidly, but vulnerability takes a long time to accrue. Since vulnerability is the root cause of disaster, the disaster is a process requiring a long time to develop. Disasters cannot be rapid-onset or sudden-onset, but are all slow-onset, emerging from vulnerability which is about humanity's decisions, attitudes, values, and activities over the long-term. For example, an earthquake quickly jolts a building, but it took decades or longer for the urban planning, building codes, and construction to manifest in such a way that infrastructure collapses and causes casualties. It is not always the individuals affected who deserve blame, but is usually certain sectors of society with decision-making power and resource control who could have acted, but chose not to.

5 Exceptions Exist

Examples of environmental phenomena and processes exist that would threaten the entire planet, making it almost impossible to reduce vulnerability substantially. Near-Earth Objects, such as comets and meteors, with the potential for striking the planet do not fall into this category because (a) the bigger they are and the faster they move, the easier it is to identify them and so the more lead-time we have to deal with them if we are monitoring for them, and (b) we have the technology and resources today to respond in time to an identified, dangerous object and then to take action to nudge it out the way. Examples of truly threatening environmental hazards are gamma ray flares and supernovae from nearby stars, ice ages, basaltic flood eruptions, and supervolcanic eruptions. In some cases, global-scale responses would be feasible to assist many of those in need to continue humanity in the face of the events, particularly for the latter three hazards. For gamma ray flares and supernovae from nearby stars, they could potentially appear without warning and wipe out life on Earth.

The above principles do not suggest that society completely understands the environment or that society can or should control the environment entirely, although modifying the environment remains within the list of options as part of dealing with potential environmental hazards. Instead, the principles suggest that society has a duty and responsibility to understand and deal with hazards and hazard influencers through focusing on vulnerability. Many techniques exist for dealing with uncertainties and unknowns, including related to the environment's behaviour and properties, as epitomised in this handbook's chapters.

If, as shown by many examples throughout the *Handbook*, society fails to act on existing knowledge and techniques while failing to continue seeking new knowledge and techniques, then blame does not rest with the environment for being 'difficult', 'dangerous', or 'hazardous'. Instead, the responsibility rests with those in society who have power and resources for choices in priorities and values – and who choose vulnerability.

In effect, societal development has been denominated. Neither a magically simple nor an off-the-shelf one-size-fits-all solution exists. It takes continual creativity, dedication, and hard work. People have a right to demand, and a responsibility to make, communities less vulnerable to environmental phenomena and processes – which is intertwined with justice, equity, and not destroying ourselves or our environment.

The previous three paragraphs have not mentioned disasters or climate change. The focus is on human–environmental relations through a development lens. It automatically incorporates all the elements of CCA, DRR, DRM, and climate change mitigation, bringing them together through a set of common principles which support the desired actions for each of these processes according to their definitions. This togetherness is what development should be, representing why DRR includes CCA, to create the better future which development envisions and strives for.

Within this realm, the past has demonstrated inspirational successes and too many frightening failures. We look forward to a future where successful development including DRR including CCA is the norm.

AFTERWORD

Youth Involvement in Disaster Risk Reduction Including Climate Change Adaptation for Sustainable Development

Lydia Cumiskey and Moa M. Herrgård

Young people are experiencing a world with climate change and imminent disasters at their doorstep along with the adoption of three international agreements designed to address these challenges: The Sendai Framework for Disaster Risk Reduction (UNISDR 2015), the Paris Agreement on Climate Change (UNFCCC 2015), and Agenda 2030 for Sustainable Development (UN 2016c). Young professionals, researchers, students, and practitioners are actively contributing to disaster risk reduction (DRR) including climate change adaptation (CCA), adopting this handbook's ethos by seeking connections across sectors, disciplines, and cultures.

A need exists for continued opportunities and resources for youth participation in decision-making, (applied) research, and project/programme creation and management to utilise youth's innovative capacity for social change and sustainable development. Not only do we as youth bring unique skillsets, but also our paradigm of shifting perspective and a systems approach are crucial for upholding the integrated approach of sustainable development. This starts with a change in the way we identify needs and curate solutions, as all areas of DRR need to integrally consider climate change as one hazard influencer, so DRR can include CCA to overcome silos and systemic barriers to resilient change (Kelman 2015). This Afterword provides some evidence of actions and perspectives of young people active in the field of DRR including CCA and promoting sustainable development throughout their work.

On a practical level, we are young scientists, engineers, technologists, and communities of practice significantly contributing to multiple areas for DRR including CCA on the ground. Young people can utilise information and communications technology (ICT) tools effectively and are creating and contributing to developing open source data and tools, such as Open Street Map initiatives collecting exposure data (see for example Fernandez and Shaw 2016) and developing mobile applications in the Code for Resilience Programme such as the Messiah application, which sends emergency alerts to pre-identified groups; see World Bank and GFDRR (2015) for more examples. Risk communication has been highlighted as an area where the potential influence of young people as resources and receivers has been underestimated, particularly for early warning information (Mitchell *et al.* 2008). We can and should play an active and activist role in developing participatory videos on DRR including CCA to raise awareness in our communities (see also Haynes and Tanner 2015). Youth and ICT are being connected from many angles; e.g., Rwanda has merged national ministries of Youth and ICT, but has not been successful in translating this change to the local government level (Ben-Attar and Campbell 2013). Young people's capacity in ICT can help to improve the digital literacy of such local governments.

We identified this important role in risk communication, raising awareness, and early warning dissemination during the Children and Youth Forum at the 3rd World Conference on DRR during the capacity building events (Cumiskey *et al.* 2015). Furthermore, young water professionals and students came together at the Conference of Youth at the climate change Conference of Parties in Paris (COP21) in 2015 for a role-playing game, the Sustainable Delta Game, where participants acted as decision-makers tasked with selecting DRR including CCA measures to deal with the uncertain and risky future – as futures have always been. Building and utilising youth capacities on ICT, risk communication, and awareness raising tools can be equally applied for wider sustainable development actions.

Access to more university programmes on DRR (many of which integrate CCA) and related disciplines is increasing the capacity and skills of young people to conduct research, formulate new research agendas, and address research gaps. Through the support of the recently established Young Scientists Platform on DRR (UN MGCY 2015) young scientists came together at the Understanding Risk Conference 2016 and at an event organised by Deltares and UNESCO-IHE to identify the research gaps for the successful implementation of the Sendai Framework for DRR within each of the four Priority Areas for Action (Deltares and UNESCO-IHE 2016). There was consensus that through current and future research, young scientists can particularly support topics such as identifying underlying drivers of risk, conducting early warning system evaluations, ecosystem-based DRR including ecosystem-based adaptation, and inclusive governance. Young researchers and professionals have increased access to capacity building events, university programmes, and networking opportunities enabling us to realise our role in supporting DRR including CCA, now and in the future.

The current generation of young people is witnessing the adoption of major international agreements in and across the fields of DRR, CCA, and sustainable development, which aim to usher in a renewed paradigm for addressing systemic, cross-cutting, currently emerging, and foresight issues. Youth actively participated in the agenda-setting and policy design of these frameworks, further contributing to the implementation of those frameworks through advocacy, and especially through meaningful youth engagement in ensuring coherence amongst the follow-up, monitoring, and review processes within the different frameworks.

The official mechanism for the follow-up and review of these policy frameworks takes place through the High Level Political Forum (HLPF) (UN 2016a). The HLPF is the space for member states and non-state actors to report on progress and best practices on various intergovernmental agreements, including the Sendai Framework for DRR and UN (2015) which is the 2030 Agenda (including Sustainable Development Goal (SDG) 13 on climate change). The aligned reporting of global frameworks contributes to the crucial aspects of a nexus approach of indicators and policy implementation, in order to create a sustainable and resilient society for all (Nilsson *et al.* 2016).

The HLPF also provides a unique platform to exchange evidence-based best practices and trends in global and local development. Young scientists and youth, with expertise in development and DRR including CCA, have contributed to enhancing the science–policy interface and in providing policy briefs for the development of the 2nd Global Sustainable Development Report, launched at the 2016 HLPF (UN 2016b). A special session on 'The Experience of Youth in using Science, Technology, & Innovation (STI) for Sustainable Development' was held during the 1st Annual Multi-Stakeholder Forum on STI for the SDGs. This youth session featured innovations which young people are developing, thus promoting the importance of engaging and supporting young people in the fields of STI to pursue the SDGs (UN 2013).

For example, the organisation Global Minimum Inc. supports young innovators to solve the challenges in their local communities and highlighted the important role for young people in supporting technology transformation tailored to local needs. The official youth constituency, the United Nations Major Group for Children and Youth, and several youth-led organisations – e.g., Water Youth Network, International Federation for Medical Students Associations, and Engineers Without Borders – are taking steps towards developing youth-led reporting as well as reporting on youth engagement and affected youth, to be included in the HLPF. Together, these practices are a good start to ensure meaningful and equitable engagement of youth in policy reviews for Agenda 2030 and the Sendai Framework for DRR, particularly paragraph 36.a.iii on youth engagement (UNISDR 2015, 36.a.iii).

These practical, scientific, and policy-related activities for DRR including CCA by youth across the world show the current success and future potential for us to enable, enact, and effect sustainable development. This handbook provides many more examples and we, as youth, are willing to get involved in them all – across disciplinary, national, sectoral, and cultural borders to bring people together. Young people are directly experiencing today's challenges, foresee the future ones, and have huge innovative capacity and need to tackle them – for our own future and beyond. Young practitioners, scientists, technologists, and community members are at the forefront of developing innovative tools and approaches that can ensure DRR including CCA is socially inclusive, addressing many development concerns simultaneously. Thus, it is recommended that continued youth engagement in DRR including CCA across science, policy, and practice is encouraged and supported by public and private sector decision-makers at all levels.

We are the future. This handbook helps us to build and shape it for successful DRR including CCA.

Acknowledgements

The authors acknowledge the support from other members of the UN MGCY Disaster Risk Reduction and Science–Policy-Interface Working Group, Alexandria Mavrodivea and Donovan Guttieres, and Water Youth Network member Alix Lerebours in reviewing and commenting on this Afterword.

References

Ben-Attar, D. and Campbell, T. (2013) *ICT, Urban Governance & Youth.* Nairobi: UN-HABITAT.
Cumiskey, L., Hoang, T., Suzuki, S., Pettigrew, C. and Herrgård, M.M. (2015) 'Youth participation at the Third UN World Conference on disaster risk reduction', *International Journal of Disaster Risk Science* 6, 2: 150–163.
Deltares and UNESCO-IHE (2016) *Young Scientists Event on Disaster Risk Reduction.* Online https://www.deltares.nl/app/uploads/2016/03/Outcomedocument_YoungScientistsEventonDRR_final.pdf (accessed 17 August 2016).
Fernandez, G. and Shaw, R. (2016) 'Urban disasters and risk communication through youth organizations in the Philippines', in R. Shaw, Atta-Ur-Rahman, A. Surjan and G.A. Parvin (eds.), *Urban Disasters and Resilience in Asia*, Waltham, MA: Butterworth-Heinemann, p. 195.
Haynes, K. and Tanner, T.M. (2015) 'Empowering young people and strengthening resilience: Youth-centred participatory video as a tool for climate change adaptation and disaster risk reduction', *Children's Geographies* 13, 3: 357–371.
Kelman, I. (2015) 'Climate change and the Sendai framework for disaster risk reduction', *International Journal of Disaster Risk Science* 6, 2: 117–127.
Mitchell, T., Haynes, K., Hall, N., Choong, W. and Oven, K. (2008) 'The roles of children and youth in communicating disaster risk', *Children Youth and Environments*, 18, 1: 254–279.

Nilsson, M, Griggs, D. and Visbeck, M. (2016) 'Map the interactions between Sustainable Development Goals', *Nature* 534, 320–322.

UN (2013) *A/RES/67/290 – Format and Organizational Aspects of the High-level Political Forum on Sustainable Development*, New York: UN (United Nations).

UN (2015) *Transforming our World: The 2030 Agenda for Sustainable Development*, New York: UN (United Nations).

UN (2016a) *E/HLS/2016/1 – Ministerial Declaration of the High-level Segment of the 2016 Session of the Economic and Social Council on the Annual Theme Implementing the Post-2015 Development Agenda: Moving from Commitments to Results*, New York: UN (United Nations).

UN (2016b) *Global Sustainable Development Report 2016*, New York: Department of Economic and Social Affairs, UN (United Nations).

UN (2016c) *Inputs to the 2016 High-level Political Forum on Sustainable Development*, New York: UN (United Nations).

UNFCCC (United Nations Framework Convention on Climate Change) (2015) *Paris Agreement*, New York: United Nations.

UNISDR (2015) *Sendai Framework for Disaster Risk Reduction 2015–2030*. Available at: http://www.preventionweb.net/files/43291_sendaiframeworkfordrren.pdf.

UN MGCY (2015) *Young Scientists Platform on Disaster Risk Reduction Concept Note*. New York: UN (United Nations) Major Group for Children and Youth.

World Bank and GFDRR (2015) *Code for Resilience: bridging Communities for Disaster Response, 2014 Apps*, Washington, D.C.: World Bank and GFDRR (Global Facility for Disaster Reduction and Recovery). Online http://codeforresilience.org/apps/ (accessed 17 August 2016).

INDEX

Locators in **bold** refer to tables and those in *italics* to figures.

Printed in the United States
by Baker & Taylor Publisher Services

Printed in the United States
by Baker & Taylor Publisher Services